Lecture Notes in Networks and Systems 1028

The series "Lecture Notes in Networks and Systems" publishes the latest developments in Networks and Systems—quickly, informally and with high quality. Original research reported in proceedings and post-proceedings represents the core of LNNS.

Volumes published in LNNS embrace all aspects and subfields of, as well as new challenges in, Networks and Systems.

The series contains proceedings and edited volumes in systems and networks, spanning the areas of Cyber-Physical Systems, Autonomous Systems, Sensor Networks, Control Systems, Energy Systems, Automotive Systems, Biological Systems, Vehicular Networking and Connected Vehicles, Aerospace Systems, Automation, Manufacturing, Smart Grids, Nonlinear Systems, Power Systems, Robotics, Social Systems, Economic Systems and other. Of particular value to both the contributors and the readership are the short publication timeframe and the worldwide distribution and exposure which enable both a wide and rapid dissemination of research output.

The series covers the theory, applications, and perspectives on the state of the art and future developments relevant to systems and networks, decision making, control, complex processes and related areas, as embedded in the fields of interdisciplinary and applied sciences, engineering, computer science, physics, economics, social, and life sciences, as well as the paradigms and methodologies behind them.

Indexed by SCOPUS, INSPEC, WTI Frankfurt eG, zbMATH, SCImago.

All books published in the series are submitted for consideration in Web of Science.

For proposals from Asia please contact Aninda Bose (aninda.bose@springer.com).

Michael E. Auer · Reinhard Langmann ·
Dominik May · Kim Roos
Editors

Smart Technologies for a Sustainable Future

Proceedings of the 21st International Conference on Smart Technologies & Education. Volume 2

Editors
Michael E. Auer
CTI Global
Frankfurt/Main, Germany

Dominik May
University of Wuppertal
Wuppertal, Germany

Reinhard Langmann
Edunet World Association e.V.
Blomberg, Germany

Kim Roos
Programme Director MScEng
Arcada University of Applied Sciences
Helsinki, Finland

ISSN 2367-3370 ISSN 2367-3389 (electronic)
Lecture Notes in Networks and Systems
ISBN 978-3-031-61904-5 ISBN 978-3-031-61905-2 (eBook)
https://doi.org/10.1007/978-3-031-61905-2

This Springer imprint is published by the registered company Springer Nature Switzerland AG
The registered company address is: Gewerbestrasse 11, 6330 Cham, Switzerland

Preface

It is a great privilege for us to present the proceedings of the 21st "International Conference on Smart Technologies & Education" (STE2024) to the authors and delegates of this event and to the wider, interested audience. The 2024 edition of STE was held under the general theme "Smart Technologies for a Sustainable Future", which was visible throughout the conference program.

The STE conference is the successor of the long-standing annual REV Conferences and the annual meeting of the International Association of Online Engineering (IAOE) together with the Edunet World Association (EWA) and the International Education Network (EduNet). Initiated in 2004, REV has been held in Villach (Austria), Brasov (Romania), Maribor (Slovenia), Porto (Portugal), Dusseldorf (Germany), Bridgeport (USA), Stockholm (Sweden), Brasov (Romania), Bilbao (Spain), Sydney (Australia), Porto (Portugal), Bangkok (Thailand), Madrid (Spain), New York (USA), Dusseldorf (Germany), Bengaluru (India), Georgia (USA), Hong Kong, Cairo (Egypt), and Thessaloniki (Greece).

This year, STE2024 has been organized in Helsinki, Finland as an onsite event supporting remote presentations, from March 6 until March 8, 2024. The co-organizers of STE2024 were the Arcada University of Applied Sciences, the International Association of Online Engineering (IAOE) together with the Global Online Laboratory Consortium (GOLC), the International Education Network (EduNet), and the EDUNET WORLD Association (EWA). STE2024 has been attracted 140 scientists and industrial leaders from more than 40 countries.

STE2024 is an annual event dedicated to the fundamentals, applications, and experiences in the field of Smart Technologies, Online, Remote, and Virtual Engineering, Virtual Instrumentation, and other related new technologies, including:

- Applications & Experiences
- Artificial Intelligence
- Augmented Reality
- Open Science Big Data
- Biomedical Engineering
- Cyber Physical System
- Cyber Security
- Collaborative Work in Virtual Environments
- Cross-Reality Applications
- Data Science
- Evaluation of Online Labs
- Human–Machine Interaction & Usability
- Internet of Things
- Industry 4.0
- M2M Concepts
- Mixed Reality

- Networking, Edge & Cloud Technology
- Online Engineering
- Process Visualization
- Remote Control & Measurements
- Remote & Crowd Sensing
- Smart Objects
- Smart World (City, Buildings, Home, etc.)
- Standards & Standardization Proposals
- Teleservice & Telediagnosis
- Telerobotic & Telepresence
- Teleworking Environment
- Virtual Instrumentation
- Virtual Reality
- Virtual & Remote Laboratories

The conference was opened by the Founding President of IAOE, Michael E. Auer, who underlined the importance to discuss guidelines and new concepts for engineering education in higher and vocational education institutions including emerging technologies in learning. In her greeting, the Rector of Arcada, Mona Forsskåhl pointed out the importance of the digitalization of education and more specifically the engineering education.

STE2024 offered an exciting technical program as well as networking opportunities concerning the fundamentals, applications, and experiences in the field of online engineering and related new technologies.

As part of the conference program, three pre-conference workshops have been organized:

1. Overcoming Traditional Boundaries of STEM Education and Enabling the Engineer of the Future
2. Logiccloud: The Next Generation Of Industrial Control
3. High-Performance Extreme Learning Machines

Furthermore, special sessions have been organized at REV2024, namely

1. Online Laboratories in Modern Engineering Education (OLMEE)
2. Human–Robot Interaction for Sustainable Development (HRI4SD)
3. Advances and Challenges in Applied Artificial Intelligence (ACAAI)

Four outstanding scientists and industry leaders accepted the invitation for keynote speeches:

1. Doris Sáez Hueichapan, University of Chile, Santiago, Chile, talked about "Energy & Water Management Systems for Agro-Development of Indigenous Rural Communities"
2. Dieter Uckelmann, HFT Stuttgart, Stuttgart, Germany, shared his valuable insights to "Why providing a comprehensive IoT education is impossible – but we should nevertheless strive to do so"
3. Roland Bent, Retired CTO Phoenix Contact GmbH & Co.KG, Germany, painted a vision for the future in his talk "The All Electric Society"

4. Hans-Jürgen Koch, Dipl.-Ing. for Communications Engineering, Executive Vice President of the Business Area Industry Management & Automation, Phoenix Contact GmbH, Germany, gave a fascinating introduction to "Innovative and collaborative automation platforms – The key for a sustainable world"

The conference was organized by the Faculty of Arcada University of Applied Sciences and Program Director Kim Roos served as the STE2024 chair. The President of IAOE, Prof. Dominik May has served as STE2024 general chair and Prof. Reinhard Langmann and Prof. Michael E. Auer served as Steering Committee Co-chairs.

Submissions of Full Papers, Short Papers, Work in Progress, Poster, Special Sessions, Workshops, Tutorials, Doctoral Consortium papers have been accepted.

All contributions were subject to a double-blind review. The review process was extremely competitive. We had to review about to 233 submissions. A team of over 100 program committee members and reviewers did this terrific job. Our special thanks goes to all of them.

Due to the time and conference schedule restrictions, we could finally accept only the best 76 submissions for presentation or demonstration.

The conference was supported by

- Phoenix Contact as Platinum Sponsor
- Air France and KLM as Diamond Sponsor
- As always Sebastian Schreiter did an excellent job to edit this book.

Kim Roos
Dominik May
Michael E. Auer
Reinhard Langmann

Committees

STE General Chair

Dominik May	President IAOE, University of Wuppertal, Germany

STE Steering Committee Co-chairs

Michael E. Auer (Co-chair)	IAOE, Austria
Reinhard Langmann (Co-chair)	EWA, Germany

STE2024 Chair

Kim Roos	Arcada University of Applied Sciences, Finland

Program Co-chairs

Erwin Smet	University of Antwerp, Belgium
Valery Varney	RWTH Aachen, Germany
Rizwan Ullah	Arcada University of Applied Sciences, Finland

Technical Program Chair

Sebastian Schreiter	IAOE, France

Workshop and Tutorial Chair

Alexander Kist	University of Southern Queensland, Australia

Special Session Chair

María Isabel Pozzo	National Technological University, Argentina

Award Chair

Andreas Pester	The British University in Egypt, Cairo, Egypt

Publication Chair and Web Master

Sebastian Schreiter	IAOE, France

Local Arrangement Chair

Maria von Bonsdorff-Hermunen	Arcada University of Applied Sciences, Helsinki, Finland

International Advisory Board

Abul Azad	President Global Online Laboratory Consortium, USA
Alberto Cardoso	University Coimbra, Portugal
Bert Hesselink	Stanford University, USA
Claudius Terkowsky	TU Dortmund University, Germany
Doru Ursutiu	University of Brasov, Romania
Hamadou Saliah-Hassane	Université TÉLUQ, Montréal, Canada
Krishna Vedula	IUCEE, India
Elio San Cristobal Ruiz	UNED Madrid, Spain
Teresa Restivo	University of Porto, Portugal
Uriel Cukierman	National Technological University Buenos Aires, Argentina

EduNet Forum Committee

Chair

Klaus Hengsbach	Phoenix Contact GmbH & Co KG, Germany

Members

Albert Alacorn	Phoenix Contact, Chile
Anja Schulz	Phoenix Contact, Germany
Christian Madritsch	Carinthia University of Applied Sciences, Austria
Christiane Kownatzki	Phoenix Contact, Germany
Edmond Wempe	Phoenix Contact, Germany
Felipe Mateos Martín	University of Oviedo
Glenn Williams	Harrisburg, USA
Hans Lindstrom	Phoenix Contact, Finland
Hernan Lopez	Phoenix Contact, Argentina
Jana Koenig	Phoenix Contact, Germany
Maren Gast	Phoenix Contact, Germany
Pascal Vrignat	University of Orleans, France
Reinhard Langmann	Edunet World Association, Germany

International Program Committee

Akram Abu-Aisheh	Hartford University, USA
Alexander Kist	University of Southern Queensland, Australia
Anastasios Economides	University of Macedonia, Greece
Andreas Pester	The British University in Egypt, Egypt
Anshul Jaswal	Pennsylvania State University, USA
Carlos A. Reyes Garcia	National Institute of Astrophysics, Optics and Electronics, Mexico
Chee Sai Stephen Bok	Institute of Technical Education, Singapore
Catherine Soh Geok Hong	Institute of Technical Education, Singapore
Christian Guetl	Graz University of Technology, Austria
Christian Madritsch	Carinthia University of Applied Sciences, Austria
Christos Katsanos	Aristotle University of Thessaloniki, Greece
Cornel Samoila	University of Brasov, Romania
Dario Assante	Universita Telematica Internazionale, Italy
David Boehringer	University of Stuttgart, Germany
Dieter Wuttke	TU Ilmenau, Germany
Erwin Rauch	Free University of Bolzano, Italy

Contents

Smart Technology and Education

A Machine Learning Framework for Improving Resources, Process, and Energy Efficiency Towards a Sustainable Steel Industry

Andrea Fernández Martínez[1(✉)], Santiago Muiños-Landín[1], Angelo Gordini[2], Luca Ferrari[2], Matteo Chini[3], Loris Bianco[3], and Mircea Blaga[4]

[1] AIMEN Technology Centre, Pontevedra, Spain
andrea.fernandez@aimen.es
[2] OPTIT Srl, Bologna, Italy
[3] Ferriere Nord SpA, Osoppo, Italy
[4] Tenaris Silcotub S.A., Calarasi, Romania

Abstract. In response to geopolitical instability, supply chain issues, and environmental concerns, initiatives like the European Green Deal highlight the need for a green transition in the EU industry. The steel sector, as an Energy-Intensive Industry, is crucial in this shift. This work introduces a Machine Learning framework for sustainability in the Steel Industry, addressing Resource, Process, and Energy efficiency with three ML algorithms. The framework, integrated into a Decision Support System, assists plant operators in the transition to a more sustainable process.

Keywords: Machine Learning · Sustainability · Steel Industry · Energy Efficiency · Process Optimization

1 Introduction

The Steel Industry is considered an Energy-Intensive Industry (EII) due to the reliance of its industrial processes on the use of high amounts of energy to provide essential materials and products to other industries e.g., construction, automotive, and other energy industries [3]. Recent geopolitical instability, supply chain issues, and climate change emphasize the urgency of new policies. These are essential for long-term competitiveness and the transition to a greener economy of the EU Industry as defined in new initiatives such as the European Green Deal. On this subject, the green transitioning of EIIs plays a key role as they account for more than half of the total energy consumption of the EU Industry [4, 12].

In modern steel-making processes, there can be distinguished two main technologies, namely Basic Oxygen Furnace (BOF) and Electric Arc Furnace (EAF). EAF, using recycled scrap steel, has been shown as more energy-efficient, aiding the steel industry's decarbonization [13]. Nevertheless, there are still several challenges hindering the green transitioning of this sector as the lack of digitization and data exploitation of the large

M. E. Auer et al. (Eds.): STE 2024, LNNS 1028, pp. 3–10, 2024.
https://doi.org/10.1007/978-3-031-61905-2_1

amounts of data that cover the steel value chain due to the sparsity of data sources and the complexity of its analysis [1]. In this context, Artificial Intelligence and Machine Learning are increasingly popular for their ability to analyze complex data patterns in applications like predictive maintenance, process optimization or even process oriented materials design [5], among others [6].

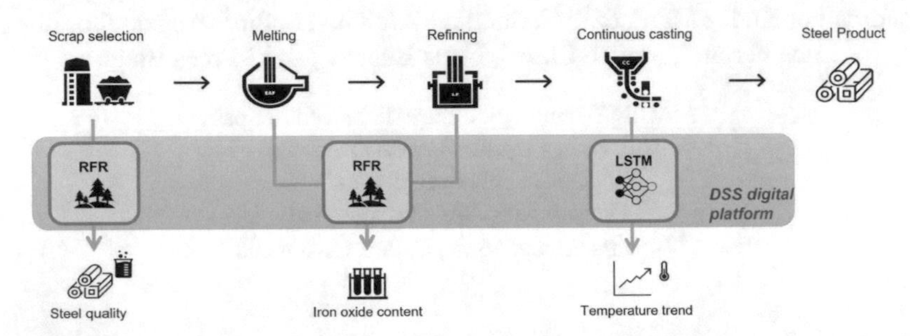

Fig. 1. Conceptual representation of the ML framework proposed. In the upper layer, schematic diagram of the phases of steel-making; in the middle layer, the ML algorithms as part of the DSS digital platform; in the lower layer, the targeted outputs of each algorithm.

This work proposes a Machine Learning (ML) framework for smart sustainability in the Steel sector, addressing resource use, melting efficiency, and energy needs during cooling. Key contributions of the ML framework include: First, a Random Forest Regressor (RFR) [2] to support the decision-making process of scraps and input materials for the steel mixture considering the final chemical composition of the steel product. Second, a RFR to forecast the composition of iron oxide in the steel slag during the melting process, which is essential for the iron content and oxygen present in the steel. Third, a Long Short-Term Memory (LSTM)-based Recurrent Neural Network (RNN) [10] to predict temperature rises in the cooling system as an indicator of the energy required.

2 Methodology

2.1 The Sustainability Challenge in the Steel Industry

The EAF-based steel-making process operates in a batch basis, referred as heats, that comprises six main operations: i) furnace charging, ii) melting, iii) refining, iv) de-slagging, v) tapping, and vi) furnace turnaround. The process starts with the selection and load of scraps in a charge bucket, followed by the introduction of the selected charge inside the furnace. After charging, the electric arc initiates the melting process, creating a molten steel pool. Oxygen injection accelerates scrap meltdown and forms steel slag. Refining operations use carbon and oxygen injection to achieve desired steel chemistry and enhance process efficiency through slag foaming. De-slagging empties the furnace of slag, followed by tapping for transferring molten steel to a ladle for molding and cooling in continuous casting (see Fig. 1).

Optimization of the Use of Resources for the Steel Mixture. EAF-based steel-making uses recycled scrap steel as its primary feedstock—material from rejected or end-of-life products. This circular approach reduces environmental impact by avoiding the use of additional resources and minimizing overall waste [3, 4]. Nevertheless, other secondary feedstock e.g., HBI, might also be incorporated as sources of clean iron to enhance the properties of the scrap charge due to the variability and frequent lack of knowledge of the properties of recycled scrap steel. To optimize melting and reduce alloy additions, our study proposes an RFR decision support model. It evaluates input material combinations for the steel composition, covering 16 materials, including diverse scrap steel types, and focusing on 7 key chemical elements, namely, Cr, Ni, Mo, Cu, Sn, V, and Pb.

Process Efficiency Through the Steel Slag Composition. Slag foaming is an essential part of the steel-making process in which foam generation is induced to generate total or partial liquid solutions i.e., slags, comprised of oxides and fluorides at the upper surface of the metal bath [7]. Foamy slag offers process benefits like increased energy efficiency by capturing heat from the arc, while also preventing metal oxidation and nitrogen incorporation. Nevertheless, there are still many limiting factors e.g., FeO content of the slag, that impede the proper control of slag foaming in industrial scenarios [7, 8]. Our study proposes an RFR, considering injections and various materials, to estimate iron oxide percentage in steel slag, crucial for slag iron content and CO gas generation in foaming.

Energy Efficiency in the Cooling Phase. Cooling systems are crucial for maintaining optimal conditions in the EAF and other components, preventing structural damage from prolonged overheating [11], and ensuring proper melting conditions. The conditions in the cooling system are directly affected by the melting process and the relationships between materials and the energy-matter within the EAF [9]. When injected materials don't properly penetrate the steel slag during melting, they can lead to temperature increases in the settling chamber and cooling system. This results in higher energy consumption. The ML framework addresses this using an LSTM Neural Network to predict temperature trends in the cooling system within 80-s time windows. It considers injections (carbon and lime), temperatures in the cooled shell and settling chamber, and the flowrate of fresh inlet water into the Water-Air Cooled (WAC) system.

2.2 Decision Support System Digital Platform

A Decision Support System (DSS) is an information system used to assist teams and organizations in complex decision-making processes that usually rely on a large number of parameters and constrains. The ML algorithms in this work are part of a DSS digital platform, facilitating information exchange via REST API. The platform comprises container images for ML models, a DSS, and a "model orchestrator" module. These container images encapsulate all dependencies, ensuring seamless interoperability and scalability. The platform is deployed on a private cloud-based server, with HTTPS protection and access control policies for secure data exchange.

3 Experimental Results

3.1 RFR for the Optimization of the Use of Resources

The dataset for the decision-support model to evaluate the combination of materials ad scraps comprised 207 samples from 11 different steel types e.g., Fe45 and Fe50. Nevertheless, the contribution of materials to the final steel quality is not considered to be steel-type-dependant.

After performing a 5-cross-validated grid-search analysis to fine-tune the hyper-parameters of the RFR, a 500-estimator-based RFR with unlimited depth of Trees and using the mean-squared-error as criterion to assess the quality of splits was trained over the 80% of the dataset (165 samples) and tested in the remaining 20% (42 independent samples) previously re-scaled within a range [0, 1] using the min-max normalization method.

The RFR predicts the percentage of the 7 targeted chemical components (Cr, Ni, Mo, Cu, Sn, V, and Pb) taking as input 16 different materials and scraps commonly used in the steel case of study. Table 1 lists the Mean Absolute Error (MAE), Mean Squared Error (MSE), and Root MSE (RMSE) obtained in the test dataset for each chemical component targeted, respectively. The results obtained confirmed that the RFR algorithm successfully achieved mapping the inputs materials with the targeted chemical components, with a maximum MAE of 0.021 for Chromium (Cr).

Table 1. Estimated MAE, MSE, and RMSE for the percentage prediction of the seven chemical components of study, respectively

	Cr%	Ni%	Mo%	Cu%	Sn%	V%	Pb%
MAE	0.021	0.014	0.008	0.035	0.005	0.000	0.001
MSE	0.001	0.000	0.000	0.002	0.000	0.000	0.000
RMSE	0.030	0.018	0.012	0.044	0.008	0.000	0.001

Figure 2 depicts the predicted targets (orange) by the RFR against the ground truth test samples (green) for the elements Mo and Pb to better visualize the error range independently. These results support our hypothesis that the RFR properly predicts the targeted chemical components with high accuracy, although there can be seen appreciable errors for the cited elements in samples close to their range limits (Mo $> 0.06\%$, Pb $< 0.001\%$). Overall, the RFR model is considered to be reliable for modelling the final expected quality of the steel based on the input materials and scraps used in the steel-making process.

3.2 RFR for Estimating the Content of Iron Oxide in the Steel Slag for Process Efficiency

For the fitting of the RFR to predict the content of iron oxide in the steel slag, an analog 5-crossed-validated hyper-parameter fine-tuning procedure was conducted considering

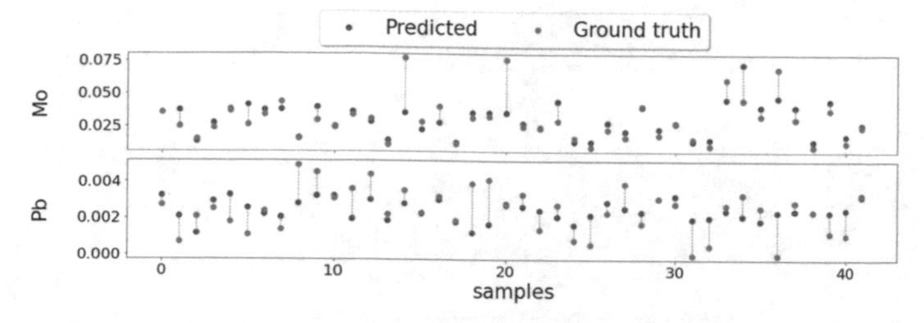

Fig. 2. Scatter plot of the predicted values (orange) vs. ground truth (green) for the seven chemical components of study

a dataset containing 4239 heat processes from November 2020 to September 2021, after performing a filtering procedure of a complete dataset containing 4714 heats to avoid introducing noise to the model due to unusual process behaviours e.g., heats during transition times in the process.

As in the previous case, 80% of the data (3391 samples) was used for training and fine-tuning, whereas the 20% of the remaining data (848 samples) was used as an independent test set to assess the model performance. The dataset was initially re-scaled in a range [0, 1] using the min-max normalization method. From the grid-search procedure, an RFR composed of 500 estimators and with the MSE as criterion for the quality of splits was found to yield the best results predicting the percentage of iron oxide in the steel slag.

Based on the results expressed in Table 2, it can be confirmed that the RFR successfully estimates the percentage of iron oxide in the steel slag with a MAE of 2.979. To better visualize the prediction error with respect to the dataset, Fig. 3 displays the predicted targets (orange) against the ground truth samples (green).

Table 2. Estimated MAE, MSE, and RMSE for the percentage of iron oxide in the steel slag

Element	MAE	MSE	RMSE
FeO%	2.979	14.654	3.828

Although the predicted values properly follow the trends of the test set as shown in Fig. 3, there can be noticed a "conservative" behaviour in the model performance when predicting extreme cases lying at the upper and lower limits of the dataset, forecasting those values towards the center of the distribution. Nevertheless, the errors obtained can be considered acceptable in the range of study [20, 48), confirming that the RFR algorithm successfully estimates the content of iron oxide in the steel slag following a complete data-driven approach.

3.3 LSTM-Based RNN for Energy Efficiency in the Cooling System

The dataset used for the development of the LSTM-based RNN comprised 50077 samples. Following best practices, the set was divided into training (72%, 36055 samples),

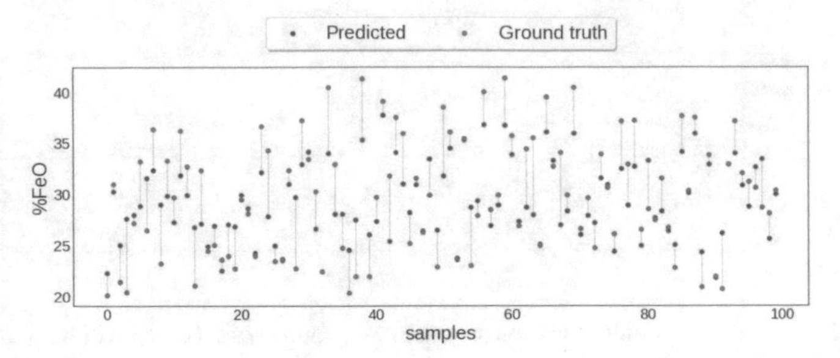

Fig. 3. Scatter plot of the predicted values (orange) vs. ground truth (green) for the percentage of iron oxide in the steel slag

validation (18%, 9014 samples), and test sets (10%, 5008 samples), respectively. The validation set was used during training to fine-tune the parameters of the network, whereas the test set was used during the final assessment of the network as an independent set.

After a fine-tuning analysis, the architecture of the final network consisted of 1-hidden layer with 8 neurons (LSTM cells), using MAE as the loss function and Adam as optimizer of the network. Due to the temporal nature of the problem, the features used as input data (see Sect. 2.1.3) were sampled in batches of 4 considering a time window of 60 s in time i.e., sampling frequency of 20 s between samples considering the last minute of data.

Figure 4 displays the evolution of 6 random samples over the fixed 80-s time-window of study from the test set (green) against the corresponding predicted trends by the LSTM (orange), manifesting the capability of the network to predict the evolution of temperatures in the cooling system in different temperature ranges (specified in the y-axis) and following different temperature patterns. Indeed, the network seems to appropriately predict the trends within a + 80-s time-window in 5 out of the 6 cases shown, that is, in all cases with the exception of the middle figure in the second row, in which the predicted trend differs from the real temperature evolution over time. Nevertheless, these results suggest that the network can successfully predict the temperature trends in the cooling system with high accuracy, especially considering the first +40-s time-window.

In order to delve in the model performance over time, the prediction error was assessed for the different time steps considered i.e., +20 s, +40 s, +60 s, and + 80 s. Table 3 expresses the MAE, MSE, and RMSE on the test set considering each time step separately i.e., errors were computed in a column-wise fashion based on the predictions of the 5008 test samples. It can be observed that the MAE increases from 0.372 in the +20 s-related predicted values to 2.174 in the +80 s prediction case, confirming our previous hypothesis that the predictive accuracy of the network decreases over time.

For visualization purposes, Fig. 5 displays the first 1000 samples of the test set (green) against their corresponding predicted values (orange) for the time windows of + 20s and + 80s, separately. The right rectangles in the sub-figures zoom in on a temperature peak, showcasing the evolving prediction errors over time. Despite the increase, these results suggest promising implications for using deep learning models in forecasting to counteract potential process issues before they occur.

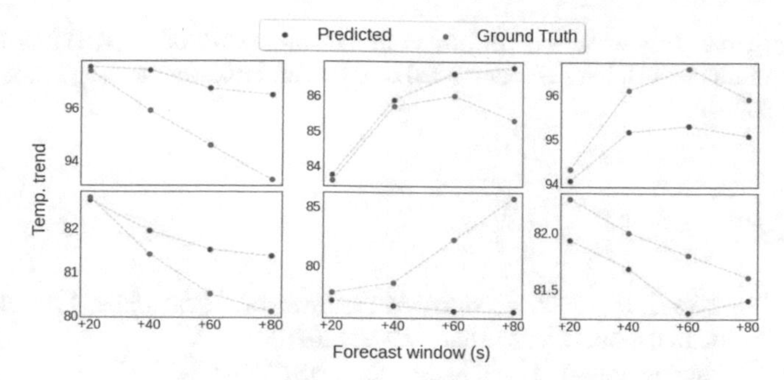

Fig. 4. Scatter plot of temperature trends in +80-s forecast-windows for predicted values (orange) vs. ground truth (green).

Table 3. Estimated MAE, MSE, and RMSE for the estimation of the temperature (T) for each time step considered.

	T+20 s	T+40 s	T+60 s	T+80 s
MAE	0.372	0.926	1.597	2.174
MSE	0.344	2.299	6.653	11.736
RMSE	0.587	1.516	2.579	3.426

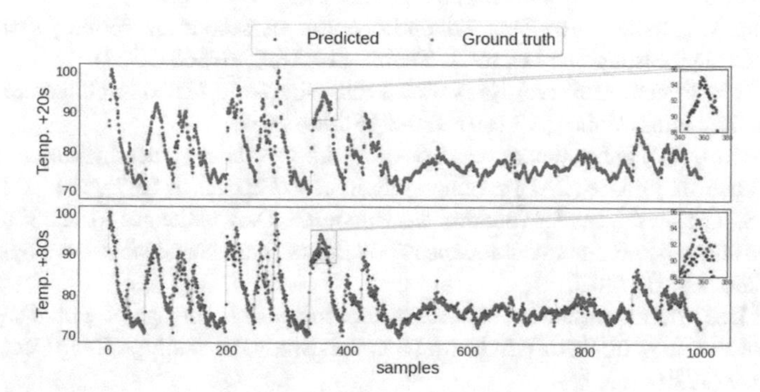

Fig. 5. Scatter plot of the predicted values (orange) by the LSTM network vs. ground truth values (green) in each specific time step considered.

4 Conclusion

The proposed ML framework for steel sustainability includes two RFRs and one LSTM-based RNN, addressing resource use, process, and energy efficiency. It integrates data across manufacturing stages, aiding operators in daily decisions and process control. As part of a DSS digital platform, it facilitates seamless interaction between ML algorithms and operators, supporting intelligent sustainability strategies.

Acknowledgments. This work was performed in the framework of the RETROFEED Project granted by the European Union's Horizon 2020 research and innovation programme under grant agreement N.869939.

References

1. Branca, T.A., Fornai, B., Colla, V., Murri, M.M., Streppa, E., Schröder, A.J.: The challenge of digitalization in the steel sector. Metals **10**(2) (2020)
2. Breiman, L.: Random forests. Mach. Learn. **45**, 5–32 (2001)
3. Fleiter, T., Herbst, A., Arens, M., Stevenson, P., Chan, Y.: Industrial innovation: Pathways to deep decarbonisation of industry. Part 1: Technology analysis. Technical report, ICF Consulting Services Limited and Fraunhofer ISI (2019)
4. de Bruyn, S., Jongsma, C., Kampman, B., Görlach, B., Thie, J.E.: Energy-intensive industries: challenges and opportunities in energy transition. Policy Department for Economic, Scientific and Quality of Life Policies (2020)
5. Gregores Coto, A., et al.: The use of generative models to speed up the discovery of materials. Comput. Methods Mater. Sci. **23**(1), 13–26 (2023)
6. Kong, J.H., Lee, S.W., Kim, S.W., et al.: Recent advances of artificial intelligence in manufacturing industrial sectors: a review. Int. J. Precis. Eng. Manuf. 111–129 (2022)
7. Tomba Martinez, A.G., López, F., Bonadia, P., Luz, A.P., Pandolfelli, V.C.: Slag-foaming practice in the steelmaking process. Ceram. Int. **44**, 8727–8741 (2018)
8. Vieira, D., Bielefeldt, W.V., De Almeida, R.A.M., Faria Vilela, A.C.: Slag foaming fundamentals - a critical assessment. Mater. Res. **20**, 474–480 (2017)
9. Pourfathi, A., Tavakoli, R.: Thermal optimization of secondary cooling systems in the continuous steel casting process. Int. J. Therm. Sci. **183**, 107860 (2023)
10. Salem, F.M.: Recurrent Neural Networks: From Simple to Gated Architectures. Springer, Cham (2022). https://doi.org/10.1007/978-3-030-89929-5
11. Vazdirvanidis, A., Pantazopoulos, G., Louvaris, A.: Overheat induced failure of a steel tube in an electric arc furnace (EAF) cooling system. Eng. Fail. Anal. **15**(7), 931–937 (2008)
12. Vögele, S., Grajewski, M., Govorukha, K., Rübbelke, D.: Challenges for the European steel industry: analysis, possible consequences and impacts on sustainable development. Appl. Energy **264**, 114633 (2020)
13. Zhu, X.: Deep decarbonisation of iron and steel industry in the age of global supply chain-issues and solutions. In: ECEEE Industrial Summer Study Proceedings, Deep Decarbonisation of Industry (2020)

Overall Writing Effectiveness: Exploring Students' Use of LLMs, Pushing the Limits of Automated Text Generation

Simon Wilbers[1(✉)], Johanna Gröpler[2], Bastian Prell[1], and Jörg Reiff-Stephan[1]

[1] Technical University of Applied Sciences Wildau, Hochschulring 1, 15745 Wildau, Germany
simon.wilbers@th-wildau.de
[2] Freie Universität Berlin, Garystraße 39, 14195 Berlin, Germany

Abstract. The advent of generative artificial intelligence for text generation, epitomized by the introduction of ChatGPT in November 2022, represents a significant shift in the academic writing paradigm. This pre-study examines how students make use of Large Language Models (LLMs) for their academic writing processes, transitioning from solitary writing to true human-machine collaboration. Participants were recruited from a workshop on LLMs and were subsequently interviewed qualitatively after two weeks of unsupervised usage. These interviews were designed using the new Overall Writing Effectiveness (OWE) framework and focused on LLMs' role in academic writing. The qualitative content of these interviews was analysed following Mayring's methodology. Findings indicate that LLMs did not substantially accelerate the writing process but enhanced the quality of the texts and redefined writing as a collaborative effort. This study not only explores the limits of automation in academic writing but also highlights how generative AI is pushing the boundaries of what is considered genuine human capabilities. This analysis opens the discussion of how to incorporate such technologies into future education curriculums.

Keywords: Academic Writing · LLMs · Text Automation · Student Experiences · AI and Academia

1 Introduction

The convergence of AI and academia has prompted an unprecedented revolution in how students approach academic writing. Particularly after November 2022, a shift from traditional methods to automated text generation tools has gained momentum [1, 2]. As we address these new literacy practices, it is critical to understand the challenges and benefits of LLMs, such as the well-discussed ChatGPT. Especially user-friendly, commercially available, or even free-of-charge online services seem to have gained popularity among students and teachers in higher education. First quantitative studies found that generative AI-tools are widely used among students, mainly for private purposes [3]. The main reasons cited are time savings and idea generation [4, 5]. However, there are also

M. E. Auer et al. (Eds.): STE 2024, LNNS 1028, pp. 11–22, 2024.
https://doi.org/10.1007/978-3-031-61905-2_2

concerns among respondents about whether they are allowed to use such tools in an academic context [4, 6]. This pre-study seeks to contribute towards an understanding of how students use LLMs for their educational writing tasks. One main focal point is addressed by the self-perceived effects on the availability, performance, and quality of both the text generated and the process of generating text [7]. We anticipated an increase in all three indicators when it comes to human-machine collaboration. To measure the impact of LLMs on the student's writing process, the concept of Overall Equipment Effectiveness (OEE) [8] has been adapted and slightly modified. The contribution is structured as follows: the journey begins with an examination of the academic writing process, identifying common challenges and the potential for technology to mitigate these frictional losses. We then introduce a novel concept, Overall Writing Effectiveness (OWE), adapting the principles of OEE from manufacturing to the context of academic writing. This approach provides a fresh perspective on assessing writing efficiency in an era of technological integration. The paper also discusses the balance between speed and quality in writing, particularly concerning the limits of automation. We delve into how AI tools may influence a writer's maximum speed and explore the potential boundaries of this technology. Our findings are grounded in a detailed data collection process, capturing diverse perspectives from participants. The results offer insightful reflections on students' perceptions of LLMs and their impact on the writing process. Finally, the paper concludes with a synthesis of these findings, contemplating the implications of LLMs in academic writing and suggesting directions for future research. This exploration aims not only to inform, but also to engage readers in a critical discussion on the evolving intersection of technology and academia.

2 The Academic Writing Process and the Overall Equipment Effectiveness Model

2.1 Potential Difficulties in the Academic Writing Process

The academic writing process consists of several steps, each of which can present challenges for the writer. Table 1 shows a simplified concept of the writing process with examples of writing difficulties that may occur [9]:

Writing is inherently a highly individualized process, where the steps involved can either be undertaken in sequence or concurrently. The writer's excessive reliance on technology is already a norm, covering everything from searching digital sources to checking spelling and even employing voice commands to type, all of which are increasingly supported by technological devices. The presence and accessibility of these tools can significantly influence the overall writing process. In this context, identifying strategies to overcome the various challenges in writing with increasingly automated tools becomes crucial.

2.2 From Overall Equipment Effectiveness to Overall Writing Effectiveness

One approach is the application of the Overall Equipment Effectiveness [8], which facilitates the measurement of challenges by quantifying availability, performance, and

Table 1. Simplified concept of the academic writing process and potential writing difficulties.

Step in the writing process	Potential writing difficulty (examples)
Finding a topic and formulating a research question	Difficulty in defining a clear, manageable scope that is neither overly broad nor excessively narrow, often leading to ambiguity in the research direction
Finding sources	Challenges in sourcing relevant material, either encountering an overwhelming abundance of information or a scarcity of adequate research on the topic
Writing	Common issues include writer's block, where ideas don't flow, or anxiety about articulating thoughts coherently and persuasively
Revising	Difficulty in critically evaluating and restructuring the draft to enhance coherence, clarity, and argument strength
Submitting	Potential last-minute concerns about the quality or completeness of the work, or technical issues in meeting submission requirements

quality. This quantifying and separating into separate categories is particularly insightful for assessing where and how AI integration, especially through the use of LLMs, can be most beneficial in enhancing the writing process.

- **Availability:** How do AI tools affect the frequency and duration of both planned and unplanned stops in the writing process? Planned stops could be a lunch break, an unplanned stop would be a technical problem interrupting the writer.
- **Performance:** Insufficient writing tools, or incompetence in using them compromising the speed of the writing process, accounting for slow cycles and small stops.
- **Quality:** The extent to which (AI) tools contribute to or mitigate defects in academic writing, including the need for rework.

An OEE [8] score of 100% would indicate optimal writing conditions: uninterrupted writing time, at maximum speed, and without the need for revisions of the written text; unlike availability and quality, which are straightforward to measure and calculate:

$$100\%\ \text{Availability} = \frac{\mathit{Operating\ (Writing)\ Time}}{\mathit{Loading\ (Planned\ writing)\ Time}}$$

where $[\mathit{LoadingTime}] = [\text{Operating (Writing) Time}]$

$$100\,\%\ \text{Quality} = \left[\text{Number of parts Texts produced}\right] - [\text{Revisions/rework}] \\ - \left[\text{scrap}\right]\ \text{where}\ [\text{Revisions/rework}] = 0 \wedge \left[\text{scrap}\right] = 0$$

Performance hinges on a well-defined planned cycle time:

$$100\%\,\text{Performance} = \frac{\textit{Planned cycle time} \times \textit{number of parts manufactured (Text written)}}{\textit{Loading (Planned writing) Time}}$$

While involved in the thought process of deducing a realistic [planned cycle time], dependencies and causalities are uncovered that also lead to a more profound understanding of the specific process and its inherent challenges. Thus, the thought process of deducing the [planned cycle time] serves as an excellent tool for fathoming the limitations of automation. For this research, it is again not the [planned cycle time] itself, but rather the process and considerations involved in its definition that offer valuable insights. These insights reveal how the limits of automation can be pushed through technological advancements, particularly with the integration of LLMs.

The OEE framework was included in the design of this pre-study not as a primarily quantitative measure, but as a qualitative structure, to understand the student's self-assessment of how LLMs benefit their academic writing.

OEE is a critical tool for recognizing inefficiencies and identifying specific losses in manufacturing processes, providing a clear focus for enhancing equipment productivity and quality output [8]. By systematically categorizing losses into availability, performance, and quality, OEE exposes the underlying causes of impaired effectiveness and reveals opportunities for achieving improvements and further automation. OEE's principles, while originating in manufacturing, are adaptable to various fields, including academia. The following illustrates this versatility by adapting and applying OEE's framework to academic writing, introducing the concept of OWE. In the conceptual transition from OEE to OWE, terminology was adapted. Nuances, changes like renaming 'Operating Time' to 'Writing Time' and 'Equipment Failure' to 'Writer's Block' improve the model's acceptance among recipients by better reflecting the specific productivity challenges of writers. In the domain of OWE, "Downtime Losses" pertain to periods when the writing process is inactive, analogous to production halts in manufacturing. These losses can be attributed to various factors: Equipment Failure in the context of OWE translates to "Writer's Block", which refers to periods when a writer is unable to produce text due to mental fatigue or lack of inspiration. This mirrors mechanical breakdowns in equipment when the tool (in this case, the writer's cognitive faculty) becomes temporarily non-functional for a set task. However, the non-tangible nature of "Writer's Block" compared to more physical "Equipment Failure" marks a distinct difference between the OEE intended for Production Systems. Focused on a Machine and its operators. Whereas OWE considers more a human-machine collaboration. Nevertheless, conventional equipment failures are also categorized as "Writer's Block", relating to the human-machine system that performs the actual writing. Organizational and logistical Disruptions remain unchanged in name from OEE, which encapsulates interruptions in the writing workflow due to poor planning, lack of resources, or other administrative setbacks (Fig. 1).

The term Setup and Adjustment remains unchanged in terminology and equates to the time a writer switches between different writing tasks or preliminary activities that are necessary to create the written output, akin to machine setup times in OEE. For Speed Losses, which represent inefficiencies in the performance rate: Short Stops are akin to

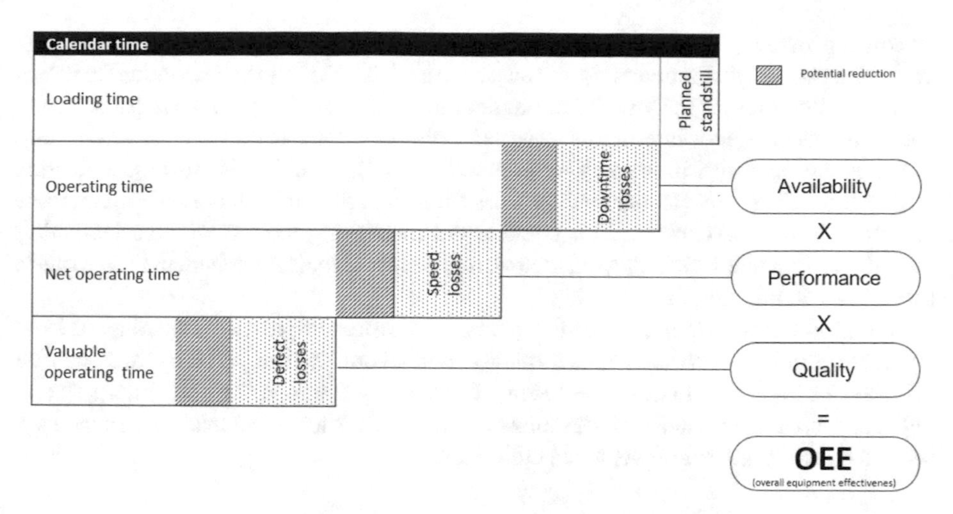

Fig. 1. Overall equipment effectiveness as defined by Nakajima [8].

"Interruptions and Distractions," where brief, unplanned breaks in writing occur due to minor distractions or momentary lapses in concentration, shorter than 3 min. These pauses, though short in duration, accumulate and lead to significant losses in writing efficiency. Reduced Speed describes times when the rate of writing is less than the optimal speed. This can be due to perfectionism, over-editing, or a lack of proficiency in the writing task at hand, resulting in slower progress than potentially achievable. Further technical problems that might occur when using writing tools usually relate to speed losses of the human-machine collaboration. Exemplarily, those could be the loading times of the LLM, and usability glitches. Lastly, Defect Losses in OWE relate to issues of output quality: Defects in Process are errors in grammar, coherence, or argumentation that require discarding the written passage. These defects degrade the quality of the written output, necessitating additional time for correction. Rework is a process wherein written content is rigorously assessed and adapted to meet the desired standards of quality. Rework is a crucial step in the refinement of the writing; however, it results in a loss of immediate output. To summarize, the OWE framework adapts to the writing process by identifying inefficiencies in writing practices, thereby identifying the potential to mitigate the inefficiencies or where to best expand the automation capabilities next.

2.3 Impact of Automation on a Writer's Technical Maximum Speed

To better assess the impact of automation on a writer's technical maximum speed, a series of steps need to be considered. **Evaluating Writing Pace** measures the writer's average speed during peak performance periods. This involves tracking the number of words or pages produced per unit of time (e.g., hour) under optimal conditions. **Assessing the Impact of Automation** as an analysis of how AI and other automated tools currently influence the writer's speed. Determine if these tools are enhancing efficiency or causing delays (for example, due to an over-reliance on editing software or AI-generated content that requires significant rework). **Identifying Performance Barriers** as factors that prevent the writer from maintaining or exceeding this peak pace. These could include frequent interruptions, writer's block, technical issues with automation tools, or inefficiencies in the writing process itself. **Optimizing Writing Conditions** so that the writing environment and process minimize disruptions and the efficient use of automation is maximized. This could entail modifying the workflow, establishing specific writing times, reducing reliance on imperfect AI tools, or enhancing the writer's proficiency in utilizing these tools.

By thoroughly examining these aspects, a comprehensive understanding of how automation affects a writer's speed, quality, and overall writing effectiveness can be achieved as well as a better assessment of achieving the technical maximum speed. OWE as well as the related considerations for the writer's technical maximum speed are used to structure the subsequent data collection.

3 Data Collection

Study participants were recruited from students at the University of Applied Sciences Wildau. In preliminary talks with students and educators alike, it became apparent that understanding and utilization of LLMs varied greatly among students. Thus, it was decided to commence the study with a workshop introducing the fundamental usage of LLMs for text generation. Drawing the participants of the study only from the pool of workshop attendees achieves better comparability, where otherwise a slightly different social media channel exposure could greatly alter the student's knowledge and abilities concerning LLMs. The workshop aimed to introduce a range of AI models suitable for higher education writing and focused on effective prompting. Primarily focusing on broadening students' capabilities with LLMs, emphasizing a comprehensive understanding of their potential and applications, while offering limited, focused training to develop specific competencies for specific LLM tools. Furthermore, the workshop was used as an acquisition tool, as students could attend free of charge. But at the end of the workshop, they were asked whether they would be interested in being interviewed about their LLM usage. 10 of the workshop participants consented and subsequently formed the pool interviewed in the follow-up interviews.

An at least two-week long period without any supervision of LLM usage followed the workshop. Naturally, for students, having assignments resulted in more intense usage of LLMs in contrast to those who did not have any significant writing tasks, or at least approaching due dates at hand at the time. The two weeks of unsupervised usage after the workshop concluded with a semi-structured interview (see Fig. 2). The leading questions for those interviews were deduced from the presented concept of OWE. Hence, those questions focused on the availability, performance, and quality of their writing with LLMs.

1. Usage: Which LLMs, specifically which commercial products, do you utilize and in what areas? Please provide us with exact examples.
2. Advantages: What advantages do you observe in your personal work when using these models? Please respond with examples.

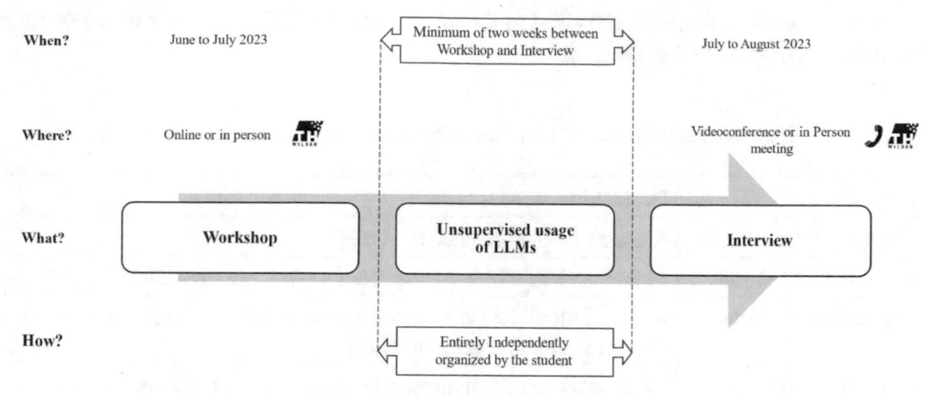

Fig. 2. Timeline of the study.

3. Disadvantages: What disadvantages do you perceive in the use of these models? Please answer with examples where possible.
4. Performance: Has the use of LLMs increased the speed of your writing process? If so, how has this been noticeable?
5. Quality: Has the use of LLMs influenced the quality of your written work? Has the quality changed? In addition, please provide examples of quality improvement or deterioration.
6. Do you have any further questions for us, or is there anything else you would like to share regarding the use of LLMs or AI in general?

The interviews were recorded and transcribed with the consent of the interviewees. The evaluation of the interviews' content was conducted through a qualitative content analysis, as per Mayring's methodology [7, 8], considering the dimensions deduced from the OWE model, which were also used in structuring the interview's leading questions. In the following section, the presented results are enriched with some excerpts of the interviews orally conducted in German, then transcribed and translated into English with careful attention to maintaining the original tone, intent, and content. Due to the inherent differences between spoken and written language, the verbatim transcriptions of participants' responses may appear unconventional but have been deliberately preserved to reflect their original expressions authentically.

4 Results

This preliminary study focused on qualitative results in order to learn more about the effects of generative AI tools on students, their academic work, and long-term perspectives. Ten students were interviewed two weeks after attending a workshop on LLMs. The study does not aim to provide statistically generalizable results but rather offer an initial early, in-depth understanding of individual ways students are engaging with LLMs in academic writing. Qualitative data, collected through interviews, are characterized by their subjective and complex nature, thereby reducing the precision achievable in statistical analysis compared to quantitative data. To provide some validation of the findings, we structured the statements retrieved in the interviews, by different levels of consensus from the participants (see Table 2).

Table 2. The Consensus Levels of Participant Responses within the Structured Interview.

Label	Description
Unanimous Consensus	All ten participants agree on a subject
Predominant Consensus	More than half of the participants (six or more) agree on a subject
Moderate Consensus	At least half of the participants (five or more) agree on a subject, but there are other diverging opinions
Isolated Perspectives	There is no agreement among the majority of participants on a specific subject

The 'Usage' section of the questionnaire revealed varied applications of LLMs among the interviewees. Participants detailed their use of commercial LLM-based applications for diverse purposes, highlighting the broad applicability and integration of these models in different work settings.

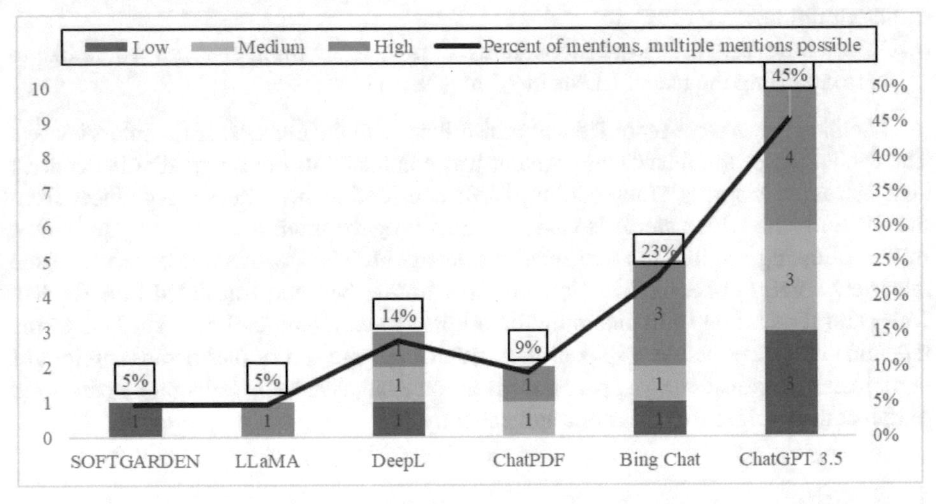

Fig. 3. Number of students using a specific tool and their intensity of usage (N = 10).

The participants were asked to name verbally the LLM tools they were familiar with or had used. It is noteworthy, while somewhat expected, that all ten students used OpenAI's ChatGPT 3.5: *"Of course, ChatGPT and somehow something from Bing, I think. Yes, and then there was DeepL."* Only five of the students, also, actively recalled Bing Chat. The Interview focused on all questions that were asked to not lead the interview more than necessary. In Fig. 3, all the tools named are depicted, where students could name as many as they wanted. The figure also shows the categorization of the ten participants into three distinct groups: Low, Medium, and High intensity of usage of LLM. This classification was derived upon analyzing their indicated usage patterns, during the unsupervised testing phase. The one group, labeled as Low, comprised three individuals who did not engage in any significant writing tasks, such as homework assignments, bachelor's, or master's thesis, during the testing phase; their interaction with LLMs was not motivated by any specific academic or professional writing requirements. The second group of Medium-intensity usage, consisted of three participants who had at least one academic writing task, such as writing a thesis, or significant writing projects like job applications during the test period. This category represents individuals who utilized LLMs for practical purposes beyond casual exploration, as expressed by the following quote: *"For some university projects, because I'm a telematics student, I have to do a lot of programming, and sometimes there are error codes, and sometimes they don't mean anything, but you can copy them and just throw them in the chat"*. The final group, of High intensity of usage, demonstrated a high intrinsic motivation to engage with LLMs. This was either due to pressing academic writing tasks (*"I am writing my master's thesis, and I actually used it to develop a bit of my introduction."*) or, in some cases, driven by curiosity or general interest in exploring the capabilities of LLMs. It is important to note that the study's participants were recruited from students who had registered for a workshop on LLMs and prompting techniques. This approach might produce a selection bias, as it inherently results in students with a pre-existing motivation to sign up for such an extracurricular workshop.

In the 'Disadvantages' segment, participants articulated challenges and limitations they experienced while using LLMs. This feedback sheds light on the potential drawbacks and areas for improvement in the application of these models. Hallucinations of LLMs were viewed as a problem as a predominant consensus of interviewees. One participant pointed out that *"Maybe sources can be given that are not correct, or maybe they are incorrectly worded."* The same was true for data privacy concerns and some uncertainty regarding the model's training data. An interesting remark was made by one participant: *"I almost forgot how to think for myself or write something myself."* This was considered as further evidence that users seem to trust the output, thus, emphasizing the importance of explaining how LLMs work before they are used.

Responses in the 'Advantages' category provided insights into the perceived benefits of using LLMs. Interviewees reported specific enhancements in their writing processes and outputs, underscoring the practical impacts of these models on their writing tasks. A Unanimous Consensus existed that the positive aspects of using LLMs exceed the downsides. All interviewees expressed the view that the use of LLMs helped to tackle procrastination (*"I think for me, it's more like that I'd probably be procrastinating a lot if I hadn't used it."*).

The notion that they can produce more text and thus the technical maximum speed is perceivably higher was supported by a predominant consensus. Given the importance of this question to the OEE and OWE framework, the interviewer laid increased emphasis on this question. When the students were asked to give tangible examples of the increased speed, most participants had trouble finding such and started questioning the assumption of an increased speed. Regarding the quality of their writing, a predominant consensus existed among the participants that using an LLM in the writing process had increased the quality of their writing. One student, for example, stated: *"If you read through it, I think the quality is better."* while this was agreed upon by other participants: *"I find it difficult [...] to formulate it well, and I definitely see an advantage in it, [...] it sounds better."* Here lies a great potential in using AI tools because having trouble academically expressing themselves may cause writer's block.

Under the 'Quality' section, participants assessed the influence of LLMs on the satisfaction of their writing. This involved sharing tangible examples where the use of these models either improved or impaired the quality of their work. Quality was defined by the participants as the quality of the final text, as well as the quantity of references reviewed. Tools such as Bing Chat or Chat PDF allowed the students to check more references during their research process and thus incorporate them into their final work.

The 'Performance' category focused on the impact of LLMs on writing speed. Responses indicated whether and how the adoption of these models has altered the efficiency of their writing process, with participants providing examples to illustrate their experiences. They had to admit that they had to double-check the output: *"But I just have to take another look at it myself and see whether what he wrote really makes sense [...] even more time required"*.

Contrary to expectations, the study found that the use of LLMs most likely did not accelerate the writing process. Participants reported needing the same amount of time for writing tasks as they did before the use of LLMs. However, the students did perceive an enhancement in the quality of their work. For instance, where they might have previously consulted five references for a paper, the use of LLMs enabled them to review and incorporate up to ten, thereby enriching their academic output.

Additionally, students reported perceiving an improvement in writing style when using LLMs, describing the AI-assisted texts to be superior to their own writing [12]. Most notably, the use of LLMs appeared to mitigate periods of procrastination or writer's block. Participants described the writing process as transforming from a solitary task to a form of "teaming up with a machine" where the LLM served as a collaborative partner. This shift was credited with reduced downtimes and making the writing process more efficient.

5 Conclusions

This exploratory pre-study, while limited in scope, offers valuable insights into the interaction between students and LLMs in academic writing, establishing a foundation for future research in a rapidly evolving field. Notably, the introduction of LLMs most likely did not expedite the writing process as previously hypothesized. This observation raises critical questions about the balance between speed and quality in AI-augmented academic writing, particularly regarding the limits of automation. Our findings indicate that the technically maximal speed of writing, or minimal cycle time, mostly remains unaffected by the use of LLMs. This challenges the traditional view of automation as a means to enhance speed, instead revealing its role of improving the quality of academic outputs within existing time frames. Such results suggest that LLMs are pushing the boundaries of what is achievable in academic writing, not by accelerating the process, but by enriching the content depth and quality. Moreover, the introduction of LLMs may have facilitated a paradigm shift in the writing process, transforming it from an individual to a collaborative human-machine endeavor. This change, marked by a decrease in procrastination and less likely occurrence of writer's block, indicates the potential of LLMs to redefine academic writing strategies and pedagogy. Furthermore, the concept of OWE emerges as a promising framework to enhance the writing process. By applying OWE, future research can identify areas where AI and LLMs can be more effectively integrated. The study suggests that exploring OWE in conjunction with LLMs could reveal untapped potential for automation in academic writing, offering new pathways for augmenting human capabilities with AI. In conclusion, this pre-study underscores the importance of continued exploration into the role of LLMs and the application of frameworks like OWE in academic contexts and puts forward evidence on which topics quantitative research ought to focus. Such research is crucial to understanding the evolving dynamics of human-machine collaboration in writing and harnessing the full potential of AI in enhancing the academic writing process.

Acknowledgements. In this exploration of LLMs in academic writing, we naturally used the support of GPT-4 and other language tools for some early drafts of our manuscript. We also extend our gratitude to the students who voluntarily participated in our interviews. This publication was made possible through the funding of the PhD program Innovation and Career Center - Integrated Engineering by the state of Brandenburg's Ministry of Research, Education and Culture (Germany), the Zukunftszentrum Brandenburg, funded by the Federal Ministry of Labour and Social Affairs (Germany) and the European Social Fund ESF, as well as co-funded by the state of Brandenburg's Ministry for Economic Affairs, Labour and Energy (Germany). Additionally, we are thankful for funding provided by EDIH. The pro_digital European Digital Innovation Hub (EDIH) has received a co-funding from the European Union's DIGITAL EUROPE Programme research and innovation programme grant agreement No 101083754.

References

1. Gendron, Y., Andrew, J., Cooper, C.: The perils of artificial intelligence in academic publishing. Crit. Perspect. Account. **87**, 102411 (2022). https://doi.org/10.1016/j.cpa.2021.102411
2. Gröpler, J.: Schreiben oder schreiben lassen? BuB **75**(7), 366–368 (2023)
3. Garrel, J. von Mayer, J.: Artificial intelligence in studies—use of ChatGPT and AI-based tools among students in Germany. Humanit. Soc. Sci. Commun. **10**(1) (2023). https://doi.org/10.1057/s41599-023-02304-7
4. Hoffmann, N., Schmidt, S.: Vorläufige Kurzauswertung der bundesweiten Studierendenbefragung "Die Zukunft des akademischen Schreibens mit KI gestalten" (2023). https://www.starkerstart.uni-frankfurt.de/142467510/kurzbericht-akademisches-schreiben-mit-ki.pdf
5. Humboldt-Universität zu Berlin: Kurzbefragung zu KI und Prüfungen (2023). https://pages.cms.hu-berlin.de/doeringn/dashboard/ErgebnisseKurzbefragungKIPruefungen_230621.pdf
6. Utami, S.P.T., Andayani, A., Winarni, R., Sumarwati, S.: Utilization of artificial intelligence technology in an academic writing class: how do Indonesian students perceive? Contemp. Educ. Technol. **15**(4), ep450 (2023). https://doi.org/10.30935/cedtech/13419
7. Washington, J.: The impact of generative artificial intelligence on writer's self-efficacy: a critical literature review. SSRN J. (2023). https://doi.org/10.2139/ssrn.4538043

8. Nakajima, S.: Introduction to TPM - Total Productive Maintenance: Total Productive Maintenance, 5th edn. Productivity Press, Cambridge (1990)
9. Kruse, O.: Keine Angst vor dem leeren Blatt: Ohne Schreibblockaden durchs Studium, 12th edn. Campus-Verlag, Frankfurt/Main (2007)
10. Mayring, P.: Qualitative Inhaltsanalyse: Grundlagen und Techniken, 10th edn. Beltz Pädagogik. Beltz Verlag, Weinheim (2008)
11. Mayring, P., Gläser-Zikuda, M.: Die Praxis der Qualitativen Inhaltsanalyse, 2nd edn. Beltz Pädagogik, Weinheim (2008)
12. Dell'Acqua, F., et al.: Navigating the jagged technological frontier: field experimental evidence of the effects of AI on knowledge worker productivity and quality. SSRN J. (2023). https://doi.org/10.2139/ssrn.4573321

Evaluating Room Occupancy with CO2 Monitoring in Schools: A Student-Participative Approach for Presence-Based Heating Control

Robert Otto[1], Myriam Guedey[1](✉), Boris Pohler[2], and Dieter Uckelmann[1]

[1] University of Applied Sciences Stuttgart, Schellingstr. 24, 70174 Stuttgart, Germany
{robert.otto,myriam.guedey,dieter.uckelmann}@hft-stuttgart.de
[2] Humboldtgymnasium Solingen, Humboldtstr. 5, 42719 Solingen, Germany
pohler@humboldtgymnasium-solingen.de

Abstract. Effective climate control in buildings, crucial for both heating and air conditioning, depends on accurate room occupancy monitoring. Adapting climate systems to actual demand can provide huge energy savings. This study employs a calculation method and CO2 data from a school building to optimize energy usage based on room occupancy. In collaboration with students, CO2 levels in various rooms were measured to determine occupancy and estimate the number of people present. The outcomes highlight the method's effectiveness and its current limitations.

Keywords: Smart Building · Presence-Based Heating Control · CO2 Monitoring · Presence Detection · People Counting · Research-Based Learning

1 Introduction

In Germany, public buildings such as schools account for a large share of the total energy consumption, with the vast majority of end energy consumption being used for the generation and provision of space heating [1]. Furthermore, a significant portion of these structures remains unrenovated and thus energy inefficient. The refurbishment of public buildings – renovation of building envelopes, implementation of insulation, and installation of modern, efficient heating systems – is considered a long-term endeavor due to a shortage of materials and skilled workers in the building industry.

Turning to immediate solutions: the behavior of building occupants also plays a crucial role in determining the actual energy demand, especially in heating. Managing proper heating regulation on a room level is challenging, particularly in large public buildings with numerous occupants. Holding users accountable for this can be difficult, because it would depend highly on the responsibility of individuals. Consequently, the implementation of an automated, presence-based heating control system in these public buildings may represent a user-friendly approach, offering the potential for substantial energy savings.

M. E. Auer et al. (Eds.): STE 2024, LNNS 1028, pp. 23–31, 2024.
https://doi.org/10.1007/978-3-031-61905-2_3

To accomplish this objective, we employed carbon dioxide (CO2) monitoring as a metric for assessing occupancy levels in various classrooms of a school. This approach involved active student participation in data collection, utilizing an open CO2 monitoring system. Complementary educational resources were provided to enhance the students' understanding of CO2 measurement principles. They were also encouraged to engage in independent data analysis, fostering their ability to discern occupancy patterns and estimate the number of individuals present in a given space.

2 Background and Relation to Others

2.1 Energy Savings Through Presence-Based Climate Conditioning

A building's occupants play a crucial role in determining its energy demand, significantly impacting overall efficiency. Prioritizing occupant comfort, especially in heating and cooling systems, is paramount. Understanding when and how a building is used is key to optimization. A study by Mylonas et al. suggests that efficient management may achieve energy savings of up to 20% in schools [2]. This relies on users being aware of and optimizing their heating and cooling systems. In professional environments, guiding educators on energy-saving practices and providing periodic reminders has been effective. In the era of the Internet of Things, cost-effective monitoring allows data collection for energy-saving building operation based on actual occupancy. Implementing automated solutions in these systems can substantially aid in energy conservation [3,4].

2.2 Presence Detection and People Counting

Presence detection is essential for understanding occupancy patterns in various spaces of a buildings. Adeogun et al. utilized a multi-sensor approach, gathering data such as temperature, humidity, CO2 levels, and sound pressure. This data was used to train a machine learning algorithm for estimating room occupancy, achieving up to 94.6% accuracy [5]. Ding et al.'s review discusses various projects that focus on presence detection and prediction in buildings. They highlight the use of Passive Infrared (PIR) and CO2 sensors as common methods, with prediction accuracies between 80% and 95% [6]. Cali et al. developed an algorithm that estimates both occupancy and the number of individuals in enclosed spaces of residential and non-residential buildings using CO2 levels, achieving approximately 80% accuracy. This approach provides a straightforward, less technologically complex solution for evaluating room usage in buildings [7].

2.3 CO2 Monitoring in Non-residential Buildings

The assessment of air quality to promote physical health and cognitive well-being is well established in building management [8]. The DIN EN 16798 standard categorizes air quality based on CO2 concentration into four distinct indoor air level categories [9]. A concentration ranging from 400 to 800 ppm signifies high-quality air, while a range of 1000 to 1400 ppm represents the lower limit for acceptable air quality. To monitor CO2 levels in non-residential buildings, various devices are available on the market. During the COVID-19 pandemic, there was increased interest in monitoring devices to enhance room conditions by signaling the need for fresh air exchange and thus reducing infection risks. Faced with a market shortage at that time, researchers developed an open CO2 gauge that visually indicates CO2 levels and can also transmit data via Wi-Fi for remote room monitoring [10]. Beyond aiding in proper room ventilation, such systems can also be used to deduce actual room occupancy and gather additional information on room usage, which is beneficial for heating control.

3 Methodology

3.1 Determining Occupancy by Gradient Slope of CO2 Levels

In a closed environment, variations in CO2 levels indicate occupancy. Increasing CO2 levels typically signal the presence of people, while stagnant or decreasing levels suggest absence. To quantify room occupancy more precisely, we use established benchmarks for individual fresh air intake and CO2 emissions during periods of low activity. This method allows for a more accurate estimation of the number of individuals in a space. The linear increase of CO2 with ongoing activity furthermore allows for the extrapolation of measurements to infer the number of occupants. Additionally, by applying mean values in a given formula, we can calculate the CO2 concentration in a room, considering its volume, interior characteristics, and occupant count. In this analysis, natural ventilation due to leaks, like unsealed windows or doors, has not been considered, as previous measurements have shown that their impact is marginal.

CO2 Behavior in a Room Without Fresh Air Ventilation. We assume that E represents the total CO2 emission measured in parts per million [ppm], and e denotes the specific emission rate, which varies based on the room's volume V_{total} and the time t, expressed in ppm per cubic meter per minute $[ppm/(m^3 \times \text{min})]$.

$$\Delta E(t) := \text{absolute } CO_2 \text{ emissions over time in [ppm]} \tag{1}$$

$$e := CO_2 \text{ specific emissions in } [(ppm \times m^3)/min \times specificPerson] \tag{2}$$

$$\overline{e} := \text{mean } CO_2 \text{ specific emissions in } [(ppm \times m^3)/min \times Person] \tag{3}$$

$$P := \text{estimated count of people} \tag{4}$$

Room with one person:

$$\Delta E(t) = \frac{e \times t}{V_{\text{total}}} \tag{5}$$

Room with several persons:

$$\Delta E(t) = \frac{P \times \overline{e} \times t}{V_{\text{total}}} \tag{6}$$

People counting by CO2 emissions:

$$P = \frac{\Delta E(t) \times V_{\text{total}}}{\overline{e} \times t} \tag{7}$$

People Specific CO2 Emissions. Table 1 presents the standard values for the fresh air demand of both adults and teenagers based on Larsen and Ziegenfuß [11]. These values are crucial for accurately calculating the volume of air that individuals breathe within a given period. Approximately 4% of the air exhaled by a person is composed of CO2. Utilizing this information, we can effectively compute the incremental increase in CO2 concentration within a room per percentage point.

Table 1. Air demand of female (F) and male (M) persons in relaxed sitting mode

	F (adult)	F (teen)	M (adult)	M (teen)
Air intake [l/min]	7.2	6	9	7.5

Equation (8) illustrates the function for calculating CO2 emissions in ppm through breathing, differentiated by gender and age. Using the mean value described in Eq. (9), it becomes possible to calculate for spaces with individuals where the specific gender and age distribution is not explicitly known.

$$e = AirIntake_{Person} \times 0.04 \times 1 \times 10^6 \tag{8}$$

$$\overline{e} = \frac{e_{Fa} + e_{Ma} + e_{Ft} + e_{Mt}}{4} \tag{9}$$

3.2 Collaborative Data Collection in a School Building

To validate the introduced method for ascertaining occupancy and estimating people count, we collected data from real spaces. In collaboration with students from a school in Solingen, Germany, several measurement series were conducted in several classrooms of the school building. Additionally, we provided the students with learning materials on the topic along with a series of tasks. The project was implemented as part of the regular curriculum in mathematics and computer science over a period of eight weeks, with two school hours per week.

Data Collection. We provided assembled CO2 gauges (Fig. 1) to the students, who were divided into five groups, each consisting of two to three students. Each group was introduced to handling and measuring with the CO2 gauge. The CO2 gauge is designed as an open kit with open firmware [12], based on the ESP8266 microcontroller. It is capable of not only displaying the measured CO2 levels on the LED ring and saving them as CSV on the EEPROM but also sending the data via Wi-Fi for remote monitoring. Furthermore, the device offers a web server with a simple user interface to read the exact measurements. The students utilized the web server and CSV to collect the data.

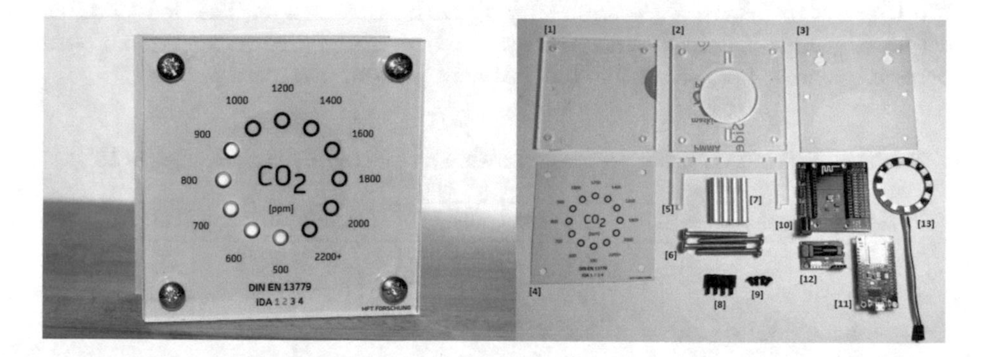

Fig. 1. The CO2 gauge kit used, displaying measured CO2 levels on a LED ring

The measurements were conducted during August 2023. Each measurement series took place in a classroom and lasted for a minimum of 15 min. During the measurements, students recorded the actual number of people in the room and noted any special occurrences, such as the times when windows or doors were opened or closed. Furthermore, they calculated the room volume. After each measurement series, they downloaded the CSV for further analysis. A total of 25 measurement series were recorded. Due to the heat during the summer season, it was not possible for the students to conduct all measurements in completely closed rooms, as it would have been ideal regarding the assumptions mentioned in Sect. 3.1.

Learning Tasks. In addition to collecting validation data, we employed a research-based learning approach, allowing the students to explore the topic and solve the problem independently. Therefore, together with the CO2 gauges, we provided the 10th-grade students with additional learning materials and a preliminary working plan. The plan comprised a series of tasks that progressively build on each other and may be skipped towards the end if the groups cannot proceed as quickly as anticipated.

The overarching learning objective is to cultivate a comprehensive understanding of the significance of CO2 monitoring, the measurement methods, the factors influencing its levels in rooms, the mathematical derivation of its slope, and the inference of the number of individuals in a room based on CO2 measurements. By project completion, students ideally not only have grasped the theoretical concepts but also have translated them into a mathematical function as well as developed a simple algorithm and thoroughly documented their work.

3.3 Outcomes

Outcomes from the Data Collection. Using student-collected data, we validated the method for estimating the number of individuals in a room from rising CO2 levels. Table 2 shows the values for these estimates, based on student measurements. The mean CO2 emission rate $\overline{e}$ was calculated at about 300 $[(ppm \times m^3)/min \times Person]$, varying with student age. This led to different accuracy levels across measurements. On average, there was an 11% absolute percentage error between estimated and actual room occupancy.

Table 2. Estimation based on measurements conducted by students (class 10B)

Measurement	10B_M1	10B_M2	10B_M3	10B_M4	10B_M5
Minutes	18	20	20	20	19
ΔE(t)	667	974	720	737	574
Volume total	189.5	189.5	189.5	189.5	189.5
$\overline{e}$	300	300	300	300	300
Persons estimated	23.41	30.76	22.74	23.28	19.08
Persons counted	26	26	21	23	23
Absolute percentage error	9.97%	18.32%	8.29%	1.20%	17.03%

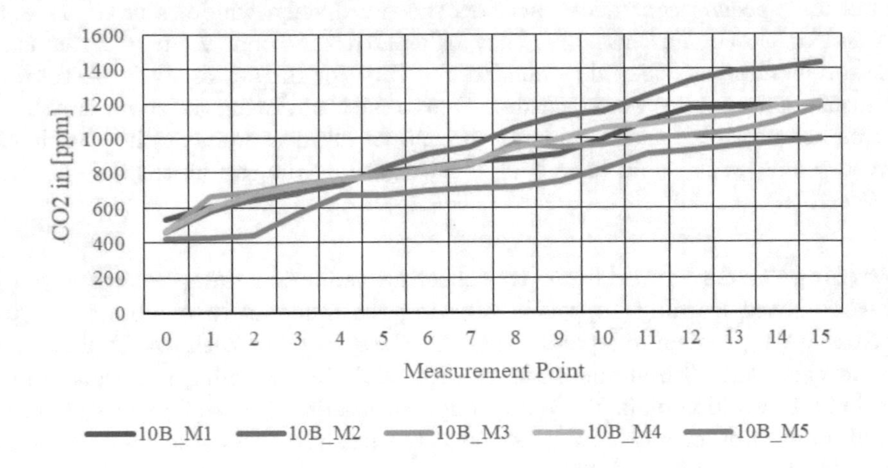

Fig. 2. Normalized measurements conducted by students (class 10B)

Figure 2 shows normalized measurements under various occupancy scenarios, demonstrating a clear correlation between occupancy and CO2 concentration in the room. The graph includes 15 data points over 18 to 20 min. This method is valid in closed environments with an increasing CO2 trend and no air ventilation caused by opening windows and doors or leaving persons. Figure 3 illustrates how open or tilted windows impact CO2 levels, showing a rapid decrease in concentration or a gentler slope compared to a closed environment.

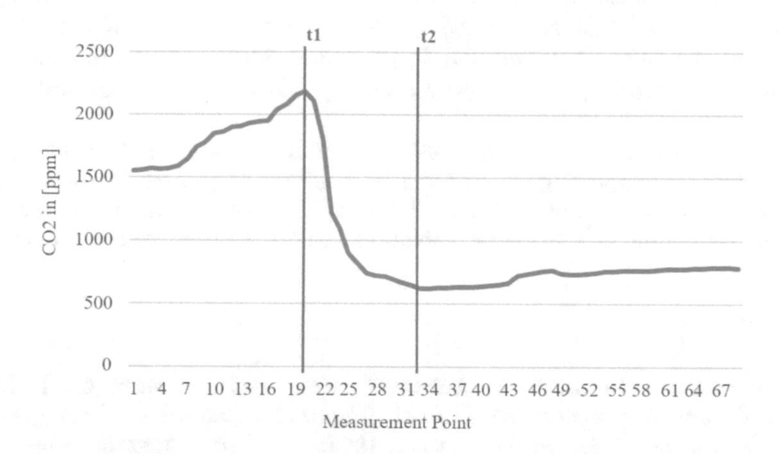

Fig. 3. Measurement in a classroom with open (t1) and tilted (t2) window event

Outcomes from the Learning Tasks. Upon evaluating the students' project documentation and considering feedback from their teacher, the research-based approach has demonstrated its effectiveness in fostering self-directed and problem-oriented learning. However, due to a reduction in school hours, the student groups were unable to completely engage with all assigned tasks. Furthermore, it was observed that the learning materials and tasks would benefit from additional guidance to support the students, rather than relying on their independent efforts alone. The project's focus on a relevant topic, applied within the students' own context, proved to enhance student engagement and interest.

4 Discussion and Outlook

The results show a clear correlation between measured parameters and room occupancy. Accurate person count estimation is feasible, with emissions per person adapting to activity levels. Measurement duration extension reduces deviations, but reliable assessment requires a slope over an extended period (>15 min), posing a limitation particularly for real-time applications. Addressing this in algorithmic implementation becomes crucial. In spaces with natural ventilation through open windows, the approach encounters limitations, necessitating detailed information on fresh air input and window openings for robust results.

Looking ahead, there is potential to optimize room utilization by leveraging heating surface areas. Information on the number of people in a room over time allows for insights into actual space utilization, which can help building operators to plan room usage more efficiently. Future efforts will focus on enhancing the accuracy of spatial estimation, recognizing the need for real-time adaptability and addressing challenges posed by varying ventilation conditions. The goal is to refine the calculation method to provide reliable insights into occupancy dynamics, contributing to an improved energy efficiency of school buildings.

Acknowledgments. The authors thank the German Federal Ministry of Education and Research for supporting the research by funding the project FH-Impuls 2016 I: Urban Digital Twins for the Intelligent City (13FH91061A). Furthermore, the authors thank the 10th-grade students of Humboldtgymnasium Solingen for their contribution.

References

1. Bigalke, U., Armbruster, A., Lukas, F., Krieger, O., Schuch, C., Kunde, J.: Der dena-Gebäudereport 2016. Statistiken und Analysen zur Energieeffizienz im Gebäudebestand, Technical report, Deutsche Energie-Agentur GmbH (dena) (2016)
2. Mylonas, G., Amaxilatis, D., Tsampas, S., Pocero, L., Gunneriusson, J.: A methodology for saving energy in educational buildings using an IoT infrastructure. In: 2019 10th International Conference on Information, Intelligence, Systems and Applications (IISA), pp. 1–7. IEEE (2019). https://doi.org/10.1109/IISA.2019.8900707, https://ieeexplore.ieee.org/document/8900707/
3. Otto, R., Guedey, M., Uckelmann, D.: Developing user- and technology-driven use cases for smart city applications. In: 16th Multi Conference on Computer Science and Information Systems (MCCSIS). Proceedings of the CGVCVIP, CSC, BIGDACI and TPMC 2022 Proceedings, Lissabon, Portugal (2022)
4. Metallidou, C.K., Psannis, K.E., Egyptiadou, E.A.: Energy efficiency in smart buildings: IoT approaches. IEEE Access **8**, 63679–63699 (2020). https://doi.org/10.1109/ACCESS.2020.2984461
5. Adeogun, R., Rodriguez, I., Razzaghpour, M., Berardinelli, G., Christensen, P.H., Mogensen, P.E.: Indoor occupancy detection and estimation using machine learning and measurements from an IoT LoRa-based monitoring system. In: 2019 Global IoT Summit (GIoTS), pp. 1–5 (2019). https://doi.org/10.1109/GIOTS.2019.8766374
6. Ding, Y., Han, S., Tian, Z., Yao, J., Chen, W., Zhang, Q.: Review on occupancy detection and prediction in building simulation. Build. Simul. **15**(3), 333–356 (2022). https://doi.org/10.1007/s12273-021-0813-8
7. Calì, D., Matthes, P., Huchtemann, K., Streblow, R., Müller, D.: CO2 based occupancy detection algorithm: experimental analysis and validation for office and residential buildings. Build. Environ. **86** (2015). https://doi.org/10.1016/j.buildenv.2014.12.011
8. Gesundheitliche Bewertung von Kohlendioxid in der Innenraumluft, Bundesgesundheitsblatt - Gesundheitsforschung - Gesundheitsschutz **51**(11) (2008). https://doi.org/10.1007/s00103-008-0707-2
9. Deutsches Institut für Normung e.V., DIN EN 16798-3:2017-11 (2017)

10. Erhart, T.: CO2-Ampel: Lüften gegen Covid-19 (2020). https://www.hft-stuttgart.de/forschung/news
11. Larsen, R., Ziegenfuß, T.: Physiologie der Atmung. In: Larsen, R., Ziegenfuß, T. (eds.) Beatmung: Grundlagen und Praxis, pp. 18–53. Springer, Heidelberg (2009). https://doi.org/10.1007/978-3-540-88812-3_2
12. Duminil, E., Otto, R., Guedey, M., Erhart, T.G., Stave, J., Käppler, M.: CO2-Ampel HFT Stuttgart (2020). https://transfer.hft-stuttgart.de/gitlab/co2ampel

Data-Driven Mobility and Transport Planning in Municipalities: Smart Solutions for Limited Resources

Erik Höhne[1], Tobias Teich[2], Oliver Scharf[2](✉), Sven Leonhardt[2], Maximilian Schlachte[2], Martin Trommer[2], Christoph Mewes[1], Manoel Kraus[2], Susan Bergelt[2], and Silvia Queck-Hänel[1]

[1] Stadt Zwickau (City of Zwickau), Hauptmarkt 1, 08056 Zwickau, Germany
[2] Westsächsische Hochschule Zwickau (University of Applied Sciences Zwickau), Kornmarkt 1, 08056 Zwickau, Germany
oliver.scharf@fh.zwickau.de

Abstract. The research paper highlights the importance of data-driven approaches in urban and mobility development for local communities. Tailor-made solutions, based on a precise understanding of local challenges, are essential, especially for smaller or financially weaker municipalities. Implementing a data-gathering infrastructure, particularly within Smart City concepts, often presents challenges due to limited technical and financial resources, as well as a lack of expertise. To address these issues, a video-based traffic detector was developed to enable flexible and minimally invasive traffic measurement. This technology differs from traditional AI models through its miniaturization and application to less powerful embedded systems. The 'Machine Learning on the edge' approach was tested in Zwickau, Saxony, Germany, where its robustness and reliability were confirmed by comparison with manual traffic counts. The field study demonstrated that reliable data collection is possible even with minimal technical resources. This contrasts with traditional traffic detection technologies, which rely on induction loops and compute-intensive systems. The novel approach offers a cost-efficient alternative, particularly suitable for minor townships with limited fiscal capacity. The study's conclusions recommend promoting data-based approaches in municipal planning, investing in cost-effective technologies like video-based traffic detectors, emphasizing scalability and adaptability, and continuously monitoring and improving data solutions. This enables the implementation of Smart City concepts even in financially challenged municipalities, contributing to the improvement of local quality of life.

Keywords: Tiny Machine Learning · video-based detection · AI model

1 Introduction

The transformation of urban mobility towards more environmentally friendly and sustainable systems is a key concern for cities and municipalities worldwide. This change is crucial to reducing air pollution and making cities more liveable overall. Digitalisation

M. E. Auer et al. (Eds.): STE 2024, LNNS 1028, pp. 32–39, 2024.
https://doi.org/10.1007/978-3-031-61905-2_4

plays a key role in this endeavour, as it has the potential to make mobility more efficient, cleaner and smarter. By using digital technologies, traffic can be controlled more effectively and traffic flows can be better managed, which helps to avoid congestion and improve overall traffic efficiency. Intelligent vehicles, which are informed about available parking spaces through real-time data, and networked mobility services, such as sharing models or on-demand services, illustrate the potential of these technologies. In addition, digitalisation in the logistics sector leads to optimised use of free capacity and reduces unnecessary empty runs [1].

Despite these advantages, it is mainly progressive pioneering municipalities that have benefited from the digital mobility revolution so far. These municipalities have in-depth knowledge of handling data, sufficient staff, advanced IT infrastructures and the necessary financial resources. Many municipalities, on the other hand, face the challenge of developing their own data infrastructures with limited financial, human and time resources and a lack of expertise. They are often reliant on expensive third-party providers, whereby the full potential of the data provided remains unutilised, partly due to a lack of specialist knowledge about the potential applications of this data [2].

In this dynamic and challenging landscape, the city of Zwickau, in close cooperation with the Westsächsische Hochschule Zwickau - University of Applied Sciences Zwickau, is pursuing an innovative approach to bridge the gap between pioneer and laggard cities. Through the targeted use of open-source technologies, the city aims to develop scalable and cost-efficient solutions that can be applied beyond individual cases. This strategy emphasises the importance of knowledge exchange and cooperation between universities and local authorities in order to overcome technical and financial hurdles. A concrete example of this is the cooperation project "Z-Move 2025" [3], which demonstrates how a data-generating infrastructure can revolutionise mobility development in municipalities even without extensive technical and financial resources and expertise. By presenting the methodology, results and implications of this approach in detail, this paper aims to make a substantial contribution to the discussion on the future of urban mobility and smart city initiatives.

2 Background and Theoretical Framework

The concept of the smart city, similar to the established concept of sustainability, embodies progressive and future-oriented visions for urban development. Despite its increasing relevance and popularity in contemporary urban planning discussions, the smart city concept lacks a generally recognised and precise definition. The term "smartness" is frequently used in academic discourse, but its meaning often remains undefined and multidimensional. Nevertheless, the concept of the smart city is characterised by two distinct features: the efficient use of resources and the comprehensive integration of digital technologies. These elements form the core of the Smart City initiative and clearly distinguish it from other urban planning approaches [4]. At its core, the smart city concept aims to achieve a harmonious fusion of urban life and cutting-edge technology, creating a dynamic and efficient urban ecosystem.

The increasing digitalisation of numerous areas of life is also creating opportunities for urban development to actively shape change and use it for sustainable urban development. Information and communication technologies play a key role in this by intelligently networking municipal infrastructures in the areas of energy, buildings, transport, water and wastewater. The basic prerequisite is access to high-performance digital infrastructures and municipal expertise regarding data responsibility and data sovereignty. Only these key components will enable a new level of urban interaction and administration that is based on efficiency [5].

In the past, municipal traffic monitoring was primarily based on traditional methods aimed at analysing urban traffic flows and improving road safety. These methods included manual observations and counts of vehicles at strategic points, supplemented by mechanical devices such as pneumatic tube counters, which enabled limited analyses at a local level. Also common were standardised, isolated technologies such as inductive loops for traffic light control and traffic counting as well as traffic cameras for congestion and accident detection. These technologies, often provided by external providers, were associated with high acquisition, licence and maintenance costs and offered only limited application possibilities. A key weakness was the lack of networking and communication between the systems, which made holistic urban traffic planning difficult. These outdated technologies tended to be inaccurate in complex traffic situations, and aspects such as data protection, security, scalability and data integration were often insufficiently considered. These limitations emphasise the urgency for continuous innovation and investment in advanced, flexible and integrated traffic monitoring systems to meet the demands of modern urban mobility [6].

The ongoing digitalisation of urban traffic management as part of smart city initiatives offers municipalities substantial advantages. Through the use of real-time data collection infrastructures supported by modern sensor technology and advanced analysis methods such as artificial intelligence (AI) and big data, city-wide traffic flows can be organised more efficiently. Not only does this lead to an optimised use of road infrastructure and a reduction in traffic congestion, but can also help to improve air quality. In addition, data-based systems promote urban planning and services by providing deeper insights into the behaviours and needs of city residents, which in turn supports informed decision-making at municipal level [1].

However, smaller and financially weaker local authorities in particular face specific challenges when implementing these technologies. The initial investment costs as well as the ongoing expenses for maintaining and updating such systems can represent a significant financial burden. Added to this is the need for technical expertise and resources, which are often limited in smaller municipalities. Furthermore, data protection and data security issues require careful consideration and measures to ensure the handling of sensitive information.

Despite these obstacles, smaller municipalities can also benefit from data-generating infrastructure. This can be achieved through gradual and targeted investments, the establishment of collaborations with other municipalities, research institutions or private partners, as well as by utilising funding and subsidies to ease the financial burden.

As part of the Z-Move 2025 project, the city of Zwickau is cooperating with the University of Applied Sciences Zwickau to work on innovative and intelligent mobility solutions. The focus here is on the development of advanced traffic monitoring technologies. These make it possible to analyse and optimise the current traffic situation live. The data obtained is also used to enable local mobility users to organise their own mobility more efficiently. A key feature of these technologies is their cost efficiency and the open and compatible interfaces. The data-based approach to improving mobility is analysed in detail below.

3 Technical Realisation and Results

Various technologies such as inductive measuring loops, laser, infrared, radar and camera-based systems are available for the automated detection of bicycle traffic, with each method having specific advantages and disadvantages [7]. The primary goal of this project was to develop a cost-efficient and energy-autonomous bicycle counter that utilises machine learning algorithms on resource-limited microcontrollers.

The first implementation of this system took place on the Zwickau Mulderadweg, an intensively used route in the immediate vicinity of the university and research site of the "ubineum" science and transfer centre. The traffic detector was designed specifically for the conditions there, which include regular use by cyclists, pedestrians and occasionally other road users.

The data collected on bicycle traffic is visualised locally on a display and also transmitted digitally via a LoRa radio interface for further analyses and public communication. Particularly noteworthy is the design of the measuring system, which enables a high degree of flexibility and self-sufficient operation at locations without an existing power supply. At the selected location, the energy supply is realised by photovoltaics in combination with a battery storage system, which ensures sustainable energy independence and at the same time guarantees flexibility and mobility (Fig. 1).

In contrast to traditional bicycle detection systems, which are based on induction coils in the road surface and require extensive structural interventions, this project aims to provide a flexible, less stationary solution. The integration of camera systems with object recognition functions represents a technically advanced alternative to bicycle detection. The latest developments in the field of Tiny Machine Learning (TinyML) enable the use of image recognition algorithms on energy-efficient microcontroller systems, allowing a simplified yet precise AI model to be implemented on a hardware-limited platform.

TinyML stands for the implementation of machine learning models on small, resource-limited devices at the edge of the network (edge devices), such as microcontrollers. This technology makes it possible to realise sophisticated applications such as image recognition and data analysis directly on site without the need for external computing power or cloud connections. The key advantages of TinyML are its energy efficiency, data protection and minimised latency, which is particularly relevant for applications in resource-constrained environments [8].

As part of this project, three different hardware systems - ESP32-CAM [9], Raspberry Pi [10] and OpenMV [11] - were evaluated in terms of their suitability for the requirements set. Although the Raspberry Pi was convincing in terms of performance, it was not found to be suitable for the desired energy-autonomous application due to its relatively high power consumption. After comparing the remaining candidates, ESP32-CAM and OpenMV, the choice fell in favour of OpenMV. This decision was based on the superior performance of OpenMV, its low power consumption and its excellent support

Fig. 1. Modern cycle traffic recording: the fully installed and functional cycle counter in action.

for machine vision applications. An OpenMV-based system was developed that is capable of image processing and machine learning algorithms directly on the edge device, enabling efficient and autonomous data acquisition (Fig. 2).

Fig. 2. OpenMV camera module in action: integrated into a customised 3D-printed mount for optimised traffic monitoring.

In a typical TinyML workflow, the process begins with the collection, preparation and transformation of data required to train machine learning models. The TinyML software Edge Impulse [12] provides a comprehensive solution for the development and implementation of such models on embedded edge devices. In particular, this software platform supports the OpenMV hardware and covers the entire TinyML workflow.

The development process for the model takes place in the Edge Impulse Studio, which is characterised by its user-friendly web interface. Within this studio, the training of the model is initiated, whereby both the progress of the training process and the performance of the developed model are continuously monitored and analysed. Once training is complete, the finished model is deployed on the OpenMV microcontroller and finally tested and applied in the field (Fig. 3).

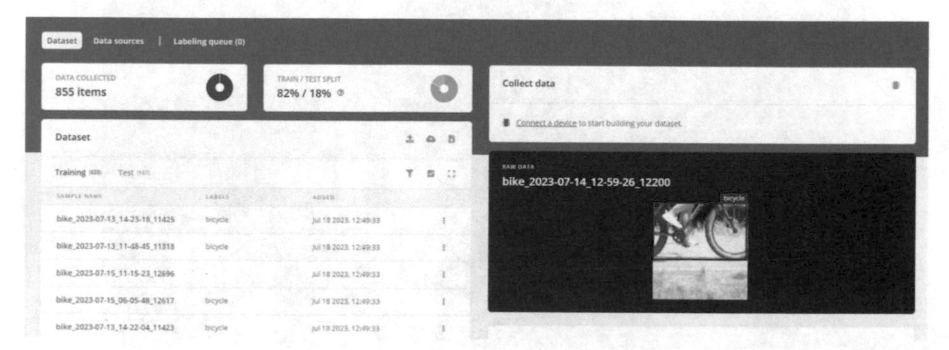

Fig. 3. Edge-Impulse Studio: Precise classification and analysis of training data for advanced traffic detection.

In order to validate the accuracy and reliability of the automated cycle traffic count, a validation procedure was carried out that included a random manual count. In this method, cycle traffic was recorded manually over a period of one hour and recorded in parallel with the automated records. This manual data collection served as a reference to check the consistency with the data generated by the automated count. The subsequent comparison of these two data sets resulted in a match rate of 91%, which can be considered a satisfactory result and underlines the effectiveness of the automated counting system.

The resource-saving and energy-efficient approach of this project opens up significant scaling possibilities. The self-sufficient and advanced design of the counting technology enables flexible application at various locations and adaptable customisation to different types of traffic. This is achieved through specific training of the system, as described in this paper. The modularity and interface compatibility of the technology allows it to be customised to the requirements of individual municipalities. This includes the provision of transparent traffic information for passers-by by means of digital displays as well as the presentation and processing of data on digital platforms.

Beyond the purely data based utilisation for transport planning, the external communication of the knowledge gained offers far-reaching potential. In an era of mobility transformation, constructive and progress-orientated communication about sustainable mobility concepts is of crucial importance. The approach used in this field test pursues two specific strategies: Firstly, cyclists receive motivating and interactive feedback via a digital display that informs them about the number of daily, monthly or annual road users. Secondly, the Zwickau mobility platform developed as part of the project enables online access to live data. This platform supports in-depth analyses through customisable dashboards and the integration of additional data sets to support a comprehensive understanding of urban mobility patterns (Fig. 4).

Fig. 4. Interactive Z-Move data platform: real-time visualisation of cycle traffic with connected counter and personalised dashboards.

4 Discussion and Conclusions

This paper underlines the central role of digitalisation in the transformation of urban mobility systems. The implementation of digital technologies, as pursued in the city of Zwickau and in the "Z-Move 2025" project, demonstrates how more efficient, cleaner and smarter mobility solutions can be realised. These developments make a significant contribution to making cities more liveable and reducing environmental pollution.

At the same time, the analysis highlights the challenges faced by smaller and financially weaker municipalities in particular. The lack of resources and expertise can make it difficult to implement digital mobility concepts. Solutions such as the use of open-source technologies and the promotion of knowledge exchange between municipalities and scientific institutions offer promising prospects here. The Z-Move 2025 project shows how cooperation and innovative approaches can enable the effective utilisation of limited resources.

The introduction of new technologies such as TinyML as part of the project represents significant progress. These technologies make it possible to carry out complex applications such as energy-efficient image recognition and analysis on the edge. The application of these technologies to microcontroller-based systems not only offers cost benefits, but also contributes to data protection and enables fast data processing.

The data and insights gained from such projects offer valuable insights for urban planning and mobility development. They enable more informed decision-making and support municipalities in developing future-oriented, sustainable and user-friendly mobility solutions. This is particularly important in the context of smart city initiatives, where the efficient use of resources and the integration of digital technologies are key objectives.

The results of this paper emphasise the importance of digitalisation for the future development of urban mobility. It becomes clear that the success of such transformation processes strongly depends on the availability of resources, knowledge and cooperation. The Zwickau case study provides an inspiring example of how innovative approaches and interdisciplinary cooperation can achieve significant progress in urban mobility development. In conclusion, the path to more sustainable and efficient urban mobility is challenging, but can be successfully realised through the use of advanced technologies and collaborative approaches.

References

1. BMVI: Digitalisierung kommunaler Verkehrssysteme. Journal (2019)
2. Höhne, E., et al.: Smarte, digitale Handlungsansätze zur Gestaltung der Mobilitätswende als Chance für Mittelstädte im ländlichen Raum – Fallbeispiel Zwickau. In: Leonhardt, S., Neumann, T., Kretz, D., Teich, T., Bodach, M. (eds.) Innovation und Kooperation auf dem Weg zur All Electric Society, pp. 143–174. Springer, Wiesbaden (2022). https://doi.org/10.1007/978-3-658-38706-8_8
3. More information. https://www.zwickau.de/de/politik/emobilitaet/region/forschungsprojekte/Z-Move.php
4. Haupt, W.: The sustainable and the smart city: distinguishing two contemporary urban visions. In: The Palgrave Encyclopedia of Urban and Regional Futures, Macmillan. IOP Publishing Ltd. (2020)
5. BBSR: Smart City Charta – Digitale Transformation in den Kommunen nachhaltig gestalten. Journal (2019)
6. FGSV: Empfehlungen für Verkehrerhebungen. Journal (2012)
7. Eder, E.: Counting systems for bicycle traffic – analysis in Austria and tests. Master thesis, pp. 12–23. Graz (2020)
8. Iodice, G.M.: TinyML Cookbook. Packt (2022)
9. More information. https://www.az-delivery.de/products/esp32-cam-modul-esp32-wifi-bluetooth-modul-inklusive-kamera%20%0d
10. More information. https://www.raspberrypi.com/products/raspberry-pi-4-model-b/
11. More information. https://openmv.io/
12. More information. https://edgeimpulse.com/

Investigating the Effect of Personal Emotional Score Display on Classroom Learning

Brainerd Prince(✉), Siddharth, Vinayak Joshi, and Rukmani Keshav

Plaksha University, IT City, Mohali, Punjab, India
brainerd.prince@plaksha.edu.in

Abstract. In this work, we present the preliminary results from an exploratory experiment in which we studied the influence of exposing students to a personal emotional score on their classroom learning. This score derived from deep learning algorithms is presented to students on a computer screen while they are learning in a classroom. With the advent of deep learning models for the classification of emotions, it is now possible to gauge emotions in real-time. Most of these algorithms depend on facial expressions to identify human emotional and attentional levels. Such algorithms are even being integrated within video conferencing platforms such as Zoom, Teams, etc. to quantify if the participants in the meeting are attentive. The assumption here is that our facial expressions reflect our emotions.

The hypotheses tested by this study is a) if such algorithms accurately reflect human emotions, and b) if the very act of showing such a metric to learners (students in a classroom) will influence their cognitive learning capabilities. We conducted the experiment in a freshman-year undergraduate class where students were asked to compile a bibliography, read the research papers and annotate them, and then write a summary. We used a state-of-the-art deep neural network trained on thousands of face images in multiple orientations to detect seven basic human emotions (neutral, happiness, sadness, anger, surprise, disgust, and fear) from the students' facial expressions. Students' facial images were captured in real-time through their laptop's camera. The experimental group of students were presented with their emotional scores and labels while they were working in the class. The control group of students were not exposed to this information. After each class, we asked all students to do a self-assessment of scores and asked for the instructor's assessment of the students' work as well. We investigated if the self-perception of their own emotional states helped students to perform better. The key outcomes of this study are threefold. First, the AI models have not reached a stage yet where they could accurately reflect human emotions using facial expressions. Second, the study shows that the self-perception of emotions provide a boost to the perception of productivity in students. Finally, this study also reveals that a higher self-perception of productivity due to being exposed to the emotion recognition AI model do not necessarily contribute to actual higher productivity.

Keywords: Deep Learning · Attention Monitoring · Classroom Learning · Emotion Detection · Affective Computing

M. E. Auer et al. (Eds.): STE 2024, LNNS 1028, pp. 40–51, 2024.
https://doi.org/10.1007/978-3-031-61905-2_5

1 Introduction

In the evolving landscape of Computer-Based Learning Environments (CBEL), the transition from traditional computer-aided instruction to the integration of deep learning technologies signifies a notable shift toward learner-centered approaches. As CBEL strives not only to yield favorable learning outcomes but also to actively engage learners in the educational process, the focus shifts to creating engaged learners. An engaged learner shows behavioral, emotional, and intellectual involvement during learning (Shen et al., 2009). Despite the central role played by emotions in different transitions, the focus of research in higher education leans towards cognitive and motivational aspect of learning (Postareff et al., 2017). However, in recent years researchers have begun to explore the role of emotions in successful learning and motivations.

This study contributes to the discourse by outlining the use of affective computing in learning process. It showcases the machine's capability to recognize learner emotions, a critical aspect of the proposed affective computing framework (Shen et al., 2009). Sundström's (2005) proposed an interesting concept called an "affective loop," which denotes an affective interaction process where emotions play a significant role in a student's interaction and involvement during learning. This affective interaction process provided a theoretical starting point for our research.

Emotion recognition emerges as a pivotal step in the pursuit of affective computing, with various efforts directed towards deciphering emotions through facial expressions, speech, and physiological signals (Cowie et al., 2001). The evolution of e-learning platforms, transitioning from Intelligent Tutoring Systems (ITSs) to Affective Tutoring Systems (ATSs), reflects the integration of emotion detection methods. These methods aim to automatically identify students' emotional states during the learning process (Craig et al., 2004). Current approaches encompass physiological signals, facial expressions, speech, and physical postures for emotional feature identification (Moridis & Economides, 2009). Notably, facial-expression-based recognition emerges as a promising avenue, given its efficacy in capturing a significant portion of emotional information conveyed by students and its non-intrusive nature compared to embodied devices. In our study, we specifically explore the implications of incorporating automatic emotion recognition using an AI model during real-time writing assignments.

The idea is to understand if there is any change in the productivity levels of learners and if they undergo adaptive emotion regulation that enhances their motivation during the process.

This paper contributes to understanding the role of cognizance of emotions facilitated by emotional feedback mechanisms within a Computer-Based Learning Environment (CBLE). This establishes a connection between how a student feels and the outcomes of task performance. The subsequent sections delineate the theoretical framework guiding this research, present empirical evidence for change in productivity and performance levels following emotion feedback in CBLEs, and delve into the broader landscape of using emotions to facilitate motivation in students.

2 Background

The exploration of the human face finds a historical milestone in Charles Darwin, who, in 1872, not only organized important studies on facial expressions but also proposed in his book "The Expression of the Emotions in Man And Animals" that emotional expressions are fundamental, universal across species, and innate without cultural variations (Miranda et al., 2015).

Building on Darwin's insights, psychologist Paul Ekman and other researchers, through substantial evidence, supported the idea that basic emotions are universally expressed innately by humans. In 1978, Ekman and Friesen introduced the Facial Action Coding System (FACS), a scientific method for measuring facial actions in humans. The FACS was later revised, with a critical Portuguese version presented by Freitas-Magalhães (2016a), featuring Action Descriptors (AD) and Movements (M) as primary categorizations of facial movements. The 44 main units, known as Action Units (AU), correspond to specific facial muscles and are studied in both the upper and lower face (Xia, 2020). The universality of six basic facial expression categories (fear, happiness, sadness, disgust, anger, and surprise) across different cultures highlights the significance of facial expressions (Ekman et al., 1987).

The applications of FACS extend across various domains, including research on the relationship between voice and facial expression, graphic arts applications of facial studies, the effects of botulinum toxin on the face, and the correlation between wrinkles and action units (Aoki et al., 2021). Moreover, FACS contributes to communication processes in classrooms, self-assessment through emotion recognition, interpersonal communication, academic endeavors, and clinical applications such as nonverbal therapeutic communication, pain assessment, and the impact of recognizing facial expressions on social interactions (Adolphs, 2002).

Through this study, we explore the correlation between emotions and facial expressions and investigate whether knowledge of emotions contributes to better productivity in an educational setting. Facial expressions are crucial for social interaction, allowing individuals to quickly infer the emotional state of their peers.

Emotion recognition, primarily from facial expressions, plays a crucial role in communication, team engagement, and productivity (Gadigi & Veerabhadrappa, 2023). As student learning outcomes are prioritized in modern education, it becomes important to understand the significance of emotions and their impact on individual and team output. Productivity is a key factor in organizational performance, and technologies have historically influenced productivity. One way to measure an employee's level of productivity is to look at the amount of work they get done in a given amount of time.

Emotion, a complex feeling state, is intertwined with motivation, influencing human effort toward goals. Employees often bring their emotions to work, and negative emotions can impact workplace dynamics. The difficulty in introspecting and providing accurate information about one's emotions underscores the importance of understanding emotional states in the workplace (Silva et al., 2022).

Rosalind Picard (1999) emphasized the significance of emotions in the computing field, encompassing both emotion detection and expression. Emotion detection is crucial for creating adaptive computer systems, enhancing productivity by aligning with users' personalities. Studies (Dryer, 1999) highlight the collaborative benefits of personality-matched interactions and the perception of computers having personalities. Longitudinal understanding of users' emotional states allows computers to adapt working styles, fostering increased productivity (Ark et al., 1999).

Automatic facial expression analysis, a field driven by machine learning, evaluates worker efficiency and motivation through emotion detection. Drawing on Paul Ekman's work on Facial Action Coding System (FACS), the methodological approaches to studying nonverbal behavior are discussed below. The Facial Action Coding System originated from an examination of the anatomical foundations of facial movement and is purportedly applicable to delineate any facial motion observed through various media, employing anatomically defined action units (Ekman, 1982).

However, recent scholars such as Lisa Feldman Barrett, critically examine the scientific basis for the widespread assumption that a person's emotions can be easily deduced from their facial expressions. These scholars critique the dominant position put forward by Paul Ekman that there human emotions can be recognized through facial expressions across cultures. Barrett questions whether the validity of its increasingly prevalent applications, such as in technology companies attempting to interpret emotions, schools instructing children on emotional expressions, federal agents training to detect emotions and predict behaviours from facial movements, and mental-health specialists diagnosing and treating psychiatric disorders based on facial expressions (Barrett et al., 2019).

Scholarship and research based on Ekman's theory support the view that specific emotions are reliably conveyed through certain facial-muscle configurations leading to six emotion categories (anger, disgust, fear, happiness, sadness, and surprise). Barrett and other scholars proposed four criteria to question if a facial expression reflects a person's emotional state: reliability, specificity, generalizability, and validity. Reviewing studies on expression production and emotion perception across various populations, they found evidence contradicting Ekman's claims. Smiles, for example, may signal submission instead of happiness (Cordaro et al., 2018). Thus, Barrett and others advocate for new multidisciplinary research methods, incorporating neuroscience and machine learning, to explore how people express emotions in everyday contexts and to develop innovative hypotheses about the nature of emotion (Barrett et al., 2019). The presented paper seriously engages with Barrett as well as Ekman even though its experiment design is based on Ekman's premises. In light of this debate, we raise the following questions:

1. Does AI models accurately reflect human emotions using facial expression data?
2. Does perceived emotional state has an impact on productivity?

3 Methodology

3.1 Research Design

There are many machine learning models utilizing feed from a camera to recognize human emotions using facial expressions. These machine learning models are usually trained on large emotion databases such as the Facial Emotion Recognition (FER) 2013 dataset which has neutral and six primary emotion categories (happy, sad, angry, surprise, fear, and disgust). FER 2013 dataset contains approximately 30,000 facial color images of different expressions with size restricted to 48×48 pixels, ranging from 600 images for disgust to about 5000 images for the other emotions.

Some AI models using facial expressions for emotion recognition are from Face++ AI, Affectiva, Faception, IBM, and Microsoft etc. (Buolamwini & Gebru, 2018).

We utilized one such AI model developed by Dr. Alexa Hagerty, Dr. Igor Rubinov, Dr. Alexandra Albert et al., called Emojify (available to use online free of cost at https://emojify.info/menu) which is able to gauge and display human emotions utilizing a laptop's camera feed. This model has the added advantage of being trained on thousands of images, therefore providing robust performance in real-time enabling us to conduct our experiments with students as they simultaneously work on a different productivity task.

This study will use AI model-based emotions detector and its effect upon the productivity (perceived and actual scores) of college students while doing an assignment. It will be a repeated measures study, where all the students will go through both the situations (control and experiment).

3.2 Participants

The study involved 22 (14 male, 8 female, mean age = 19.25 years) students from a technological university who are in their sophomore year and taking the course named Research Communication. The participants were chosen randomly and assigned into two groups. This was a crossover study where both groups went through the control and experimental conditions at different periods of time.

3.3 Procedure

1. The experiment began with a basic summary being given to students about the experiment, opening the website on their laptop and what they had to do with a short demo. (5–10 min).
2. Students were randomly divided into two groups based on tutors (V & R).
3. Students were asked to identify two research-based solutions to any real-world problem. (15 min).
4. *Experiment Group:* For solution 1 they were required to go through these steps: Bibliography (10 min) > Annotations (10 min) > Writing out the solutions (10 min).
5. After every 5 min, they were subjected to report the emotion they experienced, intensity and the emotion that the AI camera reports. They also reported on the dissonance between the emotion that they felt vs what the AI model shows. This dissonance score was on a range of 0 to 10 where 0 meant absolutely no dissonance (such as students being neutral and AI model reporting neutral) while 10 meant high dissonance (such as students being sad and the AI model reporting happy). The students were also asked to rate self-perception of their productivity on a scale of 0 (very low) to 10 (very high).
6. *Control group*: For solution 1, they also had the same task, Bibliography (10 min) > Annotations (10 min) > Writing out the solutions (10 min). This group did not fill any form related to the AI emotion recognition model except for a final form asking about their self-perception of productivity on a scale of 0 (very low) to 10 (very high).

7. Students then moved to solution 2 and the groups are switched (Control Group becomes experiment group and vice versa for solution 2).
8. Same procedure was repeated. (30 min).
9. End survey was given to every student. (5 min).
10. End of experiment.
11. Total experiment duration: 1 h 30 min.

In order to gather data on the actual productivity of students, in line with a predesigned rubric, the submissions of the students were assessed on a range of 0 (very poor) to 10 (excellent).

4 Results

(See Table 1).

Table 1. Descriptives of both final scores and perceived productivity scores for both the groups.

Descriptives									
	N	Missing	Mean	Median	SD	Minimum	Maximam	Shapiro-Wilk	
								W	P
Final Score Experiment	22	0	8.64	9.50	1.65	6	10	0.750	<.001
Perceived Productivity Experiment	22	0	4.77	5.00	3.18	0	10	0.922	0.083
Final Score Control	22	0	8059	9.00	1.56	6	10	0.797	<.001
Perceived Productivity Control	22	0	7.27	7.00	1.96	4	10	0.909	0.045

4.1 Dissonance Between AI-Reported and Self-reported Emotions

Null Hypothesis: There is no significant difference between self-reported emotions and AI reported emotions.

Alternate Hypothesis: There is a significant difference between self-reported emotions and AI reported emotions.

As the Fig. 1 depicts, a single sample T-test was done for dissonance ratings between AI-reported and self-reported emotions shows that the students reported lower level of dissonance ($M = 2.63$, $SD = 2.71$) than expected norm rating, $t(22) = 4.59$, $p < .001$.

Quite interestingly, As Fig. 2 above shows, the overall dissonance might be lower as majority of responses were neutral. Which is why a single sample T-test was also done for dissonance ratings of all the emotions excluding the neutral emotion.

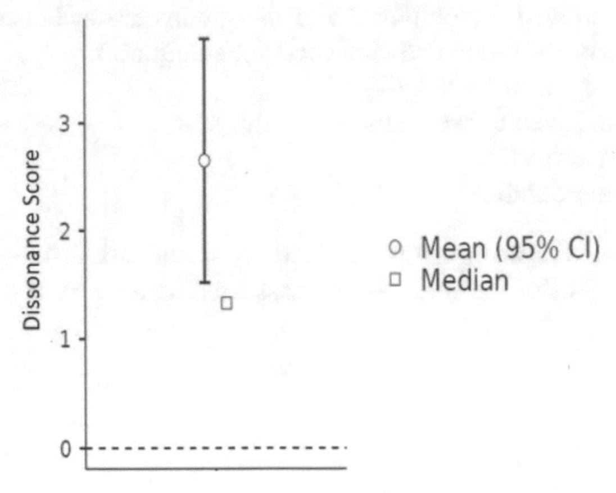

Fig. 1. Graphical depiction of reported dissonance between AI-reported and self-reported emotions.

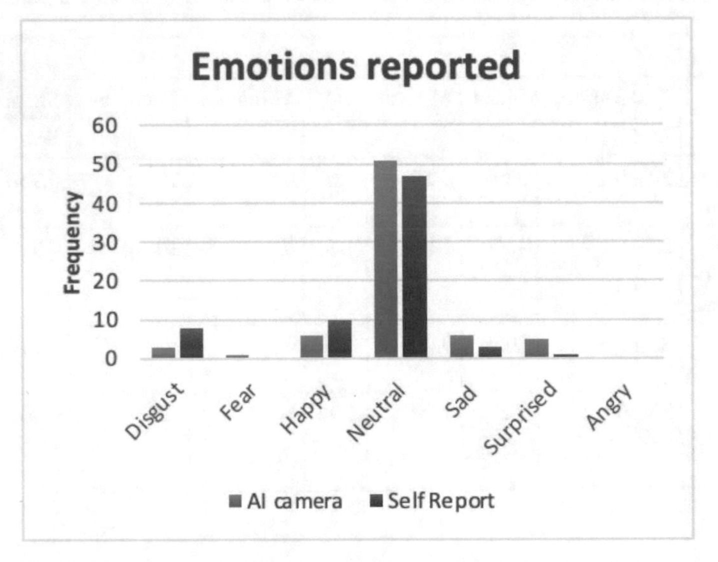

Fig. 2. Frequency of different emotions reported during the whole trial.

A single sample t-test between AI-reported and self-reported emotions excluding neutral emotion shows that the students reported much higher level of dissonance ($M = 4.91$, $SD = 3.34$) than mean dissonance when neutral emotion is also taken into account, $t(34) = 8.57$, $p < .001$.

4.2 Perceived Productivity

Null Hypothesis: There is no significant difference in perceived productivity of students between the control and experiment settings.

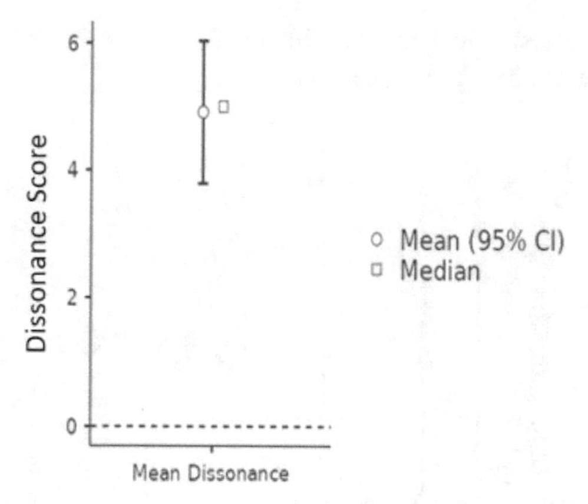

Fig. 3. Graphical depiction of reported dissonance between AI-reported and self-reported emotions excluding neutral responses

Alternate Hypothesis: There is a significant difference in perceived productivity of students between the control and experiment settings.

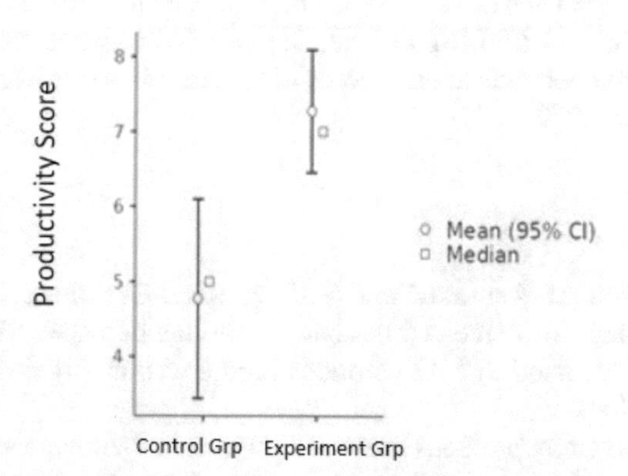

Fig. 4. Graphical Depiction of perceived productivity reported by students across control and experiment conditions.

A paired samples t-test was done to see the difference in perceived productivity scores for experiment and control trial. The results from Experiment trial ($M = 4.77$, $SD = 3.18$) and Control trial ($M = 7.27$, $SD = 1.96$) indicate that there is a statistically significant difference in the self-perceived productivity scores between two trials and students feel more productive in control trial in comparison to the experiment trial $t(22) = -3.35$, $p = 0.003$.

4.3 Actual Student Scores

Null Hypothesis: There is no significant difference in actual student scores of students between the control and experiment settings.

Alternate Hypothesis: There is a significant difference in actual student scores of students between the control and experiment settings.

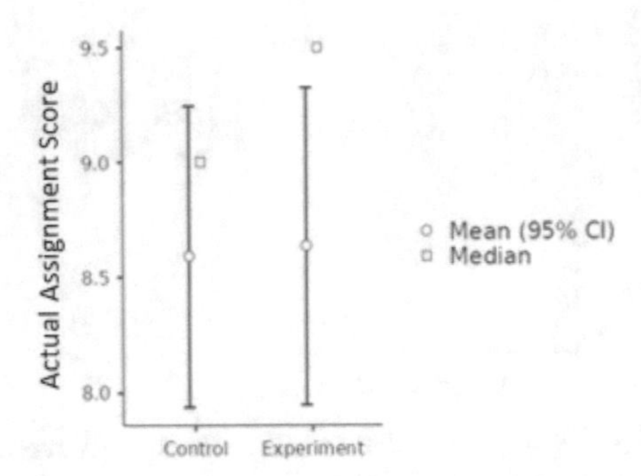

Fig. 5. Graphical Depiction of actual student scores obtained across control and experiment conditions.

A paired samples t-test (Fig. 4) was done to see the difference in actual student scores for experiment and control trial in the test. The results from Experiment trial (M = 8.59, SD = 1.56) and Control trial (M = 8.64, SD = 1.65) indicate the is no statistically significant difference between actual scores of students in experiment and control trial $t(22) = -0.09$, $p = 0.929$.

5 Discussion

Dissonance Between AI-Reported and Self-reported Emotions
A single sample t-test for perceived dissonance ratings between AI-reported and self-reported emotions revealed that AI emotions model was in slight dissonance with the self-reported emotions.

However, as shown by previous literature, majority of students while working on an assignment who experience neutral emotions, especially negative ones, are reported at low to moderate intensities makes investigating emotions extremely difficult (Harley & Azevedo 2014; Craig et al., 2008; D'Mello et al., 2013). Because of these low intensities, statistical variance is constrained, and it may be challenging to draw conclusions about student groups that are relevant based on the students' self-reported emotion profiles (Jarrell et al., 2017).

To counter this problem, we did a single sample t-test for the dissonance ratings after removing all the neutral responses. This significantly increased the mean dissonance between AI-reported and self-reported emotions which is in conjunction with previous literature.

Our discussion is focused on answering the two questions that were raised at the end of the Background Section. Our data analysis and interpretations presented above are summarized below with a view to answering these questions.
Does AI Models Accurately Reflect Human Emotions Using Facial Expression Data?

Therefore, according to our experiment, AI model-based facial emotion recognition models reflect human emotions with some dissonance (Fig. 3) in a classroom learning task. This is evidenced by a high score of dissonance between their perceived emotion and what the AI model showed their emotion to be based on their facial expressions. The effect is further amplified when only non-neutral emotional responses are taken into account. Therefore, this study's findings offer a critical counterpoint to the recent literature which wholesale aims at modeling human emotions using facial expressions through AI models. While there may be some merit in using AI models to recognize human emotions, the technology is not at the level where it can guarantee a high degree of accuracy. This study supports the argument of scholars like Lisa Feldman Barrett who emphasize that facial expressions do not accurately reflect human emotions.

Does Perceived Emotional State Has an Impact on Productivity?

This experiment enabled the students to get two sets of data on their emotions while they were in the learning process. While on the one hand, the AI model gave a set of data on their emotional states. On the other hand, the students were able to perceive their own emotional states, and compare and contrast it with those of the AI model. Through this exercise, students had a clear perception of their emotional states. The question we are raising is if having this perception has an impact on their productivity. With regard to productivity also, we have two sets of data. On the one hand, we have data with regard to students' perception of their productivity while on the other hand, we have their actual assessed productivity. From Fig. 4, we gather that exposure to their emotional state through both AI model as well as their self-perception increased the students' perceived productivity compared with the control group which only had self-perception but was not exposed to the AI model. This has huge implications for human-technology interaction. Even if the current state of AI models do not accurately reflect human emotions, their presence did have a discernable impact on students' self-perception. Therefore, it can be summarized that such AI models may have a huge role to play in activities that are primarily driven by human perceptions.

However, from Fig. 5, it is also observed that with regard to the students' performance on the classroom learning test there was no discernible difference between the experimental and control groups' results. In other words, the quality of output did not proportionally increase with the increase in self-perception of their emotional states. One reason could be that this very act of engaging with the AI model was distractive and therefore it did not present an opportunity for an increase in productivity. In this sense, the AI model was actually intrusive and was unhelpful in boosting productivity.

6 Limitations and Future Scope

The detection of emotions during learning in real-time has not received enough attention in the field of classroom learning. This lack of research effort makes it difficult to understand the evolution of emotions during learning and use of emotional feedback to design classroom learning methodologies and pedagogies that are more effective.

Therefore, it is important to conduct more research on this topic. By understanding the emotions that learners experience during different stages of learning, we can design learning systems that are better suited to their needs. For example, if we detect that a learner is feeling frustrated or overwhelmed, we can provide them with additional resources or support to help them overcome these challenges. However, the pressing need of the hour is for AI technologies to create more robust models that can effectively correlate facial expressions data with human emotions and thus overcome Barrett's critique which was also exemplified in this research.

References

Adolphs, R.: Recognizing emotion from facial expressions: psychological and neurological mechanisms. Behav. Cogn. Neurosci. Rev. **1**(1), 21–62 (2002). https://doi.org/10.1177/1534582302001001003

Aoki, H., Ohnishi, A., Isoyama, N., Terada, T., Tsukamoto, M.: FaceRecGlasses: a wearable system for recognizing self facial expressions using compact wearable cameras. In: 2021 Proceedings of the Augmented Humans International Conference, pp. 55–65 (2021). https://doi.org/10.1145/3458709.3458983

Ark, W.S., Dryer, D.C., Lu, D.J.: The Emotion Mouse. In: HCI, no. 1, pp. 818–823 (1999)

Barrett, L.F., Adolphs, R., Marsella, S., Martinez, A.M., Pollak, S.D.: Emotional expressions reconsidered: challenges to inferring emotion from human facial movements. Psychol. Sci. Public Interest **20**(1), 1–68 (2019)

Buolamwini, J., Gebru, T.: Gender shades: intersectional accuracy disparities in commercial gender classification. In: Conference on Fairness, Accountability and Transparency, pp. 77–91. PMLR (2018)

Cordaro, D.T., Sun, R., Keltner, D., Kamble, S., Huddar, N., McNeil, G.: Universals and cultural variations in 22 emotional expressions across five cultures. Emotion **18**(1), 75 (2018)

Cowie, R., et al.: Emotion recognition in human-computer interaction. IEEE Signal Process. Mag. **18**(1), 32–80 (2001)

Craig, B.M., Lipp, O.V.: The influence of multiple social categories on emotion perception. J. Exp. Soc. Psychol. **75**, 27–35 (2018)

De Silva, T.R.S., Dayananda, K.Y., Galagama Arachchi, R.C., Amerasekara, M.K.S.B., Silva, S., Gamage, N.: Solution to measure employee productivity with employee emotion detection. In: 2022 4th International Conference on Advancements in Computing (ICAC), pp. 210–215 (2022). https://doi.org/10.1109/ICAC57685.2022.10025132

D'Mello, S., Kappas, A., Gratch, J.: The affective computing approach to affect measurement. Emot. Rev. **10**(2), 174–183 (2018)

Dryer, D.C.: Getting personal with computers: how to design personalities for agents. Appl. Artif. Intell. **13**(3), 273–295 (1999)

Ekman, P.: Methods for measuring facial action. Handb. Methods Nonverbal Behav. Res. 45–90 (1982)

Ekman, P., et al.: Universals and cultural differences in the judgments of facial expressions of emotion. J. Pers. Soc. Psychol. **53**(4), 712 (1987)

Emotional Expressions Reconsidered: Challenges to Inferring Emotion From Human Facial Movements. Association for Psychological Science - APS (n.d.). https://www.psychologicalscience.org/publications/emotional-expressions-reconsidered-challenges-to-inferring-emotion-from-human-facial-movements.html. Accessed 29 Nov 2023

Gadigi, V., Veerabhadrappa, H.: International research journal of engineering and technology (IRJET) a literature review on technique of facial variance to detect the state of emotion and map productivity. Int. Res. J. Eng. Sci. Technol. Innov. **7**, 814–817 (2023)

Harley, J.M., Azevedo, R.: Toward a feature-driven understanding of students' emotions during interactions with agent-based learning environments: a selective review. Int. J. Gaming Comput.-Mediat. Simul. (IJGCMS) **6**(3), 17–34 (2014)

Jarrell, A., Harley, J.M., Lajoie, S., Naismith, L.: Success, failure and emotions: examining the relationship between performance feedback and emotions in diagnostic reasoning. Educ. Tech. Res. Dev. **65**(5), 1263–1284 (2017)

Miranda, C.R., et al.: Facial expressions tracking and recognition: database protocols for systems validation and evaluation. arXiv preprint arXiv:1506.00925 (2015)

Moridis, C.N., Economides, A.A.: Mood recognition during online self-assessment tests. IEEE Trans. Learn. Technol. **2**(1), 50–61 (2009)

Picard, R.W.: Affective computing for HCI. In: HCI, no. 1, pp. pp. 829–833 (1999)

Postareff, L., Mattsson, M., Lindblom-Ylänne, S., Hailikari, T.: The complex relationship between emotions, approaches to learning, study success and study progress during the transition to university. High. Educ. **73**, 441–457 (2017)

Shen, L., Wang, M., Shen, R.: Affective e-learning: using "emotional" data to improve learning in pervasive learning environment. J. Educ. Technol. Soc. **12**(2), 176–189 (2009)

Sundström, P.: Exploring the affective loop. Doctoral dissertation (2005)

Xia, Y.: Upper, middle and lower region learning for facial action unit detection. arXiv preprint arXiv:2002.04023 (2020)

Development of a Health Monitoring System for Students in Shelters as an Interdisciplinary Project During the Training of Masters in Automation at Ukrainian Universities

Viktoriya Voropayeva[1], Zhukovska Daria[1], Labuzova Anastasiia[1], and Anna Voropaieva[2](✉)

[1] Donetsk National Technical University, Pokrovsk, Ukraine
[2] Ivano-Frankivsk National Technical University of Oil and Gas, Ivano-Frankivsk, Ukraine
anna.voropaieva@nung.edu.ua

Abstract. Since the beginning of the full-scale invasion of the Russian Federation on the territory of Ukraine due to missile and artillery fire, the air alerts have been announced more than 31,000 times with an average duration of 58 min. Educational process in the distant from the front line areas of Ukraine is carried out in either online, offline or mixed formats, accordingly, during air raids students and school children are to go to shelters and have to spend there from 30 min to 3 h or longer.

While staying in the shelter young people and children suffer from psychological and emotional stress which can cause changes in the basic physiological indicators (pulse, blood pressure, body temperature), in particular, up to levels leading to considerable health damage. The task of provisioning of operational control of a person's vital indicators during the stay in a civil protection shelter becomes obvious, it will enable timely and appropriate response to potential threats.

At Donetsk National Technical University and Ivano-Frankivsk National Technical University of Oil and Gas the teachers offered the students of the second (master's) educational level of the automation program a task to develop a health monitoring system for vital activity indicators of people in shelters as an interdisciplinary project with practical case to consolidate theoretical knowledge.

Keywords: Health Monitoring System · Interdisciplinary Project · Decision support system

1 Introduction

Since the beginning of russia's full-scale military invasion in Ukraine, air raid alerts have been announced 28,912 times, lasting between 30 to 200 min on average [1]. During an "air raid" signal, citizens must take shelter in civil defense facilities to save their lives and health. These include fortification structures, basement or underground rooms where people can stay temporarily and be safe.

M. E. Auer et al. (Eds.): STE 2024, LNNS 1028, pp. 52–59, 2024.
https://doi.org/10.1007/978-3-031-61905-2_6

Despite the martial law, residents of Ukraine's rear area make maximum efforts to support the country's economy by working, paying taxes, making donations to the armed forces and refugees, and volunteering. The country's main infrastructure sectors, such as medical, transportation, banking, energy, education, continue to operate normally. Since September 2022, the training process in educational institutions of the country has resumed in offline, online, or in a mixed format, depending on the security situation and proximity to the front line [2]. As of January 2023, 4318 schools and 80% of higher education institutions have introduced offline and mixed format of learning [3]. Consequently, during air raid alerts, thousands of students stay in civil defense shelters, experiencing difficult emotions: sadness, fear, tension, anger. As is known, constant stress and depressed psycho-emotional state can lead to changes in basic physiological indicators (pulse, blood pressure, body temperature) and affect human physical health in adverse way, causing panic attacks, exacerbations of chronic diseases, strokes and heart attacks.

Currently, methods for monitoring the students' health are based on the results of regular medical examinations. An obvious task is to develop a real-time monitoring system for vital indicators of a person during their stay in a shelter, which will enable to respond promptly to potentially dangerous situations and take appropriate actions.

This work was supported by the European Commission [grant number ENI/2019/413-664 "EDUTIP"] and Edunet World Association (EWA).

2 Formulation of the Problem

Solving the above issues requires all the available potential, both material and human (intellectual), that is, involving student youth. During their study for a master degree in automation at Donetsk National Technical University and Ivano-Frankivsk National Technical University of Oil and Gas, the students were proposed to develop a system for monitoring the vital parameters of persons in shelters as a practical case study. This system performs the following tasks:

1. Autonomous recording of basic physiological indicators;
2. Transmission of information over supported telecommunication protocols to a data processing center;
3. Storage of personalized indicator history in dedicated database;
4. Comparative analysis of current values against regulatory or customary values for a particular person;
5. Alert in case of a hazardous health condition of a student and the need to make certain decisions.

Working on this case will enhance students' ability to "integrate knowledge from other fields, apply a systematic approach, and take into account non-technical aspects when solving engineering problems and conducting research" which is one of the necessary professional competencies according to the current standard [4]. This allows to use an interdisciplinary approach, which is very promising when teaching IT professionals [5]. Both universities already have positive experience in using practical cases to teach automation specialists for the mining industry [6], which not only improves the quality of professional training, but also stimulates students' creativity [7].

Solving problems 1) and 2) is based on the latest research in the IoT field. Modern sensors embedded in personalized gadgets allow measuring the necessary physiological parameters, providing continuous monitoring and transmitting them for further processing [8]. As part of the Internet of Things and Industrial Internet of Things course, students performed the analysis of available tools and the selection of the optimal ones. To solve the next tasks, students carried out data preparation and data initial processing using fuzzy logic rules and algorithms. They also developed a personalized decision-making model for health status using neural networks as part of the Intelligent Control Technologies course.

3 Stages of the Project Implementation

3.1 Forming the Project Teams and Developing of an Organizational Framework for Working on a Case Study

Using approaches to create and develop start-up projects from the Entrepreneurial University course, students formed teams, distributed roles, and developed a project implementation schedule. Target groups and their needs were identified, approaches to practical implementation were substantiated, desired results of each stage and the minimal valuable product (MVP) were specified.

3.2 Selecting the Means for Measuring Physiological Parameters

Recent research provides examples of the successful use of the Internet of Things (IoT) in medicine. One innovative solution is a patient health monitoring system that uses IoT technologies. In this system, a sensor collects information about the patient's health status, including oxygen saturation level, heart rate, and body temperature. The main feature of this approach is that the sensor is compact, fast and cost-effective. However, an important limitation of this system is the lack of consideration of the patient's location and environment [9]. After conducting an analysis as part of the course "Technology of the Internet of Things and Industrial Internet of Things" students offered the use of a fitness bracelet (smart watch) with built-in sensors that allow measuring pulse, blood pressure and body temperature. In addition, either the fitness bracelet or the smartphone to which it is connected must contain a GPS navigator and allow determining the location of a person in a shelter.

3.3 Creating a Personalized History of Health Indicators

As part of the "Intelligent Management Technologies" course, students developed and implemented an effective toolkit for operational monitoring of participants in the educational process in terms of health and behavior in shelters and decision-making. Objective indicators of a person's physical condition include: body temperature (BT), pulse or heart rate (HR), systolic blood pressure (S) and diastolic blood pressure (D), respiratory rate, vital capacity of the lungs, etc. For further research, the indicators available for autonomous real-time monitoring using an individual bracelet were selected, namely:

BT, HR, S and D. All these indicators are monitored during regular preventive medical examinations. Medical research and recommendations of the World Health Organization (WHO) determine the normative values of these indicators and their fluctuations, depending on gender, age, physical activity [10].

Taking into account these recommendations, it is proposed to divide the allowable range of values for each indicator into three zones, and the measurements of each zone should be transmitted to the database at a certain time interval (Fig. 1):

- green - the zone of permissible values, monitoring every 30 min;
- yellow - the zone of potential danger, monitoring every 15 min;
- red - the critical zone, monitoring every 5 min.

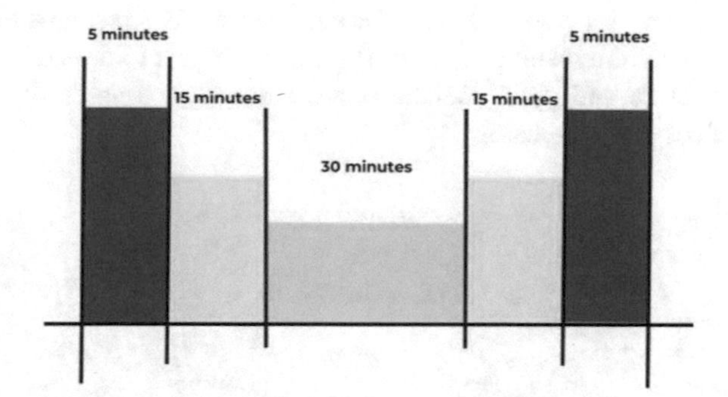

Fig. 1. Indicator monitoring zones.

The determination of zone boundaries is based on WHO recommendations for a certain age group. At the same time, each person whose condition is monitored may have their own characteristics such as chronic or current diseases that can affect the usual indicators for this person. Therefore, the students considered the use of neural networks to personalize the limit values of such zones. The collected statistical data of health status indicators of various individuals were used as input data for the neural network, which adjusts the normative values taking into account the individual indicators of the participants of the educational process according to previous and current measurements. At the output of the neural network, the personalized data for each individual was obtained. This enables further qualitative decision-making about the criticality of the deviation from the norm of a specific indicator when monitoring the physical condition of persons staying in shelters.

3.4 Developing a Set of Rules and Algorithms for Decision-Making Regarding the Critical/normal Health Condition

Decision-making during the classification of the state of each participant is carried out in the conditions of the stochastic nature of external influences, the absence of an adequate mathematical model of functioning, ambiguity, the human factor, etc. Uncertainty of the system leads to an increase in risks from ineffective decision-making. To make effective decisions under uncertainty, fuzzy logic inference methods based on IFTHEN rules are used. This method uses a fuzzy logic knowledge base, a set of IFTHEN rules that define the relationship between the inputs and outputs of the object being studied. It is recommended to use the Mamdani fuzzy rule-based models which are compiled taking into account the hierarchy of priorities of input variables established by the Saaty AHP (Saaty analytic hierarchy process) [11] using the agreed matrix of pairwise comparison.

Figure 2 shows a preliminary fuzzification model for blood pressure indicator. It is necessary to analyze not one value for this indicator, but three - S, D and their difference. In the SystolicPressure, DiastolicPressure, s-d blocks, the processes of fuzzification of the corresponding indicators take place, i.e. the conversion of the distinct values of the input variables into a fuzzy form by determining the degree of belonging of the input variable value to its zones (terms). It is worth emphasizing that zone boundaries are determined separately for each indicator for each person as a result of using the neural network. The calculation result of the mmHg unit is subject to defuzzification. Models for body temperature and pulse indicators are somewhat simpler since they require fuzzification of only one indicator.

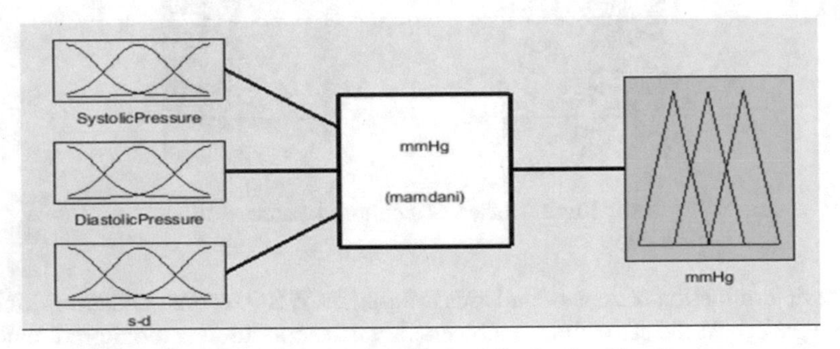

Fig. 2. Structure of the fuzzy logic model reflecting blood pressure.

After performing such fuzzification of indicators, i.e. determining the zone to which the specific value of body temperature, blood pressure and pulse falls for each person, it is necessary to draw a conclusion about their general condition. For this purpose, an integral indicator F is introduced (Fig. 3).

The model structure shown in Fig. 3 corresponds to the hierarchical system of fuzzy logical inference. The given system enables to establish a multifactorial dependence using the results of logical inference from the fuzzy knowledge bases of the previous levels, namely the results of the model from Fig. 2 and similar models for body temperature and pulse indicators.

Appropriate algorithms should be developed for decision-making on the obtained values of the integral indicator F. In order to do this, the result obtained at the output of block F in the range from 0 to 1 is divided into zones from normal [0:0.2) to critical (0.8;1) with a step of 0.2.

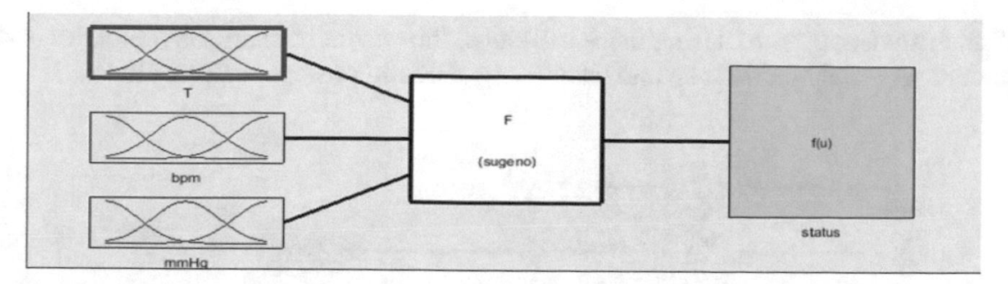

Fig. 3. Structure of the fuzzy logic model for determining the integral indicator F.

Training of the fuzzy model according to the specified rules was carried out using the method of subtractive clustering [12]. Figure 4 shows the result. The surface plot shown in Fig. 4 indicates a nonlinear relationship between input variables: T, bpm, mmHg and output variable F.

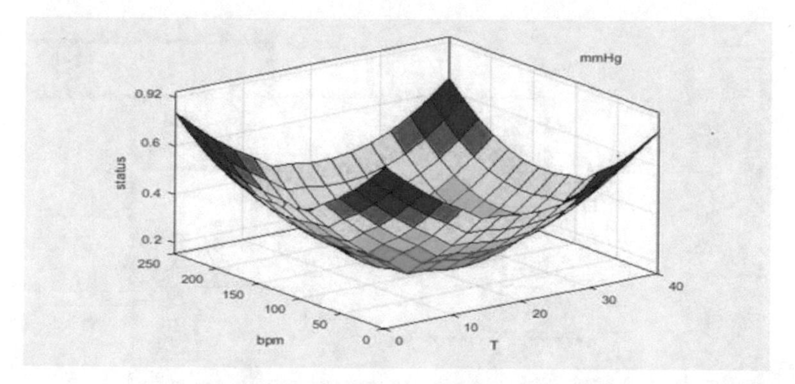

Fig. 4. Clustering of the integral health indicator.

Thus, a set of fuzzy logic models has been offered for preliminary fuzzification of input variables and clustering of output variable F, which allows interpreting input data and results obtained [13]. Implementation of this complex for each person is integrated into the Simulink environment.

Figure 5 shows an example of calculating the output variable F according to the statistical data of vital indicators when a person's condition in the shelter is within normal limits for all indicators.

In Fig. 5, the value of output variable $F = 0.1998$ indicates that this combination of input parameter values (temperature, pulse, blood pressure) allows us to conclude that the person's condition is normal. If all three indicators significantly exceed the norm, the resulting value of $F = 0.9461$ indicates a significant danger.

The proposed algorithm makes it possible to track the current vital parameters for each participant of the educational process who is currently in the shelter and display this information on the device of the person in charge (teacher, curator). Visualization is displayed in the form of a shelter map with color markings indicating the health status of each student (Fig. 6). Using these markings, the person in charge can monitor each student separately on the map and take timely action to prevent complications.

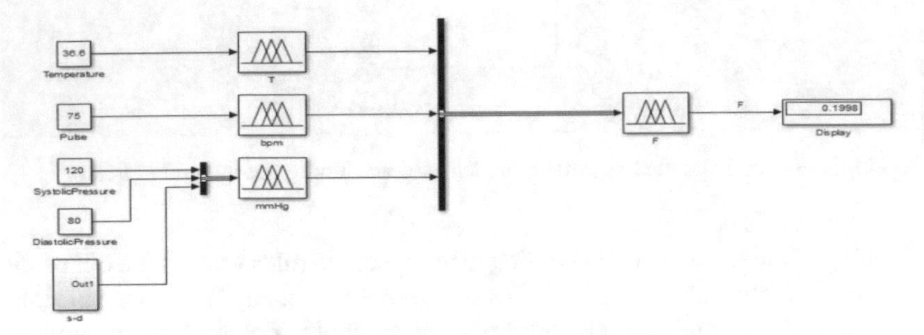

Fig. 5. Implementation of the set of models in Simulink.

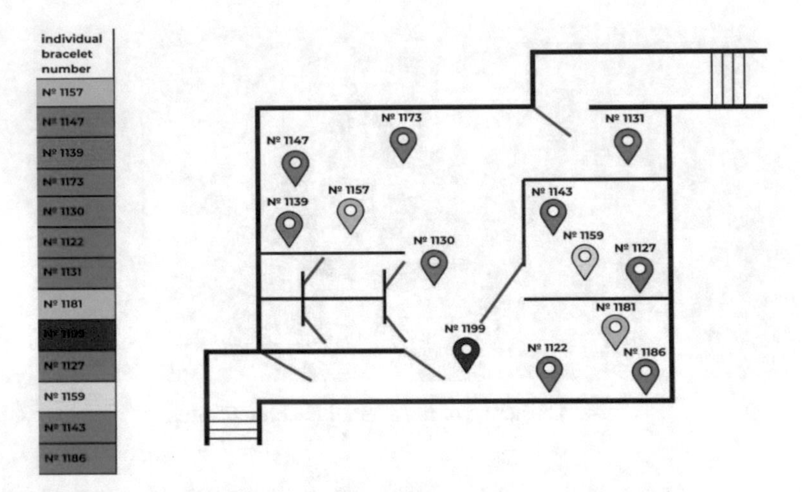

Fig. 6. Screen layout of the person in charge.

When a yellow label is detected (which indicates a potential danger), the person in charge conducts an analysis of the situation, waits and analyzes the re-measurement. In case of appearance of an orange label, the information is transmitted to the medical personnel in order to organize an immediate response and provide assistance. Medical staff take additional assessments and measures to provide assistance in accordance with established protocols. In the event of a critical situation (red label), the person in charge immediately begins actions to organize rescue operations. Rescue teams receive information about a significant deterioration in the health of a participant in the educational process, including their location and current condition. This enables them to respond quickly to any changes and take immediate action to save the participant in the educational process.

4 Conclusion

Thus, the development and further implementation of a health monitoring system for students in shelters will allow timely response or warning of serious threats to the health of students. The use of such a task in the educational program of masters in automation will improve the quality of practical training and strengthen interdisciplinary connections.

References

1. Air alert map of Ukraine. https://alerts.in.ua/. Accessed 27 Nov 2023
2. MES of Ukraine Organizational Features of the 2022/23 Academic Year. https://mon.gov.ua/ua/news/osoblivosti-organizaciyi-202223-navchalnogo-roku. Accessed 19 Nov 2023
3. Ukrainian Pravda. The Ministry of Education, Culture and Science told how many schools and kindergartens work face-to-face or remotely. https://life.pravda.com.ua/society/2023/02/16/252883. Accessed 20 Nov 2023
4. Standard of higher education master's degree in specialty 151 Automation and computer-integrated technologies. https://mon.gov.ua/storage/app/media/vishchaosvita/zatverdzeni%20standarty/2020/08/10/151-avtomatizatsiya-ta-kit-magistr.pdf. Accessed 28 Nov 2023
5. Interdisciplinary Project Based Learning Approach for Machine Learning and Internet of Things. https://ieeexplore.ieee.org/document/9280619. Accessed 16 Nov 2023
6. Voropayeva, A., Stupak, H., Voropayeva, V., Potsepaiev, V.: Case Study "automation system for mine drainage unit. In: 2020 IEEE European Technology & Engineering Management Summit, 5–7 March 2020, Dortmund, Germany, pp. 160–164 (2020)
7. Kabanets, M., Voropayeva, V., Voropaieva, A.: Pedagogical provisions of students' creativity development by means of SmartLabs. In: 20th International Conference on Remote Engineering and Virtual Instrumentation (REV2023), International Edunet World Conference (IEWC 2023), 01–03 March 2023, Porto Palace Hotel, Thessaloniki, Greece, pp. 407–413 (2023). In press
8. IoT-Based Health Monitoring System Development and Analysis. https://www.hindawi.com/journals/scn/2022/9639195/. Accessed 13 Nov 2023
9. An Overview on the Internet of Things for Health Monitoring Systems. https://www.researchgate.net/publication/312078649_An_Overview_on_the_Internet_of_Things_for_Health_Monitoring_Systems. Accessed 20 Nov 2023
10. World Health Organization (WHO). https://www.who.int/. Accessed 22 Nov 2023
11. Ikeda, H., Kawamura, Y., Tungol, Z.P.L., Moridi, M.A., Jang, H.: Implementation and verification of a wi-fi ad hoc communication system in an underground mine environment. J. Min. Sci. 505–514 (2019)
12. Karakose, M., Akin, E.: Block-based fuzzy controllers, vol. 3, no. 1, pp. 100–110. Department of Computer Engineering, University of Firat, Elazig, Turkey (2010)
13. Labuzova, A.M., Voropaieva, A.O., Voropayeva, V.Ya.: Structure and functionality of an automated subsystem for monitoring and control of technological indicators of coal mine workings, pp. 63–73. Scientific Bulletin of Donetsk National Technical University, Lutsk (2022)

Enhancing Municipal Fleet Management in Smart Cities Through 5G Integration

Salam Traboulsi(✉) and Dieter Uckelmann

University of Applied Sciences, Schellingstr. 24, 70174 Stuttgart, Germany
{salam.traboulsi,dieter.uckelmann}@hft-stuttgat.de

Abstract. This paper presents a revolutionary research initiative focused on improving municipal fleet management in smart cities by leveraging the power of 5G technology. Through adopting 5G, the municipality gains access to real-time communication, precise vehicle tracking, optimized route planning, and improved operational efficiency. Delving into the realm of 5G technology, the study examines its multiple applications in fleet management, including real-time monitoring and predictive maintenance strategies. The initial findings establish a robust foundation, laying the groundwork for forthcoming initiatives aimed at enhancing municipal fleet operations in smart cities through the seamless integration of 5G connectivity. Our objective is to amplify the impact of our work, making a substantial contribution to the continuous evolution of smart city infrastructures. Subsequent research endeavors will delve even deeper into the expansive potential of 5G, exploring its applications in fleet optimization, intelligent transportation systems, and innovative solutions for sustainable mobility.

Keywords: Smart Cities · 5G Technology · Municipal Fleet Management · Predictive Maintenance

1 Introduction

The concept of smart cities has emerged as a solution to the challenge of urbanization, improving quality of life, promoting sustainability, and enhancing efficiency for citizens [8]. These cities rely on advanced technologies to create interconnected ecosystems that optimize various urban domains, including transportation, healthcare, utilities, and public services [8–10].

This paper presents the findings of our research carried out in alignment with the objectives of the iCity project [12]. The iCity project represents a collaborative effort between academia, industry, and government agencies, aimed at creating sustainable and efficient urban environments by leveraging advanced technologies. Our research within the iCity project focuses on exploring the potential of 5G technology as a mobile communication network to enhance existing use cases and enable innovative solutions in smart cities [30]. Inspired by the groundbreaking advances made by companies like Tesla [27], which have transformed the

M. E. Auer et al. (Eds.): STE 2024, LNNS 1028, pp. 60–71, 2024.
https://doi.org/10.1007/978-3-031-61905-2_7

automotive industry with their electric vehicles (EVs) and innovative connectivity solutions, such as the Tesla Fleet Telemeters framework [28], we are striving to harness the capabilities of 5G to optimize fleet operations, improve vehicle monitoring, streamline maintenance practices, and optimize resource allocation within the municipal fleet. This paper sets the foundation of our work, exploring the potential of integrating 5G technology into municipal fleet management.

The structure of this paper is as follows: In Sect. 2, a comprehensive review of existing literature is presented, focusing on the incorporation of advanced technologies in smart cities. In addition, it highlights research gaps regarding the use of 5G technology within technical municipal services. Section 3 describes the key functional aspects of municipal fleet management, including vehicle monitoring, predictive maintenance, advanced analytics, and resource allocation optimization. Next, Sect. 4 offers an overview of 5G technology, including its evolving standards, as well as its integration in connected vehicles. Section 5 describes the methodology and evaluation process for implementing these functional aspects and assessing performance parameters, as well as presents the initial results obtained from the evaluation, highlighting the benefits and improvements brought about by the integration of 5G technology. Finally, Sect. 6 concludes the paper and discusses the prospects for leveraging 5G technology in mechanic's workshops within smart cities.

2 Literature Review

In the context of smart cities, the integration of advanced technologies has become an important area of research and development to meet the challenges of urbanization and improve the quality of life, sustainability, and efficiency for citizens [23]. Companies like Tesla have played a pivotal role in this domain by driving innovations in EVs, autonomous driving, and vehicle connectivity. Tesla's EVs have revolutionized the automotive industry, helping to reduce greenhouse gas emissions and improve air quality in urban environments. Their advances in autonomous driving have the potential to transform urban transport systems, improving efficiency and safety. Moreover, Tesla's vehicle connectivity solutions have paved the way for real-time data collection, analysis, and communication, enabling the optimization of transportation systems and overall operational efficiency within smart cities [27,28].

Smart cities harness a range of cutting-edge technologies, including the Internet of Things (IoT), Cloud Computing, and Artificial Intelligence (AI), to establish interconnected ecosystems that enhance various urban sectors, like transportation, healthcare, utilities, and public services [5,6,12,14,17,26,29,34]. Among these technologies, 5G stands out as a transformative enabler due to its high-speed connectivity, low latency, and capacity to connect a massive number of devices [15,25]. Extensive research has been conducted on the application of 5G technology in smart cities, with a focus on optimizing transportation systems through real-time traffic monitoring, intelligent routing, and vehicle-to-infrastructure communication [6,10,20,22,32].

However, limited attention has been given to the utilization of 5G technology in the maintenance and performance of municipal vehicles and their associated technical municipal services, which are essential for ensuring the operational efficiency of urban fleets encompassing public transportation, waste management, and emergency services [11,21]. Traditional approaches to vehicle maintenance often lack proactive monitoring, timely diagnostics, and efficient communication between mechanics and vehicles, leading to inefficiencies and increased downtime.

Within the iCity project, our research focuses on exploring the potential of 5G technology in municipal fleet management. By leveraging the capabilities of 5G, our objective is to enhance communication and connectivity between municipal vehicles and mechanics, enabling real-time monitoring, diagnostics, and efficient maintenance procedures. The potential of this integration lies in enhancing the overall operational efficiency and effectiveness of municipal fleets, resulting in decreased downtime and improved performance. Notably, limited existing research addresses the integration of 5G technology in municipal fleet management within the context of smart cities, highlighting the importance of our work in filling this research gap and contributing to the broader knowledge of leveraging advanced technologies in smart city infrastructure.

3 Transformative Solutions for Optimizing Municipal Fleet Management

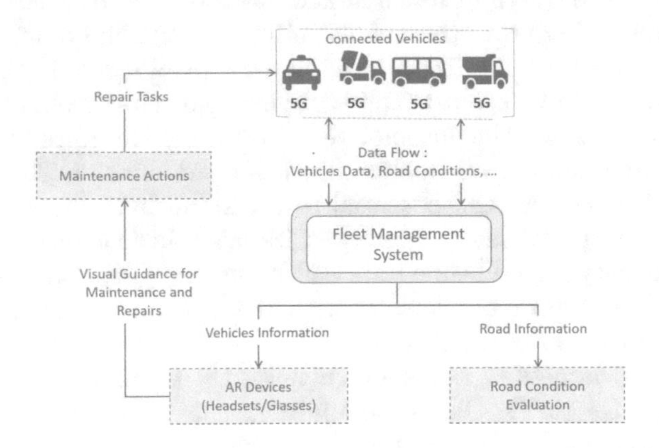

Fig. 1. Flowchart Illustrating Key Functional Aspects of Smart Fleet Management

In this section, we explore the municipal fleet management use case, a pivotal element of our project strategically crafted to address distinct needs and requirements, providing solutions to relevant challenges. The detailed exploration of the essential functions of this use case illustrates the transformative potential of 5G technology in significantly enhancing efficiency and optimizing fleet operations and maintenance practices.

In the context of smart cities, there is a need for smarter and more efficient urban environments, steering our investigation towards the integration of 5G technology to elevate municipal fleet management in the context of smart cities. The primary objective is to harness the advanced capabilities of 5G for real-time communication, efficient vehicle tracking, optimized route planning, and overall operational efficiency. This forward-looking initiative sets the stage for the evolution of fleet management practices. Our inspiration for this use case is drawn from trailblazing advancements in the automotive industry, particularly the paradigm-shifting contributions of Tesla in the realm of electric vehicles (EVs) and connectivity solutions. Additionally, the iCity project serves as a guiding framework, emphasizing the exploration of 5G technology's transformative potential within the domain of municipal fleet management.

3.1 Key Functional Aspects in Focus

Figure 1 illustrates the key functional aspects spotlighted in this use case. The "Fleet Management System" acts as the central hub for data processing and management, surrounded by "Connected Vehicles" representing the fleet seamlessly connected to the system through 5G technology. "AR Devices" equipped with AR capabilities, such as headsets or smart glasses, are utilized by mechanics for real-time maintenance guidance. These devices interact with the fleet management system, accessing the latest vehicle data to provide accurate instructions during maintenance tasks. The figure also highlights bidirectional data flow between the connected vehicles and the fleet management system, signifying the continuous exchange of vehicle information. This data flow enables real-time monitoring of vehicle performance, receipt of diagnostics data, and optimization of maintenance schedules. Additionally, the figure incorporates "Reporting Road Conditions," indicating real-time reporting from connected vehicles to the fleet management system. This capability allows for prompt response and collaboration with relevant authorities to address potential hazards.

- Connected Vehicle Monitoring and Diagnostics: Fleet vehicles seamlessly connect with the management system, enabling real-time monitoring of vehicle performance and remote diagnostics. Instant notifications about vehicle issues and remote access to detailed diagnostics data empower mechanics and fleet managers. Leveraging onboard diagnostics (OBD) data [19], the system facilitates real-time transmission of crucial vehicle information, allowing continuous monitoring of key parameters to optimize maintenance schedules.
- Augmented Reality (AR) for Precision Maintenance: Integration of AR technology provides real-time visual guidance to mechanics during maintenance tasks within municipal fleet management. Mechanics equipped with AR devices, seamlessly connected to the 5G network, access real-time visualizations, instructions, and annotations overlaid on physical vehicle components. This AR assistance significantly improves repair efficiency, accuracy, and knowledge transfer during vehicle maintenance operations.

- Enhancing Road Condition Reporting: The fleet management system is equipped to enable the reporting of road conditions, leveraging precise localization systems integrated into fleet vehicles. Fleet vehicles contribute valuable data on hazards, obstacles, or areas requiring immediate attention, optimizing street sweeping routes, enhancing safety, and improving overall operational efficiency. The capability to report road conditions facilitates prompt responses and collaboration with relevant authorities for necessary actions.

In summary, this use case not only tackles urbanization challenges but also introduces smart solutions powered by the integration of 5G technology. Inspired by industry pioneers and collaborative projects, the identified functional aspects underscore the potential of 5G to revolutionize municipal fleet management, steering towards sustainable and efficient urban environments.

4 The Power of 5G in Municipal Fleet Management

Integrating 5G technology in municipal fleet management brings numerous benefits and enhances the capabilities of the automotive industry [18]. By harnessing the potential of 5G, municipalities can optimize fleet operations, enhance vehicle monitoring, implement effective maintenance practices, and streamline resource allocation without the municipal fleet.

4.1 Overview of 5G

The 5G mobile network revolutionizes connectivity, providing faster, more flexible, scalable, and reliable technologies that connect people and devices globally. It encompasses various technologies designed to cater to diverse network requirements, enabling a wide range of use cases. These use cases are classified into three categories: enhanced Mobile Broadband (eMBB), massive Machine-Type Communication (mMTC), and Ultra-reliable Low-Latency Communication (URLLC), each with its defined Key Performance Indicators (KPIs). The development of the 5G standard is led by the 3rd Generation Partnership Project (3GPP) and evolves through progressive release phases [16]. The 5G New Radio (5G NR) was introduced in Release 15, and its architectures, are represented by 5G Non-Standalone (NSA) and 5G Standalone (SA). Release 16 enhanced the 5G SA architecture improving features such as slicing, uRLLC, and mMTC, while also enhancing positioning accuracy. Release 17 further improved IoT implementations and automation, offering support for private networks, Multi-Access Edge Computing (MEC), Massive Multiple-Input Multiple-Output (MIMO), network slicing, high-accuracy positioning, and more. These ongoing advancements demonstrate the evolution of 5G and its expanding capabilities [1–4,7,16,31].

We conducted a theoretical evaluation to assess the potential impact of integrating 5G functions into municipal fleet management, with a specific focus on anticipated benefits and improvements in key performance indicators (KPIs) associated with 5G-based functionalities. The evaluation results highlight the following benefits:

1. Integrating eMBB in municipal fleet vehicles is expected to provide faster data transfer rates, higher bandwidth, and improved network capacity. This enables seamless streaming of high-definition content, real-time monitoring of vehicle data, and rapid over-the-air software updates, enhancing the overall fleet management experience.
2. The mMTC functionality enables efficient communication among a massive number of connected devices, including municipal fleet vehicles. By integrating mMTC in the fleet, improved connectivity and reliable communication can be achieved, leading to efficient coordination of fleet operations, enhanced safety features, and optimized resource allocation.
3. URLLC plays a pivotal role in facilitating critical applications that demand stringent latency and reliability requirements. Integrating URLLC in municipal fleet vehicles is expected to significantly reduce communication latency, ensuring real-time communication between vehicles, fleet management systems, and infrastructure. This results in faster response times for fleet monitoring, incident management, and enhanced operational efficiency.
4. The integration of MEC in 5G-enabled municipal fleet vehicles enables localized processing and storage capabilities at the network edge. This allows for faster response times for latency-sensitive applications, reduced network congestion, and enhanced privacy and security through localized data processing. Fleet management systems can leverage MEC to enable real-time decision-making, predictive maintenance, and efficient resource utilization.
5. Massive MIMO technology employs numerous antennas to enhance spectral efficiency and signal quality. By integrating massive MIMO in municipal fleet vehicles, improved signal strength, increased network capacity, and reduced interference can be achieved, resulting in better connectivity and higher data throughput, especially in densely populated areas.
6. Network slicing allows for the creation of virtual networks tailored to specific use cases. By utilizing network slicing for municipal fleet vehicles, dedicated network resources can be allocated, ensuring reliable and uninterrupted communication, even in congested network conditions. This enables fleet management systems to have dedicated network slices for secure and efficient vehicle communication.
7. The integration of high-accuracy positioning technology in municipal fleet vehicles enables precise location tracking. This technology enhances applications such as precise vehicle tracking, optimized route planning, and location-based services, improving overall fleet management efficiency and safety.

4.2 Applications of 5G in Municipal Fleet Management

The exploration of 5G technology within the realm of fleet management highlights its varied applications, particularly in the realms of real-time monitoring and predictive maintenance strategies. Key areas where 5G plays a crucial role in municipal fleet management include:

1. Connected vehicle Monitoring and diagnostics: Utilizing 5G connectivity for real-time monitoring of vehicle performance, remote diagnostics, and continuous optimization of maintenance schedules using onboard diagnostics (OBD) data.
2. Augmented Reality (AR) for Maintenance: Leveraging 5G's low-latency and high-bandwidth capabilities for real-time visual guidance to mechanics during maintenance tasks using AR technology.
3. Reporting Road Conditions: Employing 5G-powered municipal fleet management systems for reporting road conditions, optimizing street sweeping routes, and enhancing overall operational efficiency.

As 5G technology is still progressing through developmental phases, technical specifications, and experiences undergo constant evolution. Immediate access to comprehensive 5G options and technologies may be somewhat limited, yet gradual advancements are foreseen. The integration of 5G in municipal fleet management, in tandem with progress in sustainable urban transportation and economic digitization, plays a pivotal role in attaining overarching objectives in the evolution of smart cities.

In the context of this work, the strategic utilization of 5G acts as a robust communication network, seamlessly transmitting data from the connected fleet to the server for subsequent analysis. The focus is on conceptual development, involving suitable hardware and a schematic software structure tailored for the identified use case. In this scenario, the role of 5G is distinctly defined as a communication network between devices. The subsequent sections will delve into practical implementation steps, outlining the achievements targeted in this initial stage of the use case. Following discussions will provide a thorough examination of concrete results and progress achieved through the incorporation of 5G technology within the specified context.

5 Data Logging and Analysis

Within our data logging and analysis methodology, we employ OBD-II system to collect metrics when the vehicle is connected. This system, utilizing the Controller Area Network (CAN) bus protocol, ensures compatibility with various vehicle makes and models. For effective OBD-II data logging and WiFi transmission, we opted for the CANedge2 from CSS electronic [13], chosen for its efficiency despite having fewer features than some alternatives.

Figure 2 illustrates the collected metrics, providing insights into crucial data regarding the vehicle's performance and operating conditions. This foundational data serves as the basis for subsequent interpretation and analysis.

In the following phase, depicted in Fig. 3, the CANedge2 device plays a central role in logging CAN data from a bus. Serving as dedicated hardware, it facilitates the recording of diverse metrics from both the Global Positioning System (GPS) and the vehicle's CAN Bus. Seamless data transfer via 5G mobile radio occurs during operation, with transmitted data sent to a self-hosted Minio server deployed on a virtual machine for efficient data exchange.

A Python script is pivotal in processing the collected data, decoding it, and writing it to the database. The data, collected in the ".MF4" format, undergoes decoding using ".dbc" files. Distinct ".dbc" files for electric and non-electric vehicles ensure human-readable data before storage in InfluxDB2 and visualization through dashboards.

Metrics for Electrical Vehicles (EVs)		Metrics for Non-Electrical Vehicles	
Metrics	**Description**	**Metrics**	**Description**
Battery State of Health (SoH)	Represents the overall health and remaining capacity of the battery.	Fuel Efficiency	Measures the vehicle's efficiency in converting fuel to mileage.
State of Charge (SoC)	Indicates the current charge level of the battery.	Engine Health Monitors	the overall health and performance of the engine.
Cell Voltages	Provides voltage information for individual battery cells.	Emissions Levels	Tracks the level of emissions produced by the vehicle.
Cumulative Charge Power	Tracks the total power consumed or regenerated by the battery.	Operational Performance	Evaluates the general performance and behavior of the vehicle.

Fig. 2. List of Crucial Metrics.

Currently, no standardized system exists for the battery management of electric vehicles, presenting challenges in accessing specific information. Efforts are underway to enhance our system's capability to extract the required information, especially concerning electric vehicles. This involves exploring additional components that can be integrated to retrieve relevant details, even from proprietary or restricted systems within electric vehicles [24,33].

Our focus on collecting crucial metrics extends to both electrical and non-electrical vehicles, developing a cost-effective methodology using the OBD II system. For electric cars, key parameters such as battery health, state of charge, temperature, cell voltages, and cumulative charge power are targeted. For non-electric cars, the emphasis extends to vital parameters such as fuel efficiency, engine health, emissions levels, and operational performance.

5.1 Data Interpretation and Visualization

To interpret collected metrics effectively, we leverage the open-source software "asammdf," offering robust dashboard capabilities and informative visualizations. Figure 4 showcases sample dashboards, providing insights into crucial aspects of vehicle performance, extracted from the vehicle's CAN Bus, including engine RPM, coolant temperature, and throttle position.

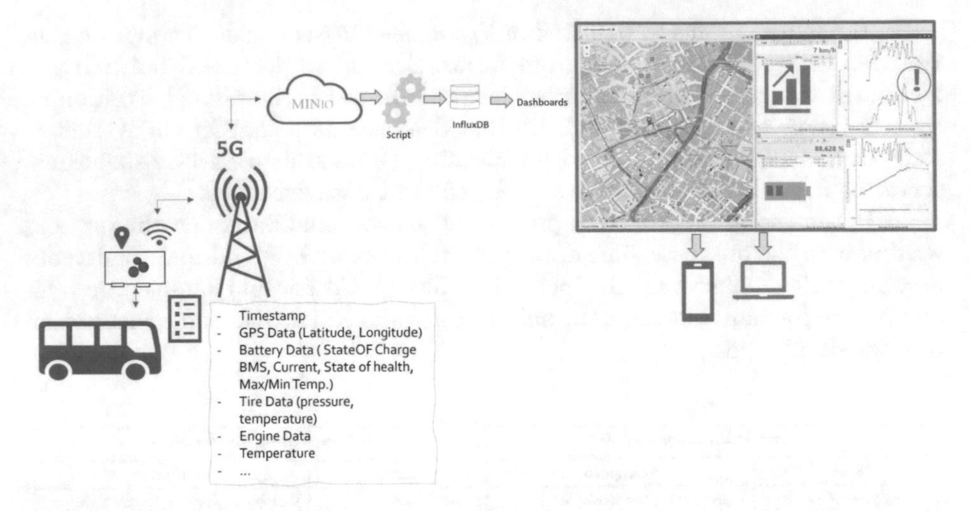

Fig. 3. Diagram Showing the Method Developed for Data Logging from Connected Vehicles.

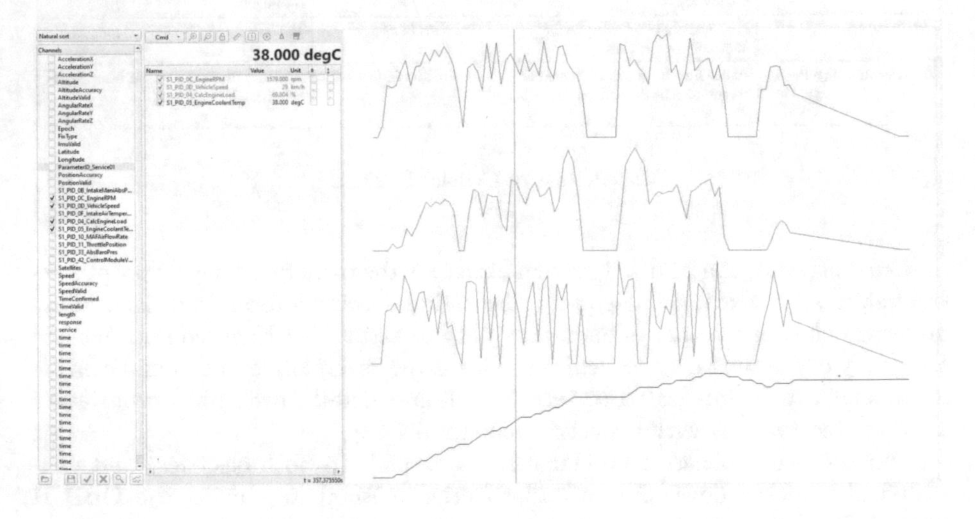

Fig. 4. Visualization of the Collected Metrics.

Analyzing these metrics equips municipal fleet managers with insights into various performance aspects, facilitating informed decision-making and optimized fleet operations. The data contributes to the development of predictive maintenance models, offering valuable insights into the impact of driving conditions on vehicle performance. Proactive maintenance planning, coupled with preventative measures, enhances the overall reliability and efficiency of the municipal vehicle fleet.

This comprehensive analysis not only provides valuable insights but also empowers fleet managers to make informed decisions, implement efficient maintenance strategies, and optimize overall fleet performance.

6 Conclusion and Future Work

The conclusion underscores the transformative impact of integrating 5G technology into municipal fleet management within smart cities. Municipalities embracing 5G experience precise vehicle tracking, optimized route planning, and overall enhanced fleet performance. This integration not only signifies a technological upgrade but also acts as a catalyst for sustainable urban development, contributing to cost savings and heightened service delivery.

Looking ahead, future work envisions strategic expansions in AR applications, refining real-time reporting algorithms, integrating advanced predictive analytics, and fostering collaboration with broader smart city infrastructure. These endeavors aim to further enhance municipal fleet management, creating a smarter, more responsive urban environment. The outlined future work aligns with the overarching goal of advancing smart city infrastructures and ensuring that municipal fleet management evolves as a pivotal component of sustainable urban development.

Acknowledgment. This work is financially supported by the Federal Ministry of Education and Research (BMBF), Germany, under the funding code 13FH9I02IA, as part of the FH-Impuls 2016 research project: 'Urban Digital Twins for the Intelligent City.' However, the primary responsibility for the content of this publication rests solely with the first author.

References

1. 5G in release 17-strong radio evolution. Technical report, 3GPP, TS 23.501 V16.1.0 (2019)
2. System architecture for the 5G system. Technical report, 3GPP, TS 23.501 V16.1.0 (2019)
3. Minimum requirements related to technical performance for IMT-2020 radio interface (s). Technical report, document ITU-R SG05 (2017). [Online]
4. 3GPP: Study on scenarios and requirements for next generation access technologies (release 16). Technical report, TR 38.913 V16.0.0 (2020)
5. Al Nuaimi, E., Al Neyadi, H., Mohamed, N., Al-Jaroodi, J.: Applications of big data to smart cities. J. Internet Serv. Appl. **6**(1), 1–15 (2015)
6. Alsaleh, A.: The impact of various communication methods: how vehicle-to-vehicle and vehicle-to-infrastructure messages affect the performance of driving smart vehicles. Available at SSRN 4339865
7. Bertenyi, B.: 5G evolution: what's next? IEEE Wirel. Commun. **28**(1), 4–8 (2021)
8. Bibri, S.E.: On the sustainability of smart and smarter cities in the era of big data: an interdisciplinary and transdisciplinary literature review. J. Big Data **6**(1), 1–64 (2019)

9. Bibri, S.E., Krogstie, J.: Smart sustainable cities of the future: an extensive interdisciplinary literature review. Sustain. Urban Areas **31**, 183–212 (2017)
10. Biswas, A., Wang, H.C.: Autonomous vehicles enabled by the integration of IoT, edge intelligence, 5G, and blockchain. Sensors **23**(4), 1963 (2023)
11. Cerasi, I.R.: The potential of autonomous and connected sweepers for smart and sustainable cities. Int. J. Transp. Dev. Integration **6**(1), 37–57 (2022)
12. Coors, V., Pietruschka, D., Zeitler, B.: iCity. Transformative Research for the Livable, Intelligent, and Sustainable City: Research Findings of University of Applied Sciences Stuttgart. Springer Nature (2022)
13. CSS Electronic: CAN Bus Data Loggers (2023). https://www.csselectronics.com/
14. Dabeedooal, Y.J., Dindoyal, V., Allam, Z., Jones, D.S.: Smart tourism as a pillar for sustainable urban development: an alternate smart city strategy from Mauritius. Smart Cities **2**(2), 153–162 (2019)
15. Dai, C., Liu, X., Lai, J., Li, P., Chao, H.C.: Human behavior deep recognition architecture for smart city applications in the 5G environment. IEEE Netw. **33**(5), 206–211 (2019)
16. Ghosh, A., Maeder, A., Baker, M., Chandramouli, D.: 5G evolution: a view on 5G cellular technology beyond 3G pp release 15. IEEE Access **7**, 127639–127651 (2019)
17. Gohar, A., Nencioni, G.: The role of 5G technologies in a smart city: the case for intelligent transportation system. Sustainability **13**(9), 5188 (2021)
18. Ji, H., Huang, H.: The integration and development trend of China's 5G technology and smart cleaning. In: Journal of Physics: Conference Series, vol. 1812, p. 012015. IOP Publishing (2021)
19. Joseph, P.C., Kumar, S.P.: Design and development of OBD-II compliant driver information system. Indian J. Sci. Technol. **8**(21) (2015)
20. Kombate, D., et al.: The internet of vehicles based on 5G communications. In: 2016 IEEE International Conference on Internet of Things (iThings) and IEEE Green Computing and Communications (GreenCom) and IEEE Cyber, Physical and Social Computing (CPSCom) and IEEE Smart Data (SmartData), pp. 445–448. IEEE (2016)
21. Li, W., Bhushan, B., Gao, J., Zhang, P.: Smartclean: smart city street cleanliness system using multi-level assessment model. Int. J. Softw. Eng. Knowl. Eng. **28**(11n12), 1755–1774 (2018)
22. Malasinghe, L.P., Ramzan, N., Dahal, K.: Remote patient monitoring: a comprehensive study. J. Ambient. Intell. Humaniz. Comput. **10**, 57–76 (2019)
23. Nagaraj, P., Lakshmanaprakash, S., Muneeswaran, V.: Edge computing and deep learning based urban street cleanliness assessment system. In: 2022 International Conference on Data Science, Agents & Artificial Intelligence (ICDSAAI), vol. 01, pp. 1–6 (2022)
24. Ramai, C., Ramnarine, V., Ramharack, S., Bahadoorsingh, S., Sharma, C.: Framework for building low-cost OBD-II data-logging systems for battery electric vehicles. Vehicles **4**(4), 1209–1222 (2022)
25. Rao, S.K., Prasad, R.: Impact of 5G technologies on smart city implementation. Wirel. Pers. Commun. **100**, 161–176 (2018)
26. Rathore, M.M., Paul, A., Hong, W.H., Seo, H., Awan, I., Saeed, S.: Exploiting IoT and big data analytics: defining smart digital city using real-time urban data. Sustain. Urban Areas **40**, 600–610 (2018)
27. Tesla: Tesla (2023). https://www.tesla.com/
28. Tesla, Inc.: Fleet telemetry (2023). https://github.com/teslamotors/fleet-telemetry

29. Traboulsi, S.: Overview of 5G-oriented positioning technology in smart cities. Procedia Comput. Sci. **201**, 368–374 (2022)
30. Traboulsi, S., Uckelmann, D.: 5G as enabler technology for smart city use cases. In: 2022 9th International Conference on Internet of Things: Systems, Management and Security (IOTSMS), pp. 1–7. IEEE (2022)
31. TU: Work plan, timeline, process and deliverables for the future development of IMT. ITU-R WP5D (2015)
32. Usman, M.A., Philip, N.Y., Politis, C.: 5G enabled mobile healthcare for ambulances. In: 2019 IEEE Globecom Workshops (GC Wkshps), pp. 1–6. IEEE (2019)
33. Wang, L., Wang, L., Liu, W., Zhang, Y.: Research on fault diagnosis system of electric vehicle power battery based on OBD technology. In: 2017 International Conference on Circuits, Devices and Systems (ICCDS), pp. 95–99. IEEE (2017)
34. Zhang, P., Zhao, Q., Gao, J., Li, W., Lu, J.: Urban street cleanliness assessment using mobile edge computing and deep learning. IEEE Access **7**, 63550–63563 (2019)

Developing a GPT Chatbot Model for Students Programming Education

Horia Alexandru Modran[1](✉), Doru Ursuțiu[1,2], Cornel Samoilă[1,3], and Elena-Cătălina Gherman-Dolhăscu[1]

[1] Transilvania University of Brașov, Brașov, Romania
{horia.modran,udoru,csam,elena.dolhascu}@unitbv.ro
[2] Romanian Academy of Scientists, Bucharest, Romania
[3] Romanian Academy of Technical Sciences, Bucharest, Romania

Abstract. While general-purpose language models, such as ChatGPT, have performed very well in various domains, the unique challenges posed by programming necessitate a specialized approach. This study focuses on the development of a specialized GPT model aimed at supporting students in their programming education. The model is designed to serve a triple purpose: assisting students in generating specific parts of their code and guiding them through the process of improving and debugging their own programs. Notably, the model incorporates a unique customization feature with moderation to ensure that the generated code does not closely mimic assignments designated by teachers. The article provides an in-depth exploration of the model's construction, detailing its distinctive features intended to empower students while preserving the academic integrity of the learning experience. This innovative contribution holds significant potential in transforming programming education, offering targeted assistance that expands individual problem-solving skills and strikes a balance between automated support and educational objectives. The research underscores the importance of maintaining fairness and integrity in the learning process while leveraging advanced technology to enhance students' programming capabilities.

Keywords: AI · Programming Education · GPT · Chatbot

1 Introduction

In recent years, the development of Artificial Intelligence (AI) technologies has significantly influenced various domains, such as education and research. One notable technology in this area is ChatGPT (Chat Generative Pre-trained Transformer), a robust large language model developed by OpenAI. This innovation opens intriguing possibilities for both engineering students and teachers, encompassing personalized feedback, higher accessibility, interactive dialogues, lesson preparation and assessments.

While ChatGPT shows proved to be useful for teachers, facilitating tasks like generating course materials and offering suggestions, and as a virtual tutor for students, assisting with queries and fostering collaboration, its current usage presents some challenges. These challenges include the generation of inaccurate or false information and

M. E. Auer et al. (Eds.): STE 2024, LNNS 1028, pp. 72–82, 2024.
https://doi.org/10.1007/978-3-031-61905-2_8

the potential to elude plagiarism detection. Teacher training and student education are imperative to address the impact of ChatGPT on the educational landscape [1].

In the ever-evolving landscape of programming education, the role of Artificial Intelligence (AI) has become increasingly prominent, offering innovative solutions to enhance the learning experience. In the current context, ChatGPT can be useful for enhancing engineering students' programming proficiency. A comprehensive literature review was carried out on this topic. To illustrate the capabilities of ChatGPT in this field, M. Rahman et al. [2] conducted diverse coding-oriented experiments employing ChatGPT, such as generating code from problem descriptions, creating pseudocode for algorithms derived from textual content and code correction. The generated codes underwent validation using an online judging system to assess their accuracy. Additionally, they conducted surveys among students and educators to gain insights into how ChatGPT supports the teaching and learning of programming.

The study conducted by Feng et al. [3] introduces a framework designed to explore the code generation functionalities of ChatGPT by analyzing crowdsourced data from Twitter and Reddit. The findings indicate that Python and JavaScript emerge as the most frequently discussed programming languages on social media, while ChatGPT finds application in diverse code generation domains, including debugging code, preparing for programming interviews and addressing academic assignments.

Other research [4] delves into the utilization of ChatGPT for solving programming bugs. The study examines the attributes of ChatGPT and how they can be harnessed to offer assistance in debugging, predicting bugs, and explaining issues to help in resolving programming challenges. It concludes by emphasizing the potential of ChatGPT as a component within a comprehensive debugging toolkit, highlighting the advantages of synergizing its strengths with those of other debugging tools.

Phung et al. [5] conducted a systematic assessment of two models, ChatGPT (based on GPT-3.5) and GPT-4, comparing their efficacy with human tutors across various scenarios. Their evaluation encompasses five introductory Python programming challenges and programs with bugs sourced from an online platform, with performance gauged through expert-based evaluation. Their findings reveal a substantial performance advantage of GPT-4 over ChatGPT (based on GPT-3.5).

Paper [6] explores the use of ChatGPT as a programming assistant in the context of programming education. The primary objective was to evaluate the impact of ChatGPT on students' understanding and performance in these fundamental problem-solving techniques. The results indicate that the use of ChatGPT as a programming assistant yielded better outcomes compared to working without any external assistance.

Our research team already proposed in a previous study a method for teaching engineers all the necessary steps for developing, validating, and deploying machine learning-based systems [7] and to integrate Artificial Intelligence and ChatGPT into Engineering Education. The current study extends the previous one and is focused on the development of a specialized Generative Pre-trained Transformer (GPT) model based on OpenAI API. Compared to existing literature and applications, the proposed model is not a one-size-fits-all solution, but rather a specialized tool finely attuned for programming education.

2 Developing a GPT Web Programming Model

2.1 Customizing the GPT Model

Developing a custom GPT (Generative Pre-trained Transformer) model using the OpenAI API for programming tasks has many key advantages. Firstly, customization is crucial for tailoring the model to the specific needs and learning objectives of engineering students. By fine-tuning the model on task-specific data relevant to the curriculum, teachers can ensure that the generated code aligns closely with the programming challenges students are likely to encounter. The customization process involves adapting the model to the nuances of programming languages, coding paradigms, and domain-specific requirements.

Moreover, compared to ChatGPT [8], a custom GPT model allows for granular control over the model's behavior, enabling instructors to align generated code with specific coding standards and best practices. This control fosters a better approach to teaching coding conventions and encourages adherence to industry-standard practices.

By integrating the model into the learning environment, teachers can observe how students interact with AI-generated code, providing data on common challenges, misconceptions, and areas where additional support might be needed. This observational data can inform future model refinements and pedagogical strategies.

The development and implementation of a custom GPT model in an educational context aligns with the principles of personalized learning and adaptive education. Scientifically evaluating the impact of personalized, AI-generated code on individual student performance and comprehension provides empirical evidence for the efficacy of this approach.

MODEL	DESCRIPTION
GPT-4 and GPT-4 Turbo	A set of models that improve on GPT-3.5 and can understand as well as generate natural language or code
GPT-3.5	A set of models that improve on GPT-3 and can understand as well as generate natural language or code
DALL·E	A model that can generate and edit images given a natural language prompt
TTS	A set of models that can convert text into natural sounding spoken audio
Whisper	A model that can convert audio into text
Embeddings	A set of models that can convert text into a numerical form
Moderation	A fine-tuned model that can detect whether text may be sensitive or unsafe
GPT base	A set of models without instruction following that can understand as well as generate natural language or code
GPT-3 Legacy	A set of models that can understand and generate natural language
Deprecated	A full list of models that have been deprecated along with the suggested replacement

Fig. 1. OpenAI API Models [9].

The OpenAI API [9] has several advantages compared to developing a GPT model due to resource efficiency, as it eliminates the need for substantial computational power and extensive training data. The API provides access to powerful, pre-trained models like GPT-4 and GPT-3.5 Turbo, saving both time and effort. Additionally, it ensures ongoing updates and improvements from OpenAI, freeing the developer from the maintenance burden associated with developing and fine-tuning a custom model. All the current models described in Fig. 1 were considered, while the GPT-4 proved to have a better grasp of complex programming tasks, an improved code quality and enhanced contextual understanding, leading to more precise and context-aware code generation.

Considering the previously presented reasons, our research team decided to develop a specialized programming GPT Model by using OpenAI GPT-4 model, named "CVTC Coding Expert". It should only be able to generate programming code, debug code and improve or refactor code. Crafting a clear and concise system message is crucial to guide the behavior of the custom GPT model. Several samples of system messages for instructing the model to focus specifically for these three tasks, while rejecting any other task, were tested, while the one that performed very well is illustrated in Fig. 2.

```
24  const instructionObj = {
25      role: 'system',
26      content: "You are an expert programming assistant specialized in generating, debugging, and improving code.  \n\
27      Please provide code snippets in response to programming-related prompts.  \n \
28      If the input contains code, your task is to understand, debug, and enhance it.  \n \
29      If the input is not related to programming tasks, respond with a message indicating that you can only assist  \n\
30      with programming-related queries.\n \
31      Do not generate or engage in content unrelated to programming.  \n \
32      Your goal is to be a valuable resource for students for code-related tasks."
33  }
```

Fig. 2. GPT Model system message.

For interacting with the model, an API Key from the OpenAI web application must be obtained beforehand. It serves as a form of authentication, allowing the model to access OpenAI's services securely and it should be included in the request header for every request sent to the OpenAI API. To interact with the OpenAI mode, the OpenA API library for JavaScript was used, which provides methods to get completion for a provided input. The API provides some finetuning parameters for setting a specific behavior for the GPT model, which are described in Table 1 [9].

Table 1. OpenAI GPT Model parameters.

Parameter	Description
Temperature	The sampling temperature, between 0 and 1. Higher values will make the output more random, while lower values will make it more focused and deterministic
Frequency penalty	The frequency penalty, between −2 and +2 (default is 0). A higher frequency penalty discourages repetitive phrases or tokens in the output. A lower frequency penalty allows for more repetition
Presence penalty	The presence penalty, between −2 and +2 (default is 0). A higher value discourages the model from including specific tokens in the output, while a lower allows for the inclusion of specific tokens
Top p	The nucleus sampling, ranging from 0 to 1. It is used to control how many of the highest-probability words are selected to be included in the generated text

Several possible values were tested for each of those three parameters, and the values that proved to be the best were the following (Fig. 3):

- Temperature of 0, as the model should be more predictable.
- Frequency penalty of 0.1, for obtaining a balanced output.
- Presence penalty of 0.1, for a focused code generation task.
- Top p of 1.0, so that the highest-probability words will be selected.

```
const response = await openai.createChatCompletion({
    model: 'gpt-4',
    messages: conversationArr,
    temperature: 0,
    presence_penalty: 0.1,
    frequency_penalty: 0.1,
    top_p: 1.0
})
```

Fig. 3. Setting the Model's parameters.

Furthermore, the model was instructed by using the few-shot prompting method. This technique is used for instructing GPT models, where a prompt consists of a few examples or demonstrations of the desired behavior to guide the model's response. This method is useful in the current scenario, as the goal is to provide specific context and examples to influence the model's output without extensive retraining. Some sample prompts provided to the model are presented in Fig. 4.

```
Prompt:
### Python Code Generation, Debugging, and Improvement

1. **Generating Code - Sinusoidal Signal:**
   - Given a frequency parameter, generate a Python function that produces a sinusoidal signal.
   - Example:
     - Input: "Generate a Python function for a sinusoidal signal with a frequency parameter."
     - Output:
       ```python
       def generate_sinusoidal_signal(frequency, duration, sampling_rate):
           import numpy as np
           t = np.arange(0, duration, 1/sampling_rate)
           signal = np.sin(2 * np.pi * frequency * t)
           return signal
       ```

2. **Debugging Code - Complex Code Snippet:**
   - Identify and fix issues in the provided Python code snippet, which implements a custom sorting algorithm.
   - Example:
     - Input: "Fix the bug in the following code:"
       ```python
       def custom_sort(input_list):
           for i in range(len(input_list)):
               for j in range(i, len(input_list)):
                   if input_list[i] > input_list[j]:
                       input_list[i], input_list[j] = input_list[j], input_list[i]
           return input_list
       ```
     - Output:
       ```python
       def custom_sort(input_list):
           return sorted(input_list)
       ```

3. **Improving Code - Unoptimized Code Snippet:**
   - Optimize the provided Python code snippet, which calculates the nth Fibonacci number using a recursive approach.
   - Example:
     - Input: "Optimize the following code for calculating the nth Fibonacci number:"
       ```python
       def fibonacci(n):
           if n <= 1:
               return n
           else:
               return fibonacci(n-1) + fibonacci(n-2)
       ```
     - Output:
       ```python
       def fibonacci(n, memo={}):
           if n <= 1:
               return n
           if n not in memo:
               memo[n] = fibonacci(n-1, memo) + fibonacci(n-2, memo)
           return memo[n]
       ```

```

**Fig. 4.** Samples prompts and answers provided to the GPT model.

## 2.2 Developing the Chatbot Web Application

A Web application was developed for interacting with the GPT model that was created, using HTML5, CSS3 and JavaScript, together with the OpenAI API library.

For displaying a chat history, an array for storing the previous conversation held in the same session was created and a JavaScript function will get the response from the OpenAI API and append it to that array (Fig. 5). The response is obtained in an asynchronous way, in order not to block the entire web applications until the response is fetched and rendered in the webpage.

```
50 function fetchReply() {
51   get(conversationInDb).then(async (snapshot) => {
52   if (snapshot.exists()) {
53   const conversationArr = Object.values(snapshot.val())
54   conversationArr.unshift(instructionObj)
55   const response = await openai.createChatCompletion({
56   model: 'gpt-4',
57   messages: conversationArr,
58   temperature: 0,
59   presence_penalty: 0.1,
60   frequency_penalty: 0.1,
61   top_p: 1.0
62   })
63   push(conversationInDb, response.data.choices[0].message)
64   renderTypewriterText(response.data.choices[0].message.content)
65   }
66   else {
67   console.log('No data available')
68   }
69
70   })
71 }
```

Fig. 5. JavaScript Function for fetching and appending OpenAI model response.

Using a JavaScript function, the web application will be able to render the responses from ChatGPT in a typewriter way. Furthermore, to store all the conversations for further use even after the browser window is closed or the web application stopped, a Firebase Realtime Database [10] was used to store the conversation history. It can store all previous conversations, while a button to delete all history was added on the main webpage. All the conversation history can be reviewed and checked by the teacher, who can change his educational methods according to the students' needs. Figure 6 shows sample conversation history of a user stored in the Firebase database.

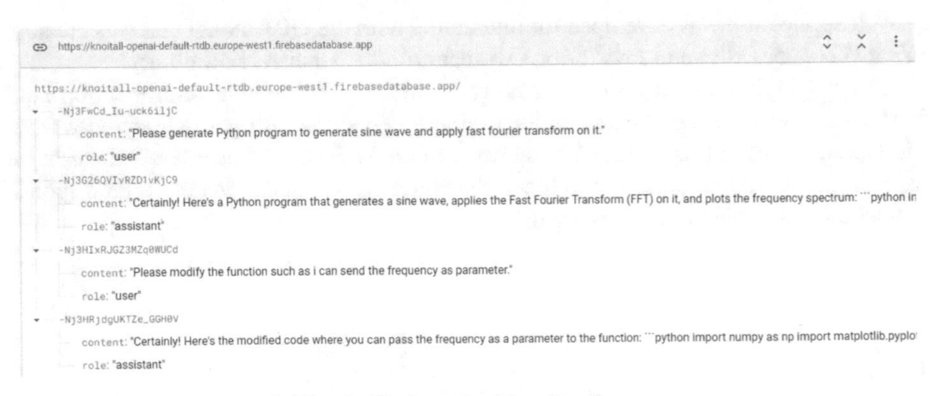

Fig. 6. Firebase Realtime Database.

3 Testing and Validating the GPT Model

Following the development of the model, it underwent thorough testing and assessment across diverse scenarios to evaluate its performance and capabilities.

The initial testing scenario aimed to verify the accurate interpretation of instructions provided to the model. The primary focus was to ensure that the model responded exclusively to queries related to programming, as illustrated in Fig. 7. This rigorous evaluation aimed to ascertain the model's precision in adhering to specific instructions and limiting its responses to the designated subject matter.

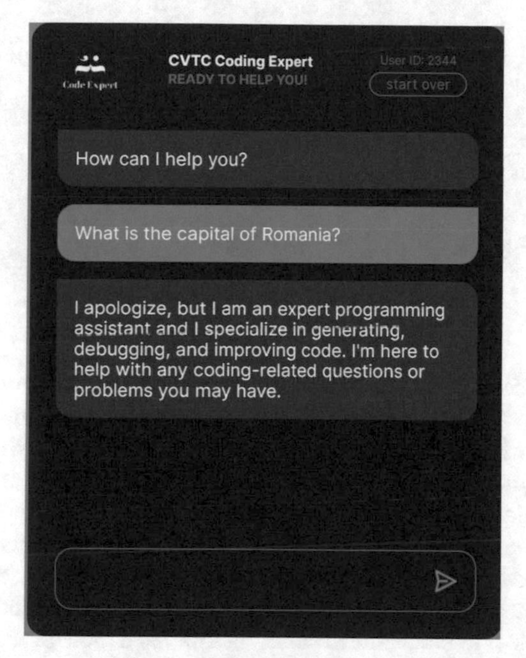

Fig. 7. Model behavior check.

In addition, it is very important to instruct students on the art of crafting effective prompts. This involves guiding them in the process of formulating queries in a manner that optimally elicits the desired information or response from the GPT Chatbot. Crafting good prompts for the OpenAI API is crucial because it directly influences the quality and relevance of the responses generated by the model. Well-constructed prompts help articulate user intentions clearly, prompting the model to provide more accurate and desired outputs. Providing students with insights into prompt construction not only empowers them to extract the information they seek, but also enhances their overall proficiency in interacting with the chatbot effectively. This educational approach contributes to a deeper understanding of how language input influences the output generated by the GPT Chatbot, fostering a more constructive and insightful engagement with the AI technologies.

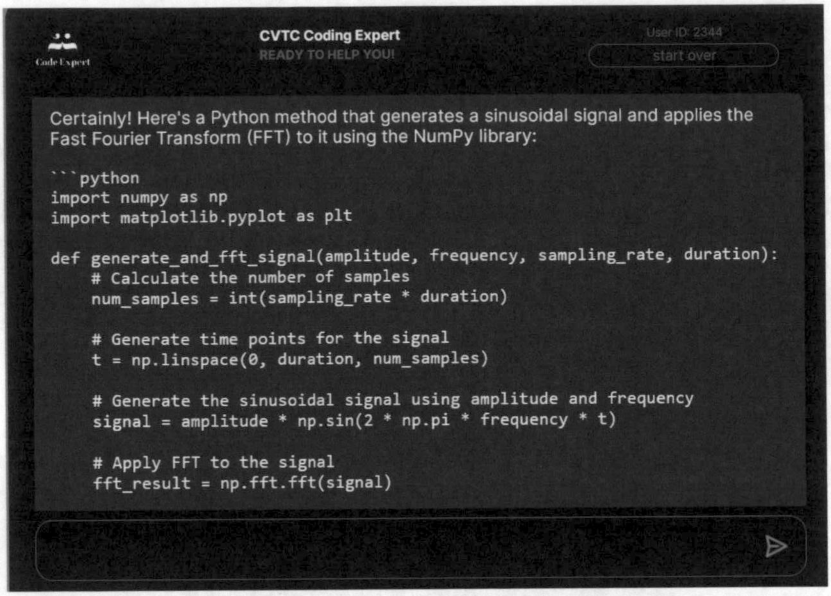

Code Expert

CVTC Coding Expert
READY TO HELP YOU!

User ID: 2344
start over

Certainly! Here's a Python method that generates a sinusoidal signal and applies the Fast Fourier Transform (FFT) to it using the NumPy library:

```
```python
import numpy as np
import matplotlib.pyplot as plt

def generate_and_fft_signal(amplitude, frequency, sampling_rate, duration):
    # Calculate the number of samples
    num_samples = int(sampling_rate * duration)

    # Generate time points for the signal
    t = np.linspace(0, duration, num_samples)

    # Generate the sinusoidal signal using amplitude and frequency
    signal = amplitude * np.sin(2 * np.pi * frequency * t)

    # Apply FFT to the signal
    fft_result = np.fft.fft(signal)
```

Fig. 8. Code Generation Example.

An exemplary demonstration of leveraging the developed GPT model in a proficient manner to generate a Python program is illustrated in Fig. 8. In this instance, the GPT model was tasked with creating a Python method for generating a sinusoidal signal and subsequently applying Fast Fourier Transform (FFT) to it. Notably, the model generated a highly efficient and parameterized method for executing this task, accompanied by detailed explanations for each step. This not only equips students with a functional Python code snippet but also offers comprehensive insights, enabling them to understand, test, and potentially modify the code to suit their specific needs. The provision of clear explanations enhances the educational value, facilitating a more profound understanding of both the code snippet and the underlying processes involved.

Another useful application of the CVTC Coding Expert Chatbot lies in its capability to correct errors in code. To leverage this functionality, it is essential to present the code snippet in a clear and well-formatted manner. Figure 9.1 provides a concrete example of utilizing the Chatbot for code correction, showcasing how a correctly formatted input can prompt the Chatbot to identify and rectify errors within the given code. Furthermore, Fig. 9.2 shows how the code improvement function facilitated by the GPT model works, focusing on the corrective mechanisms employed by the model in enhancing coding best practices and functionality. This feature not only streamlines the debugging process but also serves as an instructive tool for students, offering insights into common coding errors and their resolutions.
```

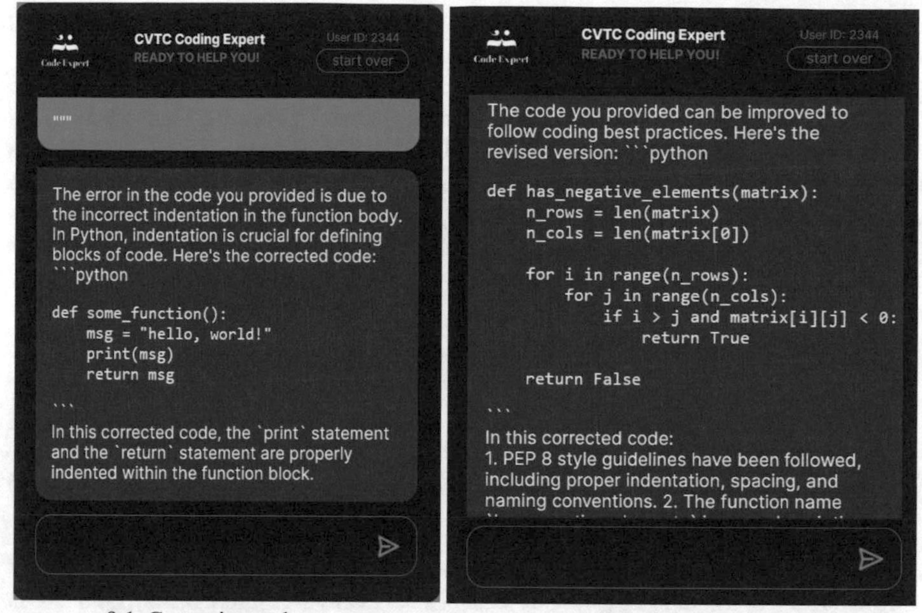

9.1. Correcting code 9.2. Code improvement

Fig. 9. Correcting Code Errors and Improving code.

4 Conclusions

While prior studies have explored the utility of general-purpose language models in various domains, the unique challenges posed by programming necessitate a specialized approach. Existing literature often falls short in addressing the specific needs of students, particularly in guiding them through the intricacies of code generation, improvement and debugging.

The current study described the development of a specialized Generative Pre-trained Transformer (GPT) model, tailored to specifically address the unique needs of students in programming education. While general-purpose language models, such as ChatGPT, have demonstrated remarkable capabilities, the imperative for a dedicated tool optimized for code generation and debugging within educational contexts becomes evident.

One of the main advantages of harnessing the OpenAI API for this specialized GPT model lies in the customization opportunities it affords. Unlike ChatGPT, which operates as a versatile language model, the proposed GPT model is finely tuned to the specific requirements of programming education. This customization, coupled with moderation features, ensures that the model provides assistance aligned with pedagogical objectives, without inadvertently generating solutions that might compromise the integrity of students' independent work.

In essence, this study seeks to contribute to the ongoing discourse on AI in education, by highlighting the advantages of a specialized GPT model based on the OpenAI API. By drawing comparisons with existing literature and applications, the aim was to underscore the nuanced benefits and potential impact of tailoring AI tools to the specific demands of programming education. Analyzing the technical aspects of model development and exploring its features, the goal is to pave the way for a more effective and ethically grounded integration of AI in the realm of programming pedagogy.

The GPT model is currently in the beta testing phase. However, there are plans to assess the effectiveness of the developed Coding Expert Chatbot in the near future. The intention is to introduce this Chatbot to engineering students at our university and to collect valuable feedback through an online anonymous survey. This comprehensive evaluation strategy aims to gauge the practical utility of the Chatbot in real educational contexts and gather insights that can help to improve it.

References

1. Lo, C.K.: What is the impact of ChatGPT on education? A rapid review of the literature. Educ. Sci. **13**, 410 (2023). https://doi.org/10.3390/educsci13040410
2. Rahman, M., Watanobe, Y.: ChatGPT for education and research: opportunities, threats, and strategies. Appl. Sci. **13**, 5783 (2023). https://doi.org/10.3390/app13095783
3. Feng, Y., Vanam, S., Cherukupally, M., Zheng, W., Qiu, M., Chen, H.: Investigating code generation performance of ChatGPT with crowdsourcing social data. In: 2023 IEEE 47th Annual Computers, Software, and Applications Conference (COMPSAC), Torino, Italy, pp. 876–885 (2023). https://doi.org/10.1109/COMPSAC57700.2023.00117
4. Surameery, N., Shakor, M.: Use chat GPT to solve programming bugs. Int. J. Inf. Technol. Comput. Eng. (IJITC) **3**, 17–22 (2023). https://doi.org/10.55529/ijitc.31.17.22. ISSN 2455-5290
5. Phung, T., et al.: Generative AI for programming education: benchmarking ChatGPT, GPT-4, and human tutors. In: Proceedings of the 2023 ACM Conference on International Computing Education Research - Volume 2 (ICER 2023), vol. 2, pp. 41–42. Association for Computing Machinery, New York (2023). https://doi.org/10.1145/3568812.3603476
6. Vukojičić, M., Krstić, J.: ChatGPT in programming education: ChatGPT as a programming assistant. InspirED Teachers' Voice **2023**(1), 7–13 (2023)
7. Modran, H.A., Ursutiu, D., Samoila, C., Chamunorwa, T.: Learning methods based on artificial intelligence in educating engineers for the new jobs of the 5th industrial revolution. In: Auer, M.E., Rüütmann, T. (eds.) ICL 2020, pp. 561–571. Springer, Cham (2021). https://doi.org/10.1007/978-3-030-68201-9_55
8. ChatGPT. https://chat.openai.com/. Accessed 13 Nov 2023
9. Open AI API Documentation. https://platform.openai.com/docs/introduction. Accessed 13 Nov 2023
10. Firebase API Reference. https://firebase.google.com/docs/reference. Accessed 13 Nov 2023

Integrated Real-Time Web-Based System of Accidental Hoisting Detection Algorithms for Hoist Lift in Harbors

Shengwang Ye[1(✉)], Guo-Ping Liu[1,2], Wenshan Hu[1], and Zhongcheng Lei[1]

[1] Wuhan University, Wuhan 430072, China
2014301470055@whu.edu.cn

[2] Southern University of Science and Technology, Shenzhen 518055, China

Abstract. The system is proposed owing to three factors, convenience, rapidness and scalability. Firstly, the data transfer and hoist lift sample and control functions are encapsulated to simplify algorithm development and verification process, which allows algorithm designers to focus on the efficiency of algorithms rather than implementation details. Secondly, the proposed system serves as the benchmark to evaluate the performance of various detection algorithms based on sampled history data. Besides, grammar errors or logical errors in the implementation of algorithms can be exposed and corrected via repeated verifications. All algorithms that pass the verification can be directly put into production without any modifications. The web-based integrated development and monitoring system is deployed in a harbor for container hoisting. The status of a three-phase asynchronous motor in a hoist process is displayed. The parameters of the monitoring algorithm are online adjusted. The effectiveness and scalability of the proposed system have been demonstrated through experiments, making it suitable for use in a variety of harbors.

Keywords: Accidental hoisting detection · Online monitor · Web-based

1 Introduction

With the development of Web technology, web-based applications have grown rapidly due to the consistent display effect on all systems [1, 2]. Container loading and unloading are executed daily in all harbors. However, along with daily tasks, accidental hoist of trucks or trains frequently occurs, which can cause loss of persons or property in harbors. Therefore, necessary algorithms are required to detect accidental hoists and improve the security of container hoisting.

Fault detection systems are applied in various applications, including wind turbine real-time remote monitoring [3], photovoltaic plants [4, 5] and the substation of petrochemical facilities [6]. The existing systems provide abundant monitoring widgets, such as line charts, labels and gauges to improve the monitoring efficiency in the process. However, the support for customized complex algorithms by users is lacking in current

M. E. Auer et al. (Eds.): STE 2024, LNNS 1028, pp. 83–89, 2024.
https://doi.org/10.1007/978-3-031-61905-2_9

platforms. The supervisory algorithms are fixed and hard to modify without professional software, which reduces the convenience of the platform.

Several platforms support online control algorithm design, such as NCSLab [7] and iLabs [8]. The existing algorithm development process is static, consisting of data process, algorithm design, offline logic check and test, and algorithm deployment. Compared with simple control algorithms, artificial intelligence (AI)-based fault detection algorithms are complicated [9, 10], which require an integrated development mechanism to improve the efficiency of development.

To help users develop and verify accidental hoisting detection algorithms rapidly, an integrated real-time web-based system is proposed. The features of the proposed system are listed below:

- Integrated development: The integrated development mechanism incorporates algorithm development, deployment and verification. Users can finish the full process in only one system. Thus, except for a web browser, no additional software is required to be installed. The technologies used are all mature open-source technologies, and there are no copyright issues or cost issues, which can be applied on all occasions.
- Online Monitor: The online monitor interface presents the live status of monitored hoists. The parameters of algorithms can be adjusted online in the hoist process without any modifications of algorithms. The same algorithm logic can be applied in various harbors.

The integrated web-based system is proposed given three factors, convenience, rapidness and scalability. Firstly, the data transfer and hoist lift sample and control functions are encapsulated to accelerate algorithm development and verification. Algorithm designers can pay attention to the efficiency of algorithms rather than the implementation details. Secondly, the proposed system serves as the benchmark to evaluate the performance of various detection algorithms based on sampled history data. Besides, grammar errors or logical errors in the implementation of algorithms can be exposed and corrected via repeated verifications. Furthermore, all algorithms that pass the verification can be directly put into production without any modifications.

2 Methodology

2.1 Distributed Web-Based Framework

The proposed system is built under the distributed framework, including the frontend, the backend server, the database server, the code server and the experiment server, as shown in Fig. 1. The proxy server accepts secure incoming connections from the internet, decrypts the requests and redirects to the target server according to the type of request items. The frontend server provides all graphical interfaces for users, such as hoist management, online monitor and algorithm design. Different operations require different privileges to some extent. The backend server stores the static information of users, hoist specifications, programs and hoist status. The code server is responsible for converting block diagrams to executed programs. The experiment server can transfer commands from users to monitor applications, and fetch real-time data from hoists. Once the algorithm is downloaded into the target host, the necessary data transfer and pre-processing procedure will be carried out automatically on the target.

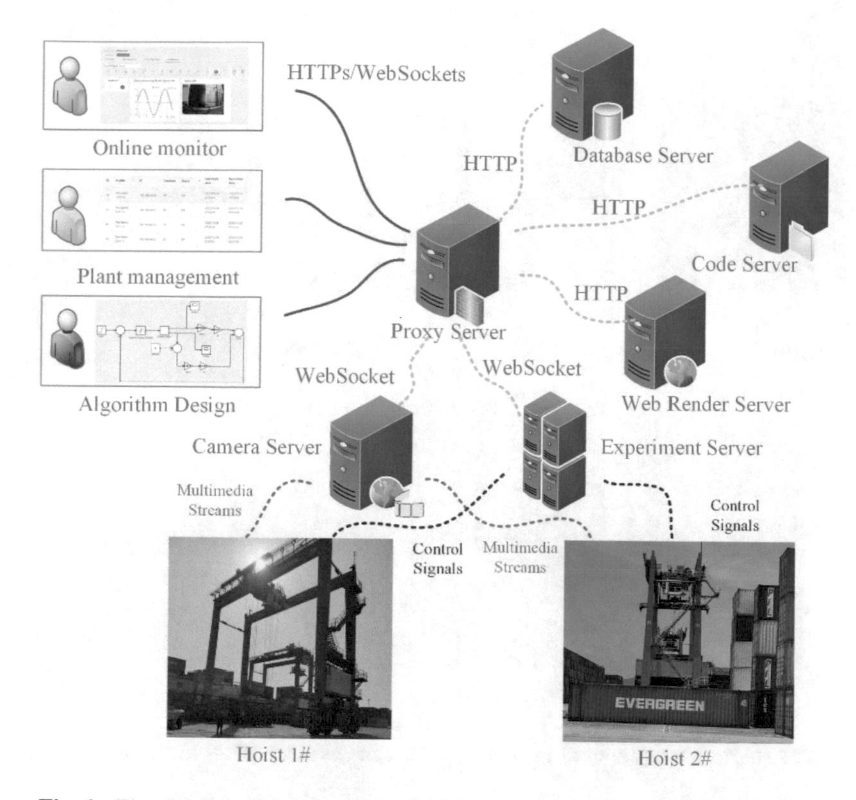

Fig. 1. The distributed web-based framework of an integrated development system in harbors.

2.2 Integrated Development Mechanism

The integrated development mechanism comprises three essential steps: algorithm development, deployment and verification, as illustrated in Fig. 2. The interconnection among these steps is visually represented in the diagram. The creation of high-performance algorithms necessitates the iterative execution of these three steps. The specific details of each step are elaborated below.

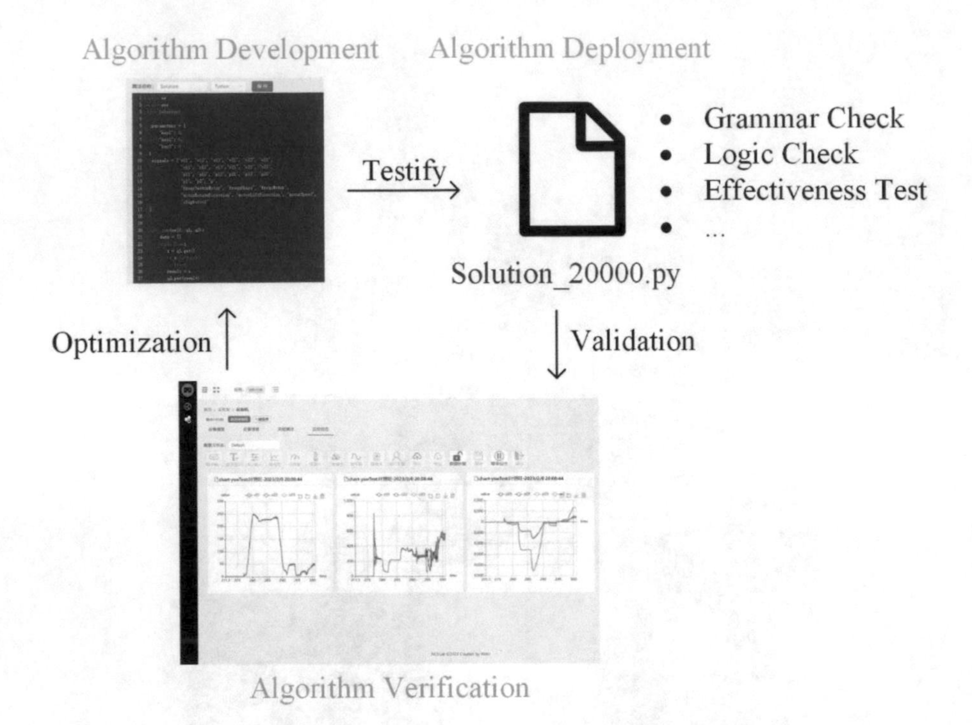

Fig. 2. The three steps of the proposed integrated development mechanism in the web-based system.

Algorithm Development

Algorithm development is the first step to implementing complex theories and powerful algorithms. Due to powerful third-party libraries and convenient input/output interfaces, Python is selected as the basic language in the backend server. All algorithms are written in accordance with Python's syntax, including necessary indents, keywords and operators. Customized algorithms can be implemented in code manners or block diagrams, accelerating the algorithm's development. Furthermore, to utilize advanced AI algorithms in production, the typical AI development environment has been installed in the backend server, including PyTorch, pandas, SciPy and other necessary libraries. Users can utilize existing functions to implement customized algorithms rapidly regardless of the details of input and output. A monitor algorithm of two three-phase asynchronous motors status is programmed as follows:

```
class Solution:
 parameters = { "key1": 0, "key2": 0, "key3": 0 }
 signals = ["v11", "v12", "v13", "v21", "v22", "v23",
      "c11", "c12", "c13","c21", "c22", "c23",
      "p11", "p12", "p13","p21", "p22", "p23",
      "p1", "p2", "p",
      "freqcVectorMotor", "freqcPhase", "freqcMotor",
      "motorRotateDirection", "motorLiftDirection",
      "motorSpeed", "slipRatio1"]

 def run(self, q1, q2):
  data = []
  while True:
   x = q1.get()
   if x is None:
    break
   result = x
   q2.put(result)
```

The algorithm is written based on an object-oriented principle, i.e., the predesigned class Solution. Each solution consists of three parts, parameters, signals and run functions. Parameters are a group of key-value pairs, which can be online modified in the monitoring process, such as the trigger threshold. Signals are an array of signals to be monitored in the monitoring process. The frontend fetches the data from the real-time application according to selected items on the web browser. The remaining parts are functions to implement algorithm logic.

Algorithm Deployment

Owing to the powerful compatibility of Python virtual environment on multiple platforms, such as Windows, Ubuntu, and Raspberry Pi OS, the system can be deployed in various terminals, including PCs, servers and embedded devices. The backend server listens on the specific ports, i.e. 20000–20009. The number of listening ports determines the number of algorithms that can be run simultaneously. Once the designer transfers the designed algorithm, the backend server receives the algorithm and renames the received file to *Solution_port.py*.

Algorithm Verification

The algorithm verification comprises offline verification and online testing. The algorithm accepts offline validation data or real-time application data as input. Offline validation data are utilized to verify the basic program logic and identify potential errors in the running process. On the other hand, real-time application data are employed to validate the effectiveness of the designed algorithm in practical production.

3 Online Experiment

3.1 Experimental Configuration

To monitor the hoist lift status, the proposed system has been installed in several harbors, where multiple hybrid model-based and data-based algorithms are deployed to detect accidental hoisting. In this article, the proposed web-based integrated system is deployed in a harbor to demonstrate the monitoring process.

3.2 Experiment Details

Take the container hoist process in a certain area as an example, an online monitor of a three-phase asynchronous motor in a hoist process is shown in Fig. 3. During the working mode, the real-time hoist lift data are continuously sampled and recorded in the database. The voltage and total power of a three-phase asynchronous motor are graphically depicted in Fig. 3. The hoist lift status and accidental hoisting detection results are packaged and presented on the browser. Once the accidental hoisting occurs, the proposed system can alarm and stop the current operation in time without human intervention. The parameters *key1* and *key2* can be adjusted online to determine the detection sensitivity or trigger threshold in different hoist processes. Furthermore, the efficiency of detection algorithms can be gradually improved with more and more data generated by daily operations.

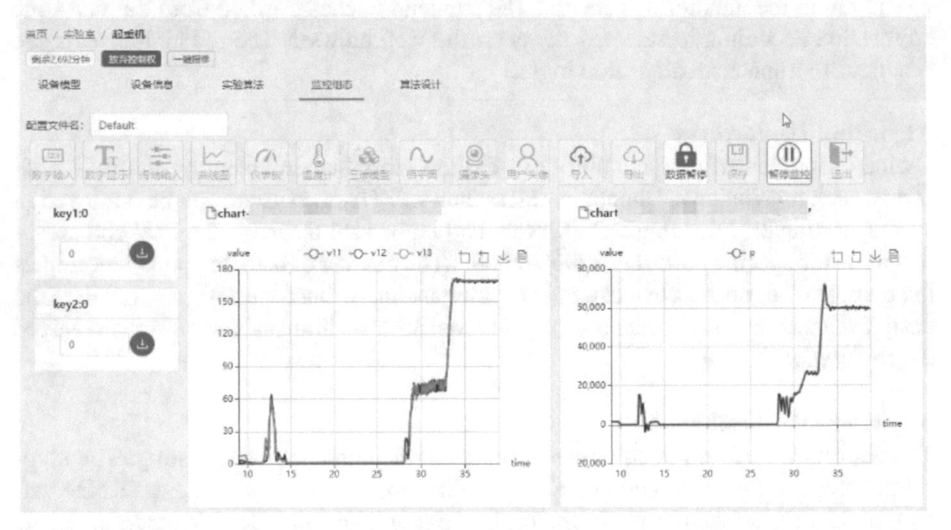

Fig. 3. An online monitor of voltages and power of a three-phase asynchronous motor in a hoist process.

4 Conclusion

The integrated Web-based system has been deployed in harbors to detect accidental hoisting of trucks. Compared with other existing systems, the proposed system can support the integrated development from algorithm design to application in practice. In the future, a camera-assistant system will be introduced into the proposed system where live video streams on the hoist lifts are available. Thus, combined with necessary video recognition algorithms, more advanced and precise technologies can be implemented with multi-dimensional data based on the proposed system.

References

1. Ye, S., Liu, G.-P., Hu, W., Lei, Z.: Design and implementation of a novel compact laboratory for web-based multiagent system simulation and experimentation. IEEE Trans. Ind. Inform. (2023). Early Access
2. Araby, A.A., et al.: Smart IoT monitoring system for agriculture with predictive analysis. In: 2019 8th International Conference on Modern Circuits and Systems Technologies (MOCAST), pp. 1–4. IEEE (2019)
3. Chakkor, S., Baghouri, M., Hajraoui, A.: Wind turbine fault detection system in real time remote monitoring. Int. J. Electr. Comput. Eng. **4**(6), 882–892 (2014)
4. Lazzaretti, A.E., et al.: A monitoring system for online fault detection and classification in photovoltaic plants. Sensors **20**(17), 4688 (2020)
5. Madeti, S.R.K.: A monitoring system for online fault detection in multiple photovoltaic arrays. Renew. Energy Focus **41**, 160–178 (2022)
6. Zhao, L., Matsuo, I.B.M., Salehi, F., Zhou, Y., Lee, W.-J.: Development of a real-time web-based power monitoring system for the substation of petrochemical facilities. IEEE Trans. Ind. Appl. **55**(1), 43–50 (2019)
7. Lei, Z., Zhou, H., Hu, W., Liu, G.-P.: Unified and flexible online experimental framework for control engineering education. IEEE Trans. Ind. Electron. **69**(1), 835–844 (2021)
8. De La Torre, L., Neustock, L.T., Herring, G.K., Chacon, J., Clemente, F.J.G., Hesselink, L.: Automatic generation and easy deployment of digitized laboratories. IEEE Trans. Ind. Inform. **16**(12), 7328–7337 (2020)
9. Jing, L., Zhao, M., Li, P., Xu, X.: A convolutional neural network based feature learning and fault diagnosis method for the condition monitoring of gearbox. Measurement **111**, 1–10 (2017)
10. Alsumaidaee, Y.A.M., Yaw, C.T., Koh, S.P., Tiong, S.K., Chen, C.P., Ali, K.: Review of medium-voltage switchgear fault detection in a condition-based monitoring system by using deep learning. Energies **15**(18), 6762 (2022)

Work in Progress: Empowering Vocational Education with Automation Technology and PLC Integration

Lizhi Song[2], Xulong Zhang[1], Xixin Wei[1], and Mingshen Fu[2](✉)

[1] Rui De International School, Taishan Road No. 1, Zhengzhou, China
[2] Beijing German-Sino Education Technology Co., Ltd., Banjing Road 96, Beijing, China
mingshen.fu@asc-bildung.de

Abstract. This paper details the development and application of a Multi-Platform PLC Working Station (MPPWS) in the context of vocational education at Rui De International School, a collaborative venture between Henan Mechanical and Electrical Vocational College and Beijing German-Sino Education Technology Co., Ltd. The MPPWS is a hands-on educational platform designed to familiarize teachers and students with various Programmable Logic Controllers (PLCs) from different brands. The project, initiated through participation in the Xplore 2023 competition, resulted in the need for a multi-brand PLC training platform. The proposed solution offers a practical teaching model for vocational education in China. The device integrates PLCs from Phoenix Contact, Inovance, and Omron, allowing simultaneous training on different platforms. Training scenarios, such as the "Automatic Mixing System for Multiple Liquids," demonstrate the device's versatility. The educational approach involves group-based training for teachers, emphasizing communication among PLCs of different brands and distributed Input/Output (IO). For students, the focus is on simpler project content with an emphasis on software and parameter settings. The expected outcomes include an enhanced knowledge base for educators, transformative shifts in vocational education methodologies, and improved alignment with industry needs. The iterative feedback process ensures ongoing improvement, maintaining the relevance and effectiveness of the solution in vocational education.

Keywords: Vocational Education · Automation Technology · PLC Integration

1 Introduction and Context

1.1 Introduction of Rui De International School

Rui De International School is a second-tier college jointly established by Henan Mechanical and Electrical Vocational College and Beijing German-Sino Education Technology Co., Ltd. (GSDET) Drawing inspiration from the German dual education model, the college combines theory and practice. It utilizes a self-developed Automation Production Line for Teaching (APLT) as the foundation to create field-specific courses, thereby constructing an integrated theory-practice education system.

M. E. Auer et al. (Eds.): STE 2024, LNNS 1028, pp. 90–98, 2024.
https://doi.org/10.1007/978-3-031-61905-2_10

1.2 Introduction to Trustee Management

Henan Mechanical and Electrical Vocational College entrusts the management and operation of Rui De International College to GSDET, employing a trustee system. Throughout the trustee-based educational process, both parties adhere to principles of equal negotiation and mutual benefit, fully leveraging their respective resource advantages to achieve shared resources and cooperative success. Henan Mechanical and Electrical Vocational College provides Reed with the necessary facilities and equipment, while GSDET deploys teaching and managerial personnel to participate in and oversee educational, training, and daily management activities at the college.

1.3 Xplore2023 Project Overview and Insights

To address the imperative of advancing sustainable vocational education and augmenting the proficiency of educators and students, our team engaged in the Xplore 2023 competition, facilitated by Phoenix Contact. Our strategic initiative involved conceiving, finalizing, and executing a comprehensive plan to develop a simulated canning equipment system. This encompassed mechanical design, the selection and procurement of standard and electrical components, non-standard part machining, mechanical assembly, electrical design, paneling, sensor integration, servo motor parameterization, and the development and debugging of PLC and HMI programs. The collaborative undertaking aimed at optimizing the cultivation and refinement of comprehensive skills, crucial in the context of vocational education in China. Emphasizing practical, hands-on problem-solving, our independent project represents a systematic approach currently lacking in Chinese vocational education. Positioned as a template for future reference, our project is tailored to facilitate sustainable learning, providing a structured framework for both educators and students throughout the vocational education phase.

Our Xplore2023 Project

Our project introduces a PLC-driven simulated canning system, emphasizing intelligence, efficiency, and safety. With over 20 sensors and a peristaltic pump, it caters to both beginners and advanced training, featuring a modular design for flexibility. The sensor array enables a comprehensive understanding of digital and analog inputs. The peristaltic pump, a pollution-free and precise industrial choice, governs canning precision. The PLC serves as the main controller, offering a universally accepted interface with a ladder diagram language for foundational PLC programming. Incorporating common electrical components, the project's modular approach facilitates the addition or removal of modules for diverse control processes. Simulating feeding, filling, sealing, and labeling, each step corresponds to an independent mechanism, progressively increasing programming complexity. The modular design ensures cost-effectiveness, maintenance ease, and scalability, with each segment operating independently.

Encountered Issues during Xplore2023 Project

During the competition, numerous difficulties arose due to limited access to information. The following is a list of some encountered problems:

- The firmware upgrade for the BTP2070W touchscreen failed, resulting in the inability to establish OPC communication with the PLC, and PLC variables couldn't be read.
- The BTP2070W touchscreen system lacked a Chinese plugin, and the absence of a Chinese installation package on the internet prevented the display of Chinese variables, leading to poor Chinese compatibility.
- Visu+ software couldn't display Chinese variables, and the compatibility between PLC Next and Visu+ software for Chinese display was poor, resulting in Chinese variable garbled text and making it challenging to use directly in the Visu+ software.
- Communication failure between Visu+ and MySQL database occurred with limited tutorial materials, and there were insufficient reference materials for troubleshooting after communication failure.
- PLC Next software lacked a "STOP" function, making it inconvenient to initialize PLC internal data during debugging, as it required a power cycle.

Due to the limited official documentation and forum information on PLC Next in Chinese, it was challenging to quickly find solutions to encountered issues.

Additionally, a significant realization was that the competition's focus on innovative project ideas was more aligned with higher education goals. In contrast, vocational education, rooted in practical production scenarios, emphasizes students' ability to independently design, execute, and troubleshoot tasks within established processes, rather than imposing excessive innovation requirements.

2 Problem Analysis

The challenges encountered during the competition, viewed from a user's perspective, primarily stem from the difficulty in acquiring relevant PLC knowledge and information. The multitude of PLC manufacturers on the market, each with its distinct features, and the variations in programming software, operational habits, and PLC programming languages within and across brands, contribute to the complexity. In vocational schools, students often focus on learning a specific mainstream PLC brand, and obtaining Chinese instructional materials for some brands can be challenging. This creates a hurdle when attempting to apply a completely new PLC to a practical project. A multi-brand PLC training platform is needed by the teachers and students from the vocational school system, as well as it's training material and manual in local language.

3 Methodology

To address these challenges, we propose designing and building an educational platform to lower the learning curve for various series, types, or brands of PLCs available in the market. The primary goal of this device is to facilitate teachers and students in getting acquainted with and mastering the use of different PLCs and their basic operations. We have selected three popular PLCs from different countries and brands, each representing a certain type. By comparing the three PLCs on the device, students and teachers can quickly adapt to new PLCs after gaining proficiency in the application of specific ones.

Drawing from our experience in the Xplore2023 competition, we intend to create a comprehensive instructional manual and corresponding training courses for this educational device as well. This will help educators become familiar with the operations and teaching processes of Phoenix Contact, Inovance, and Omron PLCs.

4 Design of the Training Device

By following the methodology we proposed, we designed a device named 'Multi-Platform PLC Working Station (MPPWS).' This device comprises three PLCs from different brands, a set of typical control input/output electrical components (such as buttons, indicators, motors, sensors, cylinders, etc.), remote IO supporting PROFINET and EtherCAT communication protocols, and a switch to connect them using Ethernet cables. A wireless router is also included, allowing operators to wirelessly connect their personal computers to a designated PLC within the same local area network. This setup aims to provide a hands-on platform for teachers and students to understand and operate various PLCs in a controlled learning environment (Fig. 1).

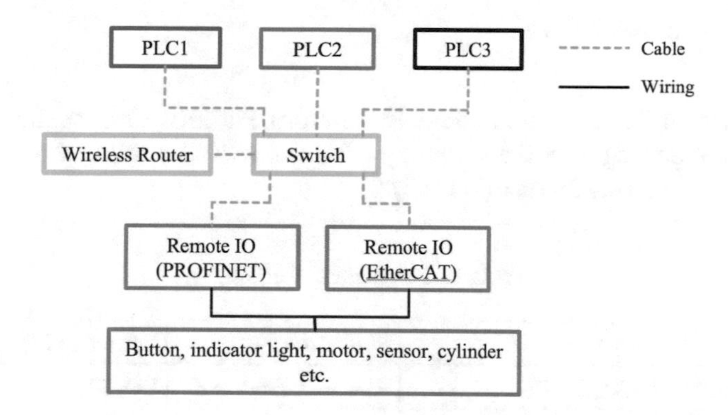

Fig. 1. Equipment Structure Topology Diagram

4.1 Features of MPPWS

The remote IO has been pre-connected with a series of typical input/output electrical components such as buttons, indicators, and motors. The remote IO is connected to the PLC via Ethernet cables and a switch. Therefore, using this device eliminates the need for additional wiring. Users only need to install the programming software for the target PLC on their personal computers and wirelessly connect to the designated PLC. Subsequently, they can carry out relevant communication and configuration operations between the target PLC and the corresponding remote IO, facilitating basic operational exercises for the target PLC. Examples include the use of target PLC programming software, writing simple control programs, controlling motor start/stop, and controlling the extension/retraction of a cylinder (Fig. 2).

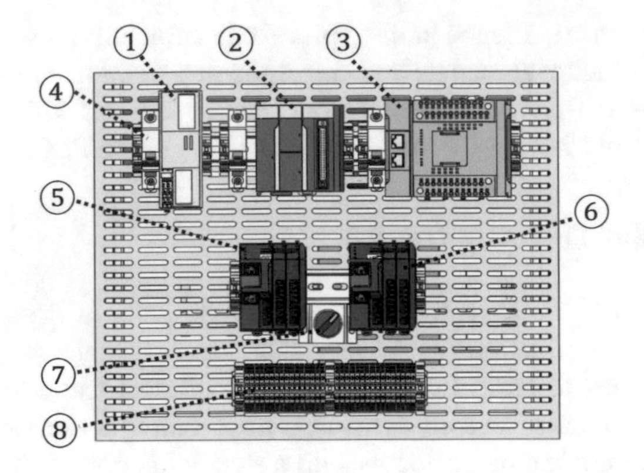

Fig. 2. Frontal Arrangement Diagram: ① PxC PLC. ② Inovance PLC. ③ Omron PLC. ④ Circuit Breaker. ⑤ Remote IO (PROFINET). ⑥ Remote IO (EtherCAT). ⑦ Rotary Switch. ⑧ Terminal Block

4.2 Component of MPPWS

PLC

The equipment utilizes three different PLCs from Phoenix Contact, Inovance, and Omron, each requiring only the necessary CPU module. The relevant parameters for the selected PLCs are as follows (Table 1):

Table 1. Parameters of PLCs.

Producer	Type	Software	Programming Language
Phoenix Contact	AXC F 2152	PLCnext Engineer	LD, ST, FBD, SFC
Inovance	AM401-CPU1608TP	InoProShop	LD, ST, FBD, SFC
Omron	NX1P2-9024DT	SysmacStudio	LD, ST, FBD, SFC

Remote IO

The controllers (adapters) of the two remote IOs support both PROFINET and EtherCAT communication protocols. Each remote IO is equipped with a set of 8-point digital input and 8-point digital output modules. The relevant parameters for the selected remote IOs are as follows (Table 2):

Table 2. Parameters of remote IOs.

Brand	Adapters	Bus Protocol	Input	Output
DECOWELL	PROFINET	PROFINET (RT, MRP)	EX-2118	EX-3118
	EtherCAT	EtherCAT (Ring, Free Run)		

The two remote IOs share a set of input/output electrical components (such as buttons, indicators, motors, sensors, etc.). During device operation for training purposes, the opening or closing of the two remote IOs is controlled through a rotary switch. The communication and configuration between the two remote IOs and the three PLCs mentioned above are conducted based on the communication protocol types they support (Fig. 3).

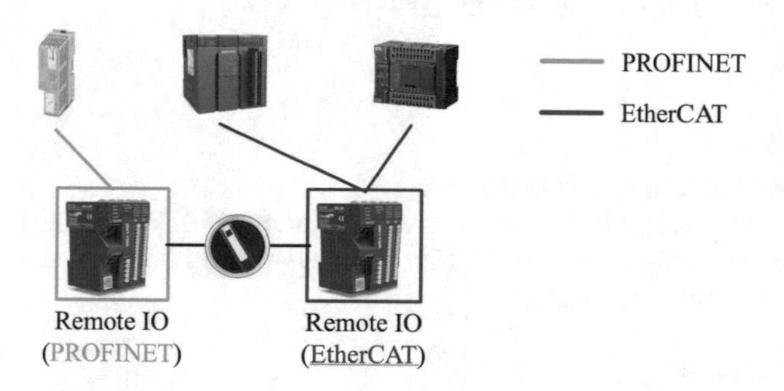

Fig. 3. Communication between PLC and Remote IO

Wireless Router.
A wireless router is installed on the device, connected to the three PLCs via Ethernet cables and a switch. This facilitates wireless connection from personal computers to the target PLC for programming and debugging. The relevant parameters are as follows (Table 3):

Table 3. Parameters of wireless router.

Model	TAS-IT-681-WF
Support	Universal 4G SIM card slot, 2.4G WiFi, 5dbi full-frequency antenna
Interfaces	1 * RJ45 port (WAN/LAN), 1 * RS485 serial port
Input Voltage	DC9~36 V (circular power connector + power terminal)

5 Application of MPPWS

Our focus is always on establishing a practical teaching model in the field of vocational education in China with value for promotion and operational feasibility. The "MPPWS" is powerful and can run various project scenarios such as beer canning production lines, automatic car wash machines, and material sorting. MPPWS meet the requirements for basic and advanced training of PLC programming. Hence, we aim to apply this solution primarily in teacher training, with the expectation of achieving the desired outcomes.

Here is one of the training projects, titled "Automatic Mixing System for Multiple Liquids," as an example.

5.1 Training Plan for Vocational Teachers Using MPPWS

Training Project Title
Automatic Mixing System for Multiple Liquids.

Training Duration
30 h

Requirements of Venue and Hardware
At least 1 set of "Multi-Platform PLC Working Station (MPPWS)". The training venue should have a projection or display platform, no fewer than 12 dedicated computers, and corresponding tables and chairs.

Teaching Methods
Lecture, discussion, group presentations, practical demonstrations.

Teaching Objective
The device can only start running when the valves A, B, and C are closed, and the start button SB1 is pressed. The specific control process is as follows:

- Open valve A, allowing liquid A to flow into the container. When the liquid level reaches SQ3, close valve A and open valve B.
- When the liquid level reaches SQ2, close valve B and open valve C.
- When the liquid reaches SQ1, close valve C, and the stirring motor begins stirring.
- After the stirring motor works for 1 min, stop stirring. Open valve D for the mixed liquid to be released.
- When the liquid level drops to SQ4, wait for 20 s, then empty the container. Close valve D for the mixed liquid, starting the next cycle.
- After pressing the stop button SB2, stop the operation only after completing the current mixing and emptying cycle (Fig. 4).

Teaching Implementation Plan
The MPPWS integrates PLCs from Phoenix Contact, Inovance, and Omron across three distinct platforms, allowing trainees to work simultaneously on different platforms to

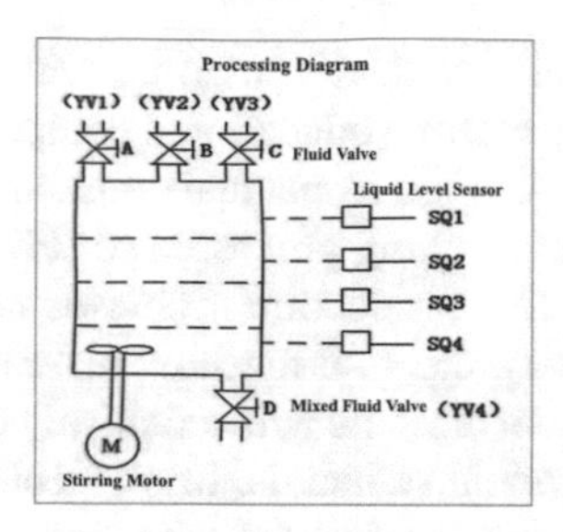

Fig. 4. Schematic Diagram

accomplish identical project tasks. Each platform is supported by corresponding documentation and worksheets. Trainees familiarize themselves with specific tasks based on documentation or instructor guidance, addressing encountered issues through group discussions led by the instructor.

Groups of 5 members, each representing a different platform, take turns testing programs. Discussions and communication within the group or with the instructor help resolve issues. After completing training on one platform, groups switch to others, ensuring exposure to all three platforms. The training is tailored for professional instructors with a PLC and programming background, emphasizing communication among PLCs of different brands and distributed IO. Corresponding instructional manuals are provided to facilitate the learning process.

5.2 Student Training Recommendations for MPPWS

In contrast to educators, students possess a less robust foundational skill set and a narrower knowledge base. Consequently, the project content and control requirements for students are relatively straightforward, emphasizing common elements for ease of comprehension. Leveraging students' existing basic programming knowledge, the focus shifts to software and parameter settings, guided through detailed documentation and worksheets.

To enhance learning, the incorporation of simple tasks involving peripherals like buttons, indicators, and motors on the test bench is recommended. This approach extends beyond programming, encompassing knowledge about the hardware peripherals of the equipment. As students may encounter various challenges during training, it is imperative to allocate time for Q&A sessions and group discussions within the class organization. Summarizing and addressing common student issues will contribute to the creation of a comprehensive Q&A document, providing valuable support for future student training.

6 Expected Outcomes

Our educational projects, enriched by practical insights, not only elevated the knowledge base of our teaching team but also triggered transformative shifts in vocational education methodologies. The focus on tailored teaching materials extends to influencing broader methodologies in vocational education settings. The objective was to create practical educational materials seamlessly aligning with the dynamic needs of vocational colleges. Independently crafting educational equipment, coupled with vocational education teachers' insights, resulted in manuals and courses. These empower teachers to navigate operations and teaching processes for Phoenix Contact, Inovance, and Omron PLCs. Concurrently, real production scenarios contribute to flexible learning cases and designated enterprise projects as teaching materials. Aligning the professional competence of vocational school teachers with technical positions in enterprises ensures graduates meet future employment standards. Implementing the project plan enhances professional skills and cognition, extending to both professional capabilities and teaching competencies.

Our team outlined a task plan based on the proposed solution, completing initial tasks. The plan includes using the equipment for a mid-January 2024 teacher training session. Subsequent refinements, based on teacher feedback, will enhance the equipment scheme. This iterative feedback process ensures ongoing improvement, maintaining the solution's relevance and effectiveness in vocational education.

A Smart Electromagnetic Lens Antenna Design for Next-Generation Location-Based Wireless Communication

Muneer Ahmed Shaikh[1(✉)], Asif Ali Wagan[1], Sarmad Ahmed Shaikh[1], Muhammad Minhaj Arshad[1], Syed Masaab Ahmed[1], and Shahzad Arshad[2]

[1] Computer Science Department, Sindh Madressatul Islam University, Karachi, Pakistan
PCS21F006@stu.smiu.edu.pk, sarmad@smiu.edu.pk

[2] National Aerospace Science and Technology Park, Rawalpindi, Pakistan

Abstract. The next-generation wireless communication systems will require smart techniques to provide network coverage, particularly location-based services. In this paper, we propose a smart electromagnetic (EM) lens antenna design that can provide better performance in next-generation wireless communication systems i.e., 6G technology. The proposed lens antenna has the property to focus the signal with high gain and narrow beamwidth in a desired direction. Moreover, the lens-assisted antenna array can steer the signal in a desired direction, using a subset of an antenna array, as a function of the position of the antenna subset i.e., as the position of the antenna subset varies, the direction of the signal varies. Hence, it is advantageous as we need to process a few antennas and achieve the location based services with low complexity and high precision. The simulated results of the proposed EM lens antenna design at 4.2 GHz frequency show that the scattering parameters, gain, and 3dB beamwidth are much improved as compared to a standard microstrip patch antenna without a lens.

Keywords: Wireless Communication · Smart Antenna Design · Smart Objects · Next-Generation Systems

1 Introduction

The deployment of fifth-generation (5G) wireless communication technology has commenced globally to propel mobile communication systems forward. However, it is projected that the current capacity will run out soon due to the growing need for higher bandwidth and the demand for automation and smart networking particularly location-based services [1]. It appears that the constantly expanding technological needs would not be entirely satisfied by current 5G mobile communication [2]. Because of this, a large number of researchers have already begun working on sixth-generation (6G) technology, which is anticipated to completely transform the digital world and incorporate new efficient techniques in the radio frequency front end (RFFE) of the cellular systems [1–3].

M. E. Auer et al. (Eds.): STE 2024, LNNS 1028, pp. 99–108, 2024.
https://doi.org/10.1007/978-3-031-61905-2_11

Currently, cellular communication systems provide limited and low data rates-based services to users with the traditional RFFE infrastructure. Furthermore, the recent advanced techniques proposed in the related literature, i.e., massive multiple-input-multiple-output (MIMO), provide better performance, particularly in enhancing data rates and capacity [4]. However, these techniques face the challenges of increased hardware and signal processing complexity due to the increased number of antenna elements in the RFFE [3, 5, 6]. Moreover, location-based services in cellular communication use information about a user's location to provide customized services and information with MIMO technology. MIMO technology can reduce the impact of signal fading and improve location tracking accuracy. The location efficiency can be also improved by employing signal processing techniques. In this regard, the authors in [7], by ignoring the hardware complexity, propose a novel signal processing technique to find the location information in terms of the direction of arrival (DoA) with high accuracy. Additionally, approaches like fingerprinting [8] and compressed sensing [9] also exist in the literature to perform the location-finding task. These approaches are able to provide better results but are generally less efficient in terms of hardware and mutual coupling.

In this way, another approach to improving the next-generation wireless communication systems' performance is to design smart antennas. The smart antennas provide better antenna parameters (i.e., high gain, high directivity, narrow 3dB beamwidth, better scattering parameters, etc.,) with less complexity and design reconfiguration ability. In this regard, a Luneburg lens antenna and its variants have been proposed in the literature to achieve high gain, overall performance, and precise control over the radiation pattern of the antenna. Planar Luneburg lens antenna has also been proposed to achieve circular polarization, narrow beam width, and low profile at Ka-band frequency [10]. The antenna is oriented at 45° in the horizontal direction. The different phase delays of the TM0 and TE0 waves in the lens result in a 90° phase difference between them, leading to the desired circular polarization. The method has Fresnel zone plates or Luneburg lenses that consist of the assembly of several materials or are based on an adequate locally variable index of refraction. Luneburg lens is attractive but usually limited by complicated manufacturing processes. Due to the high complexity in designing and methodology of the Fresnel zone plates and Luneburg lens antenna, there is another method introduced which is a two-dimensional implementation of metamaterials. Metasurfaces can easily be printed on a substrate. Therefore, their manufacturing cost may be considerably low. However, at high frequencies, dielectric materials are lossy, and fully metallic implementations are desired. One of the main limitations of the first implementations of metasurfaces was their narrow band of operation. Another technique is a multibeam geodesic lens antenna (GLA) at mmWave frequencies with relatively uniform patterns in the H-plane to reduce gain roll-off over the service area. The problem is Ray tracing is difficult to perform and mathematical complexity can be increased in the lens antenna arrays [11]. Moreover, these antennas might not be used in applications such as location-based services, particularly in next-generation artificial intelligence-enabled wireless systems due to hardware and signal processing complexity especially when employing the massive antenna arrays-based RFFE [12].

Therefore, this paper introduces a smart electromagnetic (EM) lens antenna that is able to provide better antenna parameters (i.e., high gain and narrow 3dB beamwidth) reduce the hardware and signal processing complexity, and eventually provide the location-based services. The paper is further organized as follows. Section 2 describes the proposed method for EM lens antenna design which includes EM lens geometry and integrated microstrip patch antenna at the 4.2 GHz frequency. The obtained EM simulation results are discussed in Sect. 3. Finally, the paper is concluded in Sect. 4.

2 Proposed Method

In order to improve the performance of wireless communication systems, our approach is to combine an EM lens with a microstrip patch antenna (MPA) called an EM lens antenna, as shown in Fig. 1. The EM lens has the ability to manipulate the radiation pattern of the patch antenna which in turn can enhance the antenna gain and narrow down the beamwidth, as well as focus the signal in a desired direction. By using a 3D EM simulator, we design the EM lens antenna at 4.2 GHz which is one of the frequencies of the mid-band spectrum (3.5 GHz to 6 GHz) [13]. Moreover, we consider the extended hemispherical type EM lens whose geometry includes a radius (R_l) and a section of the extended slot (L_{ex}), and made of dielectric material polylactic acid (PLA). While quarter wave transformer (QWT) fed MPA has been designed at the desired frequency of operation using FR-4 dielectric substrate. The MPA has been integrated at the bottom center of the EM lens. Moreover, by changing the position of the patch antenna, the overall radiation of the lens antenna can be controlled without significant performance degradation.

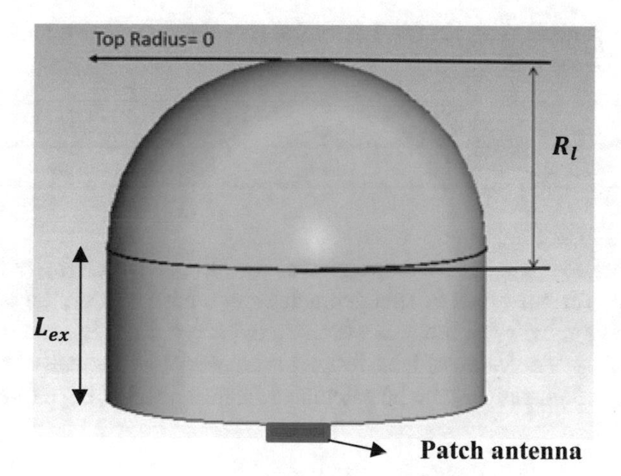

Fig. 1. The proposed EM lens antenna design.

In this way, first, a simple QWT-fed MPA, as shown in Fig. 2, has been designed using FR4 substrate. The substrate properties and the design parameters of the MPA at desired frequency have been obtained from the literature [14–16] and are summarized in Table 1.

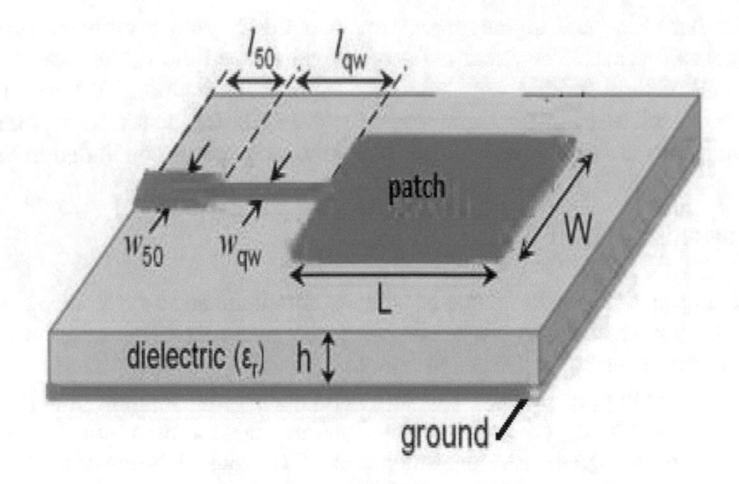

Fig. 2. Typical design of a QWT matched MPA.

Table 1. Substrate properties.

Substrate	ε_r	*h (mm)*	*tan* δ	*t (mm)*
FR4	4.4	1.57	0.02	0.035

Table 2. Design parameters of the QWT fed MPA (in millimeter – mm unit) without lens at 4.2 GHz using FR4 substrate.

Substrate	L	W	l_{qw}	w_{qw}	l_{50}	w_{50}
FR4	15.5	27.9	10.9	1.38	9.88	3.05

Similarly, an EM lens has been designed using PLA substrate ($\varepsilon_{rl} = 3.54$ and *tan* δ $= 0.011$). From various types of lens geometries, we have chosen the extended hemispherical (EHS) geometry of the lens which includes a radius (R_l) and a section of the extended slot (L_{ex}). The value of lens design parameter R_l is normally chosen from the range of $5\lambda_0$ to 2 $5\lambda_0$, and L_{ex} can be calculated from the following expression [17, 18] (Table 2).

$$L_{ex} = R_l\left(\frac{\sqrt{\varepsilon_{rl}} + 1}{\sqrt{\varepsilon_{rl}} - 1} - 1\right), \tag{1}$$

where λ_0 is the wavelength at the desired frequency (f_0). In this way, the optimized lens design parameters (from practically realization point of view) of R_l and L_{ex} considered in this work are 198.8 mm (i.e., 2.8 λ_0) and 160.63 mm (from expression (1)), respectively. Thus, by integrating the MPA at the bottom center of the lens, the EM simulation was run in order to observe the results which are discussed in the next section.

3 Results and Discussion

In this section, we discuss the observed simulated results of the proposed MPA with a lens designed at 4.2 GHz frequency and compare the results with the traditional MPA without a lens. The designs of the lens and MPA are shown above in Fig. 1 and Fig. 2, respectively, which are simulated on a 3D EM simulator. The obtained simulated return loss or scattering parameters (S11) of the MPA with and without lens at 4.2 GHz frequency, as shown in Fig. 3, are −23.43 dB and −27.68 dB, respectively. The obtained S11 in the case of the lens is below −10 dB which shows that the proposed lens antenna resonates at the desired frequency. In the case without a lens, the MPA resonates at another frequency i.e., 4.6 GHz which shows the dual-band MPA. Moreover, the radiation patterns (of the gain parameter) in 2D and 3D of the MPA with and without lens are shown in Fig. 4 and Fig. 5, respectively. The radiation pattern of the proposed lens antenna shows that the 3dB beamwidth (BW) is narrow i.e., 9.5°, as compared to the without lens case where 3 dB beamwidth is around 78.9°. Moreover, the obtained antenna parameters i.e., gain (G_a) in dBi, main lobe direction (D_l) in degrees, 3 dB beamwidth (BW) in degrees, and side lobe level (S_l) in dB), with and without lens cases are provided in Table 3. The antenna parameters of the proposed lens antenna design are much improved, i.e., $G_a =$ 16.7 dBi, as compared to traditional without lens antenna i.e., $G_a = 3.35$ dBi.

Furthermore, we have formed a linear array of MPA of 1×2 size and combined it with the lens at the center, as shown in Fig. 6. By shifting the position of the MPA array (1×2) by $d_s = \lambda_0/2$ distance to the left and right sides from the center of the lens, the beam steering capability of the proposed lens antenna structure has been observed. As we shift the antenna array to the left side from the lens bottom center, the beam is steered to the right side from the lens top center by an angular amount which is a function of the shifting distance. At different distances i.e., $d_s = \pm\lambda_0/2$, to the left ($d_s = -\lambda_0/2$) and to the right ($d_s = +\lambda_0/2$), the observed results are summarized in Table 4. In this way, the observed 3D radiation patterns are illustrated in Fig. 7(a) and Fig. 7(b) for the left and right sides shifting array, respectively. Thus, it shows that the beam can be steered in the desired location, as shown in D_l field of Table 4, without hardware and signal processing complexity. Eventually, it can enable location-based services in next-generation wireless systems with less cost and improved performance as compared to the traditional without a lens antenna array technique. Note that the same designs can be used to fabricate the structure by using a readily available 3D printer. However, the fabrication and physical measurement process is the future work of the study and therefore is beyond the scope of this paper.

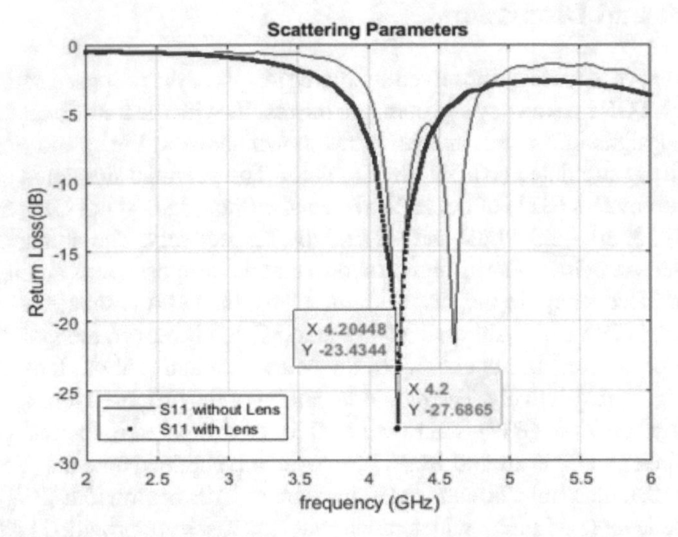

Fig. 3. Observed scattering parameters of MPA with and without lens.

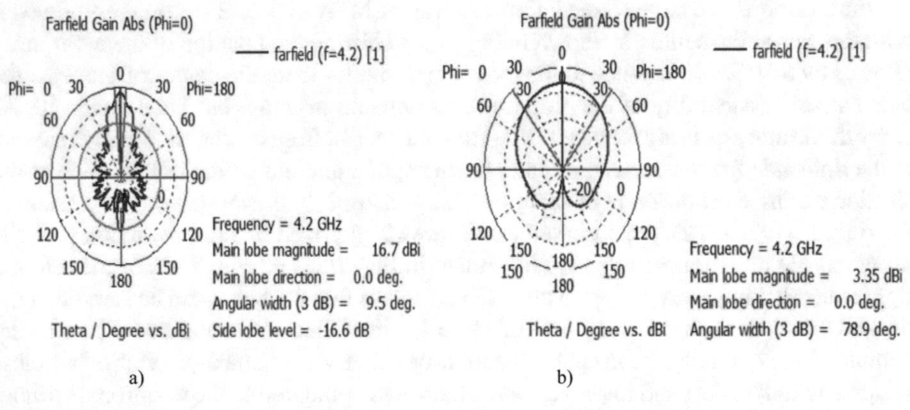

Fig. 4. Observed 2D radiation pattern of the MPA with (a) and without lens (b).

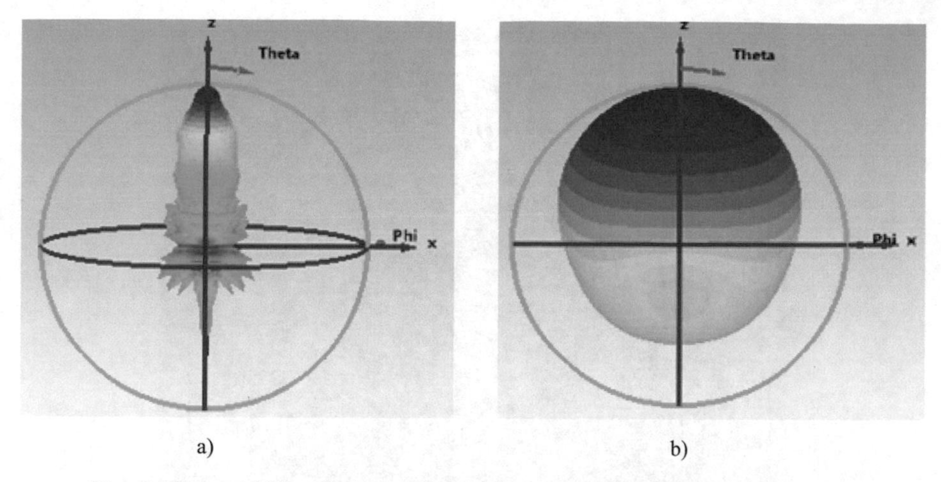

Fig. 5. Observed 3D radiation pattern of the MPA with (a) and without lens (b).

Table 3. Simulated Antenna Parameters With and Without Lens

MPA	S11 (dB)	G_a (dBi)	D_l	BW	S_l (dB)
With Lens (**Proposed**)	−23.43	16.7	0°	9.5°	−16.6
Without Lens (Traditional)	−27.68	3.35	0°	78.9°	0.0

Table 4. Simulated Antenna Parameters of MPA Linear Array (1 × 2) with Lens and without lens at 4.2 GHz frequency.

MPA Array	S11 (dB)	G_a (dBi)	D_l	BW	S_l (dB)
With Lens and $d_s = -\lambda_0/2$ (**Proposed**)	−21.51	18.51	9	12.7	−15.5
With Lens and $d_s = +\lambda_0/2$ (**Proposed**)	−21.22	18.51	−9	12.7	−15.5
Without Lens (Traditional)	−40.04	5.845	0	48.9	−21.2

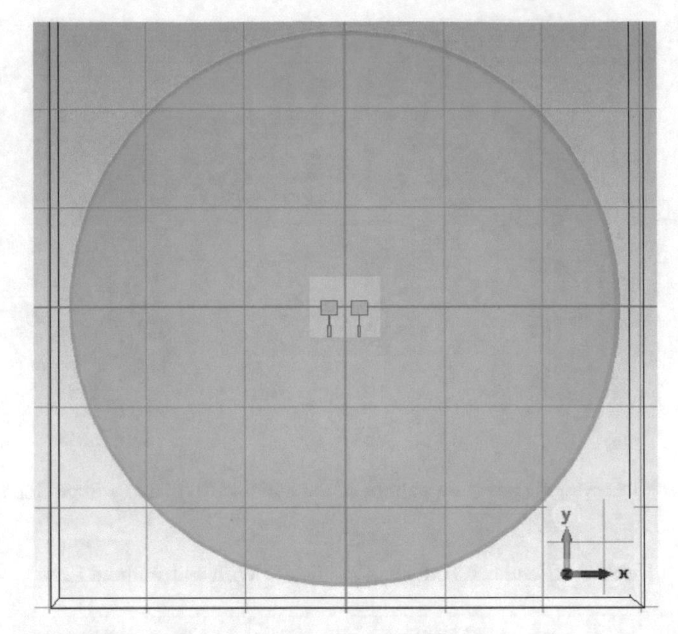

Fig. 6. Design of MPA array (1 × 2) integrated with the lens.

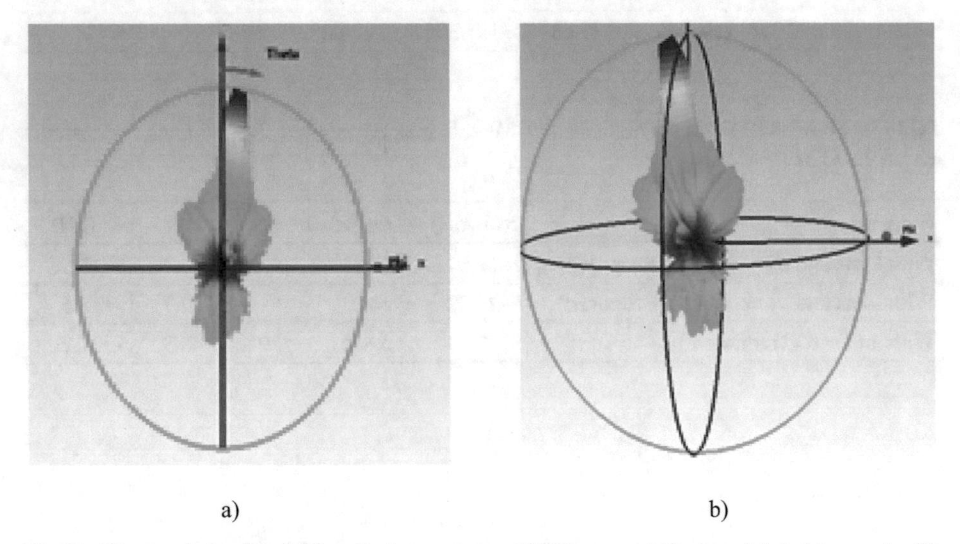

a) b)

Fig. 7. Observed simulated 3D radiation pattern of MPA array shifted to a) left ($d_s = -\lambda_0/2$), and b) right ($d_s = +\lambda_0/2$), sides from the lens center.

4 Conclusion

In this paper, we proposed a smart EM lens antenna design. We combined an extended hemispherical type EM lens with a patch antenna of FR4 substrate at 4.2 GHz operational frequency. The EM lens has the ability to focus the signal with a high gain and narrow beam in a desired direction. Therefore, the simulated proposed lens antenna provided better antenna parameters i.e., gain = 16.7 dBi, as compared to traditional without lens antenna i.e., gain = 3.35 dBi. Moreover, by changing the position of the MPA at the bottom of the lens, the beam can be steered in the desired direction. Hence, the proposed lens antenna can provide better performance when used in an RFFE of the next-generation AI-enabled wireless communication systems, particularly for the location-based services application.

Acknowledgments. The authors acknowledge the Higher Education Commission (HEC) of Pakistan for the research grant to this work under the National Research Program for Universities (NRPU) scheme with project No: 20- 15948/NRPU/R&D/HEC/2021. Also special thanks to the National Aerospace Science and Technology Park (NASTP), Pakistan, for their technical support in this project.

References

1. Yang, P., Xiao, Y., Xiao, M., Li, S.: 6G wireless communications: vision and potential techniques. IEEE Netw. **33**(4), 70–75 (2019)
2. Lu, Y., Zheng, X.: 6G: a survey on technologies, scenarios, challenges, and the related issues. J. Ind. Inf. Integr. 1–14 (2020)
3. Zhang, Z., et al.: 6G wireless networks: vision, requirements, architecture, and key technologies. IEEE Veh. Technol. Mag. **14**(3), 28–41 (2019)
4. Larsson, E.G., Edfors, O., Tufvesson, F., Marzetta, T.L.: Massive MIMO for next generation wireless systems. IEEE Commun. Mag. **52**(2), 186–195 (2014). https://doi.org/10.1109/MCOM.2014.6736761
5. Liu, Y., Zhang, S., Gao, Y.: A high-temperature stable antenna array for the satellite navigation system. IEEE Antennas Wirel. Propag. Lett. **16**, 1397–1400 (2017). https://ieeexplore.ieee.org/document/7782420
6. Zeng, Y., Zhang, R., Chen, Z.N.: Electromagnetic lens focusing antenna enabled massive MIMO. In: Proceedings of the IEEE CIC International Conference on Communications (ICCC), pp. 454–459 (2013)
7. Yang, K.Y., Wu, J.Y., Li, W.H.: A low-complexity direction-of-arrival estimation algorithm for full-dimension massive MIMO systems. In: IEEE International Conference on Communication Systems, pp. 472–476 (2014)
8. Savic, V., Larsson, E.: Fingerprinting-based positioning in distributed massive MIMO systems. In: Proceedings of the IEEE 82nd Vehicular Technology Conference (VTC Fall) (2015)
9. Gu, Y., Zhang, Y.D., Goodman, N.A.: Optimized compressive sensing based direction-of-arrival estimation in massive MIMO. In: IEEE International Conference on Acoustics, Speech and Signal Processing (ICASSP), pp. 3181–3185 (2017)
10. Segura, C.M., Dyke, A., Dyke, H., Haq, S., Hao, Y.: Flat Luneburg lens via transformation optics for directive antenna applications. IEEE Trans. Antennas Propag. **62**(4), 1945–1953 (2014)
11. Orgeira, O., León, G., Fonseca, N.J.G., Mongelos, P., Quevedo-Teruel, O.: Near-field focusing multibeam geodesic lens antenna for stable aggregate gain in far-field. IEEE Trans. Antennas Propag. **70**(5), 3320–3328 (2022)

12. Shaikh, M.A., Shaikh, S.A., Shaikh, F.Z.: Location based services for remote education and health institutes using sum-difference signal patterns. Int. J. Interact. Mob. Technol. **16**(22), 4–14 (2022)
13. 5G spectrum bands explained, Nokia (2023). https://www.nokia.com/thought-leadership/articles/spectrum-bands-5g-world/#:~:text=Mid%2Dband%20spectrum%20(1%20GHz,already%20designated%20it%20for%205G. Accessed 01 Sept 2023
14. Bansal, A., Gupta, R.: A review of microstrip patch antenna and feeding techniques. Int. J. Inf. Technol. **12**, 149–154 (2020)
15. Balanis, C.A.: Antenna Theory and Analysis and Design. Wiley, New York (1997)
16. Stutzman, W.L., Thiele, G.A.: Antenna Theory and Design. Wiley, New York (1998)
17. Saleem, M.K., Xie, M., Alkanhal, M.A.S., Saadi, M.: Effect of dielectric materials on integrated lens antenna for millimeter wave applications. Microwave Opt. Technol. Lett. **61**(4), 1079–1083 (2019)
18. Nguyen, N.T., Rolland, A., Boriskin, A.V., Valerio, G., Le Coq, L., Sauleau, R.: Size and weight reduction of integrated lens antennas using a cylindrical air cavity. IEEE Trans. Antennas Propag. **60**(12), 5993–5998 (2012)

Contribution to Characterizing Time-Dependent Handgrip Strength Profiles

Alberto Cardoso[1(✉)], Diana Urbano[2], and Maria Teresa Restivo[3]

[1] Department of Informatics Engineering, CISUC-LASI, University of Coimbra, Coimbra, Portugal
alberto@dei.uc.pt

[2] LAETA-INEGI, Faculty of Engineering (FEUP), University of Porto, Porto, Portugal
urbano@fe.up.pt

[3] LAETA-INEGI, Faculty of Engineering (FEUP), University of Porto, A3ES, Porto, Portugal
trestivo@fe.up.pt

Abstract. The maximum value of HandGrip Strength (usually named by HGS) is an important biomarker that indicates several physical conditions such as diabetes, malnutrition, sarcopenia, frailty and general functional physical capacity. It is also relevant for evaluation purposes in the context of rehabilitation, in patients suffering from hand musculoskeletal or other pathologies.

The HGS dynamometry is non-invasive, portable, easy to perform, fast and reliable, making it suitable for routine assessment in different health and sports-related areas. Besides the HGS value (maximum value), other parameters such as rate of force development and sustainability of force, obtained by measuring time-dependent HGS (HGS(t)), have also been used in different assessment contexts utilizing modern dynamometers such as the BodyGrip (now on its commercialized version, Gripwise).

A universal definition of the parameters characterizing HGS(t) curves has been proposed, and a study based on an artificial neural network model used those parameters to predict the frailty of elderly women. In this work, a new methodology was used to investigate the characteristics of HGS(t). The main idea is to identify and explore a first or second-order continuous-time transfer function, which can be used to estimate the force profile, using the Mean Squared Error and the overall percentual fit as measures of the approximation quality.

The authors consider the results to be promising for future exploitation of the model to investigate and extract the characteristics of the time-dependent HGS (HGS(t)).

Keywords: Handgrip · Handgrip Strength time profile · Model Identification

1 Introduction

In the context of the increase of the aging population, it is important to consider solutions to monitoring the health and physical conditions related to biomarkers, which can provide valuable information about disease progression, treatment responses and prediction of rehabilitation time, among others.

M. E. Auer et al. (Eds.): STE 2024, LNNS 1028, pp. 109–117, 2024.
https://doi.org/10.1007/978-3-031-61905-2_12

This monitoring process allows for timely interventions, resulting in the increase of general well-being of elderly populations. Moreover, it informs about procedures and treatments contributing to the sustainability of the health systems [1]. For example, knowing that frailty is a state of heightened vulnerability, in which minor stressors can lead to difficult situations for people and their families, biomarkers are increasingly being explored to predict frailty, particularly in the elderly population [2, 3].

One of the commonly cited biomarkers to assess frailty syndrome is HGS [4]. In reference [5], other descriptors associated with HGS(t), besides HGS were used, as inputs of an artificial neural network to predict a self-reported frailty score obtained using Tilburg tool [6]. The features considered in that work assume that the HGS(t) time profile corresponds to the response of a first-order system to a step function. The current work aims to explore the idea that individual HGS(t) matches instead the response of a second-order system to a step-function. Using the same sample of study [5], a model identification procedure was applied [7] to estimate a 1st or 2nd order continuous-time transfer function of HGS(t), with or without time delay [8].

This article is organized as follows. Section 2 presents the methodology followed to explore alternative approaches to describe the force-time profile by a continuous transfer function. Next, Sect. 3 summarizes the main results and Sect. 4 discusses the results obtained using a specific handgrip and summarizes the main prospects.

2 Methodology

This section describes the participants, the equipment used to assess HGS(t), and the methodology followed in this study.

2.1 Participants

This work uses HGS(t) tests carried out with elderly women who volunteered after providing written informed consent. Data collection took place in social, recreation, and daycare centers [5]. All assessments were conducted in the local institutions where the participants were recruited, under an established protocol between the Health School of Polytechnic of Porto and the review boards of the "Private Institution for Social Solidarity of Paços de Ferreira" and the "Social and Parish Centre of Ferreira". The sample size is of 61 community-dwelling elderly women, aged from 66 to 91 years old (averaging 76.6 $\pm$ 6.3).

2.2 BodyGrip and the Typical HGS(t) Time Profile

The BodyGrip prototype dynamometer system (Fig. 1) was used for the handgrip strength tests. This system was developed by the Instrumentation for Measurement Laboratory of the Faculty of Engineering, University of Porto. The multifunction system allows the evaluation of compression and traction applied forces of different body groups of muscles, by adapting accessories. It also permits the estimation of respective energy and power. Together with a software application developed for a PC, the BodyGrip system permits performing different calculations from measuring data provided by the device to recording them in a local or remote database [9].

This prototype accomplished a long pathway from a proof of concept at the lab, its product design and technology testing and validation based on clinical trials (as the case of results here used), and many other steps like the patent process and the technology transfer [7], to reach the production of its commercialized version, the Gripwise [10]. It assumes in the present version the potentialities of Industry 5.0 equipment and could be transformed into a smart device when a convenient amount of data will permit algorithms' development.

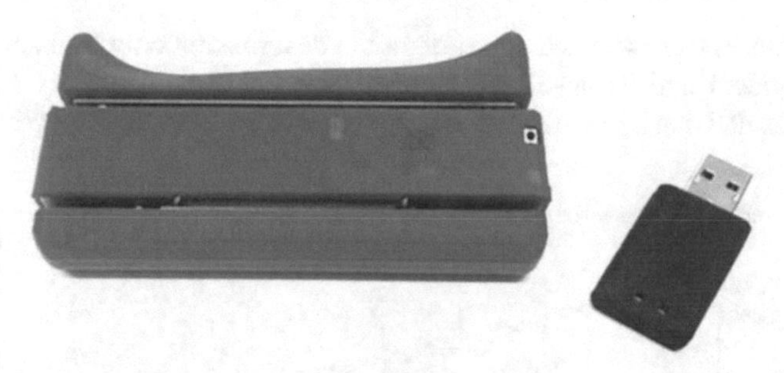

Fig. 1. The BodyGrip prototype.

The procedure protocol for recording HGS(t) data recommends that, on each test, the participants should hold the device with the dominant hand, apply his/her maximum effort, and keep exerting handgrip force for ten seconds, with constant voice incitement provided by the person conducting the tests. The force–time profile is automatically registered in the system database. It is also possible to access a graphical representation of the HGS(t) to confirm the test correctness. Figure 2 shows an example of the HGS(t) test, with a duration of 10s. It is important to notice that the time value of 10s was tuned as the convenient time interval by previous tests in a group of elderly participants.

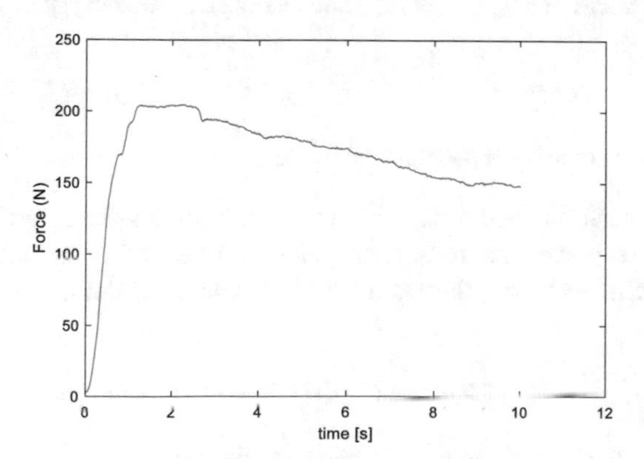

Fig. 2. Typical HGS(t) time profile.

Assuming that HGS(t) is described by the response of a 1st order system to a step function, the following universal features, shown in Fig. 3, were identified in reference [5]:

- Time T_1, corresponding to the time when the system response rises to 63.21% of its maximum value;
- Time T_2, denoting the time when the system response reaches its maximum value;
- Force $F_{\max}$, the maximum value of the system response;
- Angle α, calculated as the angle whose cotangent is given by the quotient between the normalized differences $(F_{max} - F_{final})/F_{max}$ and $(T_{final} - T_2)/T_{final}$, where T_{final} denotes the final time and F_{final} the corresponding final force.

These features were selected as an attempt to describe not only the handgrip strength of each individual and its development over time but also the ability to maintain it, characteristics that can be associated with power and endurance, respectively.

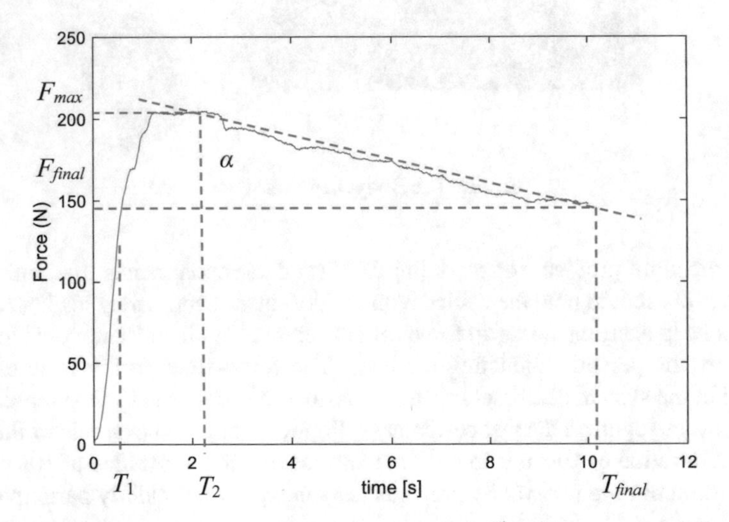

Fig. 3. Features of a typical HGS(t) considering a 1st order system modeling.

2.3 Exploring Alternative Approaches

To investigate if HGS(t) is better described by a continuous-time transfer function corresponding to the response of a first-order system or a second-order underdamped linear system, with or without time delay, to a step function, a model described by Eq. (1) is proposed.

$$F(s) = G(s)U(s) + E(s) \tag{1}$$

In this model, $F(s)$, $U(s)$, and $E(s)$ represent the Laplace transforms of the system output (the handgrip force), input, and noise, respectively, and $G(s)$ the transfer function.

The transfer function $G(s)$ is given by Eq. (2), where the polynomials $num(s)$ and $den(s)$ represent the numerator and denominator of the transfer function, and t_d is the time delay, which defines the relationship between the input and the output of the system.

$$G(s) = \frac{num(s)}{den(s)} e^{-t_d s} \tag{2}$$

The input was considered as a step function, since the test protocol requires that each individual holds, with maximum effort, the BodyGrip device and tries to maintain it during the established period, 10 s in the present case.

The estimation process was conducted using the System Identification Toolbox of MATLAB [11], seeking the optimal solution to fit the force-time profile. The results are assessed using the Normalized Root Mean Squared Error (NRMSE) to measure how well the response of the model fits the estimation data, expressed as the percentage 100 (1-NRMSE), and the Mean Squared Error (MSE).

After estimating the transfer function, the features are automatically obtained from the step system response. These features are the input of advanced methodologies that will provide biomarkers, i.e., information useful for the detection and diagnosis of different diseases, such as frailty.

3 Results

The results were obtained by applying the methodology to 61 individual tests of the sample, considering the following four possible scenarios to describe the transfer function:

- 1 zero, 2 poles and 0.1s time delay;
- 1 zero, 2 poles and without time delay;
- 1 zero, 1 pole and 0.1s time delay;
- 1 zero, 1 pole and without time delay.

To illustrate the results obtained with the application of the methodology, Fig. 4 presents one of the HGS(t) time profiles, considering a 2nd-order transfer function with 1 zero, 2 poles and 0.1s time delay and comparing the real and the estimated forces.

Assuming a normalized step input, the continuous-time transfer function, $G(s)$, of Eq. (3) was identified with a fit to estimation data of 93.343% and a MSE of 1.625. Table 1 shows the obtained features for this example.

$$G(s) = \frac{153.988s + 140.264}{s^2 + 2.057s + 1.290} e^{-0.1s} \tag{3}$$

Table 1. Features obtained with a transfer function with 1 zero, 2 poles and 0.1s time delay.

T_1 (s)	T_2 (s)	F_{max} (N)	α (°)
0.67	2.09	119.87	83.29

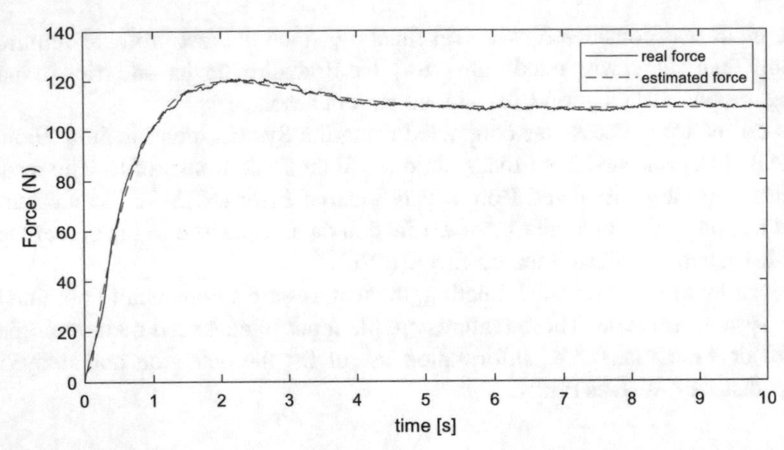

Fig. 4. Comparison of the real and the estimated force-time profiles for a 2nd-order model.

Considering another example of the HGS(t) time profile, Fig. 5 presents the result obtained with a 1st-order transfer function with 1 zero, 1 pole and without time delay.

For this case, the continuous-time identified transfer function, $G(s)$, of Eq. (4) was identified with a fit to estimation data of 75.602% and a MSE of 15.178. Table 2 shows the obtained features for this other example.

$$G(s) = \frac{-2.437s + 223.214}{s + 2.158} \tag{4}$$

Table 2. Features obtained for time profile pattern 2.

T_1 (s)	T_2 (s)	F_{max} (N)	α (°)
0.53	3.23	103.42	90.00

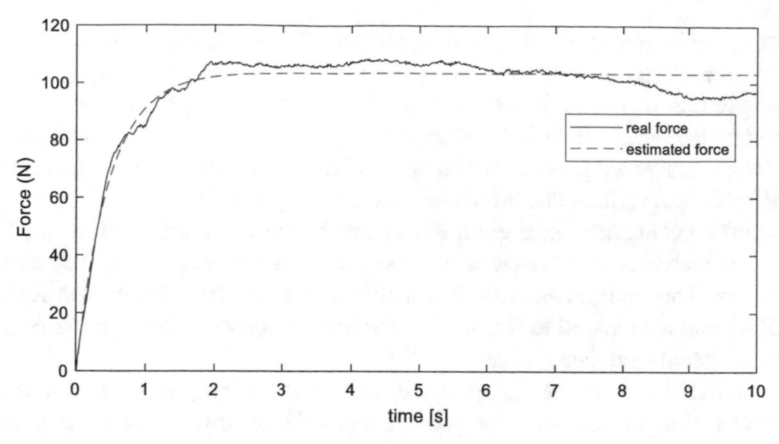

Fig. 5. Comparison of the real and the estimated force-time profiles for a 1st-order model.

4 Discussion and Prospects

This section presents a summary of the results, a discussion of the merits and limitations of the proposed methodology, and suggestions for future investigations to characterize HGS(t).

Applying the methodology to select the best approach for each sample, the performance of each solution was analyzed to fit the force-time profile, resulting in 54% for the first case, 38% for the second, 3% for the third and 5% for the fourth. These results show that the HGS(t) of the majority (92%) of the tests is best described by an underdamped second-order transfer function, with or without time delay, and only a few (8%) of the tests are better described by a first-order transfer function, with or without time delay.

Globally, the best approaches provided a fit to the estimation data with a mean of 81% and a standard deviation of 9%, and the MSE with a mean of 25.64 and a standard deviation of 28.96.

Analyzing the results obtained for all tests, it can be said that the proposed methodology to select the transfer function that best describes the force-time profile shows in general very good performance.

However, the assumption that each individual maintains the maximum effort during the defined test time is not verified in all cases. In some tests, the user's action might not be well described by just one input step.

A major limitation of the study concerns the small sample size and the fact that it only includes female participants.

Moreover, it does not provide alternative universal parameters to those proposed in [12] that could be relevant for frailty prediction.

In future studies, it could be worth exploring an input consisting of a sequence of steps with variable amplitude. Moreover, to better investigate the person's ability to maintain force, the HGS(t) evaluation could be performed considering different time intervals.

Also, to verify the consistency and reproducibility of this approach, data should be collected from a larger and diverse group of participants, including children, adolescents, and young adults, male and female, and tested with this type of modeling.

Based on the same principle of reference [12], an investigation would be then possible, to define a universal set of descriptors of HGS(t), that could be used by researchers from different areas where this biomarker plays an important role.

The current commercialized equipment named Gripwise includes iOS and Android apps, and a WebApp, and provides a cloud platform for data storage on two European servers. This equipment complies with the General Data Protection Regulation (EUGDPR) and is prepared to integrate with medical software through the HL7/FHIR standard for healthcare data exchange [10].

The availability of bigger samples will enable the appropriate conditions for the development of algorithms that can provide valuable information about physiological processes, health condition evolution and response to treatments, among other purposes.

Acknowledgments. This work was supported by the Portuguese Foundation for Science and Technology (FCT), I.P./MCTES, through national funds (PIDDAC), within the scope of Project LAETA - UIDB/50022/2020 and Project CISUC R&D Unit - UIDB/00326/2020.

References

1. Clegg, A., Young, J., Iliffe, S., Rikkert, M.O., Rockwood, K.: Frailty in elderly people Lancet **381**(9868), 752–762 (2013). https://doi.org/10.1016/S0140-6736(12)62167-9. Epub 8 February 2013. Erratum in: Lancet 19 October 2013, **382**(9901), 1328 (2013). PMID: 23395245; PMCID: PMC4098658
2. Hao, Q., Zhou, L., Dong, B., Yang, M., Dong, B., Weil, Y.: The role of frailty in predicting mortality and readmission in older adults in acute care wards: a prospective study. Sci. Rep. **9**(1), 1207 (2019)
3. Apóstolo, J., et al.: Predicting risk and outcomes for frail older adults: an umbrella review of frailty screening tools. JBI Database Syst. Rev. Implement Rep. **15**(4), 1154–1208 (2017). https://doi.org/10.11124/JBISRIR-2016-003018. PMID: 28398987; PMCID: PMC5457829
4. Saedi, A.A., Feehan, J., Phu, S., Duque, G.: Current and emerging biomarkers of frailty in the elderly. Clin. Interv. Aging **14**, 389–398 (2019). https://doi.org/10.2147/CIA.S168687. PMID: 30863033; PMCID: PMC6388773
5. Urbano, D., et al.: Handgrip strength time profile and frailty: an exploratory study. Appl. Sci. **11**(11), 5134 (2021)
6. Zamora-Sánchez, J.J., et al.: The Tilburg frailty indicator: a psychometric systematic review. Ageing Res. Rev. **76**, 101588 (2022). https://doi.org/10.1016/j.arr.2022.101588. Epub 10 February 2022. PMID: 35150901
7. Restivo, M.T., Quintas, M., da Silva, C., Andrade, T., Santos, B.: Device for Measuring Strength and Energy. US20180249940A1, 08-12-2018, B2, utility patent (2018)
8. Khoo, M.C.K.: Model identification and parameter estimation. In: Physiological Control Systems: Analysis, Simulation, and Estimation, pp. 225–288. IEEE (2018)
9. Bolton, W.: System Response, Chapter 10 of Instrumentation and Control Systems, 3rd edn. Newnes (2021)
10. Andrade, T., Restivo, M.T., Urbano, D.: Handgrip strength time profile: from BodyGrip to Gripwise. In: IEEE Xplore Proceedings of the Experiment@ International Conference 2023, 5–7 June 2023 (2023, in press)

11. MathWorks, 2023. Linear Model Identification, System Identification Toolbox of MATLAB. https://www.mathworks.com/help/ident/linear-model-identification.html. Accessed 04 Dec 2023
12. Urbano, D., Restivo, M.T., Amaral, T.F., Abreu, P., Chousal, M.D.F.: An attempt to identify meaningful descriptors of handgrip strength using a novel prototype: preliminary study. Information **11**(12), 546 (2020)

Virtual and Remote Laboratories

Wireless Remote Control Systems of Marine Diesel Engine

Vitalii Nikolskyi(✉), Mykola Slobodianiuk, Maksym Levinskyi, and Mark Nikolskyi

National University «Odessa Maritime Academy», Odesa, Ukraine
prof.nikolskyi@ukr.net

Abstract. A wireless system for energy equipment remote control in the Technical operation of fleet department laboratory of the National University «Odesa Maritime Academy» was developed. An emergency diesel generator from the Kohler company, specially adapted for marine conditions, was used in the project. The main goal of the study was to determine the technical capabilities of the developed wireless remote control scheme, which was integrated into the automatic control system of the ship's emergency diesel generator. The work justified and proved the need to modernize the control system in conditions of full-scale war and to provide electricity during missile attacks, when the national energy company turns off the power in order to reduce the emergency failure of its equipment, which is capable of displaying the technological process in real time with the help of mobile devices. The developed structural scheme of the modernized system of monitoring and control of emergency diesel generator parameters was tested in laboratory conditions. The necessary equipment was also selected, in particular, a programmable logic controller with a built-in Web server from the Phoenix Contact company using the PLCNEXT ENGINEER software environment. The tests that were conducted proved that the selected equipment is able to effectively display in real time the processes occurring during the operation of the emergency diesel generator.

Keywords: Wireless Remote Control System · Remote Control and Monitoring System · Kohler · Phoenix Contact · EduNet · PLCnext Control · TATU

1 Introduction

The nature of the development of the scientific and technological process consists of constant changes, scientific research, technical innovations and interaction between the scientific community, industry and society. Scientific and technical development is based on successive solutions - from simple to complex, from small possibilities to large ones. The specified movement indicates the satisfaction of the needs of remote control systems, namely the transmission of information flows using the wireless method.

At the current stage of development the wireless method in remote control systems has relevant advantages over wired systems, namely:

M. E. Auer et al. (Eds.): STE 2024, LNNS 1028, pp. 121–132, 2024.
https://doi.org/10.1007/978-3-031-61905-2_13

- wireless method allows you to avoid a wired connection, simplifying and improving the infrastructure and installation;
- wireless remote control systems have greater flexibility in positioning and moving equipment, which is important when designing remote control systems;
- quick installation of wireless systems contributes to their widespread implementation;
- wireless systems are usually easier to expand, upgrade and adapt to growing needs.

In addition, the use of the wireless method in remote control systems allows:

- reduce the cost of service;
- to improve the stability of the connection;
- expand the limits of accessibility;
- to reduce the cost during the modernization of the energy facility infrastructure;
- to bring innovation and reduce the impact on the environment.

In this regard it is important to emphasize the promising outlook of the wireless method development in remote control systems. A wide range of scientific research, which is presented in publications [1, 2], confirms the relevance of research topics devoted to the application of the wireless method of transmitting information flows in systems of remote control and monitoring of energy facilities.

2 Analysis of Literary Data and Statement of the Problem

During the implementation of the TEMPUS 544010-TEMPUS-1-2013-1-DE-TEMPUS-JPHES project [3], Ukrainian partners created TATU mobile smart labs (TSL) (Fig. 1) and training materials for them [4–13] with the support of the International Educational Network EduNet. This network is organized and financed by Phoenix Contact, Germany. This provided an opportunity for students to study advanced technologies in the field of wireless control and control systems.

Starting from 2020, University began preparations for the implementation of the obtained results in the educational process of the educational and scientific institute "Engineering". A special emphasis was placed on the educational machinery and boiler of the Department of technical operation of the fleet, which uses power equipment while teaching the discipline "Maintenance, diagnostics and repair of ship's technical equipment".

However, due to russia's illegal aggression against Ukraine and missile attacks on the city of Odessa by the cruiser "moscow", the work of educational laboratories was suspended for the purpose of protection and safety of students and teachers. After repelling the attack and liberating Zmiiny Island, the laboratory continued its educational and scientific work. However, as a result of the strikes on the energy facilities of Ukraine and the long-term absence of power supply (from 10 to 37 h), a decision was made to use laboratory equipment to power the educational building. Therefore, the task was adjusted to ensure not only parameter control, but also the possibility of remote control of the emergency diesel generator using both mobile and network Internet.

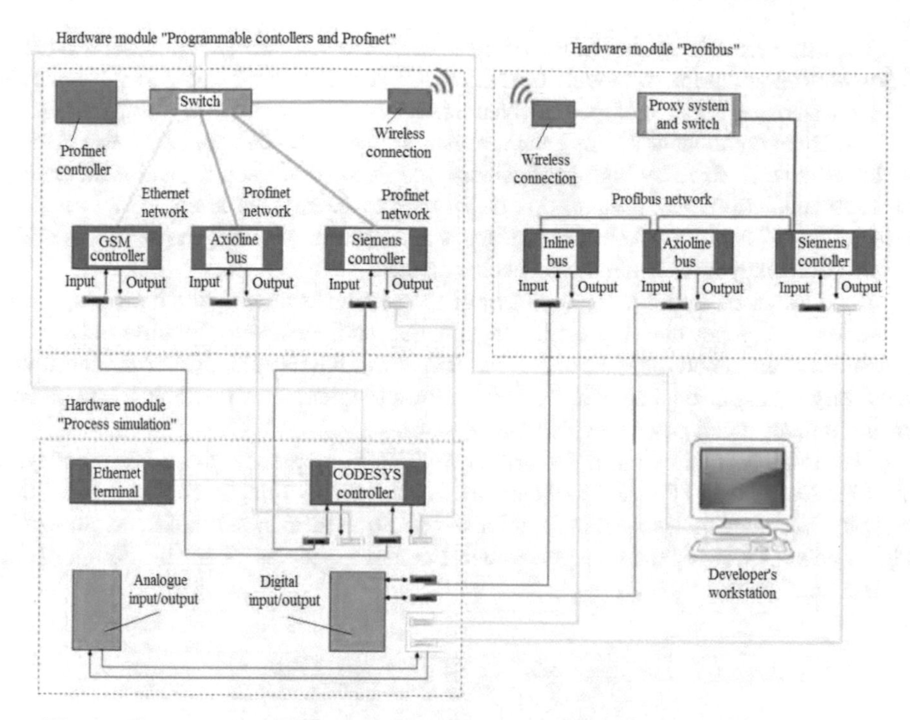

Fig. 1. The structure of TSL hardware modules and their connection to computer [5].

3 Research Goals and Objectives

The conducted research aimed to determine the technical capabilities of the wireless remote control scheme, which was developed and integrated with the automatic control system of the ship's emergency diesel generator.

To achieve the goal, the following tasks were solved:

- determination of the technical feasibility of functional expansion of the automatic control system of the ship's emergency diesel generator with modules for wireless transmission of information flow to remote monitoring systems;
- establishment of technical features and possibilities of operation of the assembled modular devices of the emergency diesel generator information flow transmission wireless system.

4 Materials and Methods of Research of Ship's Diesel Generator Wireless Remote Control Systems

4.1 Equipment that Was Researched in the Experiment to Determine the Possibilities of Functional Expansion of Its Work

The study of the wireless remote control system was carried out using the educational laboratory equipment of Technical Operation of the Fleet Department, National University "Odesa Maritime Academy".

To ensure reliable power in the marine environment, an emergency diesel generator from the Kohler company, model 50EOZD with a capacity of 50 kW, was chosen. This choice was made because of the high characteristics and reliability of this generator.

The main component of this generator is a John Deere 4045TFM75 diesel engine, which is characterized by high performance and durability. With a rotation frequency of 1500 revolutions per minute, this diesel engine guarantees a stable and efficient production of electricity, even at high loads. Its optimal working power ensures the highest efficiency with minimum fuel consumption.

In addition, an important component of the system is automatic control, which is part of the emergency diesel generator. It guarantees uninterrupted operation and reliable activation of the generator in the event of a main power failure. This system demonstrates reliability during critical situations, which allows the generator to quickly turn on and ensure uninterrupted power supply to the system.

Due to the combination of the Kohler 50EOZD emergency diesel generator with the John Deere 4045TFM75 diesel engine and a reliable automatic control system, this complex becomes indispensable for reliable operation in marine conditions. Shown in Fig. 2 general view of the emergency diesel generator serves as an illustration of the system components.

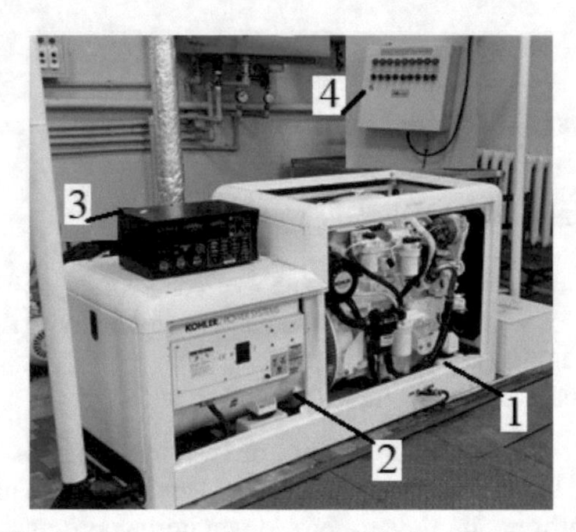

Fig. 2. Kohler 50EOZD emergency diesel generator [2]: 1 – diesel; 2 – generator; 3 – control panel; 4 – generator load control panel.

In the modern control system of an emergency diesel generator, the presence of two key components is determined: the control panel (Fig. 3, a) and the wired remote control panel (Fig. 3, b).

The control panel (Fig. 3, a) allows the operator to interact with the generator on the spot, providing convenient and safe control of parameters and operations. It is an integral part of the system, which contains the main tools and interfaces for reliable control of the generator in local mode.

The remote control panel (Fig. 3, b) opens up wide possibilities for remote monitoring and control of the generator. It is connected using wires, which ensures stable data transmission over long distances. This remote allows operators to monitor the status of the generator and take appropriate measures even from a distance, providing an effective response to changes in the situation. The distance to which the remote control can be moved is determined by the length of the wires.

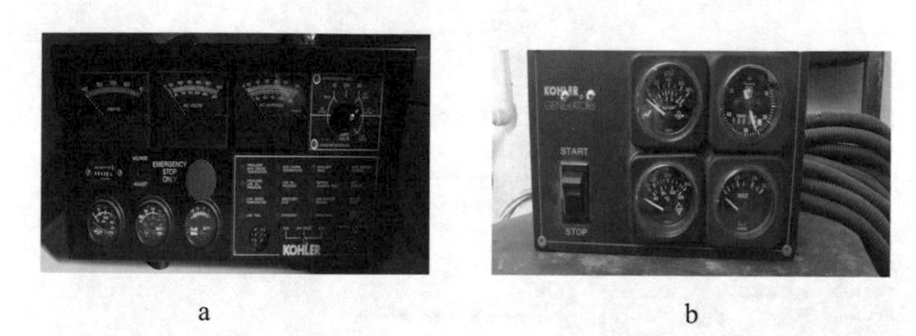

Fig. 3. Emergency diesel generator control system: a – control panel; b - wired remote control.

The emergency diesel generator control system can technically be equipped with an additional device for wireless transmission of emergency diesel generator operating parameters. At the stage of the experimental study, the control of the wireless parameters of work that are transmitted will be checked through the remote control of the emergency diesel generator.

Unfortunately, this emergency diesel generator does not have the ability to transfer basic parameters to the remote control wirelessly.

It should be noted that in certain situations on ships, there is no possibility of integrating an emergency diesel generator into the ship's SCADA system due to their autonomy. Such generators work independently. However, these circumstances do not remove the need for remote control.

When designing and implementing various types of generator systems on ships, it is important to provide the possibility of remote monitoring and control to ensure the efficient operation of the equipment, regardless of whether they are integrated into the general SCADA network or operate as stand-alone devices.

4.2 Methodology for the Composition of Modular Devices of the Wireless System for Transmitting the Information Flow of an Emergency Diesel Generator

The development of a wireless information flow transmission system was embodied in the form of a structural diagram (Fig. 4). The diagram shows the remote control and parameter monitoring system of the emergency diesel generator. This approach allows you to clearly display the components and relationships of the system, which facilitates the perception of its functions and structure.

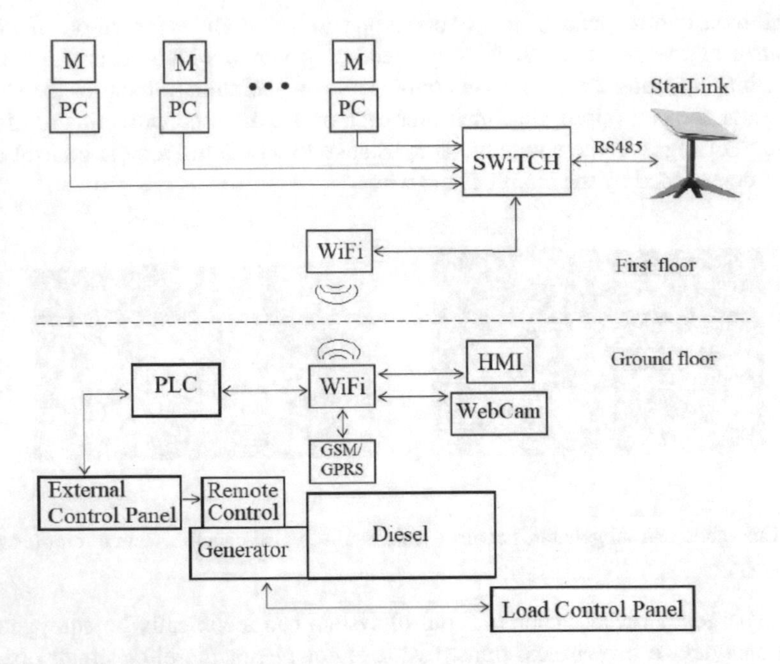

Fig. 4. Structural diagram of the wireless system for transmitting the information flow regarding the control of parameters of the emergency diesel generator: M – monitor; PC - personal computer; PLC – programmable logic controller; HMI - human-machine interface; SWiTCH – network switch; WebCam – web camera; WiFi – wi-fi router; GSM/GPRS – mobile Internet access controller; RS485 – local network of the University.

The structural diagram of the wireless information flow transmission system for controlling the parameters of the emergency diesel generator has the following organization. An emergency diesel generator is located on the graund floor (Fig. 5), and a computer classroom is located on the first floor of the educational laboratory (Fig. 6).

WiFi bridge technology was chosen to ensure communication between these premises. In particular, a wireless connection has been established that allows data to be transmitted directly between the generator control system and the computer classroom, providing a convenient and efficient exchange of information.

In addition, a speed sensor was used to accurately control the speed of the diesel engine, which was installed on the crankshaft pulley. This allows the system to accurately monitor the engine speed and respond to any changes in engine operation.

The PLC is connected to the control system of the emergency diesel generator using wires that pass through the remote control. An important condition is compliance with electromagnetic compatibility requirements to ensure stable and reliable operation of the entire system.

Fig. 5. Exterior view of the engine and boiler training department (ground floor).

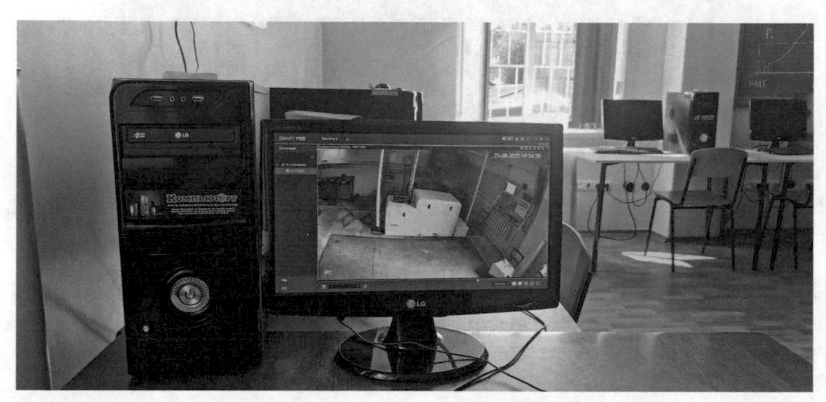

Fig. 6. A fragment of a computer class with a display of graphic information through a Web camera (first floor).

4.3 The Result of the Wireless Control and Control System Device Layout According to the Parameters of the Emergency Diesel Generator

When choosing the equipment for the creation of emergency diesel generator parameters control and remote control wireless system, preference was given to the products of the Phoenix Contact company (Germany). This company is a partner of the University through participation in the EduNet program, which has been ongoing since 2010. As part of this cooperation, two classrooms were equipped, and the teachers of the specialized department completed an internship at the Phoenix Contact company.

At the first stage of the latest development, equipment released in the 2010s was used to process information from the speed sensor, in particular, the Frequency transducer unit – MCR-F-UI-DC. This block made it possible to measure the frequency of rotation of the diesel engine. A programmable logic controller ILC 150 with a built-in Web server for displaying the state of the technological process and a GSM communication channel was also used. However, it turned out that when trying to connect to the controller using a smartphone or iPhone, no information was displayed.

For this reason, a new generation of equipment with a PLCnext Control AXC F 2152 controller was purchased (Fig. 7). Its Web server works on the basis of the HTML 5 language, which made it possible to reproduce indicators on mobile devices of various manufacturers.

Fig. 7. Remote control system based on the PLCnext Control AXC F 2152 controller.

This change made it possible to provide more flexible and convenient access to information about the state of the system, which was especially important for connecting from different devices.

5 Resulting Technical Capabilities of the Developed Wireless System of Diesel Generator Parameters Control and Monitoring

When choosing the architecture, additional functions that may be needed in the future were taken into account. In particular:

- AXC F 2152 PLC Next controller (Item number 2404267);
- AXL F BP SE6 Basic module Axioline Smart Elements (Item number 1088136);
- AXL SE AI4 I 4-20 Axioline Smart Elements analog input module (Item number 1088062);
- AXL SE AI4U0-10 Axioline Smart Elements analog input module (Item number 1088104);
- AXL SE A04 U 0-10 Axioline Smart Elements analog output module (Item number 1088126);
- AXL SE CNT1 Functional Module Axioline Smart Elements (Item number 1088131);
- AXL SE Dll6/1 Axioline Smart Elements discrete input module (Item number 1088127);
- AXL SE DO 16/1 Axioline Smart Elements discrete output module (Item number 1088129);
- UNO-PS/1AC/24DC/60W: UNO power supply, (Item number 2902992).

6 Discussion of the Research Results Regarding the Technical Capabilities of the Emergency Diesel Generator Parameters Wireless Control and Monitoring System

PLCNEXT ENGINEER software [14] with article number 1046008 was used for programming the system components.

This tool provided an opportunity to create, configure functional blocks and connections between them, simplifying the process of programming and integration of all system components.

Figure 8 shows the appearance of the modernized system of emergency diesel generator remote control.

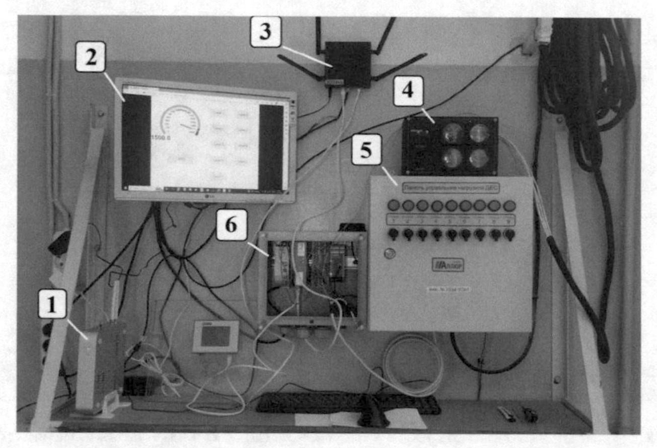

Fig. 8. Appearance of the remote access system 1 – personal computer; 2 – monitor; 3 – wi-fi router; 4 – wired remote control; 5 – generator load control panel; 6 – parameters control and monitoring system which is based on the programmable controller AXC F 2152 PLC Next controller.

Figure 9 shows a detailed fragment of the interface, which is used to control the mode of operation of the control channel. This window form provides the functionality to start the emergency diesel generator by pressing the Start button. In addition, this interface displays the crankshaft speed of the emergency diesel generator. Buttons Level 1,…, Level 9 are responsible for connecting the load - a system of rheostats, which can be replaced by a power system for classrooms. The generator stops automatically 5 min after the load is removed.

Since the generator is started from the +24 V battery, this voltage can provide power to the controller during the absence of mains electricity.

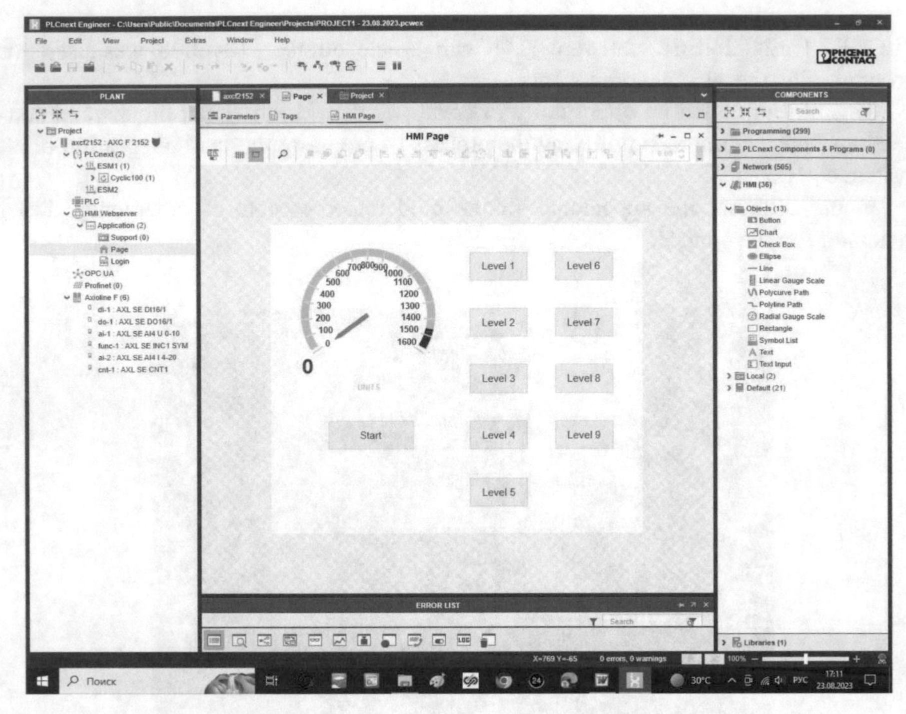

Fig. 9. The interface used in the project in the PLCNEXT ENGINEER software environment.

This interface provides the operator with a convenient and intuitive way to control the process, allowing the generator to be started and stopped quickly and efficiently as needed. In addition, the display of crankshaft revolutions allows the operator to observe the speed of the engine in real time, which can be important for monitoring its optimal functioning and maintaining safe parameters.

This interface is an important component of the system, as it allows the operator to effectively interact with the generator control system and monitor its operation at the level of speed parameters.

With the help of a Web camera installed in front of the generator and the remote control system, it is possible to monitor the situation around the emergency diesel generator using the DMSS App software [15]. Figure 10 shows the image on a smartphone (a) and a computer monitor screen (b) in the classroom on the first floor.

a b

Fig. 10. Image of the emergency diesel generator environment: a – on a smartphone; b – on the computer monitor screen.

7 Conclusions

The technologies used made it possible to modernize the control and monitoring system of emergency diesel generator parameters with the possibility of remote control with simultaneous display of the necessary parameters on students' mobile devices and personal computers.

Mastering wireless technologies and operating information flows allows students to flexibly configure both the necessary functions and the interface depending on the conditions of use.

The applied hardware provides opportunities for the future depending on the change of tasks that will need to be solved in the future. Thanks to the equipment of the Phoenix Contact company, high reliability of the wireless control system and immunity to interference are ensured.

References

1. Gorb, S.I.: Technical support for the training of ship engineers in automation systems with programmable controllers. Autom. Ship Tech. Means Sci. Tech. **22**, 39–46 (2016)
2. Golikov, V.A.: Modernization of the remote control system and control of the emergency diesel generator of the initial machine-boiler department. Ship Power Plants Sci. Tech. Collect. **44**, 64–70 (2020)
3. About the project (tatu.org.ua) Homepage. https://tatu.org.ua/. Accessed 20 Aug 2023
4. Makarov, O.: TATU SMART LAB. Kerivnitstvo koristuvach (2017)
5. Gorb, S.I.: Programming controllers in the integrated development environment: training manual. practice (2017)
6. Klyuchnik, I.: Software security CoDeSys. Module 1 (2017)
7. Shaporin, R.: Ethernet pronunciations for Profinet. Module 2 (2017)
8. Voropaeva, V.: Ethernet Promises for Modbus. Module 2 (2017)
9. Voropaeva, V.: Proxy server to Profibus network. Module (2017)
10. Klyuchnik, I.: Bezdrotov technologies. Module 2 (2017)
11. Zakhimovsky, L.M.: Curation by processes of real time. Module 3 (2017)

12. Shaporin, R.: Introduction to OPC technology. Module 4 (2017)
13. Kaghazchi, H., et al.: Trainings in automation technology for Ukraine: TATU study book. In: Madritsch, C., Werth W. (eds.) EU (2017)
14. Programming software - PLCNEXT ENGINEER Homepage. https://www.phoenixcontact.com/uk-ua/produkcija/programming-software-plcnext-engineer-1046008. Accessed 20 Aug 2023
15. DMSS HD - Apps on Google Play Homepage. https://play.google.com/store/apps/details?id=com.mm.android.DMSSHD&hl=en_US. Accessed 20 Aug 2023

Altitude's Impact on Photovoltaic Efficiency: An IoT-Enabled Geographically Distributed Remote Laboratory

Andrés Gamboa, Alex Villazón(✉), Alfredo Meneses, Omar Ormachea, and Renán Orellana

Universidad Privada Boliviana (UPB), Cochabamba, Bolivia
{andresgamboa1,avillazon,alfredomeneses1,oormachea,renanorellana}@upb.edu

Abstract. In the context of online education, remote laboratories play a crucial role in practical application of theoretical knowledge. Even though there are some remote laboratories to study the solar efficiency of photovoltaic (PV) panels, they are limited either to a single specific location or restricted to simulated conditions. In this paper, we focus on understanding the behavior of PV solar panels under diverse conditions, including altitude, Ultraviolet-A influence, temperature, and solar radiation, which are factors that cannot be accurately simulated and require real-world experimentation. We present the development of a distributed PV Solar Remote Lab deployed at three different cities and altitudes, emphasizing Internet of Things (IoT) technology for real-time data collection and experiment control. Our PV Solar Remote Lab web platform incorporates a specialized current-voltage (I-V) tracer system for accurate efficiency calculations. We obtained encouraging preliminary results on altitude's influence on PV efficiency. Our remote lab implementation leverages IoT, providing valuable insights into the impact of altitude on PV efficiency and promoting a deeper understanding of solar energy systems. We also contribute to online education by offering a user-friendly platform for practical PV experiments in diverse real-world conditions.

Keywords: Remote Labs · IoT · Online Education · PV Efficiency

1 Introduction

Online education breaks the borders of education, making learning resources accessible, and available at a global level. An important complement to online education are remote remote laboratories, which are essential to put into practice the obtained theoretical knowledge. Especially in the area of energy, it is important to understand the behavior of the photovoltaic (PV) solar panels under different conditions, such as altitude, Ultraviolet-A (UVA) influence, temperature and solar radiation, which cannot be simulated without real-world experimentation.

M. E. Auer et al. (Eds.): STE 2024, LNNS 1028, pp. 133–144, 2024.
https://doi.org/10.1007/978-3-031-61905-2_14

There are few existing remote laboratories in the photovoltaic field [1–4]. One of the most prominent is HeliosLab [3], which consists of four PV panels positioned on the roof of a laboratory at the University of West Attica in Athens - Greece. This remote lab allows carrying out experiments using real PV panels connected either in series or parallel, with the possibility of changing the tilt angle of the PV panels with respect to the horizontal. In this remote lab the data can be recorded in real conditions, as well as observing a live streaming of the experiment via a web camera. While HeliosLab facilitates experimentation and research, it is limited to a specific physical location and altitude. This limitation complicates the learning process for students interested in understanding how different altitudes, UVA exposure, and geographic location affect the efficiency of photovoltaic panels.

The remote laboratories described in [1,2,4] give the possibility to make experiments with current-voltage (I-V) curves with variable radiation and load values, and also modify the setup to use the whole PV module or just a portion of it; both have simulation of radiation with high power luminaries, so the practitioners are limited to non real-world conditions where many other factors can affect the results.

To overcome these limitations, we designed and developed a distributed PV Solar Remote Laboratory, specialized in the study of the influence of the UVA on PV panel efficiency, depending on the altitude.

The contributions of this paper to the online education, and remote laboratories community can be outlined as follows:

- The development of a PV Solar Remote Lab user-friendly platform, allowing the students and researchers to perform multiple experiments in different real-world conditions, with three fully operational identical setups deployed at different locations and altitudes.
- The application of Internet of Things (IoT) to facilitate real-time data collection and experiment control, enhancing the efficiency and accuracy of the experiments.
- The development of a specialized I-V tracer embedded system to calculates I-V curves and PV efficiency accurately.
- Preliminary results on the influence of altitude on PV efficiency.
- A dataset of historical data collected from our network of identical pyranometers, that is fully accessible to practitioners for in-depth analysis, through real-time graphs, deployed in our remote lab web platform.

2 Measuring the Solar Resource at Different Altitudes

The solar spectrum has a spectral component formed by the UVA part defined by the range 315 nm–400 nm, Visible (VIS) from 400 nm–700 nm and the Infrared (IR) part from 700 nm–4000 nm [5]. The temperature at the Sun's surface is what characterizes the solar spectrum. From all the received radiation passing through the atmosphere, approximately 6.4% is UVA radiation, 48% VIS and 45.6% IR [6].

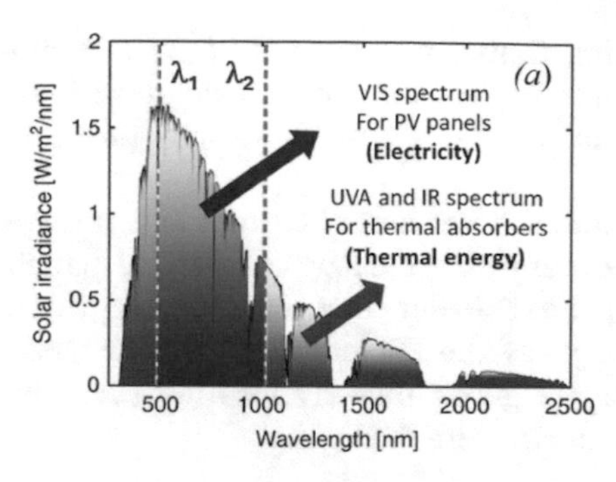

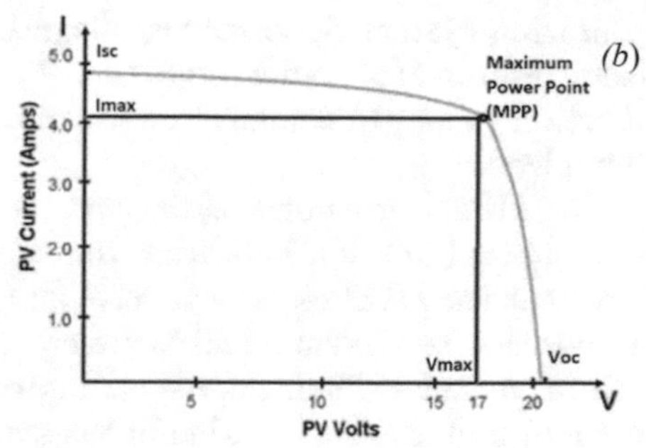

Fig. 1. (a) Intensity of the solar radiation in function of the wavelength and electricity/thermal applications according to spectral range. (b) Characteristic current-voltage (I-V) curve of a PV panel.

In order to measure the energy efficiency of PV panels, the intensity of the solar spectrum must also be measured. Depending on the altitude, the solar intensity increases, however this increase is not proportional depending on each part of the solar spectrum, e.g., the UVA increases more than the IR part [7]. In that sense, the energy efficiency of PV panels can change if we take into account the altitude above sea level, and not only the latitude.

Figure 1(a) shows the intensity of solar radiation as a function of wavelength [8], The range of $\lambda_1 = 500\,\text{nm}$ to $\lambda_2 = 1000\,\text{nm}$, is typically used for the production of electricity using PV panels, whereas the UVA and IR spectrum, is often used for thermal energy, e.g., for heating water.

3 Energy Efficiency of Photovoltaic Panels

Measuring the efficiency of PV panels is critical to performance assessment in converting solar energy into electricity. As the world embraces solar energy, accurate efficiency assessments provide information on economic viability, system optimization, and technological advances. Key parameters such as sunlight intensity and duration influence these measurements.

3.1 I-V Curves in Photovoltaic Panels

The I-V curve, which represents the relationship between current (I) and voltage (V) in a PV panel, illustrates the combinations of voltage and current at which the panel operates under varying conditions. The exact shape of the I-V curve can vary depending on the type of PV panel and environmental conditions. Variations in temperature, solar radiation, and shading can significantly affect the shape of the curve, thus affecting the efficiency of the system.

In PV systems, I-V curves are fundamental tools for evaluating the electrical characteristics of solar panels [9,10]. Understanding I-V curves is crucial for optimizing system performance, diagnosing faults, and evaluating overall efficiency. Figure 1(b) shows a typical I-V curve has a characteristic shape with an inflection point (Maximum Power Point or MPP), where the output power is maximized.

The MPP is the point on the curve where the product of current and voltage is maximized ($I * V$), indicating the maximum power output of the PV panel. Note that the MPP requires to have an optimal tilt angle of the PV panel w.r.t. the position of the sun. This therefore requires the PV panel to be positioned with an optimal tilt angle, depending on the geographically location (i.e., the latitude) and also the position of the sun during the daylight.

3.2 Measurement of the Energy Efficiency of a Photovoltaic Panel

Measuring the efficiency of a PV panel involves assessing how effectively it converts sunlight into electrical energy. The most common method for determining the efficiency of a PV panel is through the calculation of the power output relative to the incoming solar radiation. Equation (1) can be used to calculate the efficiency of a photovoltaic panel [11], where n correspond to the efficiency of the solar PV panel, $W(I)$ the power generated by the PV panel with a solar radiation intensity I (i.e., $W(I)$ is equivalent to the MPP obtained from the I-V curve), I the intensity of the solar radiation, and A the effective irradiation area.

$$n = \frac{W(I)}{I * A} * 100\% \tag{1}$$

For the measurement of the efficiency of PV panels, a fast I-V curve plotting system is necessary. In [12], the authors present circuit designs that allow the I-V curves to be plotted in a fast way. Based on this work, we designed and implemented a I-V tracer Printed Circuit Board (PCB) for our PV Solar Remote Lab (see Sect. 5.3).

4 Overview of the PV Solar Remote Lab

We have developed a user-friendly web platform that unites three fully operational PV Solar Remote Lab "kits" deployed in different cities, each one located at different altitudes. These kits provide a robust infrastructure for conducting altitude-related PV efficiency experiments and also serve to provide experimental evidence to students for understanding the influence of altitude and the increase of the UVA radiation on the efficiency of PV panels, which cannot be calculated or simulated.

In the architecture of the platform shown in Fig. 2, we have two main parts: the software infrastructure (shown on the top) and the hardware setup of the kits (shown on the bottom).

The PV Remote Lab kits include:

- the PV panel,
- the motor to change the tilt angle of the PV panel,

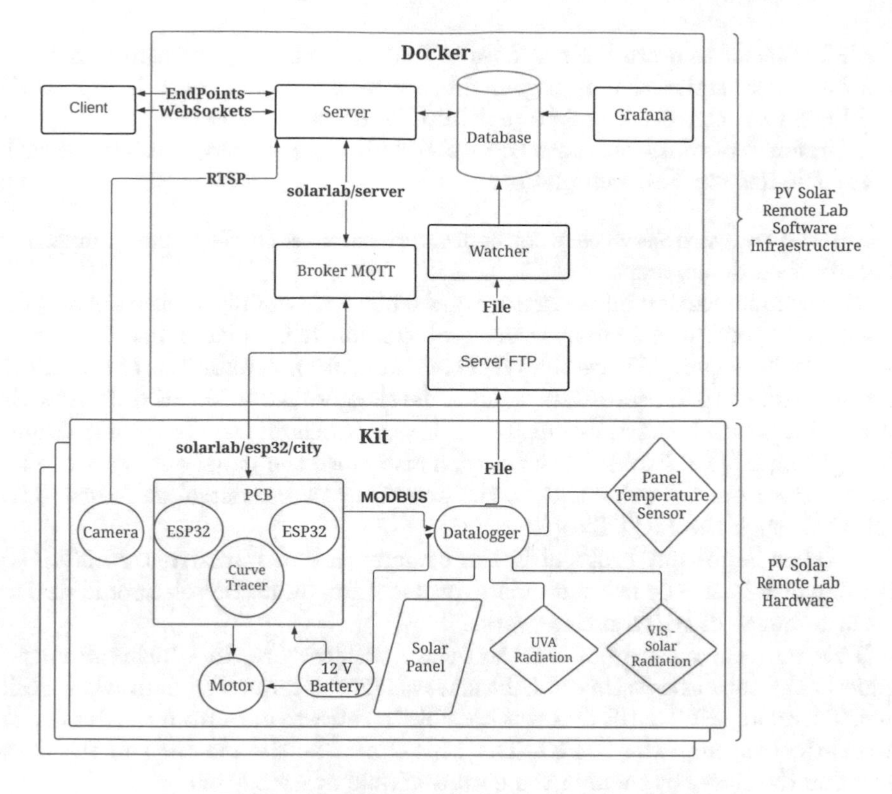

Fig. 2. The PV Remote Lab Architecture.

- a camera for real-time monitoring,
- three different sensors (a UVA radiation sensor, a VIS-IR radiation sensor, and a temperature sensor placed on the back of the PV panel),
- a datalogger to retrieve data from the sensors,
- a home-made PCB with the I-V curve tracer and two ESP32[1] (one to collect data from the I-V curve tracer and the other to control the motor), and
- a 12 V battery to store the solar energy of the PV panels, and to provide energy to the PCB, the datalogger, and the motor.

The software infrastructure consists of:

- a web server implementing the whole Remote Lab platform,
- a web client providing all the interactive interfaces to view the experiment in real-time and remotely control the PV panels,
- a database to store user-defined experiments and sensors data,
- an MQTT[2] Broker to send control messages between the server and the kits,
- a datalogger to retrieve data from the sensors,

[1] https://www.espressif.com/en/products/socs/esp32.
[2] https://mqtt.org/.

- a FTP server to store the raw data of the solar intensity measurements,
- a "watcher" script that is triggered every time that new data is sent to the FTP server, convert it and store it into the database, and
- a Grafana[3] platform instance to visualize all historic solar measurements in our PV Remote Lab web platform

All the software services were containerized in a Docker[4] environment to simplify their deployment.

The communication infrastructure of the platform mainly consists of MQTT, which is a popular communication protocol for IoT devices based on publish/subscribe topics. We use the Wi-fi communication capabilities of the ESP32 microcontroller to retrieve data from a datalogger. After the ESP32 gets the data, it has to send it in real-time to the server, publishing to the specific "topic" through the MQTT Broker. The user can also move the panel and start the I-V curve tracer from the web client, so the web client sends control messages to the ESP32 through the MQTT topics.

Another important protocol in the communication infrastructure is MODBUS[5], which is the de-facto standard protocol for industrial electronic devices in a wide range of buses and networks.

This protocol consists of a master-slave architecture, in which the master requests the data and waits until the slave sends the requested data when available. Therefore, MODBUS makes it possible to retrieve data from the datalogger in real-time through the ESP32. The ESP32 acts as the master and the datalogger as the slave, by making the data available every 500 ms.

5 Implementation and Deployment of the Remote Lab

To implement the PV Solar Remote lab, we had the following requirements: a) Replicate the PV Remote Lab kits as nearly identical as possible to allow reliable data comparison between each setup with identical components, b) Control and monitor in real-time the status of the PV panel current tilt position, to determine the optimal angle w.r.t. the sun, and c) Develop a user-friendly web platform where users can put in practice theoretical knowledge about PV efficiency.

5.1 Assembly of the PV Solar Remote Lab Kits

In a preliminary stage, we installed all three kits in a single geographical location to test the PV module motion control systems, and also to calibrate the sensors, in order to ensure that all the collected data (under the same condition and location) is as similar as possible. Figure 3(a), shows the three PV Solar Remote Lab kits that were assembled and deployed in the same location.

---

[3] https://grafana.com/.
[4] https://docker.com/.
[5] https://modbus.org/.

Fig. 3. The baseline PV Remote Lab setup with (a) three PV Solar Remote Lab kits deployed in the same location, and (b) the internal of the fiberglass box with the I-V PCB, the datalogger, and the battery.

As seen in Fig. 3(b), the electronic motor control circuits, the datalogger, the battery, the PCB for obtaining the I-V curves, and other components were placed in a special fiberglass box, which are specially designed for outdoors usage and to avoid creating interference with Wi-Fi signals.

In addition to the components of the fiberglass box, each kit consists of a 75 Wp PV panel (i.e., the peak power of the PV panel), a DC electric motor, a metal structure that allows the PV panel to be tilted from 0° to 90°, and a thermocouple on the back of each PV panel to measure the temperature.

We also installed on top of the metal structure of the 3 kits, identical digital thermopile pyranometers (CS320)[6] and UVA sensors (SU-200)[7] to measure the solar irradiance.

5.2 The Web Platform

The web-client was developed in the React[8] framework, with REST API (Application Programing Interface) on NextJS[9] to interact with the backend. We used the MySQL[10] database, a robust, reliable solution, and easy configuration. The IP-Camera is embedded in the web interface thanks to a websocket stream of the video captured from RTSP protocol. We also created a responsive design of each web page, so the web platform can be used from computers, tablets, and smartphones, thus increasing its accessibility.

The PV Solar Remote Laboratory web-client interface is shown in Fig. 4. Practitioners can move the PV panel to their required tilt angle, perform several

[6] https://www.campbellsci.com/cs320.
[7] https://www.apogeeinstruments.com/content/SU-200.pdf.
[8] https://react.dev/.
[9] https://nextjs.org/.
[10] https://www.mysql.com/.

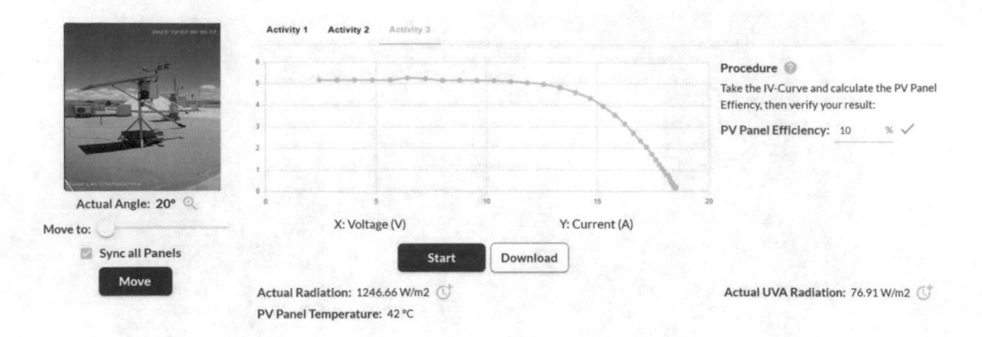

Fig. 4. Our PV Solar Remote Lab Web Platform User Interface.

experiments. The interface allows also to draw the corresponding I-V curve to calculate different factors (i.e., optimum tilt angle, MPP, and PV panel efficiency) in the activities proposed on the platform for practitioners. Also, practitioners can save experimental results and download data for further analysis. In the right side of Fig. 4, we can observe a field where the practitioner must indicate the calculated PV panel efficiency. This value is automatically and internally computed in the platform using Eq. (1), to validate the value that the practitioner calculates.

In the example, the PV panel is of 10%, which corresponds to the efficiency with a radiation of 1246.66 W, MPP of 76.68 W, and PV panel area equal to 0.61 m^2. Our PV Solar Remote Lab platform includes several training activities, where the practitioner needs to make some calculations to ensure the correct understanding of the theory.

Note that the web-client interface includes the possibility to synchronize the movement of all the PV panels, to perform experiments in all the three cities at the same time.

All the dataset of our network of pyranometers can be seen through our Grafana instance, which shows the data in time-line charts, facilitating its visualization and filtering in the way practitioners need.

5.3 The PCB Development

We developed a PCB based on the capacitance method [12] to obtain fast and high-precision I-V curves in real-time. The PCB, shown in Fig. 5, serves a dual role: adjusting the electric motor based on the panel's tilt angle and facilitating I-V curve tracing by disconnecting from the charge regulator and connecting to the dedicated circuit. The PCB includes the two ESP32 microcontrollers, which are programmed in Micropython[11], to retrieve sensor data from the datalogger with MODBUS, asynchronously receive control requests through MQTT, move the PV panel to the correct tilt angle, and take the I-V curve.

[11] https://micropython.org/.

Fig. 5. The developed PCB with the I-V tracer and electronic control system.

The successful execution of these functions highlights the impact of the integration of IoT, which enables the real-time retrieval and transmission of data and handles asynchronous control actions ensuring precise synchronization of the PV panel tilt angle and of the experiments performed by the user.

6 Results and Discussion

We have successfully deployed three fully operational PV Solar Remote Lab kits in our University campuses in three different cities in Bolivia (Santa Cruz de la Sierra, Cochabamba, and La Paz), each one located at distinct altitudes (400, 2558, and 3625 m.a.s.l. respectively). The software infrastructure was deployed in the main University campus, and all the network connections were handled through a Virtual Private Network (VPN). Our distributed laboratory provides a robust infrastructure for conducting altitude-related PV efficiency experiments. For the measurement of the efficiency of the PV panels, an electronic circuit was developed to obtain fast and high-precision I-V curves in real-time.

We have developed a user-friendly web platform that serves as a single-entry point for remotely controlling our network of PV systems distributed across various geographical locations and altitudes.

Figure 6(a) shows the intensity of the integral radiation (385 to 2105 nm), while Fig. 6(b) shows the UVA intensity of the solar radiation (305 to 390 nm). In both cases the y-axis represents units of $\mathrm{W\,m^{-2}}$, and the x-axis represents time units. The historical data can be accessed in our web platform and displayed through the Grafana instance integrated in our platform.

The web platform enables real-time monitoring and control of experiments, ensuring seamless and synchronized data collection. In Fig. 7, the I-V curves were obtained at an altitude of 2558 m.a.s.l., with a solar intensity of $964.06\,\mathrm{W\,m^{-2}}$ and multiple tilt angles with respect to the horizontal.

To manage access to the remote laboratory, we have integrated it with a centralized Booking System [13] which ensures that practitioners have exclusive access during their experiments, improving their learning experience.

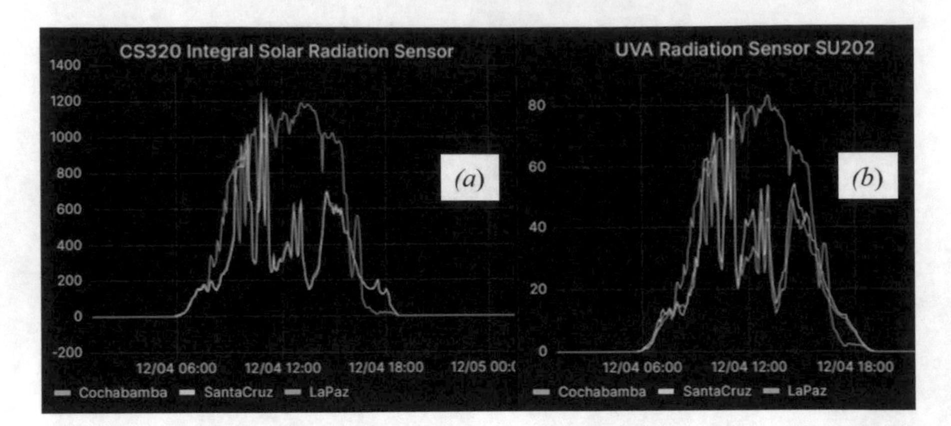

Fig. 6. Plots made with Grafana in our PV Remote Lab platform, showing (a) the intensity of integral solar radiation, and (b) intensity of UVA radiation, measured on 04/12/2023 and an altitude of 400 (yellow), 2558 (green), and 3625 (blue) m.a.s.l.

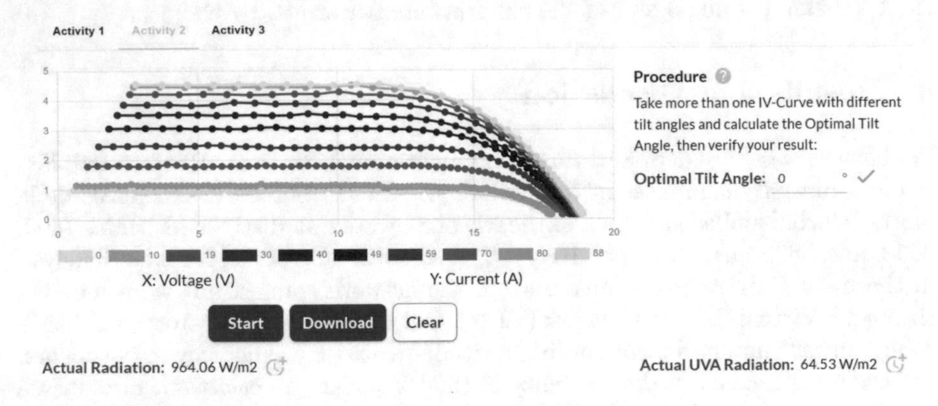

Fig. 7. I-V Curves on different tilt angles.

We obtained preliminary results on the influence of UVA radiation at different altitudes and determined the PV efficiency by comparing two identical PV panels at the same geographical position. In order to compare the efficiency of the PV panels under the same solar radiation conditions and at the same time, one was covered with a filter to eliminate the UVA solar influence, and the second didn't have any filter. At an altitude of 2558 m.a.s.l. at 11:00 a.m. an efficiency of 11% was measured for the PV panel without filter and an efficiency of 11.8% for the panel with filter. This showed us that the solar UVA component has a negative impact on the PV panel's efficiency. The temperature was measured for both panels, obtaining 63.5 °C for the panel with filter and 68.0 °C for the panel without filter. We were able to measure the energy efficiency of PV panels as a function of UVA intensity, where this solar spectral component increases as a function of altitude. We therefore confirmed that the temperature of PV panels decreases if the UVA radiation is filtered, thus increasing its efficiency.

We also meticulously crafted on-line learning material for educators and students, to facilitate training using a "flipped classroom" approach, thus enhancing the quality of education by promoting active engagement with course content and encouraging interactive learning experiences. This material is combined with hands-on activities for practitioners to fill out answer fields in the platform that can be automatically reviewed to know if their result is wrong or correct.

Finally, all data collected from students' experiments under different conditions and user-defined measuring parameters are saved and serve as a valuable dataset for further in-depth analysis.

7 Conclusions

In this paper, we describe the development of a fully operational distributed PV Solar Remote Laboratory, deployed at various altitudes. Our PV Solar Remote Lab provides several advantages, including a deeper understanding of how altitude influences the PV panel efficiency.

The integration of our remote laboratory in online education improves the active learning experience, through simultaneous measurements with identical PV panels, but with different solar irradiation conditions, different altitudes and geographical positions. This bridges the gap between theory and practice of PV efficiency in different real-world configurations, allowing practitioners to identify the factors that contribute to maximum efficiency of PV systems.

The application of IoT technology has been instrumental for the proposed PV Solar Remote Laboratory, by enabling real-time data collection, remote control, and synchronization of experiments. The proposed PV Solar Remote Laboratory software and hardware components are available as open-source[12], and can be further replicated in other countries to increase the diversity of geographical conditions, allowing a much deeper analysis not only for students but also for researchers in the area.

Acknowledgement. This work was partially funded by the Erasmus+ Project "EUBBC-Digital" (No. 618925-EPP-1-2020-1-BR-EPPKA2-CBHE-JP).

References

1. Herrera, R.S., Márquez, M.A., Mejías, A., Tirado, R., Andújar, J.M.: Exploring the usability of a remote laboratory for photovoltaic systems. IFAC-PapersOnLine **48**(29), 7–12 (2015). https://doi.org/10.1016/j.ifacol.2015.11.205
2. Das, S.: Development of low-cost remote online laboratory for photovoltaic cell and module characterization. In: 2019 ASEE Annual Conference & Exposition, Tampa, Florida, June 2019. ASEE Conferences. https://doi.org/10.18260/1-2-32657
3. Zimmer, T., Billaud, M., Fylladitakis, E.D., Axaopoulos, P.: HelionLab: a remote photovoltaic laboratory. In: 2015 3rd Experiment International Conference (exp.at'15), pp. 52–55, June 2015. https://doi.org/10.1109/EXPAT.2015.7463213

[12] https://github.com/eubbc-digital/PVSolarRemoteLab.

4. Gonthier, L., Billaud, M., Lacoste, D., Zimmer, T.: Remote photovoltaic outdoor solar lab. In: 2017 27th EAEEIE Annual Conference (EAEEIE), pp. 1–5, June 2017. https://doi.org/10.1109/EAEEIE.2017.8768678
5. Cleveland, C., Morris, C.: Handbook of Energy, Volume I: Diagrams, Charts, and Tables. Elsevier Science, Amsterdam, January 2013. ISBN ISBN-10: 008046405X; ISBN 13: 978-0080464053
6. Brune, D., Hellborg, R., Persson, B.R., Pääkkönen, R., Shapiro, J.: Radiation: at home, outdoors and in the workplace. Am. J. Phys. **71**(2), 189–190 (2003). https://doi.org/10.1119/1.1522706
7. Romero, F., et al.: Development of a system of solar radiation monitoring based on a broad-spectrum spectrometer. Investig. Desarro. **1**(11) (2011). ISSN 2518-4431
8. Huang, G., Wang, K., Markides, C.N.: Efficiency limits of concentrating spectral-splitting hybrid photovoltaic-thermal (PV-T) solar collectors and systems. Light Sci. Appl. **10**(1), (2021). https://doi.org/10.1038/s41377-021-00465-1
9. Aranda, E.D., Gomez Galan, J.A., de Cardona, M.S., Andujar Marquez, J.M.: Measuring the I-V curve of PV generators. IEEE Ind. Electron. Mag. **3**(3), 4–14 (2009). https://doi.org/10.1109/MIE.2009.933882
10. Tsuno, Y., Hishikawa, Y., Kurokawa, K.: Translation equations for temperature and irradiance of the I-V curves of various PV cells and modules. In: 2006 IEEE 4th World Conference on Photovoltaic Energy Conference, vol. 2, pp. 2246–2249, May 2006. https://doi.org/10.1109/WCPEC.2006.279619
11. Emery, K.A., Osterwald, C.R.: Solar cell efficiency measurements. Solar Cells **17**(2), 253–274 (1986). https://doi.org/10.1016/0379-6787(86)90016-5
12. Vargas Bautista, J.P.: Development of a control system to measure the efficiency and characteristic curve I-V in real time of a solar PV system using Labview® and Arduino. Investig. Desarro. **1**(15), 49–64 (2015)
13. Villazon, A., Ormachea, O., Orellana, A., Zenteno, A., Fransson, T.: Work in progress: a booking system for remote laboratories – the EXPLORE energy digital academy (EEDA) case study. In: Auer, M.E., Langmann, R., Tsiatsos, T. (eds.) Open Science in Engineering. REV 2023. LNCS, vol. 763, pp. 341–348. Springer, Cham (2023). https://doi.org/10.1007/978-3-031-42467-0_31

RHLab Interoperable Software-Defined Radio (SDR) Remote Laboratory

Marcos Inoñan[1], Zhiyun Zhang[1](✉), Pedro Amarante[1], Pablo Orduña[2], Rania Hussein[1], and Payman Arabshahi[1]

[1] University of Washington, Seattle, WA 98195, USA
{minonan,zzyzzy42,pedroa2,rhussein,paymana}@uw.edu
[2] LabsLand, San Francisco, CA 94114, USA
pablo@labsland.com

Abstract. Educational remote laboratories, dedicated to radio frequency (RF) communications, represent an innovative effort aimed at bridging the gap between theory and practice. This advancement is even more empowered by the advent of Software-Defined Radio (SDR) technology, which allows students to create a wide array of prototypes for RF communication systems. The SDR community provides highly adaptable hardware that can be easily customized with just a few lines of code. Additionally, the availability of SDR hardware is extensive, offering various features tailored to users' specific needs. One of the key advantages of the flexibility inherent in SDR technology is the shared libraries among its devices. This feature fosters interoperability which is the ability of different SDR devices or components to work together effectively. In this context, the Remote Hub Lab (RHL/RHLab) has been working to incorporate this functionality into the RHL-RELIA remote lab project. RHL-RELIA stands as a remote wireless communication lab, with its initial iteration based on the ADALM-PLUTO SDR. However, as it evolves, RHL-RELIA is incorporating other SDR devices, such as the Red Pitaya, which introduces unique features not found in the ADALM-PLUTO. This degree of diversification empowers students with a broad spectrum of choices, allowing them to select the SDR device that aligns most closely with their individual interests and educational requirements.

Keywords: GNU Radio · Interoperability · Software-Defined Radio · Remote Laboratory · Embedded Systems

1 Introduction

In the curriculum of electrical and computer engineering courses, laboratories are integrated to offer students hands-on experience across a diverse range of subjects [1]. Traditionally, students receive physical lab kits and are granted access to lab facilities to execute their assignments [2]. However, in recent years, the introduction of remote labs has emerged as a viable alternative, offering

M. E. Auer et al. (Eds.): STE 2024, LNNS 1028, pp. 145–156, 2024.
https://doi.org/10.1007/978-3-031-61905-2_15

students convenient access to equipment at any time, eliminating geographical restrictions, fostering collaboration among peers, and improving accessibility for students with disabilities [3]. This transition also contributes to promoting equitable access in education [4].

Remote laboratories have attracted considerable interest in educational research since the 1990s, as indicated by the numerous definitions found in the academic literature [5–7]. It is crucial to distinguish between "remote labs" and "virtual labs". Virtual labs are computer-based applications simulating non-physical environments, whereas remote labs enable users to access and control physical equipment from a remote location through computer and communication infrastructure [8]. The educational efficacy of remote and virtual laboratories has been validated [9–13] with their development closely tied to the role of internet technology.

In Electrical and Telecommunication Engineering, courses in wireless communication that employ RF techniques are integral to the curriculum. Traditional communication laboratories equip students with a kit of components for their lab assignments [10] including RF devices for signal transmission and reception, along with essential components like antennas and cables. SDR technology simplifies this process, reducing setup time and offering flexible hardware that can be modified with code. SDR enhances versatility, applicable in various educational settings for different types of communication [14]. This approach positively impacts education, enriching learning experiences and fostering research and collaboration among developers and instructors aligning with high standards of educational quality [15,16].

In the context of SDR remote laboratories for educational purposes, numerous initiatives provide educational solutions, often focusing on a single type of SDR device. Nevertheless, the SDR technology encompasses a diverse range of devices available in the market, each differing in several aspects. Additionally, SDR devices share a substantial portion of libraries and software tools, ensuring compatibility with a common set of resources. This particular feature fosters the inclusion of the concept of interoperability that permits the interaction with two or more devices from the same SDR family. In this direction, we[1] [17] and LabsLand[2] involved in the development of the RHL-RELIA project [18,19] are now incorporating the Red Pitaya SDR[3] as a second SDR device alongside the ADALM-Pluto[4] to provide an interoperability option, offering users a wide range of SDR selections.

The paper's structure is as follows: Sect. 2 offers an overview of interoperability concepts in communication remote labs, along with insights into projects featuring remote labs equipped with this characteristic. Section 3 details the design of the RHL-RELIA system, integrating new SDR devices. Section 4 explains the current progress of the project. Finally, Sect. 5 summarizes the article, draws conclusions, and explores avenues for future work.

---

[1] https://rhlab.ece.uw.edu.
[2] https://labsland.com.
[3] https://redpitaya.com/.
[4] https://wiki.analog.com/university/tools/pluto.

2 Background

The concept of interoperability within SDR remote labs pertains to the capability of incorporating diverse devices that collaborate on the same platform, allowing users to enhance the versatility of the lab environment [20]. This feature's advantages in a remote lab empower students to select the appropriate SDR hardware based on their specific needs. In the realm of RF devices, several key parameters include:

- Frequency Range: It covers the range of frequencies relevant to transmit or receive. Different SDRs are designed for specific frequency bands of operation.
- Sampling Rate: It determines how quickly the SDR can digitize analog signals. Higher sampling rates are beneficial for capturing high-frequency signals.
- Bit Resolution: It defines the number of bits in each sample. Higher resolution provides a more accurate representation of the analog signal.
- Dynamic Range: It indicates the SDR's ability to handle signals of varying amplitudes. A large dynamic range permits to detection of both weak and strong signals.

A prominent SDR project that offers interoperability is REDHAWK[5], a software package facilitating the design, development, deployment, management, and upgrading of network-enabled SDRs in real time. Initially starting as the OSSIE project [21], REDHAWK aims to provide portability in the development of software radio applications, given the necessity of running the same software on different hardware for SDR applications.

REDHAWK's primary contribution lies in providing a platform that renders the hardware transparent to the programmer, thereby reducing the time and cost of code development [22]. However, its utilization necessitates specialized software and dedicated hardware, precluding its remote use and limiting its potential as an educational tool with equitable access [19].

Adopting a similar approach, we incorporated interoperability into the RHL-RELIA project. RHL-RELIA, an implementation of the MELODY model, serves as an agnostic technological model providing design considerations for educational remote labs controlling SDR devices [23].

In its initial version, RHL-RELIA utilized the ADALM-Pluto SDR, known for its affordability, versatility, and community support, offering students practical experiences aligning well with modern educational approaches emphasizing real-world applications in the field of software-defined radio and communications. To enhance interoperability, RHL-RELIA introduced the Red Pitaya SDR device as a second component.

The Red Pitaya, a high-performance tool with commendable hardware and an open-source software project, supports exploration, learning, and the development of various applications [24]. Red Pitaya has gained widespread popularity as a versatile learning platform since its launch. It has been embraced by amateurs, educators, and professionals alike for diverse project creations [25–29].

[5] https://redhawksdr.org/.

This device has found application in several educational remote labs, such as the Gheorghe Asachi Technical University of Iaşi, which developed a linear electronic circuit based on Red Pitaya for electrical engineering courses. This project leverages Red Pitaya's features, including flexible hardware and peripherals like Ethernet, to remotely manipulate and visualize a signal wave originating from a breadboard [30].

Other projects also utilize Red Pitaya as a core component in educational remote labs. For instance, the National Institute of Technology, Malang, Indonesia, developed an electronics telecommunication remote lab involving Red Pitaya, functioning as an oscilloscope, signal generator, spectrum analyzer, and multitester device [31].

Compared to the ADALM-PLUTO, the Red Pitaya SDRlab 122-16 offers several additional features, as Table 1 demonstrates. Firstly, it boasts a more powerful programmable System-on-Chip SoC: the Zynq Z-7020. This upgrade from the Z-7010 within the PlutoSDR results in a significant enhancement, with the Z-7020 offering 164% more FPGA logic cells, 133% more BRAMs, and 175% more DSP slices. These enhancements ensure that the SDRlab 122-16 is considerably more future-proof in terms of computational power. Secondly, the SDRlab 122-16 provides a greater number of input and output channels, enabling the execution of more intricate programs. Thirdly, Red Pitaya offers a broader range of connectivity options. Although it shares the USB 2.0 port with the PlutoSDR, it extends its capabilities by including an Ethernet port. This additional feature proves highly beneficial, particularly for wireless connections to a host computer. Moreover, the Red Pitaya SDRlab 122-16 fills the frequency range gap left by the PlutoSDR, reaching frequencies as low as 200 kHz. Lastly, the SDRlab 122-16 boasts higher input and output resolutions compared to the PlutoSDR.

Table 1. Red Pitaya SDRlab 122-16 & ADALM-PLUTO Specifications

	SDRlab 122-16	ADALM-PLUTO
SoC	Zynq Z-7020	Zynq Z-7010
Transceiver	AD9767 & LTC2185	AD9363
Transmitter Channels	2	1
Receiver Channels	2	1
Transmitter RF Range (Hz)	300k–60M	325M–3.8G
Receiver RF Range (Hz)	300k–550M	
Output Resolution (bits)	14	12
Input Resolution (bits)	16	
Connectivity	USB 2.0 & 1 Gb Enthernet	USB 2.0
SD Card Port	Yes	No
Digital IOs	16	No

3 System Design

The architecture of RHL-RELIA in interoperability mode is designed to accommodate a variety of SDR devices. In a general scenario, the SDR device must feature either a USB 2.0, USB 3.0, or Gigabit Ethernet interface connector to enable connection with a cost-effective Raspberry Pi. A comprehensive block diagram is presented in Fig. 1, illustrating the user's remote access to the RHL-RELIA system through a server. Within this interface, students can select their preferred SDR station.

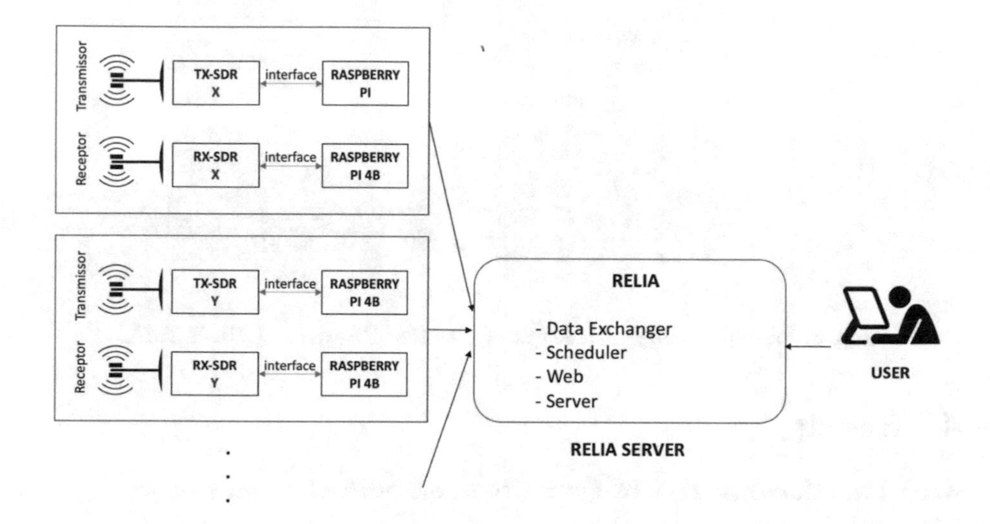

Fig. 1. RHL-RELIA general interoperability architecture.

In the specific hardware setup with Red Pitaya, the configuration encompasses two Raspberry Pis and two Red Pitayas. Each Red Pitaya is linked to an individual Raspberry Pi through an Ethernet interface (utilizing a USB connection in the case of ADALM-Pluto), and both devices are equipped with dedicated Ethernet ports. Following this, the two Raspberry Pis are interconnected with the RHL-RELIA server. A block diagram illustrating these interconnections is presented in Fig. 2.

In difference from other previous remote lab projects with Red Pitaya elaborated in Sect. 3, RHL-RELIA utilizes the GNU Radio Companion (GRC)[6] environment in order to configure Red Pitaya. This installation requires a modification in the Red Pitaya's firmware.

[6] https://wiki.gnuradio.org/index.php/Guided_Tutorial_GRC.

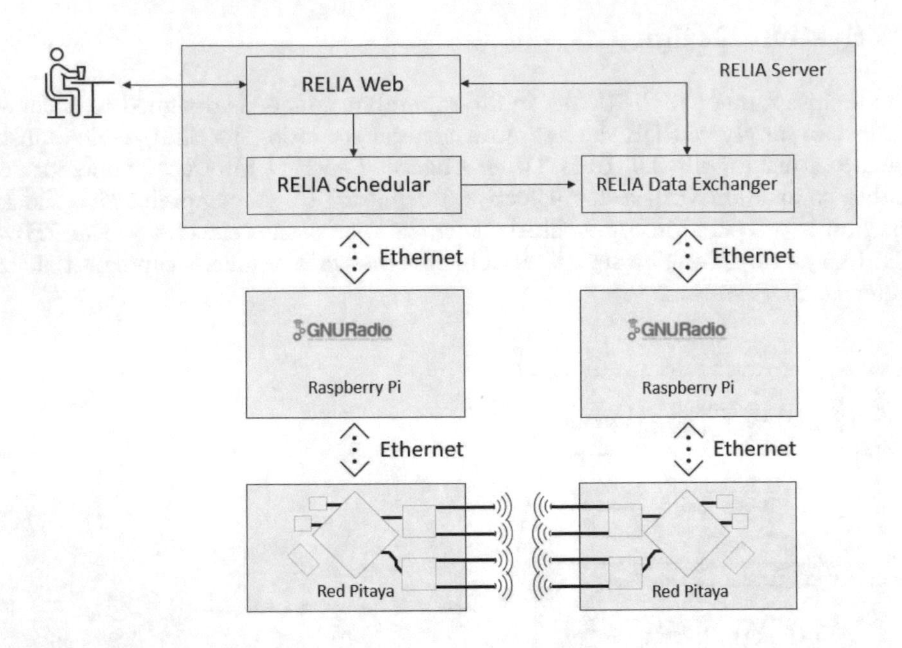

Fig. 2. Interoperability framework of two Red Pitayas in RHL-RELIA.

4 Results

4.1 Functionality Test of Four Channels with Custom Image

Prior to initiating interoperability between two Red Pitayas, we conducted initial testing of the device by installing a custom image with GRC libraries on it. A loop antenna is connected between a transmitter and receiver ports on the Red Pitaya, depicted in Fig. 3. Subsequently, the Red Pitaya was controlled by a Raspberry Pi, and programs were executed to assess functionality. The Red Pitaya successfully transmitted signals specified by GRC programs and effectively received them through its receiver.

4.2 Interoperability Between Two Red Pitayas

We then moved on to perform interoperability between two distinct Red Pitayas. The connection is shown in Fig. 4. First, a single tone transmission was performed in order to test the accuracy of the frequency transmitted. Given that each Red Pitaya incorporates an internal oscillator, even when configured with the same frequency value, slight frequency shifts are expected. While these shifts may be insignificant in certain applications, estimating their values remains crucial for more complex experiments.

The results of this basic experiment are illustrated in Fig. 5. Figure 5a shows the experiment of transmitting a signal frequency of 50 kHz modulated on a

Fig. 3. Loop test using a single Red Pitaya.

Fig. 4. Hardware setup for performing interoperability between two Red Pitayas.

carrier frequency of 50 MHz. Given the communication type is In-phase (I) and Quadrature (Q), it becomes imperative to examine each signal component. This constellation representation serves as a potent tool for both the analysis and visualization of digital signal characteristics in communication systems. The results are illustrated in Fig. 5b.

The second experiment focuses on Binary Phase Shift Keying (BPSK), a digital modulation scheme employed in telecommunications for transmitting binary data ('0's and '1's) by manipulating the phase of the signal [32]. This experiment holds significance as it constitutes one of the fundamental components in telecommunication courses, providing students with insights into more advanced transmission techniques.

Figure 6 displays the results of the experiment involving the random transmission of a sequence of 1's and 0's. The design of the experiment in GRC can be observed in Fig. 6a, with the corresponding results presented in Fig. 6b.

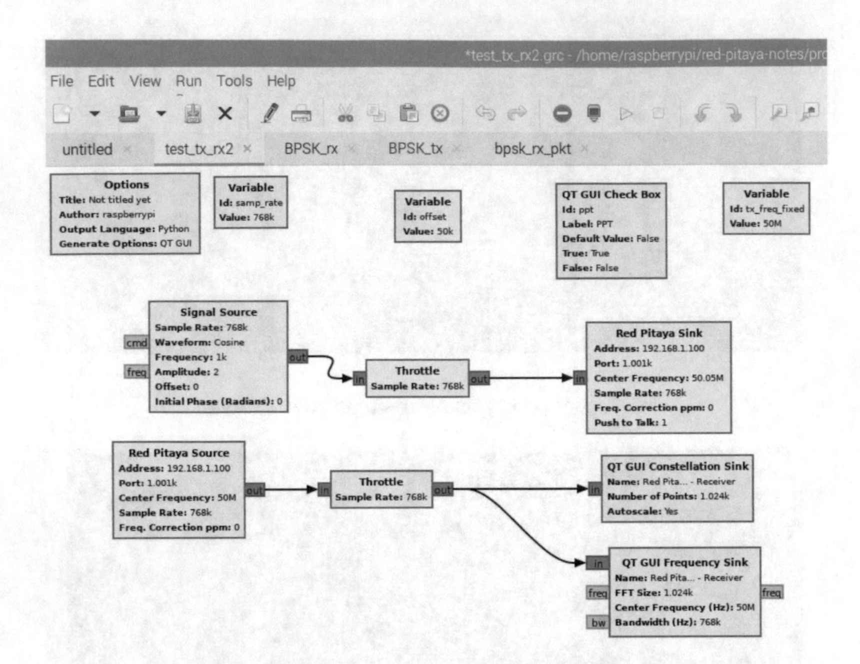

(a) RHL-RELIA single frequency tone transmission reception test - GRC.

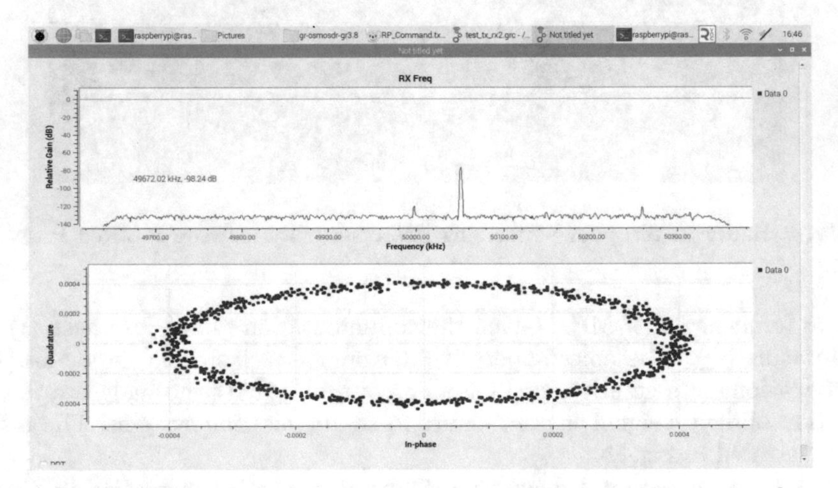

(b) RHL-RELIA single frequency tone transmission reception test - plot.

Fig. 5. Single tone test results.

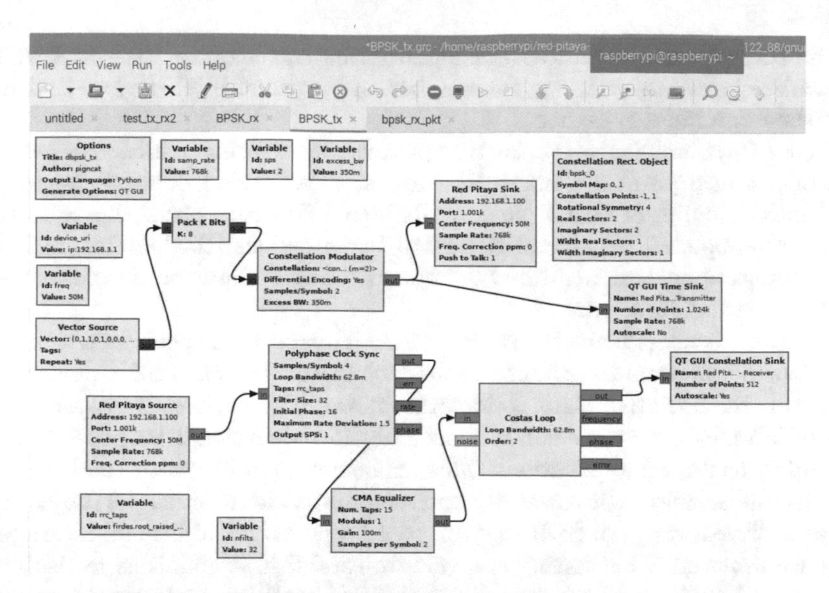

(a) RHL-RELIA BPSK transmission reception test - GRC.

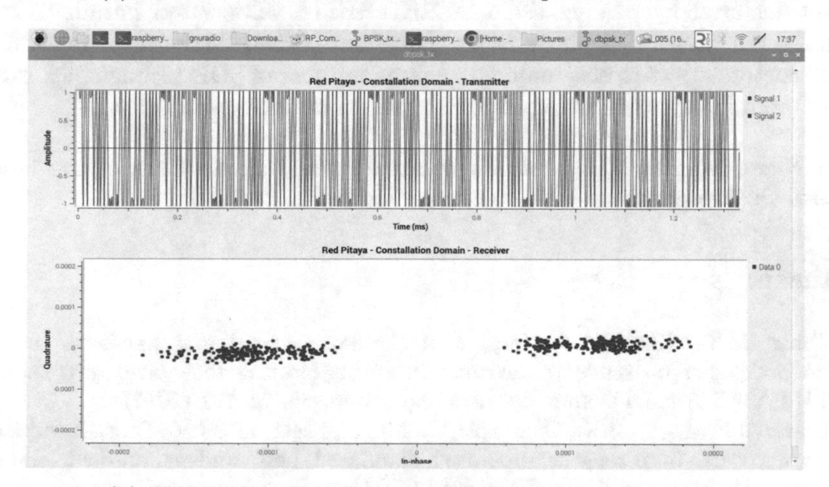

(b) RHL-RELIA BPSK transmission reception test - plot.

Fig. 6. BPSK test results

5 Conclusion and Future Work

This paper demonstrates the integration of the Red Pitaya SDR into the RHL-RELIA remote lab, introducing an interoperability feature. This functionality will allow students to select the most suitable SDR device based on the specific requirements of a communication project. Each device possesses strengths in different application domains. Furthermore, this interoperability presents a cost-effective advantage, as educational institutions can acquire RHL-RELIA and customize the lab by selecting the number and types of SDR devices that align with their specific requirements.

Red Pitaya enhances the lab's versatility by offering additional features. Notably, it incorporates a micro SD card slot, expanded onboard memory, a substantial number of GPIO pins, and onboard Ethernet ports. These features extend the application scope of RHL-RELIA to various RF domains, including radar applications, and facilitate the implementation of advanced functions such as partial reconfiguration.

In future developments, the project aims to introduce a specialized web interface that allows users to exploit the distinct features of each SDR device fully. Moreover, the initiative plans to integrate more SDR devices, transforming the remote lab into a resource suitable not only for educational purposes but also for professional-level applications. This expansion is in line with the MELODY model, emphasizing both versatility and scalability. However, as the project integrates a diverse range of SDR devices on a larger scale, additional interference tests are essential. This testing is crucial to establish mechanisms for isolation, mitigating interference between different SDR stations. Isolation is predominantly achieved by placing them in RHL-RELIA 3D-printed Faraday Cages, mounted on PhaseDock plates in the LabsLand Prism4 structure. As the lab's infrastructure grows to accommodate a large number of SDR technologies, ensuring reduced interference between SDR stations becomes a priority.

Acknowledgements. This work is supported by the National Science Foundation's Division Of Undergraduate Education under Grant No. 2141798.

References

1. Taher, M.T., Khan, A.S.: Impact of simulation-based and hands-on teaching methodologies on students' learning in an engineering technology program. In: 2014 ASEE Annual Conference and Exposition, pp. 24–701 (2014)
2. Corter, J.E., Esche, S.K., Chassapis, C., Ma, J., Nickerson, J.V.: Process and learning outcomes from remotely-operated, simulated, and hands-on student laboratories. Comput. Educ. **57**(3), 2054–2067 (2011)
3. Auer, M.E.: Virtual lab versus remote lab. In: 20th World Conference on Open Learning and Distance Education. Citeseer (2001)
4. Hussein, R., Maloney, R.C., Rodriguez-Gil, L., Beroz, J.A., Orduna, P.: RHL-BEADLE: bringing equitable access to digital logic design in engineering education. In: 2023 ASEE Annual Conference and Exposition (2023)
5. Feisel, L.D., Rosa, A.J.: The role of the laboratory in undergraduate engineering education. J. Eng. Educ. **94**(1), 121–130 (2005)
6. Restivo, M.T., Mendes, J., Lopes, A.M., Silva, C.M., Chouzal, F.: A remote laboratory in engineering measurement. IEEE Trans. Industr. Electron. **56**(12), 4836–4843 (2009)
7. Johri, A., Olds, B.M.: Situated engineering learning: bridging engineering education research and the learning sciences. J. Eng. Educ. **100**(1), 151–185 (2011)

8. Balamuralithara, B., Woods, P.C.: Virtual laboratories in engineering education: the simulation lab and remote lab. Comput. Appl. Eng. Educ. **17**(1), 108–118 (2009)
9. Aitor, V.M., García-Zubía, J., Angulo, I., Rodríguez-Gil, L.: Toward widespread remote laboratories: evaluating the effectiveness of a replication-based architecture for real-world multiinstitutional usage. IEEE Access **10**, 86298–86317 (2022)
10. Hussein, R., Wilson, D.: Remote versus in-hand hardware laboratory in digital circuits courses. In: 2021 ASEE Virtual Annual Conference Content Access (2021)
11. Heradio, R., De La Torre, L., Galan, D., Cabrerizo, F.J., Herrera-Viedma, E., Dormido, S.: Virtual and remote labs in education: a bibliometric analysis. Comput. Educ. **98**, 14–38 (2016)
12. Bencomo, S.D.: Control learning: present and future. IFAC Proc. Vol. **35**(1), 71–93 (2002)
13. Ma, J., Nickerson, J.V.: Hands-on, simulated, and remote laboratories: a comparative literature review. ACM Comput. Surv. (CSUR) **38**(3), 7-es (2006)
14. Sinha, D., Verma, A.K., Kumar, S.: Software defined radio: operation, challenges and possible solutions. In: 2016 10th International Conference on Intelligent Systems and Control (ISCO), pp. 1–5. IEEE (2016)
15. Inonan, M., Paul, A., May, D., Hussein, R.: RHLab: digital inequalities and equitable access in remote laboratories. In: 2023 ASEE Annual Conference and Exposition (2023)
16. Hussein, R., Guo, M., Amarante, P., RodriguezGil, L., Orduña, P.: Digital twinning and remote engineering for immersive embedded systems education. In: Frontiers in Education (FIE) Conference, USA. IEEE (2023)
17. Hussein, R., et al.: Remote Hub Lab - RHL: broadly accessible technologies for education and telehealth. In: Auer, M.E., Langmann, R., Tsiatsos, T. (eds.) REV 2023. LNNS, vol. 763, pp. 73–85. Springer, Cham (2023). https://doi.org/10.1007/978-3-031-42467-0_7
18. Inonan, M., Chap, B., Orduña, P., Hussein, R., Arabshahi, P.: RHLab scalable software defined radio (SDR) remote laboratory. In: Auer, M.E., Langmann, R., Tsiatsos, T. (eds.) REV 2023. LNNS, vol. 763, pp. 237–248. Springer, Cham (2023). https://doi.org/10.1007/978-3-031-42467-0_22
19. Inonan, M., Orduña, P., Hussein, R.: Adapting a remote SDR lab to analyze digital inequalities in radiofrequency education in Latin America. Revista Innovaciones Educativas (2023, in press)
20. Becker, J.K., Starobinski, D.: Snout: a middleware platform for software-defined radios. IEEE Trans. Netw. Serv. Manage. **20**(1), 644–657 (2022)
21. Snyder, J., McNair, B., Edwards, S., Dietrich, C.: OSSIE: an open source software defined radio platform for education and research. In: International Conference on Frontiers in Education: Computer Science and Computer Engineering (FECS 2011). World Congress in Computer Science, Computer Engineering and Applied Computing, Las Vegas, NV (2011)
22. Banaszak, T.: Redhawk for Vita 49 Development in Open Radio Access Networks. Ph.D. thesis, Purdue University (2020)
23. Inonan, M., Hussein, R.: Melody: a platform-agnostic model for building and evaluating remote labs of software-defined radio technology. IEEE Access **11**, 127550–127566 (2023). https://doi.org/10.1109/ACCESS.2023.3331399
24. Ştan, C.M., Neacşu, D.: A remote laboratory for linear electronics based on the Red Pitaya board. In: 2022 International Conference and Exposition on Electrical and Power Engineering (EPE), pp. 113–117. IEEE (2022)

25. Durvaux, F., Durvaux, M.: SCA-Pitaya: a practical and affordable side-channel attack setup for power leakage–based evaluations. Digit. Threats **1**(1), March 2020. https://doi.org/10.1145/3371393
26. Matusko, M., et al.: Fully digital platform for local ultra-stable optical frequency distribution. Rev. Sci. Instrum. **94**(3), 034716 (2023). https://doi.org/10.1063/5.0138599
27. Shindin, A.V., Moiseev, S.P., Vybornov, F.I., Grechneva, K.K., Pavlova, V.A., Khashev, V.R.: The prototype of a fast vertical ionosonde based on modern software-defined radio devices. Remote Sens. **14**(3), 547 (2022). https://doi.org/10.3390/rs14030547
28. Hertlein, A.F.: Measurements on the dynamics of dissipative, comb-driven solitons in femtosecond enhancement cavities p. 93 (Feb 2023), https://cloud.physik.lmu.de/index.php/s/StEQTwAoHsCPgYY?dir=undefined&path=%2F&openfile=52207912
29. Siffer, W.: https://content.redpitaya.com/blog/a-home-lab-with-red-pitaya
30. Stan, C.M., Scripcariu, L.: Development of secure remote connection for the electronics laboratory based on Red Pitaya board. In: 2023 13th International Symposium on Advanced Topics in Electrical Engineering (ATEE), pp. 1–6. IEEE (2023)
31. Limpraptono, F.Y., Nurcahyo, E., Faisol, A.: The development of electronics telecommunication remote laboratory architecture based on mobile devices. Int. J. Online Biomed. Eng. (iJOE) **17**(03), 26 (2021)
32. Sacher, W.D., et al.: Binary phase-shift keying by coupling modulation of microrings. Opt. Express **22**(17), 20252–20259 (2014)

RHL-RADAR Remote Laboratory

Marcos Inoñan(✉), Matt Reynolds, and Rania Hussein

University of Washington, Seattle, WA 98195, USA
{minonan,msreynol,rhussein}@uw.edu

Abstract. An Educational Remote Laboratory is a software and hardware solution that enables students to access real equipment from any location. In the literature there is a wide range of remote laboratories in many fields (e.g., robotics, electronics, physics, chemistry). However, few experiences have been developed for educational Radar purposes. Radar, which stands for "Radio Detection and Ranging," is a technology that utilizes radio waves to detect distant objects to measure physical parameters like the range, angle, or velocity. Most Radar systems are composed of a transmitter, receiver, antennas and a computer to apply signal processing methods of estimation. This paper introduces the Radar remote laboratory developed by the Remote Hub Lab (RHL). The laboratory involves the creation of a remotely controllable Radar system prototype designed to estimate the rotational velocity of a fixed structure, with initial results presented. This educational tool is oriented toward undergraduate and graduate engineering students, offering them a remote platform to supplement their Radar courses by actively engaging in real-world Radar problem-solving exercises.

Keywords: GNU Radio · Software Defined Radio · remote laboratory · embedded systems · Radar

1 Introduction

Remote laboratories in engineering education curricula have become increasingly popular due to their low cost, 24/7 availability, and the absence of geographic proximity restrictions [1,2]. This educational approach not only fosters collaboration among peers but also significantly improves accessibility for students with disabilities [3]. Within the electrical engineering department, an extensive array of remote labs has been introduced as a valuable alternative to traditional labs, providing students with hands-on experimentation opportunities to complement theoretical knowledge [4,5]. One notable advantage of remote labs in electrical engineering is that parameters like electrical current and electromagnetic waves are not visible or audible, eliminating the need for sound or video transmission [6]. In the case of electromagnetic waves, which play an important role due to the propagation of electromagnetic energy in the radio frequency (RF) spectrum is fundamental to designing and implementing effective wireless communication systems.

M. E. Auer et al. (Eds.): STE 2024, LNNS 1028, pp. 157–168, 2024.
https://doi.org/10.1007/978-3-031-61905-2_16

In this context, the RHL[1] group [7] and LabsLand[2] initiated the development of Remote Engineering laboratory for Inclusive Access (RHL-RELIA), a remote laboratory for wireless communication utilizing Software Defined Radio (SDR) devices. This laboratory offers the advantage of a low-cost and open-source philosophy, allowing students to access it through a web browser, thereby avoiding the expenses associated with specialized software licenses [7,8]. RELIA was developed as a implementation of MELODY, a model tailored explicitly for SDR remote labs. MELODY is characterized by its technology-agnostic and open-source approach [9].

Following with the idea of developing remote educational tools based on the MELODY model, RHL group worked on the development of a RHL-RADAR remote laboratory with the idea of offering it in Radar classes where students can manipulate remotely a radar system to experiment and apply Radar configuration and signal processing techniques. Typically in the traditional Radar courses, students apply signal processing techniques in the plane of simulation and they might build a hands-on project. However, these activities often demand a significant background in in electrical engineering and electromagnetic to build and install a Radar system. With a remote laboratory students have the Radar system installed and focus mostly in the experiment's configuration and the Radar signal processing which is beneficial from the cost-time development which is essential in classes and it can motivate students from other specialities to take Radar classes since they don't to have to handle with other backgrounds required.

This paper introduces RHL-RADAR, which implements a Continuous Wave (CW) Radar application capable of estimating the rotational velocity of a fixed structure. The obtained results are subsequently compared with those from a tachometer used for measuring rotational speed. The structure of the article is as follows: Sect. 2 offers a review of Radar tools in the education field. Section 3 provides an overview of the CW Doppler Radar, offering a mathematical analysis that elucidates the micro-Doppler effect generated by rotating structures. It also incorporates technical details about the antennas and digital receiver employed in the experiment. Section 4 presents and discusses the outcomes derived from the CW Radar, drawing comparisons with measurements obtained from the tachometer. Finally, Sect. 5 summarizes the primary findings of the article and outlines the conclusions and future work derived from the study.

2 Background

In a remote laboratory, students can remotely access actual equipment situated in the university, setting it apart from virtual laboratories that typically involve simulations rather than tangible equipment [5]. The composition of a remote laboratory includes both software and hardware elements, and owing to the advancements in internet technology, it now allows for the execution of intricate experiments closely mirroring traditional learning experiences [10,11].

[1] https://rhlab.ece.uw.edu.

[2] https://labsland.com.

While the literature encompasses a diverse array of remote laboratories across various fields, such as robotics, electronics, physics, and chemistry [12], there is a notable scarcity of experiences specifically designed for Radar applications. A Radar system utilizes radio waves to detect, locate, track, and identify objects, finding widespread applications in military, aviation, meteorology, navigation, and traffic control [13]. In the educational context, students pursuing electrical engineering often enroll in Radar classes, typically as elective courses or at the graduate level.

One compelling Radar project is offered by the University of Oklahoma (OU), presenting an integrated interdisciplinary approach to prepare students in meteorology Radar, facilitating the acquisition of quantitative weather data. OU's initiative features a laboratory with hands-on hardware setups and signal processing, complementing an extensive Radar curriculum spanning 10 courses [14]. Despite containing multiple Radar systems, this laboratory requires the physical presence of students for full operation.

Another intriguing Radar project is developed by the Massachusetts Institute of Technology (MIT) Lincoln Laboratory, where a low-cost Radar system has been designed for students participating in a Massive Open Online Course (MOOC). Valued at $530, the system is available for online ordering, guiding students through the collection and processing steps to record their own data. This project's distinctive feature lies in its emphasis on allowing students to work with and interpret real-world data, a significant departure from standard sample files [15]. This useful tool requires that a student buy and build their own Radar system and test the results at home while is taking an online class.

The objective of the RHL-RADAR project is to develop a new Software Defined Radio (SDR) remote laboratory featuring a specialized peripheral-a fan designed for testing Radar applications [16]. While RHL-RADAR is currently in the development phase, its characteristics align well with explaining the MELODY model in this context. This alignment is crucial for evaluating MELODY using a prospective remote laboratory. RHL-RADAR aims to provide students with the convenience of conducting experiments from a distance, eliminating the need for direct manipulation of physical hardware or manual configuration of Radar parameters.

3 System Design

3.1 Continuous Wave Radar

Continuous Wave (CW) Radar transmits a continuous radio frequency signal at a fix frequency without any interruption. This type of Radar is able to estimate the velocity of a target by measuring a portion of the signal reflected back to the Radar receiver. The receiver then analyzes the frequency shift, known as the Doppler shift, between the emitted signal and the received signal.

CW Radar is typically used in applications such as speed detection, however, CW Radar can't estimate range resolution because it does not utilize pulse timing to measure distances.

One application of continuous wave (CW) Radar is to estimate the rotational velocity of objects. In this particular application, the frequency of the reflected signal received by the Radar undergoes modulation, resulting in an effect known as micro-Doppler. The CW Radar system designed for this purpose is illustrated in Fig. 1.

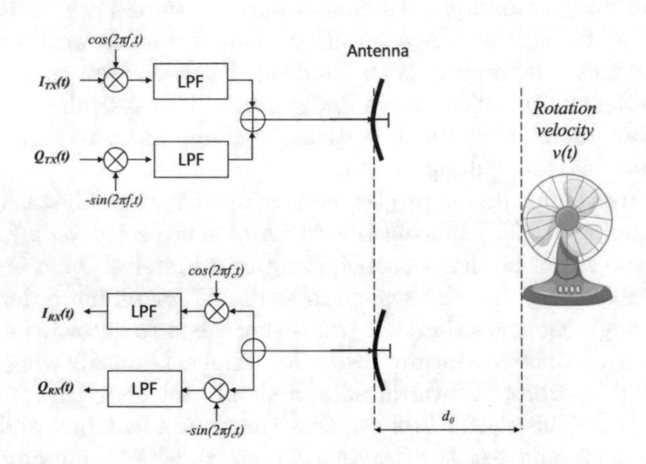

Fig. 1. CW Doppler Radar experiment diagram.

3.2 Modeling of the Rotating Structure

The mathematical analysis of the received signal in Doppler Radars encountering a rotating structure with blade length L involves complex modeling due to the requirement of azimuth and pitch angles of the incident wave. However, in the case of a fixed rotating structure where the Radar is aligned with the center, the analysis can be simplified. In this scenario, at time $t = 0$, a scattering point P_0 positioned at an initial rotation angle ϕ_0 begins rotating counterclockwise with an angular velocity of Ω. The movement is depicted in Fig. 2.

At a given time t, the point P undergoes rotation and is denoted as $P(t)$, with the rotation angle represented as $\phi_t = \Omega t + \phi_0$. The distance from the Radar to the point P can be expressed as:

$$R_p(t) = \sqrt{(R_0 + x_t)^2 + y_t^2} \tag{1}$$

$$R_p(t) = \sqrt{(R_0 + L\cos\phi_t)^2 + (L\sin\phi_t)^2} \tag{2}$$

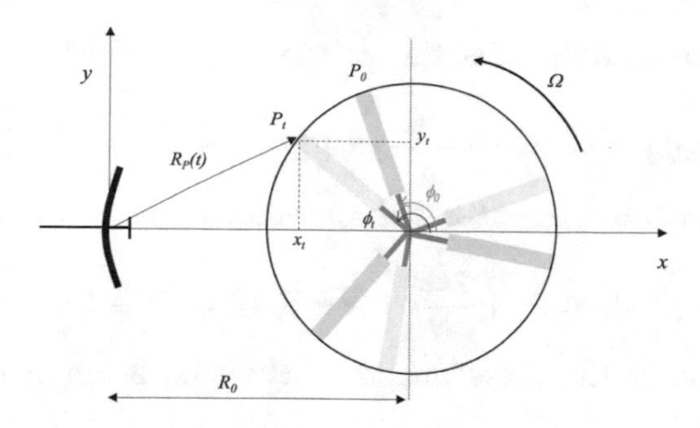

Fig. 2. Rotating structure diagram.

$$R_p(t) = \sqrt{R_0^2 + L^2 + 2R_0L\cos(\phi_t)} \tag{3}$$

$$R_p(t) = \sqrt{R_0^2 + L^2 + 2R_0L\cos(\Omega t + \phi_0)} \tag{4}$$

In the far field distance can be reduced to:

$$R_p(t) = R_0 + L\cos(\Omega t + \phi_0) \tag{5}$$

According to the Radar theory, the echo of a scattering is attenuated and delayed. The is delayed is represented by:

$$\tau(t) = \frac{2d_0}{c} + \frac{2v(t)}{c} \tag{6}$$

where d_0 is the distance between the Radar and the target and $v(t)$ is the velocity of the target. However, because it is a static infrastructure then $v(t)$ is 0. Therefore for this specific case the echo returned is:

$$s(t) = A_r \exp\{-j2\pi f_{TX}(t - \tau(t))\} \tag{7}$$

where A_r is the amplitude of the received signal and f_{TX} is the frequency of transmission. Replacing

$$s(t) = A_r \exp\{-j2\pi f_{TX}(t - \frac{2R_p(t)}{c})\} \tag{8}$$

$$s(t) = A_r \exp\{-j2\pi f_{TX}t + j\frac{4\pi R_p(t)}{\lambda}\} \tag{9}$$

$$s(t) = A_r \exp\{-j2\pi f_{TX}t + j\frac{4\pi[R_0 + L\cos(\Omega t + \phi_0)]}{\lambda}\} \tag{10}$$

Thus, the base-band signal at the scattering point P_t is:

$$s_p(t) = A_r \exp\{j\frac{4\pi R_0}{\lambda}\} \exp\{j\frac{4\pi L \cos(\Omega t + \phi_0)}{\lambda}\} \tag{11}$$

For a fan with N blades, the initial rotational angle of each blade is:

$$\phi_k = \phi_0 + \frac{2k\pi}{N} \qquad k = 0, 1, \ldots, N-1 \tag{12}$$

The expression for the phase function of the echo at the tip of the blade k is as follows:

$$\phi_k(t) = \frac{4\pi L}{\lambda} \cos(\Omega t + \phi_0 + \frac{2\pi k}{N}) \tag{13}$$

As it is know the frequency features of the blade are represented by the Doppler frequency shift is the derivative of the phase:

$$f_{D,k}(t) = \frac{1}{2\pi} \frac{d\phi_k(t)}{dt} \tag{14}$$

Then resolving Eq. 14 taking from Eq. 13

$$f_{D,k}(t) = -\frac{2\Omega L}{\lambda} \sin(\Omega t + \phi_0 + \frac{2\pi k}{N}) \tag{15}$$

From Equation it can be deducted that the rotational speed modulates the instantaneous Doppler frequency as a sinusoidal curve [17]. Therefore, the maximum Doppler is when

$$f_{Dmax} = \frac{2\Omega L}{\lambda} = \frac{2v}{\lambda} \tag{16}$$

Figure 3 illustrates the simulation of micro-Doppler characteristics for rotating 5 blades at a speed of 900 RPM, which simulates real maximum velocity $v3$ of the fan.

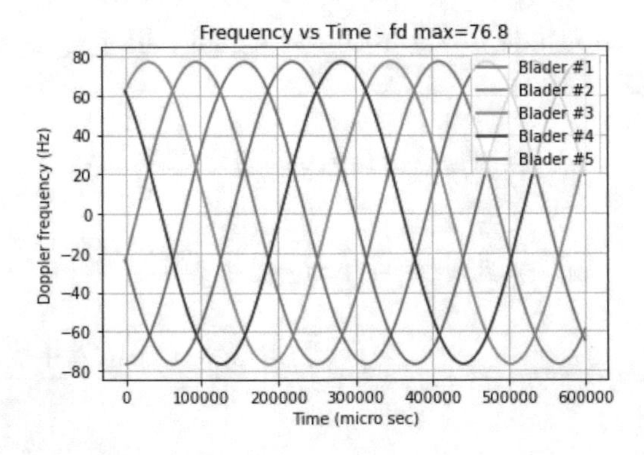

Fig. 3. The micro-Doppler features of a fan with 5 blades.

The radial velocity (v_r) of a target can be determined by measuring the frequency shift (fd) [13]. This provides information about the target and its components. Hence, estimating the RPM (rotations per minute) of a rotating structure, like rotating blades, can be achieved by observing the micro-Doppler effect generated by the rotational motion. The relationship between the observed frequency shift and the RPM of the rotating structure can be expressed as:

$$f_d = \frac{RPM_{fan} \times N}{60} \tag{17}$$

where f_d is the frequency shift and N is the number of blades of the fan.

3.3 Antenna

In Radar applications, it is essential to use an antenna that emits radio waves predominantly in a specific direction, allowing for concentrated transmission of energy in that particular direction. For this experiment, a Log Periodic Antenna is chosen because it exhibits a directional radiation pattern, offers high gain ($9 - 10dBi$), and is capable of operating across a wide range of frequencies. The specific antennas employed are the HG72710LP-NF models from L-com, which have a frequency range of 698–960 MHz and 1710–2500 MHz. These antennas feature a horizontal beam width of 78 degrees and a vertical beam width of 56 degrees. Additional specifications can be found in [18].

3.4 Digital Receiver

The ADALM-Pluto from Analog Devices [19] is a digital receiver that offers a wide range of features and capabilities that make it suitable for Radar signal processing. Its frequency range, high dyanamic range and capability of working in full-dupplex and its programmable nature allows for flexible configuration and adaptation to various Radar scenarios.

Furthermore, the inclusion of ADALM Pluto in the GNU-Radio Companion (GRC) software libraries simplifies the utilization of its powerful processing capabilities [20]. With its USB 2.0 data transfer capability, it becomes an affordable device that can be employed in conjunction with low-cost embedded systems like Raspberry Pi. These characteristics collectively make ADALM Pluto an attractive choice for Radar applications, particularly in educational settings.

4 Results

Figure 4 illustrates the Radar prototype system. The experiments utilized a fan as the target object with three distinct velocities: v1, v2, and v3. In this context, v1 represents the lowest velocity, v2 corresponds to the middle, and v3 signifies the maximum velocity. To enhance reflectivity, the blades of the fan were coated with aluminum. Additionally, the configuration parameters detailed in Table 1.

One crucial benefit of SDR technology is the ability to perform the mixing process for signal conversion to baseband digitally. Additionally, the conjugate multiplication can be carried out during data acquisition, providing a significant advantage over its analog counterpart. Consequently, this expedites the development time of RF prototypes.

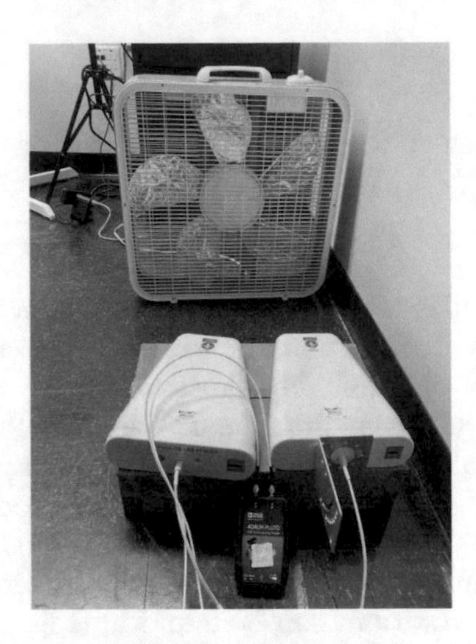

Fig. 4. CW Radar system installed.

Table 1. CW Radar Configuration.

Parameter	Value
Frequency	2.4 GHz
Tx Gain	0 dB
Rx Gain	60 dB
Sample Rate	2 MHz
Data Type and Resolution	I/Q 64bits
Low Pass Filter cut-off frequency	500 Hz

The outcomes derived from the Radar experiment are depicted in Fig. 5, showcasing a spectrogram. A spectrogram visually represents the spectrum of frequencies in a sound or signal, illustrating how they change over time [21].

Initially, the velocities may not be clearly visible in the Spectrogram. However, after applying a low-pass rectangular window filter, the Spectrogram in Fig. 6 reveals three distinct velocities: $v3$, $v2$, and $v1$. The corresponding spectra and RPM measurements for these velocities are presented in Figs. 7, 8, and 9, respectively.

By extracting a specific time slice from the data, it becomes possible to obtain a more precise estimate of the Doppler frequency, which can then be compared

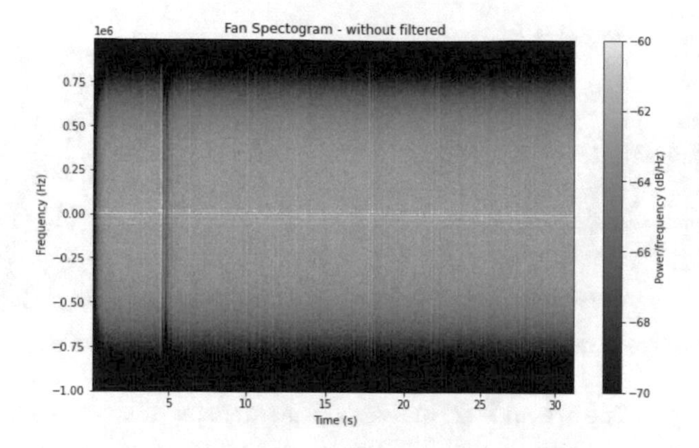

Fig. 5. Raw data Spectrogram of the rotating structure experiment.

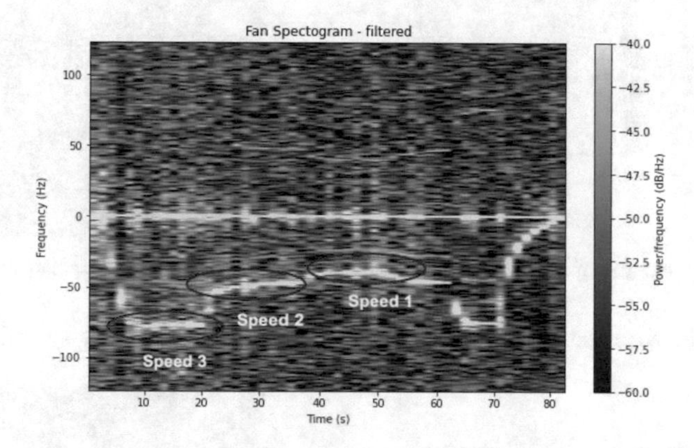

Fig. 6. Filtered data Spectrogram of the rotating structure experiment.

with the measurements obtained from the tachometer. The comparison between the two values can be performed using Eq. 17, allowing for a quantitative assessment of their agreement.

The results demonstrate a strong correlation between the two measurements, as indicated by the low error values ranging between 1–3%. Table 2 provides a summary of the three measurements obtained.

Table 2. Comparison of results from Radar and Tachometer's measurement.

Tachometer (RPM)	Radar (Hz)	Error (%)	Std Dev	Power (dB/Hz)
900.4 (v3)	76.29–77.25	1.57–2.92	4.92	-3dB
611.5 (v2)	49.59–50.54	0.74–2.68	3.88	-3dB
530.8 (v1)	44.18–45.57	0.94–3.08	3.44	-3dB

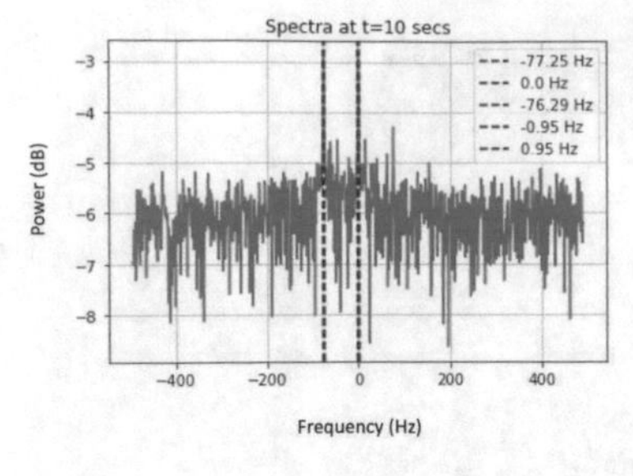

(a) Spectrum of the fan - v3.

(b) RPM of Fan's speed - v3.

Fig. 7. Spectra and tachometer's measurement of velocity - v3.

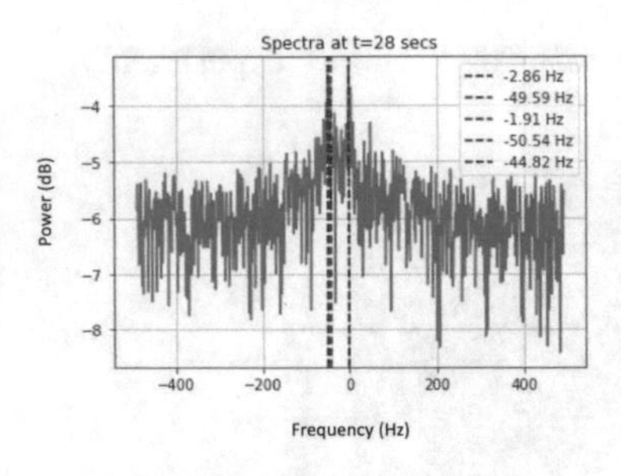

(a) Spectrum of the fan - v2.

(b) RPM of Fan's speed - v2.

Fig. 8. Spectra and tachometer's measurement of velocity - v2.

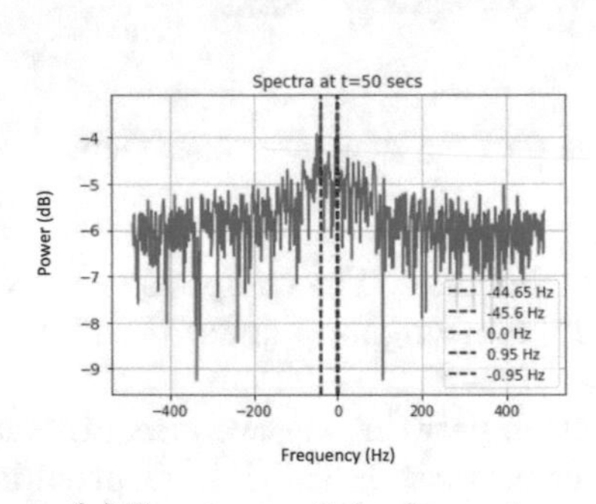

(a) Spectrum of the fan - v1.

(b) RPM of Fan's speed - v1.

Fig. 9. Spectra and tachometer's measurement of velocity - v1.

5 Conclusions and Future Work

In summary, this article introduces a prototype of a Remote Laboratory dedicated to Radar experiments, featuring a configurable Continuous Wave (CW) Radar system.

The results reveal a significant correlation between Radar and tachometer measurements, with a margin of error ranging between 1–3%, signifying a reasonably accurate estimation.

The use of affordable components, such as the Adalm Pluto (priced at $230) and Raspberry Pi 4B (priced at $120), enhances the accessibility and value of Radar experiments in Engineering Education.

The integration of this prototype into a remote SDR laboratory empowers students to conduct Radar experiments from any location, thereby augmenting the flexibility and accessibility of the learning process.

Moreover, there exists potential for expanding the scope of Radar experiments, including the exploration of CW Radar for Multi-Target estimation or the investigation of Frequency-Modulated Continuous Wave (FMCW) Radar for estimating both range and Doppler parameters.

As part of future plans, this remote laboratory can be implemented in an ultra-concurrent mode, a type of remote laboratory allowing students to interact with pre-recorded real data. This mode proves advantageous when facing infrastructure challenges, making it impractical to provide a dedicated remote laboratory station for each student.

Acknowledgements. This work was supported by the National Science Foundation's Division Of Undergraduate Education under Grant 2141798.

References

1. Hussein, R., Wilson, D.: Remote versus in-hand hardware laboratory in digital circuits courses. In: 2021 ASEE Virtual Annual Conference Content Access (2021)
2. Hussein, R., Guo, M., Amarante, P., RodriguezGil, L., Orduña, P.: Digital twinning and remote engineering for immersive embedded systems education. In: Frontiers in Education (FIE) Conference. IEEE, USA (2023)
3. Grout, I.: Supporting access to stem subjects in higher education for students with disabilities using remote laboratories. In: Proceedings of 2015 12th International Conference on Remote Engineering and Virtual Instrumentation (REV), pp. 7–13. IEEE (2015)
4. Inonan, M., Paul, A., May, D., Hussein, R.: RHLab: digital inequalities and equitable access in remote laboratories. In: 2023 ASEE Annual Conference & Exposition (2023)
5. Orduña, P.: Transitive and Scalable Federation Model for Remote Laboratories. Ph.D. thesis, Universidad de Deusto, Bilbao, Spain, May 2013. https://morelab.deusto.es/people/members/pablo-orduna/phd_dissertation/
6. Gustavsson, I.: Remote laboratory experiments in electrical engineering education. In: Proceedings of the Fourth IEEE International Caracas Conference on Devices, Circuits and Systems (Cat. No. 02TH8611), pp. I025–I025. IEEE (2002)
7. Hussein, R., et al.: Remote hub lab - RHL: broadly accessible technologies for education and telehealth. In: 20th Annual International Conference on Remote Engineering and Virtual Instrumentation REV 2023 (2023)
8. Inonan, M., Chap, B., Orduña, P., Hussein, R., Arabshahi, P.: RHLab scalable software defined radio (SDR) remote laboratory. In: 20th Annual International Conference on Remote Engineering and Virtual Instrumentation REV 2023 (2023)

9. Inonan, M., Hussein, R.: Melody: a platform-agnostic model for building andevaluating remote labs of software-defined radio technology. IEEE Access **11**, 127550–127566 (2023). https://doi.org/10.1109/ACCESS.2023.3331399
10. Orduña, P., Irurzun, J., Rodriguez-Gil, L., Garcia-Zubia, J., Gazzola, F., López-de Ipiña, D.: Adding new features to new and existing remote experiments through their integration in weblab-deusto. Int. J. Online Eng. (iJOE) **7**(S2), 33 (2011)
11. Inonan, M., Orduña, P., Hussein, R.: Adapting a remote sdr lab to analyze digital inequalities in radiofrequency education in latin america. Revista Innovaciones Educativas (2023). in press
12. Orduna, P., et al.: Generic integration of remote laboratories in learning and content management systems through federation protocols. In: 2013 IEEE Frontiers in Education Conference (FIE), pp. 1372–1378. IEEE (2013)
13. Skolnik, M.I.: Radar Handbook. McGraw-Hill Education, New York (2008)
14. Palmer, R., et al.: Weather radar education at the university of Oklahoma: an integrated interdisciplinary approach. Bull. Am. Meteor. Soc. **90**(9), 1277–1282 (2009)
15. Kolodziej, K.E., et al.: Build-a-radar self-paced massive open online course (MOOC). In: 2019 IEEE Radar Conference (RadarConf), pp. 1–5. IEEE (2019)
16. Fu, X.: Radar and Radar Imaging of Cooperative and Uncooperative Modulated Targets. Ph.D. thesis (2019)
17. Kumawat, H.C., Bazil Raj, A.: Extraction of doppler signature of micro-to-macro rotations/motions using continuous wave radar-assisted measurement system. IET Sci. Meas. Technol. **14**(7), 772–785 (2020)
18. Wireless antenna 698-960 mhz/1710-2700 mhz 10 dbi log periodic antenna n-female. https://www.l-com.com/wireless-antenna-698-960-mhz-1710-2700-mhz-10-dbi-log-periodic-antenna-n-female. Accessed 14 May 2023
19. Analog Devices: Adalm-pluto [analog devices wiki] (2021). https://wiki.analog.com/university/tools/pluto
20. Guided Tutorial GRC. https://wiki.gnuradio.org/index.php. Accessed 05 June 2023
21. Fulop, S.A., Fulop, S.A.: The fourier power spectrum and spectrogram. Speech Spectr. Anal. 69–106 (2011)

Preliminary Evaluation of RHL-RELIA Post-development

Marcos Inoñan[1(✉)], Pedro Amarante[1], Bruno Diaz[1], Nattapon Oonlamom[1], Kiana Peterson[1], Pablo Orduña[2], Candido Aramburu Mayoz[3], and Rania Hussein[1]

[1] University of Washington, Seattle, WA 98195, USA
{minonan,pedroa2,bidt,genieoon,ksimone,rhussein}@uw.edu
[2] LabsLand, San Francisco, CA 94114, USA
pablo@labsland.com
[3] Universidad Pública de Navarra, Campus Arrosadia, 31006 Pamplona, Spain
candido@unavarra.es

Abstract. Leveraging the Remote Hub Lab (RHL/RHLab) research group's work on democratizing educational access, RHLab, in partnership with LabsLand, has developed the Remote Engineering Lab for Inclusive Access (RELIA/RHL-RELIA). This initiative introduces an educational tool specifically designed for communication courses, aiming to promote learning experiences in technical education. RHL-RELIA, built upon Software Defined Radio (SDR) technology, offers a novel approach to manipulating radio-frequency (RF) hardware through programming, aiming to promote the learning experience in wireless communication. This study presents an initial evaluation of RHL-RELIA post its development, focusing on its usability and educational impact, particularly for students with limited experience in RF technology. The assessment involved three engineering students testing the system, providing feedback on its interface design and effectiveness in facilitating remote experimentation. The preliminary findings suggest that RHL-RELIA is successful in improving the efficiency of completing assignments and making complex wireless concepts more accessible. Future work will include a comprehensive evaluation of RHL-RELIA within a classroom environment, involving a more diverse student population to assess its adaptability and overall educational impact. Future studies will employ mixed methods, including surveys and interviews, to gain a deeper understanding of RHL-RELIA's role in enhancing engineering education.

Keywords: GNU Radio · Software Defined Radio · remote laboratory · embedded systems · wireless communication

1 Introduction

Remote laboratories are gaining popularity in higher education due to their flexibility, allowing students to access them anytime and from anywhere. Furthermore, the challenges associated with ensuring reliable access to hardware in

M. E. Auer et al. (Eds.): STE 2024, LNNS 1028, pp. 169–179, 2024.
https://doi.org/10.1007/978-3-031-61905-2_17

traditional hands-on labs make remote labs an attractive alternative for educational institutions [1–5]. At the same time, remote labs contribute to equitable access, expanding opportunities for students from lower-income and underrepresented minority backgrounds, especially through community colleges [6–8].

Building on the Remote Hub Lab (RHL/RHLab) [9] research group's efforts in democratizing educational access, the group continues to contribute to remote labs development, harnessing technologies like Software Defined Radio (SDR). With the aim of making advanced engineering education more universally accessible, RHLab, in collaboration with LabsLand [10], introduced the Remote Engineering Lab for Inclusive Access (RELIA) for telecommunication courses that uses SDR technology. RHL-RELIA is characterized by its affordable hardware, use of an open-source framework, and a user-friendly web interface [9,11,12]. These features are central to ensuring that diverse student populations have the opportunity to engage in advanced technical learning.

RHL-RELIA is a remote laboratory that employs SDR devices for wireless signal transmission. The primary advantage of SDR technology is its capability to facilitate various forms of wireless communication (Analog, Digital, Cellular, etc.) using a single hardware platform, which can be modified exclusively through code. ADALM-PLUTO was selected as the SDR for RHL-RELIA due to its cost-effectiveness, robust hardware, and strong community support. Additionally, RHL-RELIA provides students with the convenience of remote access through a standard web browser, eliminating the need for specialized software and reducing barriers related to prior knowledge. In this paper we present a preliminary evaluation of RHL-RELIA through the perspective of a three undergraduate students who embarked on this task as an independent study, possessing no prior knowledge in the topic area.

2 Background

In Engineering courses, having a laboratory section for practical application of learned concepts is a common practice. These labs are traditionally designed to be solved with physical hardware, often requiring significant financial resources to provide materials for all students. However, financial constraints may lead to a shortage of equipment if funds are insufficient. In addressing this challenge, remote laboratories aim to provide a solution while enhancing accessibility for users [13].

SDR educational labs effectively address many of the challenges associated with traditional labs. The SDR community consistently enhances both hardware devices and software tools, emphasizing their relevance and potential impact across various disciplines. The transition from the complexity of setting up SDR devices in the past to the current user-friendly interfaces reflects a notable trend in the ongoing efforts of the SDR community. A prominent example is the open-source graphical user interface, GNU Radio Companion (GRC), which functions as a graphical layer on the standard GNU Radio programming environment implemented in Python [14]. This interface offers an intuitive user experience comparable to Matlab's Simulink [12]. In the realm of RF education based on GRC, there are illustrative developments. For instance, Emona TIMS project offers a comprehensive package that includes hardware, software solutions, and modular experiments [15]. However, its reliance on specific EMONA hardware may limit scalability for providing flexible laboratory access to a larger student body. Another notable project is the FORGE initiative, employing 16 SDR devices for heightened flexibility and scalability [16]. Nonetheless, none of these labs currently allow students to access the hardware remotely [17], establishing the RHL-RELIA project as a pioneering initiative in this aspect. Evaluation of remote labs in the literature has included many efforts and different methodologies [1,13,18–21]. These diverse evaluation methods will collectively guide the interpretation of initial results and subsequent findings from RHL-RELIA.

In the field of engineering, especially in ECE, communication emerges as a noteworthy subject for the application of remote labs. Given the pervasive impact of modern communication systems in our daily lives, the effective instruction of communication to engineering students becomes crucial to meet industry demands [22]. The efficacy of Software-Defined Radio (SDR) in education has witnessed a significant surge in affordability and performance since the early 2000s [23]. For the successful integration of SDR into the design of communication systems, certain prerequisites must be met, including affordable SDR hardware, availability of SDR software, compatibility between SDR hardware and robust technical computing software, and established SDR-based engineering undergraduate curricula [24]. The successful realization of these crucial points, facilitated by advancements in SDR technology and the utilization of tools such as the Universal Software Radio Platform (USRP) hardware with the GNU Radio software framework in undergraduate courses, has proven effective for various radio engineering assignments [22]. The ongoing evolution of SDR tools remains pivotal in shaping the future trajectory of SDR in engineering education.

The preliminary study we are presenting in this paper is anchored in the perspective of students, examining their experiences as they undertake assignments aligned with the traditional Communication curriculum and reference conventional lab materials. In this study the RHL-RELIA lab is assessed based on its technological attributes and the perceived educational benefits from students. It is noteworthy that this study represents an initial exploration, and future publications are planned to conduct a more comprehensive investigation. Subsequent studies will employ mixed methods, pre and post surveys, and interviews for a larger student cohort.

3 RHL-RELIA Remote Laboratory

RHL-RELIA comprises components that enable users to remotely access and configure any SDR station located within an educational institution. As an implementation of the MELODY model [17], RHL-RELIA serves as an agnostic technological framework, offering design considerations for educational remote labs that control SDR devices. The block diagram illustrating the structure of RHL-RELIA is depicted in Fig. 1. Key features of RHL-RELIA include:

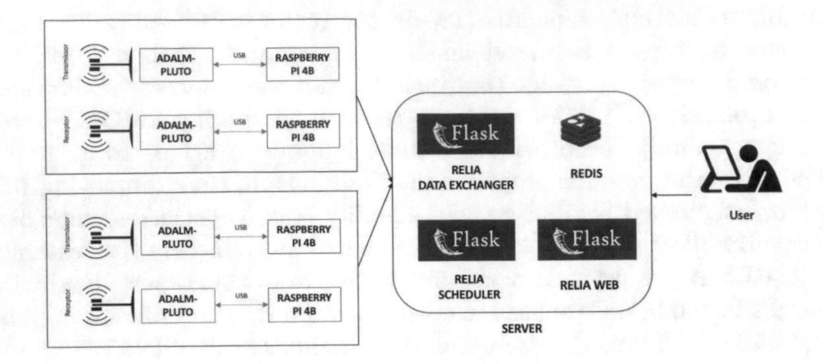

Fig. 1. RHL-RELIA block diagram

- Open source: The development code is accessible for installation by any collaborator or developer.
- Scalability: Due to its architecture and cost-effectiveness, leveraging the ADALM-PLUTO priced at $400 per unit-significantly lower than the USRP devices commonly used in literature-RHL-RELIA demonstrates flexibility in scaling the number of remote units. This scalability not only enhances accessibility but also allows for the accommodation of a greater number of units.
- Web access: RHL-RELIA can be accessed through any web browser, streamlining its usability, reducing the learning curve, and ensuring compatibility with various operating systems.

The features of RHL-RELIA suggest its potential as a useful tool for educational institutions requiring remote access and configuration of SDR stations. With its technology-agnostic framework and notable capabilities, RHL-RELIA could serve as a foundational resource for the integration into communication courses. This perspective is based on its current functionalities and observed benefits, positioning it as a candidate for consideration in future educational applications.

4 Assessment Methodology

Examining both the operational intricacies of the system and its educational implications, the methodology in this study aimed to uncover essential insights. Feedback for the evaluation was collected from three undergraduate students who served as early testers, lacking prior experience in RF topics. From an accessibility standpoint, our decision was to allow students with or without education in communication theory. We aimed to ensure that students testing the lab have a comparable level to those who will eventually use it, thereby democratizing access to this lab.

For the initial assignment, students acquaint themselves with the lab environment. They follow guidance to install the necessary software and engage in two exercises. The first exercise involves verifying correct system installation through simulation, followed by remote transmission and reception using Amplitude Modulation (AM) with the ADALM-PLUTO. Moving on to the second assignment, students undertake two experiments. The initial experiment focuses on calibrating the ADALM-PLUTO by adjusting the frequency values of the transmitter or receiver. The second experiment involves executing digital transmission using Binary Phase Shift Keying (BPSK). These assignments necessitate students to quantitatively analyze signal quality and transmission efficiency, offering a practical complement to theoretical coursework. To ensure a comprehensive comparison, the testers completed both assignments using RHL-RELIA, accessing ADALM-PLUTO remotely in one scenario and physically accessing ADALM-PLUTO in another.

Regarding the operational intricacies, we gathered feedback from students on the following:

- User interface: The user interface ensures that students can easily navigate and interact with the lab platform, enhancing the efficiency of conducting experiments.
- Technology dependence: Understanding the degree of dependence on technology helps assess the accessibility of the remote lab. Additionally, compatibility with various devices and operating systems is crucial.
- Lab availability: Lab availability refers to the accessibility of the remote lab at different times and from various locations. Moreover, assessing lab availability involves understanding how resources, such as equipment and simulation tools, are managed.

Regarding educational implications, students provided their impressions on the following:

- Time Efficiency: Explores the time-related aspects of using RHL-RELIA, focusing on efficiency and the learning curve. Understanding how long students take to complete assignments and become comfortable with the system is crucial in evaluating its practicality in an educational setting.
- Learning Independence: Assesses the level of autonomy students experienced while working with RHL-RELIA. Measuring the degree of independent work versus the need for external help provides insight into the user-friendliness and intuitiveness of the lab system.
- Interest in subject matter and overall satisfaction: Focuses on assessing students' interest in wireless communication, a field in which they were initially novices, following their use of RHL-RELIA. Additionally, it seeks to gauge their overall satisfaction with the learning experience provided by the lab.

5 Results

In this section, we present the findings from our preliminary evaluation of the RHL-RELIA as experienced by three engineering students. These students, who were novices in the field of wireless communication prior to using RHL-RELIA, provided insights into the lab's usability, access, and its educational implications.

User Interface. The user interface (UI) of RHL-RELIA presents noticeable differences when compared to traditional lab systems. RHL-RELIA is built with web-specific libraries, providing a distinct user experience from the Graphic User Interface (GUI) used with ADALM-PLUTO, which is derived from the broader SDR community. Illustrations of both interfaces are showcased in Figs. 2 and 3. Our findings indicate diverse preferences among students for the RHL-RELIA's UI versus that of conventional labs. Notably, several students highlighted a unique feature of RHL-RELIA: the system allows a 20-second window of access for each student. This functionality is particularly beneficial in situations where the demand for resources exceeds the available units, thereby demonstrating RHL-RELIA's ability to cater to a larger group of users efficiently. Table 1 summarizes students' answers.

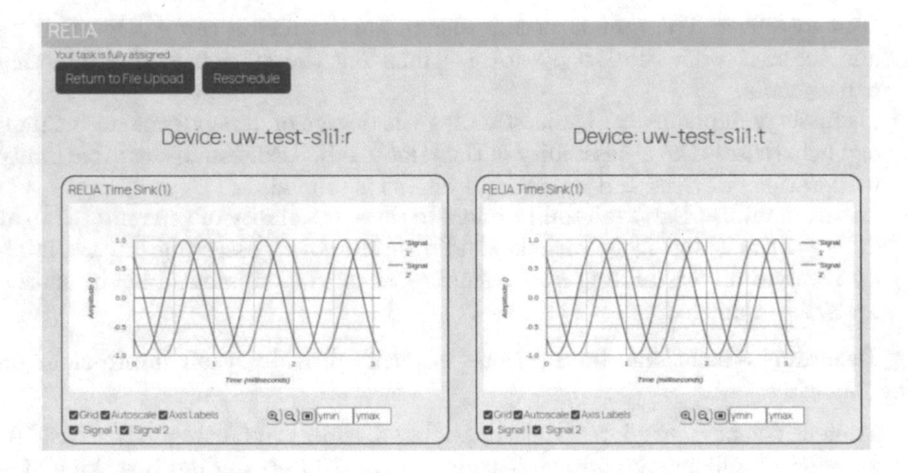

Fig. 2. RHL-RELIA web user interface

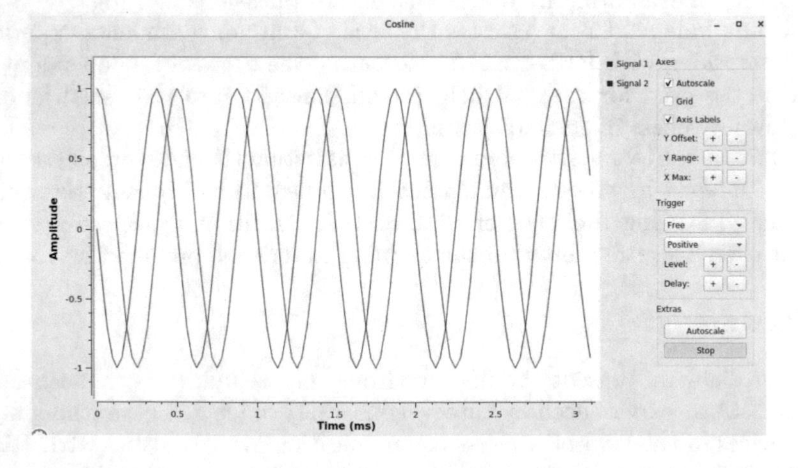

Fig. 3. GRC user interface

Table 1. Students' Impression Over the User Interface

	RHL-RELIA	Traditional
Student 1	*"The interface is friendly, but it would be more effective with prolonged use for manipulating signal graphs".*	*"The configuration process was complex, and it doesn't function on certain operating systems".*
Student 2	*"A timer of how long it is used will be helpful".*	*"Access to a more robust user interface for signal analysis and readability".*
Student 3	*"It simplifies the process but is difficult handle long period results".*	*"I can use it for as long as I need".*

Technology Dependence. We analyzed the technological components and tools necessary for configuring both RHL-RELIA and ADALM-PLUTO in a traditional lab setting. The emphasis is on familiarizing students with the requisite technology for effective engagement in these lab environments. RHL-RELIA has been designed to minimize the need for specialized components, thus prioritizing equitable access [19]. This design contrasts traditional labs, which often demand the installation of specific software on the user's computer. By reducing these technological barriers, RHL-RELIA aims to offer a more streamlined and accessible user experience, catering to a broader student base. The outcomes of the students' responses are depicted in Table 2.

Table 2. Students' Impression on the Technology Dependence

	RHL-RELIA	Traditional
Student 1	*"Requires only a device compatible with GNURadio and later any device with internet access can be used".*	*"A computer that has USB ports is needed".*
Student 2	*"Operates through the web, eliminating the need for expensive devices and dealing with varying connections to different computers. It is straightforward and accessible on all devices".*	*"GNURadio poses a challenge in terms of usage across different devices. Traditional labs, with actual hardware, add an extra layer of work".*
Student 3	*"Requires only a device compatible with GNURadio and internet access".*	*"A device with USB support is needed, and modern devices often lack USB ports, necessitating extensions".*

Availability and Accessibility. Students shared their insights regarding the accessibility of both RHL-RELIA and traditional lab setups. RHL-RELIA stands out for its 24/7 availability, albeit with a limited duration for each user to accommodate more students and reduce waiting times. In contrast, traditional labs provide complete and unrestricted access for individual students, as they have direct and personal access to the hardware. The students' responses to this aspect are presented in Table 3.

Table 3. Students' Impression Over the Availability and Accessibility

	RHL-RELIA	Traditional
Student 1	*"Accessible from anywhere once configured".*	*"Allows playing, pausing, and taking one's time but requires carrying hardware at all times".*
Student 2	*"Accessible from any device, ability to save and load previous files".*	*"Lack of features like saving files, requires connection to specific devices".*
Student 3	*"No need to carry devices, access anytime but 30-second limit can be pressuring for analysis".*	*"Allows taking time, despite the burden of carrying devices".*

Educational Implications: We gauged the potential educational impact of the lab through questions on time efficiency, learning independence, interest in subject matter and overall satisfaction. Table 4 summarizes student's input. Together, these categories offer a holistic view of RHL-RELIA's effectiveness in not only educating students but also in fostering an engaging and autonomous learning environment, thereby enhancing their interest and satisfaction in the subject matter. This multi-faceted approach underscores the lab's potential in revolutionizing educational experiences in engineering disciplines.

Table 4. Students' Educational implications

Question	Student 1	Student 2	Student 3
How long did it take you to complete the lab assignment using RHL-RELIA?	2 h	1 h	4 h
How much time did you spend learning to use the RHL-RELIA system before you felt comfortable with it?	20 min	1 h	2 h
What percentage of the lab were you able to complete independently, without external help?	75%	100%	80%
Has your experience with RHL-RELIA increased your interest in the field of wireless communication?	Somewhat	Yes	Yes
Overall, how satisfied are you with your learning experience using the RHL-RELIA lab?	Overall satisfied	8 [on a scale 1–10]	Very satisfied
Would you recommend the RHL-RELIA lab to other students learning about wireless communication?	Yes	Yes	Yes

The collective feedback from students shows a promising picture of the lab's efficacy. Completion times for lab assignments, ranging from 1 to 4 h, alongside the learning curve of 20 min to 2 h, indicate a flexible system that accommodates varying student abilities and learning speeds. The high degree of independence in lab completion, with 75% to 100% of tasks done without external help, reflects RHL-RELIA's success in fostering self-sufficiency and intuitive learning. Additionally, the increase in students' interest in wireless communication and the high levels of overall satisfaction, with scores as high as 8 out of 10 and strong endorsements for recommending the lab to peers, underscore the system's ability to engage students effectively and enhance their educational experience. These findings, albeit from a limited sample, suggest that RHL-RELIA is a valuable tool, capable of not only imparting technical knowledge but also stimulating interest and ensuring student satisfaction in the learning process.

While the scope of our data is limited in quantity, the insights gathered from the responses of the three students provide a meaningful glimpse into the effectiveness of the RHL-RELIA. Despite the small sample size, the qualitative nature of this preliminary feedback is valuable, offering a foundational understanding of how RHL-RELIA impacts student interest in wireless communication and their satisfaction with the learning experience. These initial findings, though not extensive, are instrumental in guiding future research directions and in making incremental improvements to the RHL-RELIA system. They serve as an encouraging indication of the potential benefits and educational contributions of RHL-RELIA, setting the stage for more comprehensive studies in the future.

6 Conclusions and Future Work

This initial assessment of the RHL-RELIA remote lab system has provided promising insights, particularly in its potential to enhance engineering education methodologies. The system underwent testing by three engineering students with no prior experience in software-defined radio. This deliberate choice of novice testers aimed to collect preliminary feedback before the system undergoes a more comprehensive evaluation in a classroom setting. The observed interface improvements and technology-agnostic features highlight RHL-RELIA's potential to make complex wireless communication concepts more accessible. Despite being tested in a lab setting and not yet in a classroom environment, the feedback from these students indicates a notable improvement in the efficiency and ease of completing assignments.

Building upon these preliminary findings, the next phase for RHL-RELIA entails a thorough assessment within a classroom environment. Engaging a broader and more diverse student population in this context will facilitate a more comprehensive evaluation of the system's adaptability and educational impact. This pivotal phase will specifically address the assessment of the system's user-friendliness and scalability for larger student groups, generating valuable insights for further refinements. Subsequent publications will delve into a more extensive study, employing mixed methods, pre and post surveys, and interviews conducted for a larger student cohort, providing a deeper understanding of RHL-RELIA's impact on engineering education.

Acknowledgements. This work was supported by the National Science Foundation's Division of Undergraduate Education under Grant #2141798.

References

1. Garcia-Zubia, J., et al.: Empirical analysis of the use of the VISIR remote lab in teaching analog electronics. IEEE Trans. Educ. **60**(2), 149–156 (2016)
2. Li, S., Wang, H., Rodriguez-Gil, L., Orduña, P., Hussein, R.: FPGA meets breadboard: integrating a virtual breadboard with real FPGA boards for remote access in digital design courses. In: Auer, M.E., Bhimavaram, K.R., Yue, X.-G. (eds.) REV 2021. LNNS, vol. 298, pp. 144–151. Springer, Cham (2022). https://doi.org/10.1007/978-3-030-82529-4_15
3. Orduña, P.: Transitive and Scalable Federation Model for Remote Laboratories. Ph.D. thesis, Universidad de Deusto, Bilbao, Spain, May 2013. https://morelab.deusto.es/people/members/pablo-orduna/phd_dissertation/
4. Orduña, P., Irurzun, J., Rodriguez-Gil, L., Garcia-Zubia, J., Gazzola, F., López-de Ipiña, D.: Adding new features to new and existing remote experiments through their integration in WebLab-Deusto. Int. J. Online Eng. (iJOE) **7**(S2), 33 (2011)
5. Bose, R.: Virtual labs project: a paradigm shift in internet-based remote experimentation. IEEE Access **1**, 718–725 (2013)
6. Atienza, F., Hussein, R.: Student perspectives on remote hardware labs and equitable access in a post-pandemic era. In: 2022 IEEE Frontiers in Education Conference (FIE), pp. 1–8. IEEE (2022)
7. Hussein, R., Maloney, R.C., Rodriguez-Gil, L., Beroz, J.A., Orduna, P.: RHL-BEADLE: bringing equitable access to digital logic design in engineering education. In: 2023 ASEE Annual Conference and Exposition (2023)
8. Inonan, M., Orduña, P., Hussein, R.: Adapting a remote SDR lab to analyze digital inequalities in radiofrequency education in Latin America. Revista Innovaciones Educativas (2023, in press)
9. Hussein, R., et al.: Remote Hub Lab - RHL: broadly accessible technologies for education and telehealth. In: Auer, M.E., Langmann, R., Tsiatsos, T. (eds.) REV 2023. LNNS, vol. 763, pp. 73–85. Springer, Cham (2023). https://doi.org/10.1007/978-3-031-42467-0_7
10. Orduña, P., Rodriguez-Gil, L., Garcia-Zubia, J., Angulo, I., Hernandez, U., Azcuenaga, E.: LabsLand: a sharing economy platform to promote educational remote laboratories maintainability, sustainability and adoption. In: 2016 IEEE Frontiers in Education Conference (FIE), pp. 1–6. IEEE (2016)
11. Zhang, Z., Inoñan, M., Orduña, P., Hussein, R., Arabshahi, P.: RHLab interoperable software-defined radio (SDR) remote laboratory. In: Proceedings of the 21st International Conference on Smart Technologies & Education (STE), Helsinki, Finland, 6–8 March 2024. Paper accepted for presentation (2024)
12. Inonan, M., Chap, B., Orduña, P., Hussein, R., Arabshahi, P.: RHLab scalable software defined radio (SDR) remote laboratory. In: Auer, M.E., Langmann, R., Tsiatsos, T. (eds.) REV 2023. LNNS, vol. 763, pp. 237–248. Springer, Cham (2023). https://doi.org/10.1007/978-3-031-42467-0_22

13. Hussein, R., Wilson, D.: Remote versus in-hand hardware laboratory in digital circuits courses. In: 2021 ASEE Virtual Annual Conference Content Access (2021)
14. Guided Tutorial GRC. https://wiki.gnuradio.org/index.php. Accessed 05 June 2023
15. May, D., Morkos, B., Jackson, A., Hunsu, N.J., Ingalls, A., Beyette, F.: Rapid transition of traditionally hands-on labs to online instruction in engineering courses. Eur. J. Eng. Educ. **48**(5), 842–860 (2023)
16. Abreu, P., Barbosa, M.R., Lopes, A.M.: Robotics virtual lab based on off-line robot programming software. In: 2013 2nd Experiment@ International Conference (exp.at 2013), pp. 109–113. IEEE (2013)
17. Inonan, M., Hussein, R.: Melody: a platform-agnostic model for building and evaluating remote labs of software-defined radio technology. IEEE Access **11**, 127550–127566 (2023). https://doi.org/10.1109/ACCESS.2023.3331399
18. Paul, A., Inoñan, M., Hussein, R., May, D.: Exploring diversity, equity, and inclusion in remote laboratories. In: 2023 ASEE Annual Conference and Exposition (2023)
19. Inonan, M., Paul, A., May, D., Hussein, R.: RHLab: digital inequalities and equitable access in remote laboratories. In: 2023 ASEE Annual Conference and Exposition (2023)
20. Hussein, R., Guo, M., Amarante, P., RodriguezGil, L., Orduña, P.: Digital twinning and remote engineering for immersive embedded systems education. In: Frontiers in Education (FIE) Conference, USA. IEEE (2023)
21. Gustavsson, I., Zackrisson, J., Håkansson, L., Claesson, I., Lagö, T.: The VISIR project–an open source software initiative for distributed online laboratories. In: Proceedings of the REV 2007 Conference, Porto, Portugal (2007)
22. Petrova, M., Achtzehn, A., Mähönen, P.: System-oriented communications engineering curriculum: teaching design concepts with SDR platforms. IEEE Commun. Mag. **52**(5), 202–209 (2014)
23. Tuttlebee, W.H.: Software Defined Radio: Enabling Technologies. Wiley, New York (2002)
24. Wyglinski, A.M., Orofino, D.P., Ettus, M.N., Rondeau, T.W.: Revolutionizing software defined radio: case studies in hardware, software, and education. IEEE Commun. Mag. **54**(1), 68–75 (2016)

RHLab: Towards Implementing a Partial Reconfigurable SDR Remote Lab

Zhiyun Zhang[1](✉), Marcos Inoñan[1], Pablo Orduña[2], and Rania Hussein[1]

[1] University of Washington, Seattle, WA 98195, USA
{zzyzzy42,minonan,rhussein}@uw.edu
[2] LabsLand, San Francisco, CA 94114, USA
pablo@labsland.com

Abstract. Software-Defined Radio (SDR) remote labs permit students to experiment with real wireless communication, designing Radio Frequency (RF) systems with minimal code adjustments. This feature allows them to create RF prototypes remotely in a fast way allowing them to complement their theory of communication classes. While SDR hardware suffices for most basic applications, some demand extensive Signal Processing stages that surpass the capabilities of standard SDR equipment. SDR devices are controlled by reprogrammable digital logic devices like FPGA which have some limitations in terms of capabilities/price factor. For this case Partial Reconfiguration (PR) emerges as a solution, leveraging to use the resources of these devices more efficiently. In the conventional approach, modifying FPGA designs required users to undertake the laborious process of resynthesizing, implementing, and programming the entire FPGA. Consequently, this procedure is time-consuming and impedes users' progress. However, with partial reconfiguration, users only need to resynthesize and program the specific portions or slices of the FPGA that necessitate modification. However, it necessitates a specialized understanding of FPGA design, involving the creation of modifiable regions. This paper takes initial strides towards establishing a remote laboratory for students to explore wireless communication concepts, harnessing PR for SDR devices.

Keywords: Software-Defined Radio · Remote Laboratory · Partial Reconfiguration · Embedded Systems

1 Introduction

In traditional Science, Technology, Engineering, and Mathematics (STEM) courses, students typically engage in hands-on laboratory experiments using physical lab kits provided by the instructional team in physical laboratories [1]. However, this conventional approach has several drawbacks. First, it is not inclusive for students with disabilities, as it requires them to physically attend school, which may pose risks to their physical and mental well-being [2]. Secondly, this method can lead to equipment damage, such as short circuits or the

M. E. Auer et al. (Eds.): STE 2024, LNNS 1028, pp. 180–192, 2024.
https://doi.org/10.1007/978-3-031-61905-2_18

dropping of devices [3]. Lastly, it places a greater financial burden on educational institutions, as they need to purchase and maintain a large number of lab kits [4]. Unfortunately, these lab kits are often underutilized, as not all students use the devices simultaneously, which fails to capitalize on the potential for greater flexibility.

These disadvantages were largely addressed through the advent of remote laboratories [5,6]. Following the emergence of the internet, numerous educational institutions have shown an interest in creating laboratories that grant students access to physical devices remotely [7]. Despite initial skepticism, remote laboratories have demonstrated their effectiveness. Numerous engineering educators and students have reaped the benefits of these remote facilities [8,9]. In a study conducted during a college-level FPGA course at the University of Washington, it was found that students exhibited improved analytical skills and achieved higher overall scores when utilizing a remote hardware laboratory, as compared to a hands-on laboratory. The flexible access to remote laboratories also facilitated a greater number of students in completing their work [10].

Software-Defined Radios, commonly known as SDRs, are radio systems that can be configured and controlled via software programming [11,12]. In contrast to traditional hardware-based radios, which are constrained by fixed internal components determining specifications like frequency range, bandwidth, and sampling rate, SDR devices provide substantial advantages in terms of flexibility, reconfigurability, and cost-effectiveness. They can be readily updated by reprogramming their software, and by replacing many hardware components with software-driven solutions, SDR devices often exhibit improved power efficiency and affordability [13,14]. Over the past two decades, we have observed remarkable progress in SDR devices. They have grown more robust and capable while simultaneously reducing in size and cost [15].

As a result, a growing number of educational institutions have incorporated telecommunication courses into their curriculum, aiming to provide students with expertise in areas such as radio frequency (RF), wireless communication, and software-defined radio (SDR) [16]. Additionally, many of these courses mandate a laboratory section to give students a full learning experience. Addressing the limitations associated with conventional in-person labs, the Remote Hub Lab (RHL/RHLab)[1] group [17] and LabsLand[2] have developed the RHL-RELIA project [18]. This project, based on the MELODY model [19], seeks to establish a remote laboratory facilitating student access to popular SDR devices, enabling convenient and flexible usage from any location and at any time [17,20,21].

Building upon the achievements of the existing remote laboratory, where users were empowered to program ADALM-PLUTO (PlutoSDR) through a web interface, we are advancing the platform by introducing additional features. Specialized courses are being developed to harness the capabilities of this remote lab, including the creation of practical applications such as a radar remote laboratory. Moreover, we are in the final stages of integrating another cost-effective SDR

[1] https://rhlab.ece.uw.edu.
[2] https://labsland.com.

device, the Red Pitaya, into the project. This paper covers our effort to incorporate partial reconfiguration into the existing RHL-RELIA system. This step is aimed at optimizing the use of these more economical FPGAs and enabling students to explore the potentials of partial reconfiguration [22].

The paper is structured into the following sections: Sect. 2 provides an overview of partial reconfiguration using Zynq SoCs and explores its feasibility within the SDR domain. Section 3 summarizes our approach to implementing PR within an established hardware platform. Section 4 showcases our current progress in this endeavor. The final section, Sect. 5, offers a conclusion of the paper and outlines the forthcoming steps.

2 Background

Partial reconfiguration (PR) refers to the process of updating specific portions or slices of the FPGA while leaving other sections unchanged[3]. This allows for flexibility and adaptability in SDR programs, enabling improved performance and functionality [23–25].

The majority of inexpensive SDRs are equipped with Zynq 7000 System-on-Chips (SoCs), which belong to the Xilinx family of SoCs comprising a Processing System (PS) and a Programmable Logic (PL). The PL component is built around an FPGA, offering high-performance real-time computing capabilities. However, SDRs in the price range of $200 to $400 commonly utilize Zynq Z-7010 or Z-7020 SoCs, both of which are cost-optimized members within the Zynq-7000 SoC family. They come with a limited number of logic cells and Look-Up Tables (LUTs). For instance, the Z-7010 SoC features 28,000 logic cells and 17,600 LUTs, which are only about 1/10 of the logic cells and 1/10 of the LUTs found in the mid-range Z-7035, which boasts 275,000 logic cells and 171,900 LUTs[4].

Inexpensive SDRs often deplete the FPGAs rapidly if not used efficiently, given that most signal processing applications are resource-intensive and consume a significant number of processing logic cells and LUTs. Partial reconfiguration provides a solution to this issue by allowing specific sections of an FPGA to be updated and modified, adapting to the current operational mode as needed[5]. The effectiveness of partial reconfiguration has been demonstrated in the Software-Defined Radio domain. A study using the USRP E310 SDR revealed that performing partial reconfiguration is over four times faster than a full configuration of the entire FPGA. In their signal processing design, partial reconfiguration takes 33 ms, while a full reconfiguration lasts 143 ms [26].

Dynamic Partial Reconfiguration (DPR) stands out as a specialized form of Partial Reconfiguration (PR) that empowers engineers to modify specific portions of an FPGA while it is actively processing data [27–29]. In a study conducted by a research team at Cairo University, an assessment of DPR in wireless

[3] https://www.xilinx.com/products/design-tools/partial-reconfiguration.html.

[4] https://www.xilinx.com/products/silicon-devices/soc/zynq-7000.html##productTable.

[5] https://www.xilinx.com/products/design-tools/partial-reconfiguration.html.

communication technologies like WiFi, Bluetooth, and LTE revealed that DPR can significantly reduce FPGA slice utilization and minimize power consumption. The findings indicate that DPR contributed to a 10.19% reduction in total area and a substantial 76.71% decrease in average power consumption when compared to system designs lacking DPR [30].

PR has found applications in remote laboratories [31]. In a particular project, researchers engineered a scalable remote System-on-Chip (SoC) virtual laboratory designed to offer training support and exercises for students delving into digital and embedded system design. Their utilization of partial reconfiguration aimed to enable "hardware multitasking", a feature that empowered their remote lab to concurrently execute up to four distinct FPGA designs from students [32].

Despite extensive research, we discovered a gap in remote laboratories utilizing the unique capability of partial reconfiguration for SDR projects. One of our objectives is to facilitate advanced FPGA-focused students in learning the intricacies of designing PR modules, while simultaneously offering radio frequency students the opportunity to delve into the creation and synergy of various components within a radio system.

3 System Design

This section provides an overview of the current RHL-RELIA system setup, elucidating its structure and components, in Sect. 3.1 and our plan to integrate partial reconfiguration into the existing architecture in Sect. 3.2.

3.1 RHL-RELIA System

The RHL-RELIA system comprises a user-friendly web interface, a server overseeing the allocation of SDR combinations per session, and a physical lab housing all hardware components such as SDRs, Raspberry Pis, and servers. Through this setup, students can upload their SDR designs from GNU Radio. Subsequently, the server uploads these programs onto Raspberry Pis, which then program the connected SDRs via USB cables. Once configured, users can observe and analyze real-time outputs and actively engage with the SDRs [20].

3.2 Integrating Partial Reconfiguration into the Architecture

To integrate partial reconfiguration into the RHL-RELIA network, there are two distinct architectural models that facilitate this feature: one based on a user interface (UI) and another using a command-line approach.

User Interface (UI)-Based Architecture. The first approach centers on integrating the Vivado Design Suite, an intuitive user interface application developed by Xilinx[6]. This tool empowers users to craft FPGA programs tailored for their System-on-Chips (SoCs), encompassing the Z7000 series among others.

[6] https://www.xilinx.com/products/design-tools/vivado.html.

Students are required to craft their partial reconfigurable FPGA programs using Vivado on their personal computing devices. This process encompasses project creation, design implementation, and synthesis. Upon error-free completion, students can generate bitstreams primed for uploading onto FPGA boards across different SDRs. Subsequently, these bitstreams can be uploaded onto RHL-RELIA's web user interface. Within Vivado's hardware manager window as depicted in Fig. 1, students can select and upload their preferred bitstreams, either complete programs or partial programs, onto remote SDRs. The resultant output is promptly accessible for viewing on the website.

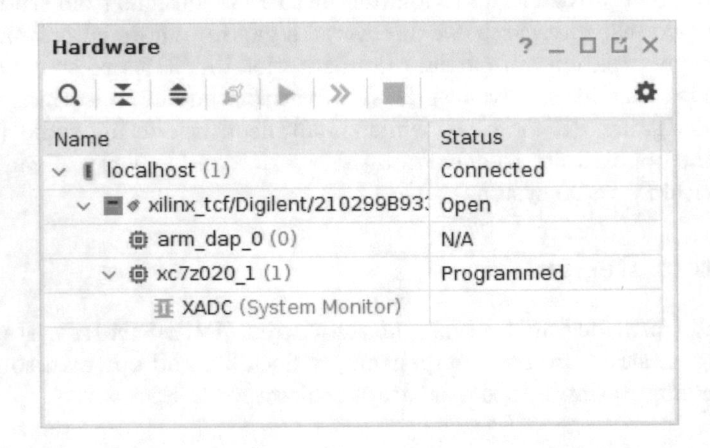

Fig. 1. Hardware manager window in Vivado Design Suite 2020.1.

UI-Based Hardware Setup. The hardware setup for UI-based partial reconfiguration involves an X86 computer, a JTAG programmer compatible with Zynq FPGAs, an SDR device, and a Raspberry Pi. The X86 architecture-based computer is essential since Vivado exclusively operates on this architecture[7]. The JTAG programmer receives bitstreams from the X86 computer, performing the task of uploading and programming the Zynq FPGA [26,33]. Its reliability is crucial; any incorrect FPGA programming often requires a hard reset, involving a complete power cycle to restore the device to its default mode. Once the successful upload of bitstreams is accomplished, the SDR device transmits data to a Raspberry Pi, which then displays the output on the RHL-RELIA website. Figure 2 provides a visual illustration of this interconnected configuration.

Command Line-Based Architecture. The second approach focuses on integrating the command prompt within the existing RHL-RELIA system to facilitate partial configuration through Linux commands. Students gain the ability to

[7] https://support.xilinx.com/s/question/0D52E00006hprrpSAA/vivado-on-arm-linux?language=en_US.

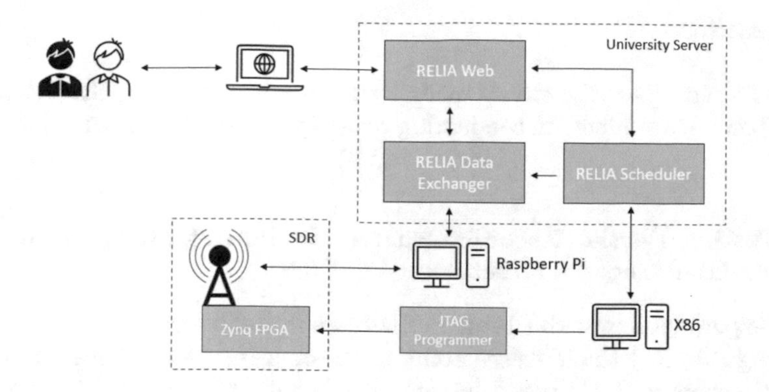

Fig. 2. RHL-RELIA architecture for UI-based FPGA PR.

upload both full and partial bitstreams onto the RHL-RELIA website and subsequently onto the FPGAs using specific commands. Although less intuitive than the UI-based interface, this method offers cost reductions in hardware setup.

Command-Based Hardware Setup. The setup cost for a command-based architecture is lower compared to a UI-based architecture due to the exclusion of an X86 computer and JTAG programmer. While both devices serve as vital bridges between the Vivado Design Suite and the FPGA, their necessity diminishes when a UI is not utilized. A command-based architecture solely necessitates a Raspberry Pi and an SDR device. The Raspberry Pi serves dual functions as both a programmer and data collector: it programs the SDR's FPGA via Secure Shell Protocol (SSH), a widely-used network protocol, and retrieves data from the SDR, displaying it on the RHL-RELIA website. The command-based architecture is shown in Fig. 3.

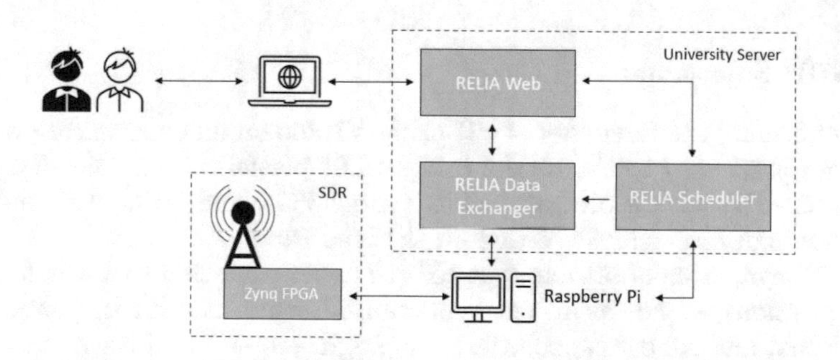

Fig. 3. RHL-RELIA architecture for command line-based FPGA PR.

4 Results

This section outlines the strides made by us in advancing the RHL-RELIA system to facilitate students in conducting real-time partial reconfiguration on SDR devices.

4.1 Testing Partial Reconfiguration: Evaluating Blackboard for UI-Based Architecture

Before experimenting with PR on an SDR, we initially assessed the feasibility of executing PR using the UI-based architecture on an economical evaluation board called Blackboard. This board integrates a Zynq 7007 SoC and incorporates a USB 2.0 transceiver capable of programming the FPGA[8].

The assessment starts with two basic calculator programs: a 4-bit calculator and an 8-bit calculator. Both programs allocate a region for PR. This segment can perform either addition or subtraction. Both full and partial bitstreams are uploaded onto the Zynq FPGA utilizing the hardware manager within Vivado. Initial findings are depicted in Table 1.

Both programs yield a complete bitstream of approximately 2 MB and partial bitstreams averaging around 100 KB. This leads to an average of 5.1% reduction in file size compared to a complete program with embedded child (partial) bitstreams. Consequently, the upload time for a partial bitstream is only 50.2% to 57.7% of that required for a full bitstream.

Table 1. Blackboard PR Program Size & Upload Time Comparison

	4-Bit Calculator			8-Bit Calculator		
	Full Bitstream	Child 1	Child 2	Full Bitstream	Child 1	Child 2
File Size (KB)	2036	104	104	2083	106	106
Upload Time (s)	4.09	2.12	2.36	4.40	2.34	2.21

4.2 SDR Selection

After confirming the feasibility of PR using Vivado on an economical evaluation board, we proceeded to choose the initial SDR platform. Our selection criteria favored an SDR with a programmable Zynq FPGA and an active community. After thorough research, we settled on the Red Pitaya.

Red Pitaya, often abbreviated as RP (to avoid confusion with PR for partial reconfiguration), stands as an open-source SDR renowned for its compact size, affordability, and robust capabilities. The engineers at Red Pitaya are not only open to exploring new concepts but also provide valuable assistance. Two RP

[8] https://www.realdigital.org/hardware/blackboard.

models gained our attention: STEMlab 125-14[9] and SDRlab 122-16[10]. In comparison to the entry-level model, the STEMlab 125-14, the slightly upgraded SDRlab 122-16 boasts a more powerful Zynq 7020 SoC. This advanced model offers a substantial increase in specifications, including 204% more logic cells, 133% more block RAMs, and 175% more DSP slices. Consequently, the SDR-lab 122-16 became our primary choice. For a detailed comparison between the STEMlab 125-14 and SDRlab 122-16, refer to Table 2.

Table 2. Red Pitaya STEMlab 125-14 & SDRlab 122-16 Specifications

	STEMLab 125-14	SDRlab 122-16
SoC	Zynq 7010	Zynq 7020
Connectivity	USB 2.0, 1 Gb Ethernet	
Number of RF Input Channels	2	
Number of RF Output Channels	2	
RF Input Bandwidth	DC - 60 MHz	300 kHz–550 MHz
RF Output Bandwidth	DC - 60 MHz	300 kHz–60 MHz
Input Sampling Rate (MS/s)	125	122.88
Output Sampling Rate (MS/s)	125	122.8
Input Resolution (bit)	14	16
Output Resolution (bit)	14	

4.3 Analysis: SDR PR via Vivado vs. Command-Line Approach

Next, we conducted PR experiments on the Red Pitaya, employing Vivado and command-line approaches to assess the feasibility of the previously proposed architectures. The first subsection provides an overview of the PR outcomes on a Red Pitaya SDRlab 122-16 utilizing Vivado, while the second subsection outlines the results achieved through command-line PR execution.

SDR PR with Vivado Hardware Manager. The process involved initiating an SDR project using a manufacturer-provided program and modifying it[11]. Subsequently, Vivado's hardware manager facilitated the upload of generated bitstreams onto the Red Pitaya. The hardware setup comprised an X86 Windows computer, a JTAG-HS3 programmer cable, and a Red Pitaya SDRlab 122-16. The JTAG-HS3 is an inexpensive programmer cable compatible with most Zynq FPGAs, including the Z7020 on the Red Pitaya[12]. It is available at approximately

[9] https://redpitaya.com/stemlab-125-14/.
[10] https://redpitaya.com/sdrlab-122-16/.
[11] https://redpitaya-knowledge-base.readthedocs.io/en/latest/learn_fpga/4_lessons/top.html#lessons.
[12] https://digilent.com/shop/jtag-hs3-programming-cable/.

60 USD. In contrast, the official Platform Cable USB II programmer costs nearly 300 USD (the prices are of January 2024).

The primary limitation of the JTAG-HS3 is its pinout disparity. While it has a total of 14 pins, there are only six JTAG pins on the Red Pitaya. Consequently, a 14-to-6-pin converter becomes necessary. However, these converters are pricey compared to non-specialized alternatives. To economize, we devised a converter by assembling a breadboard and jumper wires, detailed in Fig. 4.

Upon programming the Red Pitaya using a PR program, data on file sizes and resource usage were gathered and summarized in Table 3. Opting for PR rather than generating complete programs containing both child programs resulted in a file size reduction exceeding 5%.

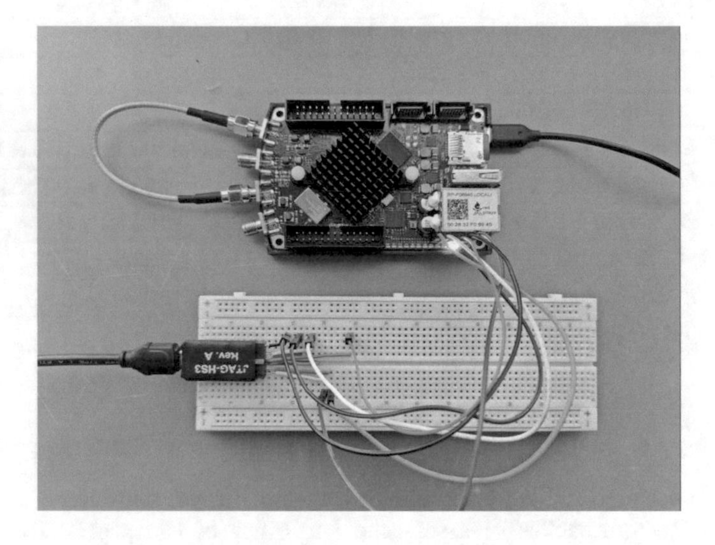

Fig. 4. JTAG-HS3 and Red Pitaya connection.

Table 3. Red Pitaya PR Program Size Comparison

	Program 1			Program 2		
	Full Bitstream	Child 1	Child 2	Full Bitstream	Child 1	Child 2
File Size (kB)	2096	106.3	106.3	4046	209.8	209.8

Additional specifics regarding resource usage for a specific SDR project are outlined in Table 4. The child program's LUT usage accounts for approximately 8.2% of the static program and utilizes about 16.9% of FFs. These figures closely align with the LUT and FF savings observed when transitioning from a non-PR program to a PR program with the same functionality.

Table 4. Red Pitaya PR Program Resource Usage

	Full Program with Empty PR Slice	Child 1	Child 2
LUT	4975	409	411
FF	5359	903	903
BRAM	32.0	0	0

SDR PR with Command Lines. We also conducted tests on the architecture using command lines, which presented a less intricate hardware setup compared to using Vivado. We connected a Red Pitaya to a Raspberry Pi 4B using an Ethernet cable.

Upon receiving the bitstreams, the Raspberry Pi initiates an SSH connection into the Red Pitaya using its IP address and transfers all necessary bitstreams onto it. Uploading partial bitstreams onto the FPGA involves asserting a PR flag. We faced challenges when the FPGA froze after uploading partial bitstreams using this approach. The FPGA becomes non-responsive after modifying the PR regions until it is reprogrammed using a full bitstream. Upon consultation with a Red Pitaya engineer, it was suggested that the issue might stem from using an outdated image.

5 Conclusion and Future Work

This work documents our investigation into integrating PR within the RHL-RELIA remote laboratory. Such a feature represents a high level of specialization in FPGA expertise, not common at the undergraduate level but valuable in graduate classes and professional projects. It underscores the inherent multidisciplinary nature of electrical engineering, necessitating collaborative efforts between FPGA and RF engineers.

Two distinct architectural models were introduced and assessed: a UI-based system and a command line-based approach. The detailed hardware setups and experimentation revealed the potential of PR in reducing upload times onto Zynq FPGAs while significantly optimizing resource utilization.

The UI-based approach configures the FPGA by an X86 computer separate from Raspberry Pis. This provides a high degree of flexibility, allowing changes during SDR execution and fostering a more professional project environment.

Conversely, the command-line method configures remote SDRs entirely by a Raspberry Pi, offering cost savings but lacking the flexibility of the UI-based approach. Users can only upload programs to FPGAs using commands.

As future work, while the Red Pitaya satisfies most requirements for an educational SDR device, exploring options for interoperability among different types of SDRs could enhance the system's capabilities. Due to challenges encountered in the command line-based approach, we will collaborate closely with Red Pitaya engineers to resolve these issues and ensure its functionality.

Acknowledgements. This work is supported by the National Science Foundation's Division Of Undergraduate Education under Grant No. 2141798.

References

1. Wang, L., Wang, J.: Design of laboratories for teaching mechatronics/electrical engineering in the context of manufacturing upgrades. Int. J. Electr. Eng. Educ. **59**(3), 251–265 (2022). https://doi.org/10.1177/0020720919837856
2. Grout, I.: Supporting access to STEM subjects in higher education for students with disabilities using remote laboratories. In: Proceedings of 2015 12th International Conference on Remote Engineering and Virtual Instrumentation (REV), pp. 7–13 (2015)
3. Love, T.: Addressing safety and liability in stem education: a review of important legal issues and case law 1. Technol. Stud. **39**, 28–41 (2013)
4. Wei, C.: Research on university laboratory management and maintenance framework based on computer aided technology. Microprocess. Microsyst. 103617 (2020). https://www.sciencedirect.com/science/article/pii/S014193312030764X
5. Hussein, R., Maloney, R.C., Rodriguez-Gil, L., Beroz, J.A., Orduna, P.: RHL-BEADLE: bringing equitable access to digital logic design in engineering education. In: 2023 ASEE Annual Conference and Exposition (2023)
6. May, D., Morkos, B., Jackson, A., Hunsu, N.J., Ingalls, A., Beyette, F.: Rapid transition of traditionally hands-on labs to online instruction in engineering courses. Eur. J. Eng. Educ. **48**(5), 842–860 (2023). https://doi.org/10.1080/03043797.2022.2046707
7. Xu, Z., Chen, W., Qu, D., Hei, X., Li, W.: Developing a massive open online lab course for learning principles of communications. In: TALE, pp. 586–590. IEEE (2020)
8. Schnieder, M., Williams, S., Ghosh, S.: Comparison of in-person and virtual labs/tutorials for engineering students using blended learning principles. Educ. Sci. **12**(3), 153 (2022). http://dx.doi.org/10.3390/educsci12030153
9. Schnieder, M., Ghosh, S., Williams, S.: Using gamification and flipped classroom for remote/virtual labs for engineering students, February 2022. https://repository.lboro.ac.uk/articles/conference_contribution/Using_gamification_and_flipped_classroom_for_remote_virtual_labs_for_engineering_students/19188251
10. Hussein, R., Wilson, D.: Remote versus in-hand hardware laboratory in digital circuits courses. In: 2021 ASEE Virtual Annual Conference Content Access. ASEE Conferences, Virtual Conference, July 2021. https://peer.asee.org/37662
11. Blossom, E.: GNU radio: tools for exploring the radio frequency spectrum. Linux J. **2004**, 4 (2004)
12. Tato, A.: Software defined radio: a brief introduction. In: XoveTIC Congress 2018. XoveTIC 2018, MDPI, September 2018. http://dx.doi.org/10.3390/proceedings2181196
13. Şorecău, M., Şorecău, E., Sârbu, A., Bechet, P.: Real-time statistical measurement of wideband signals based on software defined radio technology. Electronics **12**(13), 2920 (2023). http://dx.doi.org/10.3390/electronics12132920
14. Perotoni, M.B., Ferreira, L., Maniçoba, A.: Low-cost measurement of electromagnetic leakage in domestic appliances using software-defined radios. Revista Brasileira de Ensino de Física **44**, e20220009 (2022). https://doi.org/10.1590/1806-9126-RBEF-2022-0009

15. Collins, T., Getz, R., Wyglinski, A., Pu, D.: Software-Defined Radio for Engineers (2018)
16. Hussein, R., Guo, M., Amarante, P., RodriguezGil, L., Orduña, P.: Digital twinning and remote engineering for immersive embedded systems education. In: Frontiers in Education (FIE) Conference, USA. IEEE (2023)
17. Hussein, R., et al.: Remote Hub Lab - RHL: broadly accessible technologies for education and telehealth. In: Auer, M.E., Langmann, R., Tsiatsos, T. (eds.) REV 2023. LNNS, vol. 763, pp. 73–85. Springer, Cham (2023). https://doi.org/10.1007/978-3-031-42467-0_7
18. Inonan, M., Paul, A., May, D., Hussein, R.: RHLab: digital inequalities and equitable access in remote laboratories. In: 2023 ASEE Annual Conference and Exposition (2023)
19. Inonan, M., Hussein, R.: Melody: a platform-agnostic model for building and evaluating remote labs of software-defined radio technology. IEEE Access **11**, 127550–127566 (2023). https://doi.org/10.1109/ACCESS.2023.3331399
20. Inonan, M., Chap, B., Orduña, P., Hussein, R., Arabshahi, P.: RHLab scalable software defined radio (SDR) remote laboratory. In: Auer, M.E., Langmann, R., Tsiatsos, T. (eds.) REV 2023. LNNS, vol. 763, pp. 237–248. Springer, Cham (2023). https://doi.org/10.1007/978-3-031-42467-0_22
21. Inonan, M., Orduña, P., Hussein, R.: Adapting a remote SDR lab to analyze digital inequalities in radiofrequency education in Latin America. Revista Innovaciones Educativas (2023, in press)
22. Vipin, K., Fahmy, S.A.: ZyCAP: efficient partial reconfiguration management on the Xilinx Zynq. IEEE Embed. Syst. Lett. **6**(3), 41–44 (2014)
23. Bucknall, A.R., Fahmy, S.A.: Runtime abstraction for autonomous adaptive systems on reconfigurable hardware. In: 2021 Design, Automation & Test in Europe Conference & Exhibition (DATE), pp. 1616–1621 (2021)
24. Bucknall, A.R., Shreejith, S., Fahmy, S.A.: Network enabled partial reconfiguration for distributed FPGA edge acceleration. In: 2019 International Conference on Field-Programmable Technology (ICFPT), pp. 259–262 (2019)
25. Bucknall, A.R., Shreejith, S., Fahmy, S.A.: Build automation and runtime abstraction for partial reconfiguration on Xilinx Zynq UltraScale+. In: 2020 International Conference on Field-Programmable Technology (ICFPT), pp. 215–220 (2020)
26. Grassi, S., Convers, A., Dassatti, A.: FPGA partial reconfiguration in software defined radio devices. In: Proceedings of the GNU Radio Conference, vol. 5, no. 1 (2020). https://pubs.gnuradio.org/index.php/grcon/article/view/68
27. Bucknall, A.R., Fahmy, S.A.: ZyPR: end-to-end build tool and runtime manager for partial reconfiguration of FPGA SoCs at the edge. ACM Trans. Reconfig. Technol. Syst. **16**(3), June 2023. https://doi.org/10.1145/3585521
28. Vipin, K., Fahmy, S.A.: FPGA dynamic and partial reconfiguration: a survey of architectures, methods, and applications. ACM Comput. Surv. **51**(4), July 2018. https://doi.org/10.1145/3193827
29. Pham, K., et al.: Moving compute towards data in heterogeneous multi-FPGA clusters using partial reconfiguration and I/O virtualisation. In: 2020 International Conference on Field-Programmable Technology (ICFPT), pp. 221–226 (2020)
30. Hosny, S., Elnader, E., Gamal, M., Hussien, A., Khalil, A.H., Mostafa, H.: A software defined radio transceiver based on dynamic partial reconfiguration. In: 2018 New Generation of CAS (NGCAS), pp. 158–161 (2018)
31. Somanaidu, U., Telagam, N., Kandasamy, N., Nanjundan, M.: USRP 2901 based FM transceiver with large file capabilities in virtual and remote laboratory. Int. J. Online Eng. **14**, 193–200 (2018)

32. Machidon, O., Machidon, A., Cotfas, P., Cotfas, D.: Leveraging web services and FPGA dynamic partial reconfiguration in a virtual hardware design lab. Int. J. Eng. Educ. **33**, 865–876 (2017)
33. Hassan, A., Ahmed, R., Mostafa, H., Fahmy, H.A.H., Hussien, A.: Performance evaluation of dynamic partial reconfiguration techniques for software defined radio implementation on FPGA. In: 2015 IEEE International Conference on Electronics, Circuits, and Systems (ICECS), pp. 183–186 (2015)

A First Critical Analysis of a Hybrid Teaching Approach Using Remote Laboratories

Bastien Vincke[1,2], Cédric Vanhoolandt[3,4], Bruno Darracq[1(✉)], and Pascal Aubert[1]

[1] IUT d'Orsay, Univ. Paris-Saclay, 91400 Orsay, France
cartable-distant.iut-orsay@universite-paris-saclay.fr
[2] SATIE - CNRS UMR 8029, Univ. Paris-Saclay, 91400 Orsay, France
[3] Chaire de Recherche-action sur l'innovation Pédagogique, Univ. Paris-Saclay, 91400 Orsay, France
[4] Institut de Recherches en Didactiques et Education de l'UNamur, Université de Namur, 5000 Namur, Belgique

Abstract. Numerous technological solutions are available to enable remote lab work. These solutions are now mature enough to be used on a massive scale in university courses. These solutions make it possible to envisage pedagogical scenarios that combine face-to-face and distance learning (hybridation). In this article, we present a hybrid pedagogical sequence as part of an applied physics course. Qualitative and quantitative analysis are presented, evaluating the added value of this hybrid pedagogical approach.

Keywords: Remote Laboratories · Hybrid teaching

1 Introduction

Remote laboratories (RL) have seen a huge interest in recent years [1–3]. The development of technological knowledge and skills is made possible by the remote availability of physical equipment. However, our teaching methods need to evolve to consider the possibility of hybrid teaching, including face-to-face and distance learning.

In higher education, online activity platforms were particularly useful during the covid-19 pandemic, as they enabled learning to continue when teaching was stopped suddenly for health reasons. Technologies have emerged to facilitate remote access to RL and student monitoring [4–6]. Different remote actions to make RLs as close as possible to real laboratories have also been set up which is a key point in teaching practical skills such as applied physics. In a relatively short space of time, the financial investment of the universities and the involvement of the teaching teams have enabled the emergence of innovative learning tools for instance in electronics, acoustics and material physics.

Due to increasing student numbers in higher education, student populations are becoming more diverse. This trend has also led to create more flexible modes of education and personalized learning trajectories. Moreover, in order to offer flexibility in terms of time and place, the need of an instructional approach that combines face-to-face and

M. E. Auer et al. (Eds.): STE 2024, LNNS 1028, pp. 193–200, 2024.
https://doi.org/10.1007/978-3-031-61905-2_19

online laboratory activities is increasing. Particularly in applied physics teaching, the design of hybrid learning using the RL resources in response to student diversity is then an interesting way forward. Nevertheless, it is crucial to look first at the perceptions and learning of the students who are at the heart of this design.

2 Contribution

During the second semester of the academic year 2022–23, a cohort of students was monitored during a hybrid pedagogical sequence in applied physics curricula at Paris-Saclay University. It includes a set of lectures, tutorials and labs, as well as remotely accessible resources to enable students to supplement the knowledge and skills they could acquire at home. Each lab included a large face-to-face part and a small part to be completed remotely alone or in groups, outside academic class hours. The students were given a short questionnaire on the homework and on students perception and global satisfaction.

A first analysis of validity of the questionnaire has been carried out. Two separate analyses were carried out. The first was based on the student general perception of the remote lab homework. The second was based on the strengths and weaknesses they expressed on the pedagogical approach. Both analysis reveal some factors that teachers may tackle to foster student satisfaction with the tool.

3 Hybridization of a Course

3.1 Description of the Remote Lab

The remote lab platform is a numerical environment to work alone or with others on workstations at IUT Orsay at univ. Paris-Saclay (France), available 24 h a day, and to access to a remote catalogue of lab work [9].

Historically, this remote lab platform was created during the March-June 2020 lockdown period to ensure the continuity of theoretical and practical teaching at the Physical Measurements department of the IUT Orsay [7]. Today, as students and teachers have returned to classrooms, the remote lab is becoming a hybrid solution, complementary to traditional classroom teaching, which allows users to access, individually or in groups, their digital university environment as well as a certain number of practical works similar to those carried out in educational laboratories.

The use of the remote lab is diversified:

- Use of heavy software environments (difficult to install and maintain on individual computers);
- Carrying out experiments in support of lectures or classrooms;
- Revision for lab exams;
- Training of new teachers;
- Continuing education;
- Accessibility to practical work for students with disabilities;
- Sharing of practical work within the framework of international collaborations.

The hardware and software structures were described in detail in [1].

3.2 Description of the Computer Programming, Instrumentation Course

The course was designed for 1st year undergraduate students of the Physical Measurements curriculum between april and june 2023. The students have a scientific background. The course includes 5 lectures (6h), 8 small classes (10 h) coupled with 8 labs (22 h), theory exam (1.5 h) and a practice exam (2 h).

The target skills are: analog and digital signals acquisition, design data processing algorithms and learning of a programming language.

The learning objectives are: understand the caracteristics of an acquisition card, evaluate its metrological characteristics, understand boolean, combinatorial and sequential logic.

3.3 Example of a Hybrid Pedagogical Sequence

For several years now, we chose to combine tutorials with labs. Each student has his own workstation and a dedicated equipment. The pedagogical sequence consists in setting-up and programming a multifunctional data acquisition device (myDAQ) [8] connected to a dedicated circuit related to the lab topic.

The hybrid pedagogical sequence is focused on the concepts of data acquisition (sampling frequency, range, resolution). The lab work concerns the characterisation of an RC circuit and more specifically the configuration of the acquisition parameters to optimise the measurement of the time constant.

All fundamental concepts and some application cases were taught face-to-face. An additionnal homework lab is proposed using the remote lab platform. The students had 3 weeks to complete this homework. During this period, the students can interact with the teachers by e-mail or face-to-face. A video tutorial was made available to help students [https://youtu.be/S-sFaDjib8s]. The evaluation consists in filling out a document containing the programming of the acquisition and the measurement processing (Fig. 1).

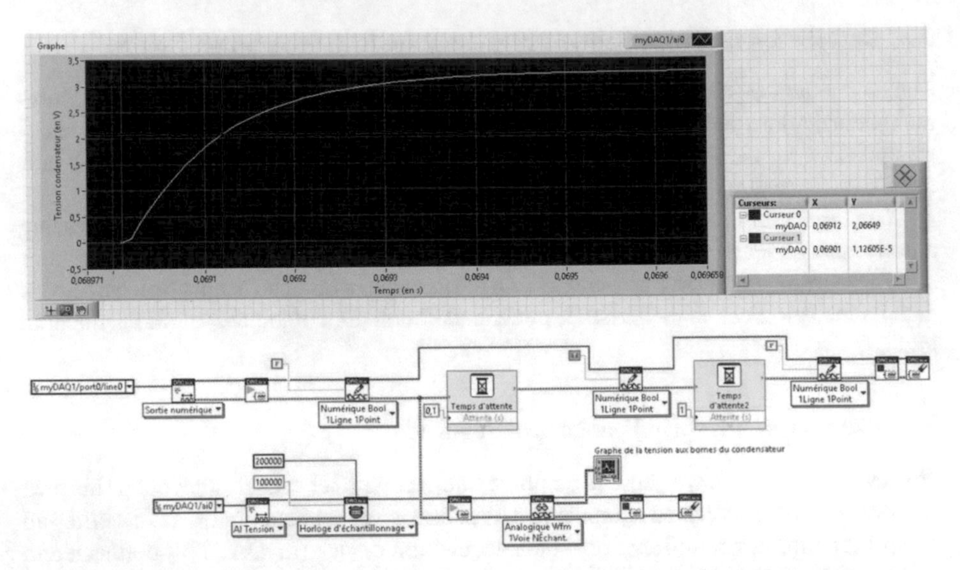

Fig. 1. Example of a document completed by students (front panel and block diagram)

4 Evaluation

4.1 Participants

Participants consisted of 132 French talking students (15% females, M $\pm$ SD = 18.9 $\pm$ 0.9 years) from IUT Orsay at univ. Paris-Saclay (France). All students were taught using the same pedagogical approach.

4.2 Data Collection and Analysis

Data on students' perceptions were collected by administering an exploratory survey on students perceptions (ESSP). The part of the questionnaire focusing the remote lab homework contains ten questions and was administered once in may 2023. This moment was chosen to reflect the state of students' perception on this topic, after the homework assignment and before the final exam of the course. The students' perceptions were collected in accordance with national and international norms governing the use of educational research participants.

The questionnaire used a five-level Likert-type scale from *strongly disagree* to *strongly agree* with a neutral answer (i.e. *neither agree nor disagree*). It was written in French-language and adapted on the pedagogical context of this study. Some open-ended questions were also added focusing the strengths and weaknesses of the pedagogical approach.

Participants filled out the questionnaires online and remotely on a voluntary basis. A total of 86 responses (i.e. 70.5% of the students) were collected, whose 80 were totally completed. About ten minutes were needed to complete the questionnaire. To ensure anonymity, no personal data (i.e. name and gender) were asked to the students and the experimenters only had access to the anonymous data.

On the basis of these data, three levels of analysis have been carried out:

- A *quantitative* analysis on the scales items. Descriptive statistics and correlational analysis were performed;
- A *qualitative* analysis on the open-ended questions. We performed a category analysis based on content analysis;
- A *confirmatory* analysis. This can be done by computing the internal validity of the questionnaire.

Data on academic results were collected by the evaluation of a homework (may 2023), a theory exam and a practice exam (june 2023). All those assessments are marked out of 20. They were evaluated by the regular teachers without any feedback from any result of this study.

4.3 Results and Discussion

The academic results of the students are shown in Table 1.

Table 1. Academic results (M ± SD) by students gender.

Results	Female	Male	Sample
Homework results (/20)	15.9 ± 3.1	12.3 ± 7.3	12.9 ± 6.9
Practical results (/20)	12.9 ± 2.1	13.1 ± 2.6	13.0 ± 2.5
Theory results (/20)	9.3 ± 2.9	9.7 ± 3.5	9.6 ± 3.4

A difference based on gender does not stand up to statistical testing for the practical results and the theory results. A Student's t test for independent samples is used to compare the homework results nonetheless. This test shows that the results depend significantly on gender ($t = -3.60$; $df = 65.68$; p-value $< .001$) as shown in Fig. 2.

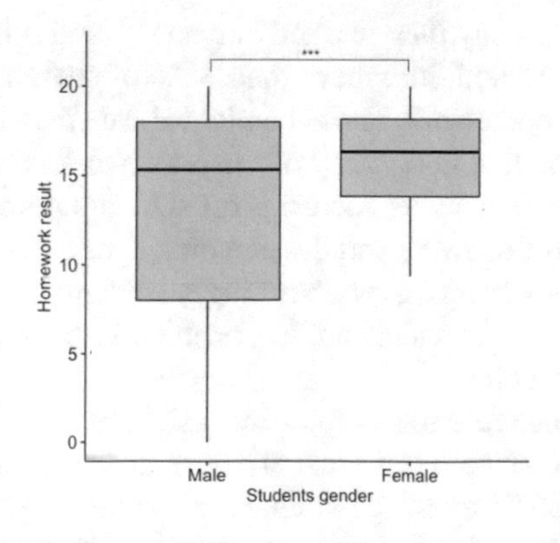

Fig. 2. Distribution of the homework results by students gender.

Table 2 shows a correlation matrix between academic results. All correlations can be seen as moderate to good. It indicates consistency between results. A student who has completed the homework successfully is therefore to succeed more likely in practice and in theory.

Table 2. Correlation matrix between academic results. Pearson coefficients were computed. Statistical significance is indicated (***: $p < .001$).

	Practice results	Theory results
Homework results	0.37 (***)	0.50 (***)
Practice results		0.54 (***)

The ESSP contains two questions considered as cornerstone. The first one is about the work time of the students. Another is about their general perception about the homework. Results are shown in Fig. 3 (a) and (b).

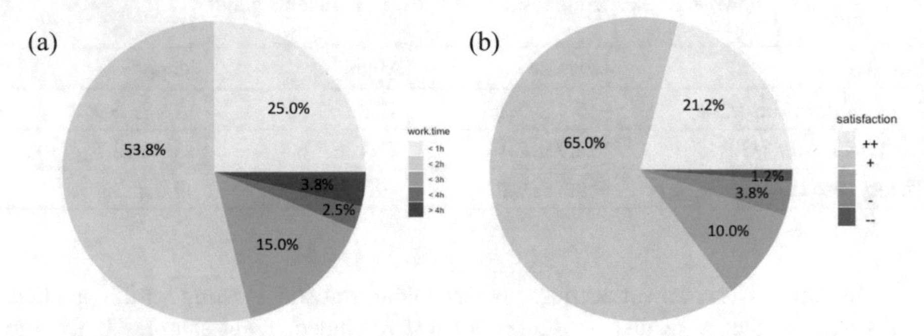

Fig. 3. Pie chart of the students' (a) work time and (b) general satisfaction about the homework (from --: completely unsatisfying to ++: completely satisfying).

Nearly 70% of students say they worked between 1 and 3 h on this topic, which is the expected time order. In addition, more than 85% of students feel satisfied or completely satisfied with this homework. Those results indicate that the students have worked seriously on this topic and that the quality of the homework has met their expectations.

Then ESSP proposes four items focusing on students perceptions. Table 3 shows the ranking by levels of the answers and the percentage of answers in favor (Fav) of the homework. This homework helped overwhelmingly students to get down to work. A few of them had connection and/or understanding problems. A majority found the support they needed to fix their problem.

Open-ended questions were asked focusing assets of this approach. No limit was imposed to students who were free to describe the main strength and weakness from their point of view. In the 86 questionnaires, more strenghts (N = 77) were described than weaknesses (N = 45). They were categorized by the researchers on the basis of their content as show in Table 4.

Table 3. Answers to items focusing on students perceptions (from --: *strongly disagree* to ++: *strongly agree*). Answers in favor (Fav) are highlighted in grey.

Item (N=86)	--	-		+	++	Fav (%)
I think this homework helped me to get down to work.	0	5	10	41	24	75.6
I had connection problems with this homework assignment.	22	26	12	13	7	55.8
I had understanding problems with this homework assignment.	17	33	17	11	2	58.1
If I had a problem, I got the support I was looking for.	0	1	31	28	20	55.8

Table 4. Categorizations of the strengths and weaknesses.

Category	Strenght N (%)	Weakness N (%)
Support for differentiation/ personalization of learning	39 (50.6%)	9 (20.0%)
Plenty of pedagogical tools	15 (19.5%)	14 (31.1%)
Remote activities	15 (19.5%)	2 (4.4%)
Quality of the access	7 (9.1%)	19 (42.2%)
Overall	77 (100%)	45 (100%)

Support for differentiation/personalization of learning is the main strength. The variety of tools (i.e. LabVIEW, videos, correction of homework…) is appreciated by the learners. In one hand, for many students there is enough information to understand clearly what is requested. In the other, for some of them, the work is too directed. The opportunity to work at their own pace, at home and/or to take the time they need is truly highlighted. From students' point of view, it's a way to "learn better" and to deepen some topics.

Some learners point out the quality of the access (i.e. easy to connect and clear instructions) while a significant number of them experienced connection problems (i.e. all resources were busy at the time they connected or they were disconnected from the platform in an unexpected way). This kind of problems affect probably the general satisfaction of those students as described below.

Eventually a pivot category appears to be the plenty of the pedagogical tools. Both a strength and a weakness, many pedagogical tools can be seen as… too many for a student. A right balance between the number of pedagogical tools seems necessary to enhance the students general satisfaction. It also appears to be a lever for the personalization of learning which is at the heart of this hybrid teaching approach.

5 Conclusions Perspectives

The conclusions point to the need of a high didactical quality of the pedagogical RL resources. Results also highlight that teacher's involvement have a significant influence on student's perception of learning and satisfaction. As far as didactical resources concern in the topic of RL, it seems that hybrid teaching approach may lead to improvement of both remote and face-to-face labs.

Acknowledgement. The authors would like to thank the *Chaire de recherche-action sur l'innovation pédagogique* supported by Univ. Paris-Saclay for its help in analyzing this pedagogical approach.

References

1. Darracq, B., Vincke, B., Aubert, P.: A platform for remote laboratories in applied physics. In: Proceedings of the 20th International Conference on Remote Engineering and Virtual Instrumentation, pp. 1035–1045, Thessaloniki (2023)
2. Cubillo, L., Dormido, S., Pedro Sanchez, J.: What remote labs can do for you. Phys. Today 48–53 (2016)
3. Tawfik, M., et al.: Virtual instrument systems in reality (VISIR) for remote wiring and measurement of electronic circuits on breadboard. IEEE Trans. Learn. Technol. **6**(1), 60–72 (2013)
4. Hardward, V.J., et al.: The iLab shared architecture: a web services infrastructure to build communities of internet accessible laboratories. Proc. IEEE **96**(6), 931–950 (2008)
5. Broisin, J., Venant, R., Visal, P.: Lab4CE : a remote laboratory for computer education. Int. J. Artif. Intell. Educ. **27**(1), 154–180 (2017)
6. Leproux, P., Barataud, D., Bailly, S., Nieto, R.: LABENVI, un dispositif pour les travaux pratiques à distance. Interfaces numériques **2**(3), 453–467 (2013)
7. Vincke, B., Darracq, B., Rodriguez, S., Reynaud, R., Tonnerre, B.: Cartable Distant: Un environnement numérique complet pour l'enseignement pratique à distance. In: La Revue 3EI, no. 100, pp. 1–6 (2020)
8. Klinger, T., Kreiter, C., Pester, A., Madritsch, C.: Low-cost remote laboratory concept based on NI myDAQ and NI ELVIS for electronic engineering education. In: IEEE Global Engineering Education Conference (EDUCON), pp. 106–109, Porto (2020)
9. Cartable Distant Homepage. https://webapps.iut-orsay.fr/cartable-distant/

Lego-Based Remote Robotics Lab: Enhancing Didactic Engagement and Learning

Dario Assante(✉), Barbara Loletti, and Daniele Pirrone

Università Telematica Internazionale Uninettuno, Corso Vittorio Emanuele II 39, 00186 Rome, Italy
d.assante@uninettuno.it

Abstract. This paper introduces the concept of a Lego-based remote robotics lab for enhancing didactic engagement and learning. It discusses the need for practical, hands-on experiences in robotics education and presents the Lego-based remote lab as a solution. Leveraging a customised software implementation, which allows interfacing Lego Mindstorm and Raspberry Pi, students are enabled to remotely control Lego-based robots. The advantages of this approach, such as overcoming geographical limitations and promoting collaboration, critical thinking, and practical application of concepts, are highlighted. The paper also addresses the implementation and evaluation of the lab, emphasizing positive outcomes in student engagement, motivation, and learning.

Keywords: Remote laboratory · Robotics · STEM education

1 Introduction

The concept of distance learning in the 21st century and the use of the modular robotics as Lego MindStorm/Spike kits [1] has taken hold in the education. The combination of these key factors, along with the growing market for remotely programmable devices, played a crucial role in the development of the remote laboratory (RL).

The academic community collaborated to develop RL, incorporating both hardware and software interfaces with the aim of improving innovative pedagogical tools. These tools can be used to create new skill improvement through well-known software tools like Simulink and Easy Java Simulations [2], helping students understand automatic control theory. Furthermore, with the onset of the COVID-19 pandemic focused on the remote environment instead of the laboratory. There are many high-cost kits on the market, however, in-house kits have been developed such as the one described in [3] that focuses on two important pillars of instrumental analysis: spectroscopy and chromatography. The implementation and integration of RL can be extended to the study of renewable energy such as the government's master plan in [4] where they have included design and integration into a curriculum model.

The constant increase of the Internet of Things (IoT) [5] shows the potential to radically transform various areas, including education, in particular the realization of the multi-purpose and low cost RL [6]. The main actors of these laboratories are usually

M. E. Auer et al. (Eds.): STE 2024, LNNS 1028, pp. 201–207, 2024.
https://doi.org/10.1007/978-3-031-61905-2_20

a (i) Raspberry PI [5], that acts as a hardware writing Python code to drive the general-purpose input/output (GPIO) port [7], as server with Apache 2 server [8–10], and as link between the sensors and actuators and the personal computer [11], (ii) Lego Mindstorm [12].

The Lego-based remote robotics lab adopts a comprehensive approach that seamlessly integrates Lego Spike and Raspberry Pi, leveraging the power of custom software implementation [13]. Each robotic experiment created with Lego MindStorm or Spike entails a central hub, connected with an array of sensors and actuators [14]. Traditionally, the hub would be programmed using Bluetooth or a USB cable, employing the dedicated Lego app. However, this work takes innovation to new heights by introducing a specially designed socket that facilitates the direct communication between the Lego hub and the Raspberry Pi via USB [15].

In this novel setup, the Raspberry Pi assumes the role of a server and web interface, enabling remote control of the Lego hub. This configuration empowers learners to engage with and manipulate the Lego robots from a distance. To enhance the student's experience, the remote-control functionality is thoughtfully designed to provide flexibility in interaction.

2 Lego Mindstorm EV3

The Fig. 1 shows the control hub of a Lego Mindstorm EV3 and EV4, respectively: both come with and external USB port [16]. In particular, the main hardware features of the EV3 Programmable Brick is a 32-bit ARM9 processor, namely Texas Instrument's AM1808, part of the 32-bit ARM RISC processor family. It also features a single core, and a single thread per core, with a clock speed of 300 MHz. The memory has 64MB of DDR RAM, 16MB of flash memory and 126 KB of EEPROM. However, it is possible to insert a MicroSD with a maximum capacity of 32 GB. On the other hand The Brick has a firmware pre-installed within its 16MB flash memory, where its possible load in Python language, Matlab/Simulink script [17] both are used to encode commands that will only be sent from the computer to the Brick via Bluetooth protocol.

Fig. 1. The central processing unit (the "hub") of a Lego Mindstorm EV3 (left) and EV4 (right).

In this paper, we used the ev3dev2 operating system that can be run directly from the Micro-SD that can be inserted inside the Brick, it is an operating system based on

Debian Linux, with the particularity that it can be run on different Lego Mindstorm platforms, including the BrickPi, which is a simple Raspberry Pi used as a simple Brick [18]. In this operating system, there is a low-level driver framework to be able to control sensors, motors and much more, as well as the fact that it supports many out-of-the-box (OOTB) scripting languages for programming in Python through predefined libraries.

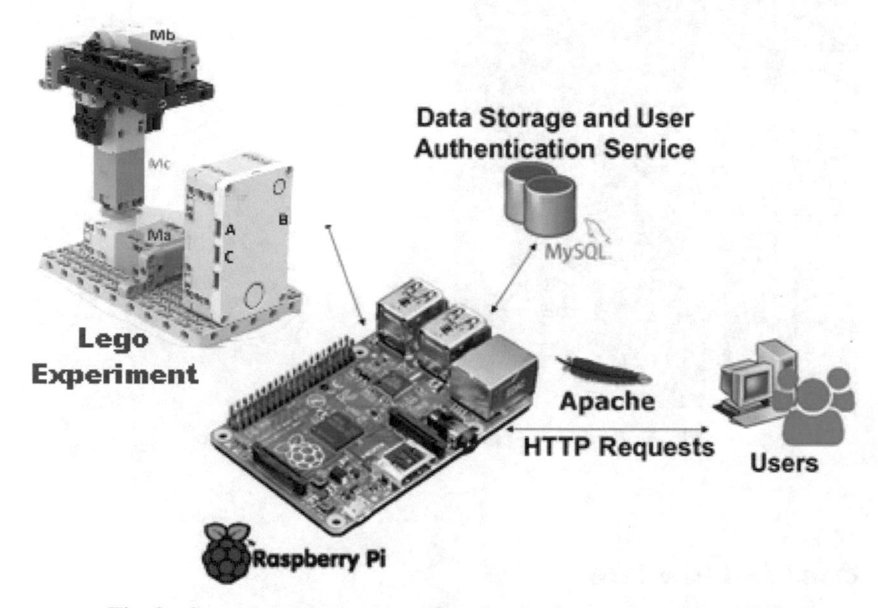

Fig. 2. Control and management functional scheme adopted from [18]

In summary, as shown in Fig. 2, the web server hosted inside a Raspberry Pi3 B+ board, exploits a client-server architecture [19–21]. Thus, the commands selected by the user is sent to the web server which, in turn, sends them to the robot through the use of a socket written in Python.

3 Web Server Interface

The web interface is performed with PHP and Python language for server and socket side respectively, while the client side is performed via JavaScript. In addition, the graphics interface is performed via HTML and CSS style sheet. The web interface allows the user access through their own credentials, which are created in the registration session (for sake of simplicity it is not shown). Once registered, the user login and it interacts with the robot during 20-min session. Therefore, client-side interface has the following characteristics to:

- Register the user.
- Make Login/logout.
- Have limited sessions up to 20 min.

- Consult an accurate list of commands that can be sent to the robot.

While, the server side receives commands from the user, then process them via socket to the robot (Fig. 3).

Fig. 3. First page of the web application for the login

4 Front End Interface

The user makes the login then he can choose two modes the Standard Mode (see the Fig. 4) and the Expert mode (see the Fig. 5.). The first choice is made for the beginner users where he can send command to the robot by pressing four buttons as shown in Fig. 4 performing the following four tasks:

- Forward: The robot moved forward for 2 s.
- Backward: The robot moved backward for 2 s.
- Left: Turn the robot 90 degrees to the left.
- Right: Turn the robot 90 degrees to the right.

The application when executing the above task it writes the respective command inside a "TextBox", simultaneously. In this way, the user learns the code-instruction relative to the task performed. Furthermore, there is button when it is pressed it starts streaming a webcam to check the outcome of the commands given to the robot.

The expert user can select Expert-mode and it directly write the instructions in a "TextBox", then it send the commands when the 'Submit' buttons is pressed. Moreover, there is a dedicate web page namely "Lego Mindstorm" where it is possible find a detailed description of the main commands and the corresponding reference links.

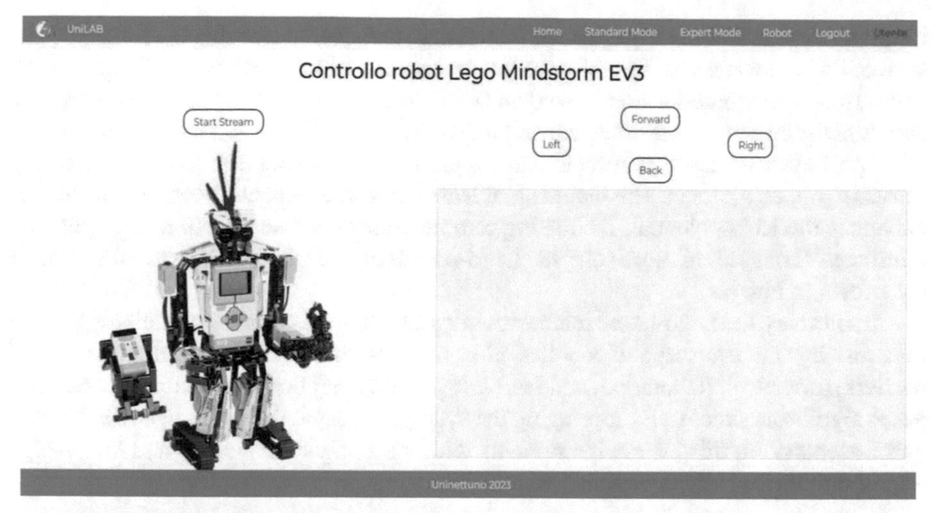

Fig. 4. Standard mode interface for the beginner user, while (b) show the expert user where it can write de code in a box

Fig. 5. Expert user mode, where it can write de code in a box

5 Conclusion

The Lego-based remote robotics lab presents a highly promising approach that holds tremendous potential to elevate didactic engagement and enrich the learning experience within the realm of robotics education. By transcending geographical limitations, this lab opens up boundless possibilities for collaboration, enabling students to actively apply theoretical concepts in practical contexts. The implementation and subsequent evaluation of this lab have yielded consistently positive outcomes, demonstrating notable improvements in student engagement, motivation, and overall learning outcomes. At the core of this innovative lab lies the seamless integration of Lego Spike and Raspberry Pi. This efficient combination equips students with invaluable skills and knowledge that are essential for thriving in the ever-evolving field of robotics. By harnessing the capabilities of Lego Spike, students gain hands-on experience in constructing and programming intricate robotic systems. The inclusion of Raspberry Pi as a central component further enhances the lab's potential, facilitating remote control, server functionality, and web interfaces. Examples of more complex Lego-based remote labs referring to robotic arms are shown in Fig. 6.

In summary, the Lego-based remote robotics lab represents a transformative approach that enriches the landscape of robotics education. Its ability to transcend geographical barriers, promote collaboration, and facilitate practical application of concepts has garnered significant success. By leveraging the synergistic capabilities of Lego Mindstorms and Raspberry Pi, this lab equips students with the foundational skills and knowledge required to thrive in the dynamic field of robotics.

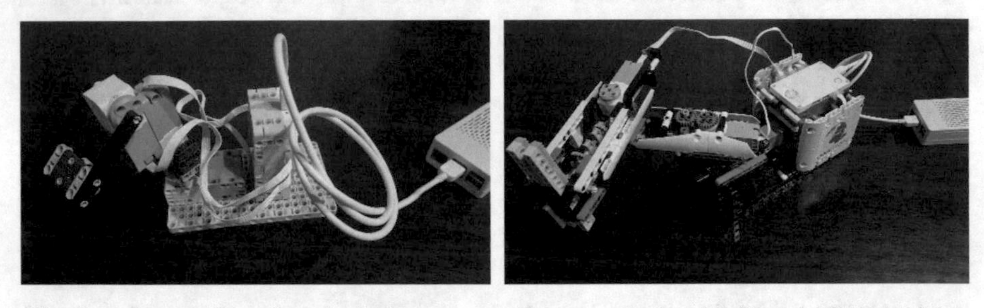

Fig. 6. Advanced Lego-based robotic arms experiences: articulated (left) and SCARA (right)

Acknowledgement. This work has been co-funded by the European Commission within the project (R-IoT-C)4VET – *Developing Innovative, Modern and Hands-On Digital Teaching Materials with a Focus on Robotics, Cloud and IoT for VET* (2021-1-HR01-KA220-VET-000034642).

References

1. Afari, E., Khine, M.S.: Robotics as an educational tool: impact of Lego Mindstorms. Int. J. Inf. Educ. Technol. **7**(6), 437–442 (2017)
2. Fabregas, E., Farias, G., Dormido-Canto, S., Dormido, S., Esquembre, F.: Developing a remote laboratory for engineering education. Comput. Educ. **57**(2), 1686–1697 (2011)
3. Miles, D.T., Wells, W.G.: Lab-in-a-Box: a guide for remote laboratory instruction in an instrumental analysis course. J. Chem. Educ. **97**(9), 2971–2975 (2020)
4. Pastor, R., et al.: Renewable energy remote online laboratories in Jordan universities: tools for training students in Jordan. Renew. Energy **149**, 749–759 (2020)
5. Maksimović, M.: IOT concept application in educational sector using collaboration. Facta Universitatis, Ser. Teach. Learn. Teach. Educ. **1**(2), 137–150 (2018)

6. Pirrone, D., Fornaro, C., Assante, D.: Open-source multi-purpose remote laboratory for IoT education. In: 2021 IEEE Global Engineering Education Conference (EDUCON), Vienna, Austria, pp. 1462–1468 (2021)
7. Zárate-Moedano, R., Canchola-Magdaleno, S.L., Arrington-Báez, A.A.: Remote laboratory, based on raspberry Pi, to facilitate scientific experimentation for secondary school students. Int. J. Online Biomed. Eng. (iJOE) **17**(14), 154–163 (2021)
8. Ariza, J.Á., Galvis, C.N.: RaspyControl Lab: a fully open-source and real-time remote laboratory for education in automatic control systems using Raspberry Pi and Python. HardwareX **13**, e00396 (2023)
9. Fernández-Pacheco, A., Sergio M., Manuel C.: Implementation of an Arduino remote laboratory with raspberry Pi. In: 2019 IEEE Global Engineering Education Conference (EDUCON) (2019)
10. Kwinana, P.M., Nomnga, P., Rani, M., Lekala, M.L.: Real laboratories available online: establishment of ReVEL as a conceptual framework for implementing remote experimentation in South African higher education institutions and rural-based schools – a case study at the university of fort hare. In: Auer, M., May, D. (eds.) Cross Reality and Data Science in Engineering. REV 2020. AISC, vol. 1231, pp. 128–142. Springer, Cham (2021). https://doi.org/10.1007/978-3-030-52575-0_10
11. Djordjević-Kozarov, J., Miodrag A.: Remote Laboratory Development for E-Learning in the Field of Electronic Measurement
12. Wu, T., Peter, A.: Investigating remote access laboratories for increasing pre-service teachers' STEM capabilities. J. Educ. Technol. Soc. **22**(1), 82–93 (2019)
13. Tweedale, J.W.: Using Lego EV3 to explore robotic concepts in a laboratory. Int. J. Adv. Intell. Paradig. **21**(3–4), 330–347 (2022)
14. Chou, P., Ru-Chu, S.: Engineering design thinking in LEGO robot projects: an experimental study. In: Huang, Y.M., Cheng, S.C., Barroso, J., Sandnes, F.E. (eds.) Innovative Technologies and Learning. ICITL 2022. LNCS, vol. 13449, pp. 324–333. Springer, Cham (2022). https://doi.org/10.1007/978-3-031-15273-3_36
15. Benedettelli, D.: The LEGO MINDSTORMS Robot Inventor Activity Book: A Beginner's Guide to Building and Programming LEGO Robots. no starch Press (2021)
16. Matyushchenko, I., Zvereva, E., Lavina, T.: Development of algorithmic thinking by means of Lego Mindstorms Ev3 on robotics. In: 2020 Ural Symposium on Bio-medical Engineering, Radioelectronics and Information Technology (USBEREIT), Yekaterinburg, Russia, pp. 444–447 (2020)
17. Pribilova, K., Gabriska, D.: Use of Lego Mindstorms EV3 MATLAB/Simulink with a focus on technical education. In: 19th International Conference on Emerging eLearning Technologies and Applications (ICETA) (2021)
18. Cornelissen, L.: Simulating Lego Mindstorms EV3 robots using unity and python. In: Radboud University Nijmegen (2019)
19. Assante, D., Capasso, C., Veneri, O.: Internet of energy training through remote laboratory demonstrator. Technologies **7**(3), 47 (2019)
20. Assante, D., Caforio, A., Flamini, M., Romano, E.: Smart education in the context of industry 4.0. In: 2019 IEEE Global Engineering Education Conference (EDUCON), pp. 1140–1145 (2019)
21. Assante, D., Flamini, M., Romano, E.: Open educational resources for industry 4.0: supporting the digital transition in a European dimension. In: 2021 IEEE Global Engineering Education Conference (EDUCON), pp. 1509–1513 (2021)

Enhancing Accessibility for Real-Time Remote Laboratories: A Web-Based Solution with Automated Validation and Access Control

Boris Pedraza, Alex Villazón(✉), and Omar Ormachea

Universidad Privada Boliviana (UPB), Cochabamba, Bolivia
{borispedraza1,avillazon,oormachea}@upb.edu

Abstract. The use of remote laboratories nowadays offers multiple advantages like accessibility, cost-effectiveness, flexibility, and safety. However, ensuring a secure and uniform interface for a remote laboratory remains a challenge. In this paper, we present the development of the Real-Time Remote Lab Bridge Server (or simply *BridgeServer*), a tiny web server that provides secure and transparent access to remote labs, without the necessity of a third-party software or password sharing. Its key features are the serve of a web-based remote desktop session created with locally stored credentials, secure file downloads, session time administration, and automated access control thanks to the validation with an Application Programming Interface (API) of a Booking System. This solution maintains control, employs consistent credentials for all users, and prevents direct access to lab equipment control software. We provide a real-world use case that exemplifies the flexibility of our solution simplifying access to a third-party software remote lab without any need for code modifications. We successfully solved access challenges in an international network of real-time remote labs deployed in Latin America, Asia, Africa, and Europe, particularly for Windows-based lab control software.

Keywords: Remote labs · real-time access · web-based remote access control

1 Introduction

Remote labs offer numerous benefits, including increased accessibility, cost savings, flexibility, safety, scalability, and the opportunity for real-time data collection and analysis [1,2]. There are two main types of remote labs [3,4]: ultra-concurrent remote labs, that are based on a prerecorded experience carried out at a real lab (i.e., allowing several users to run remote experiments concurrently), and real-time remote labs, where users actually remotely operate and interact with real lab equipment (i.e., requiring exclusive access to the remote lab equipment to prevent interferences). Ultra-concurrent remote labs are simpler

M. E. Auer et al. (Eds.): STE 2024, LNNS 1028, pp. 208–219, 2024.
https://doi.org/10.1007/978-3-031-61905-2_21

to deploy and implement, whereas real-time remote labs require more complex infrastructure as mentioned in several studies made over the years [5,6]. Nevertheless, no simple solution is available if the control software (e.g., in SCADA systems) does not provide native remote support through a web interface, is Operating System (OS)-dependent (e.g., only works on Windows), or uses proprietary software that cannot be changed or adapted.

Our solution, called "Real-time Remote Lab Bridge Server" or "*BridgeServer*" does not require any modification of the lab control software (open source or proprietary), and prevents sharing passwords thanks to time-slot session validation with a reservation or booking system. Furthermore, it provides a user-friendly interface, coupled with automated functionalities to grant access, time management, live video stream, remote file access, and the automatic termination of a session once the allocated time has elapsed. The contributions of this paper are:

- A simple yet powerful web-based solution that creates a remote display session connection with a remote lab control software in a transparent and straightforward manner. No password or extra user interaction is needed.
- A system that provides automated access control to the remote lab with time-session control and access to a filesystem location. All of these features are achieved thanks to the validation of a session with a booking system.

The rest of the paper is structured as follows: Sect. 2 describes different remote display technologies to access remote labs, Sect. 3 depicts the architecture and implementation of the *BridgeServer*, Sect. 4 describes a use case where the *BridgeServer* was used with a third-party remote lab. In Sect. 5 we describe the results. Section 6 concludes the paper.

2 Related Work on Remote Display to Access Labs

The implementation of remote display serves as a viable strategy to facilitate the use and interaction with a remote lab. Notably, various applications based on both open and proprietary remote display protocols already exist for this purpose. We can mention open protocols such as the Remote Framebuffer Protocol (RFP)[1] used in Virtual Network Computing (VNC) [7], the Simple Protocol for Independent Computing Environments (SPICE) [8] used in QEMU-KVM Virtualization[2], Promox Virtual Environment[3], and the oVirt virtualization management platform[4]. Among proprietary protocols we can mention the Remote Desktop Protocol (RDP) [9] used in Windows Remote Desktop, TeamViewer[5], AnyDesk[6], and others. Note that Microsoft's RDP is a proprietary protocol that

[1] https://datatracker.ietf.org/doc/html/rfc6143.
[2] https://www.qemu.org/.
[3] https://www.proxmox.com/en/proxmox-virtual-environment/.
[4] https://www.ovirt.org/.
[5] https://www.teamviewer.com/.
[6] https://anydesk.com/en.

is based on UIT-T T.120 [10] family protocols, but the open specifications are released, allowing the implementation of third-party clients.

Unfortunately, despite the availability of these types of applications, their usage for real-time remote labs reveals inherent limitations. First, using them involves a cumbersome multi-step process, which includes software installation on the client side, sharing passwords, and manual session setups. Second, such intricacies not only limit user interaction but also pose security risks as the controlling computer is fully accessible, creating potential vulnerabilities. Third, most of these solutions cannot be fully automated, requiring users to manually grant access to the computer to be remotely accessed. Finally, most of these solutions are OS-dependent. In contrast, Web-based technology provides OS-independent solutions that simplify the deployment and the user interaction, because no client software needs to be installed, and a single interface can be used for any OS. There are Web-based versions of different remote display protocols, such as the ApacheGuacamole[7] which supports RDP and VNC protocols, noVNC[8] an open source VNC client implemented in JavaScript, Azure Bastion[9] which supports RDP, or Chrome Remote Desktop[10].

Related work on Remote Desktop and Web-based solutions for real-time remote labs, include [11], a Guacamole-based clientless system which limits control access to users in specific slots of time, but the control is not fully automated. Also, in [12] an architecture design for a web-based remote lab is proposed for optoelectronic engineering. However, the solution only works for a single context and specific lab requirements, thus not providing flexibility for other use cases. Finally, a web page interface for the Distance Lab remote lab system is presented in [13], allowing users to access different remote labs. One drawback of the system is that, while some labs are available to the public, others require manual permission from the lab owner to get access.

The identified shortcomings in current Remote Desktop and Web-based solutions highlights the necessity for a more efficient and secure approach. The proposed *BridgeServer* aims at overcoming these challenges through a user-friendly, secure, and automated solution for real-time remote lab interactions.

3 A *BridgeServer* for Real-time Remote Labs

The *BridgeServer* is a tiny web server that acts as a gateway between the internal network where the laboratory equipment is located, and the public internet. The *BridgeServer* not only provides remote display capabilities through the VNC protocol over a web page, but also includes session management through a Booking system, remote access to files, live video streaming from cameras inside the laboratory, and prevents sharing any password. Before describing the architecture of the *BridgeServer*, we will briefly describe the Remote Lab Booking System that is used.

[7] https://guacamole.incubator.apache.org/.
[8] https://novnc.com/info.html.
[9] https://azure.microsoft.com/en-us/products/azure-bastion/.
[10] https://remotedesktop.google.com/.

The Remote Lab Booking System: We have developed Book4RLab [14], a generic Remote Lab Booking System (in the following referred simply as "the Booking System"), that can be integrated with any external remote lab. Since this paper is out of the scope of a detailed description of the Booking System, we only provide a general overview of its key elements:

- The Booking System provides a teacher and student web interface. The teacher interface allows to define the available time slots, the duration of each time slot, whereas the student interface allows to make a reservation.
- The Booking System does not share any password with the remote lab for which the booking slots are reserved.
- The validation is made through an access key that is generated by the Booking System, and is passed as an URL parameter for access validation.
- The generated URL points to the remote lab platform that requires to parse the request, extract the access key, and validate it against the Booking System Application Programming Interface (API).
- If the validation is successful (i.e., if the reserved time slot corresponds to the exact time that the request is performed), all the details of the reservation are returned, otherwise, an empty message is returned.

This generic Booking System requires an extension of the remote lab web platform, to validate the reservation. Therefore, the integration of the Booking System with the *BridgeServer* was straightforward, i.e., the parsing of the request and validation of the access key is performed, before granting access to the web interface of the remote lab.

3.1 Architecture of the *BridgeServer*

The architecture was designed based on the specific needs for any real-time remote lab to operate, namely: (a) a way to access the control software that typically runs on a dedicated computer that is connected to a lab equipment (e.g., through a hardware component or a network connection); (b) a way to download the data generated by the lab equipment, which is often a CSV or Excel file, that is generated in the computer where the control software is running; (c) a way to observe the equipment through a live video streaming, typically using different types of digital cameras, to see in real-time the equipment; and (d) a way to allow remote users to access the lab, only during a specific time slot, avoiding several users to operate the equipment concurrently with potential interference. The *BridgeServer* operates on a carefully designed system architecture that includes various components (see Fig. 1). To better understand the inner workings of our solution, let's explore the key elements of the system architecture.

- A tiny web server which provides the core functionality of the *BridgeServer*
- A simple and user-friendly web client that is served by the *BridgeServer* that contains the components for remote desktop, time management, file management, and live video streaming.

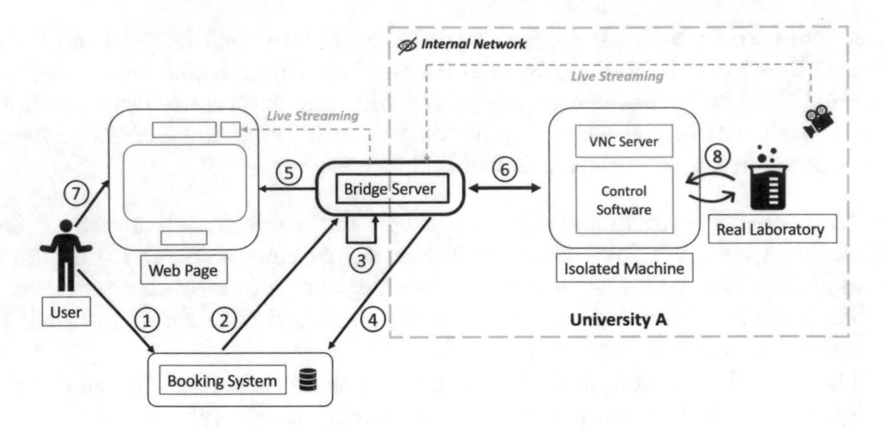

Fig. 1. The *BridgeServer* architecture.

- A Booking System that allows users to reserve time slots to have exclusive access to the remote lab. The *BridgeServer* validates the access against the Booking System, before granting access to the remote lab.
- An isolated computer within the internal network running the control software for the lab equipment.
- A real laboratory equipment interacting with the control software.
- A set of cameras to live video stream the experiments.

Basic System Setup: Since the *BridgeServer* acts as a gateway between users outside the internal network and the lab equipment, it needs to be deployed on a server that has both a public IP and connectivity to the host where the control software is operating, i.e., to have either two network interfaces or port forwarding feature. The Booking System needs to be accessible to validate the reservation and its deployment is totally independent to the *BridgeServer*. Finally, the computer running the control software should be isolated inside the internal network, to avoid any security breach. This can be done, limiting the connection within the internal network only to accept connections from the *BridgeServer* and limiting the actions that the user can do in this computer. It is important to note that the computer running the control software for the remote lab, cannot be accessed from the outside. This ensures that only the users that have a valid reservation can access this computer remotely, and therefore, users are not required to have an account or password on the *BridgeServer* and the isolated machine. This prevents users sharing passwords, and potentially accessing the remote lab at the same time.

Step-by-Step Operation: We describe the step-by-step sequence of operations within our system (see circled numbers in Fig. 1), offering a comprehensive understanding of how components interact and tasks unfold as follows:

1. Each user registers to the Booking System and reserves an exclusive time slot. The Booking System generates a link pointing to the *BridgeServer* and passing as argument the unique access key for validation.

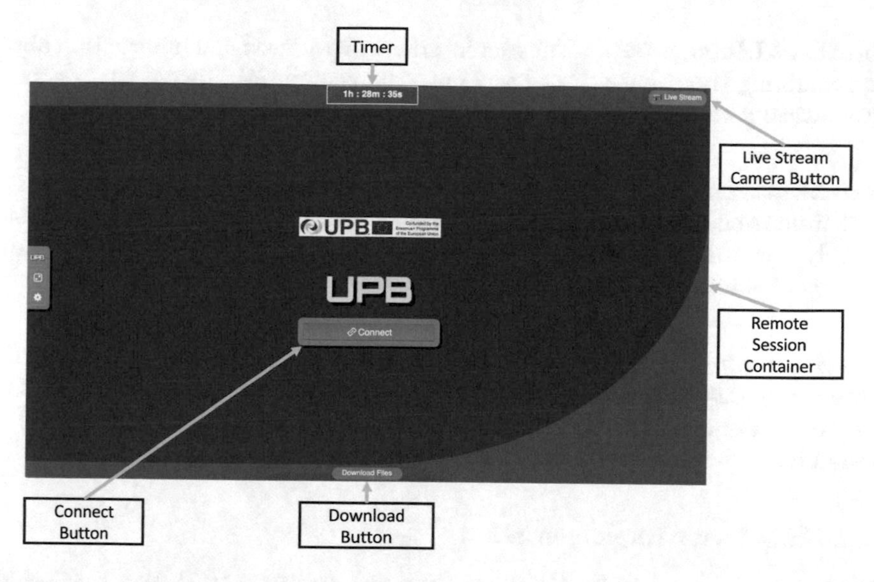

Fig. 2. *BridgeServer*'s default client interface, which includes the timer, the Remote Desktop holder (with the "Connect" button), the download files button, and the live stream button.

2. The generated link is used to access the *BridgeServer* and an HTTPS request is sent.
3. The *BridgeServer* parses the request and extracts the access key.
4. The *BridgeServer* queries the Booking System Validation API, to verify if the reservation is valid or not. If the validation fails, the *BridgeServer* returns an error page, asking to provide a valid reservation.
5. If the reservation is valid, the *BridgeServer* serves a web page which contains holders for the time management indicating the remaining time of the reservation, the remote desktop, a button to download the files generated in the computer with the control software, and the live video stream of the cameras. A "Connect" button triggers the connection to the isolated computer (see Fig. 2).
6. When the user clicks on "Connect", the connection is established between the *BridgeServer* and the isolated computer that is running a VNC server. The user does not need to provide any password.
7. The web page now contains all the information to access the remote lab. The VNC session is forwarded through the *BridgeServer* so that the user can interact with the remote lab control software, and see the live streaming from the cameras.
8. All the interactions done by the user are reflected in the real laboratory equipment that can be remotely operated to gather data.

Time Slot Management: The user interface provides visual information about the remaining time that a user has to use the remote lab, before the session is automatically closed once the allocated time runs out.

Data Management: The *BridgeServer* handles the access control that enforces strict limitations on the remote file system access. We ensure that remote users can only access and download precise paths within the isolated computer running the control software of the remote lab software.

Live Streaming: The *BridgeServer* connects to the cameras that are specified in the configuration file. These URLs are commonly used in cameras supporting different streaming formats, such as MJPEG video streams via HTTP or RTSP (Real-Time Streaming Protocol) [15].

3.2 *BridgeServer* Implementation

We combined several technologies and tools to develop and deploy the *BridgeServer*, ensuring the overall functionality and performance of the system. These technologies include:

- **Node.js**: Used for the core server and the web page. It is a popular server side open-source, cross-platform JavaScript runtime environment that enables the development of scalable and high-performance applications. The Node.js server was used to process the HTTP request, extract the access key, and query the Booking System Validation API. It also handles the timer and the closing of the connection with the VNC server.
- **noVNC**: Used to handle the remote session in the website. It uses websockify[11] that provides Web Sockets support for any application/server which allows for secure and efficient remote access to the system. We slightly modified parts of the noVNC code to integrate the timer, file download, and live streaming features.
- **TightVNC**: Used as a VNC server, providing a reliable and efficient solution for remote desktop access and control.
- **Docker**: Used to containerize the solution, enabling efficient deployment and management of the system across different environments. All the services of the *BridgeServer* can be deployed with Docker, which allows the *BridgeServer* to run on any OS.
- **Nginx**: Functioning as a reverse proxy, Nginx enhances security, load balancing, and performance. It efficiently manages incoming traffic and directs it to the appropriate server, enhancing the responsiveness of the *BridgeServer*.
- **Certbot**: Implemented for the adoption of SSL certificates, Certbot automates the process of obtaining and renewing certificates. It ensures secure communication by encrypting data exchanged between users and the *BridgeServer*, adding an extra layer of protection to sensitive information.

[11] https://github.com/novnc/websockify.

Also, we provide a configuration file (`server.env`) for customizable settings of the *BridgeServer*. The values of the configuration file are:

- `PASSWORD`: The VNC session password.
- `VIEW_ONLY_PASSWORD`: The VNC session view only password.
- `BOOKING_URL`: The URL of the Booking System API.
- `VNC_SERVER_IP_ADDRESS`: The VNC server's machine IP address.
- `VNC_SERVER_PORT`: The VNC server's machine port.
- `REFERER`: The complete URL of the *BridgeServer* web page.
- `CAMERA_URL`: The URL of the cameras' live video streams.

4 *BridgeServer* Use Case with an External Control Service

In this section, we explore the dynamic evolution of our system for a specific real-world use case. For this, we collaborated with University of São Paulo (USP) from Brazil, within an international network of remote labs. We enhanced the accessibility to USP's Refrigeration and Air Conditioning System Remote Lab [16] using our *BridgeServer*. While the core architecture remains unchanged, some adjustments were necessary to handle this particular use case demonstrating the system's inherent flexibility.

USP's remote lab uses proprietary control software, requiring the installation of a client native Windows application, based on Microsoft's .NET ClickOnce [17] technology. The users needed to run the application only through the Microsoft Edge browser, which executed the ClickOnce protocol, and downloaded a native Windows application that finally connected to the remote lab server. This limited the usage of the remote lab to Windows users, preventing users with other OS (e.g., Linux, macOS), and also preventing users from using other browsers. Furthermore, there was no way to limit the time users were connected to the remote lab, as no automatic reservation was possible, and also the link and credentials were shared with all the users, increasing the risk of concurrent access and interference.

Before using the *BridgeServer*, handling the access to the USP's remote lab, was done using a Google Calendar shared with the students (as a rudimentary booking system), where students had to request a session, forcing the remote lab manager to create a new user for each user requesting access, invalidate all the other accounts, and manually send this information via email. All this manual process was error-prone and almost impossible to manage. Also, the password was sent in plain text via email, increasing security risks.

The adapted architecture can be seen in Fig. 3. The workflow remains almost the same. The steps 1 to 6 are maintained and the steps 8 and 9 in this use case are similar to steps 7 and 8 in the core architecture. The main change is in step 7: "The isolated machine is not connected to the laboratory equipment within the local network, but executes ClickOnce third-party client control software that is installed, and remotely connects to the actual remote lab".

We used a Windows 10 Virtual Machine (VM) as the isolated computer, where all the accesses were totally restricted. Only one application should run

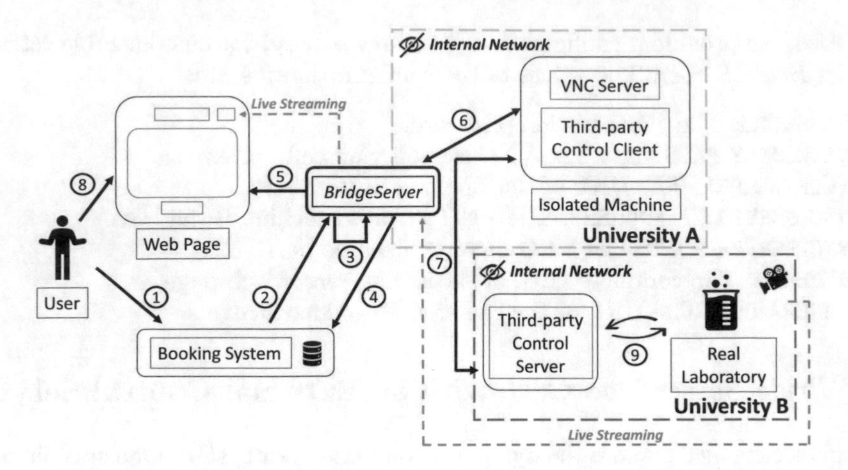

Fig. 3. Adapted *BridgeServer* architecture to support third-party remote lab software.

(i.e., the third-party ClickOnce application) without allowing the user to access any other application, get access to the network nor the file system. To make the restrictions we created a personal Group Policy Object (GPO) for the user and then modified specific policy sectors like the management of the Control Panel, the access to the settings, the taskbar interaction, the windows components and the system in general. Also, we completely disabled the access to the Start menu/screen and disable the possibility to use predefined Windows hot-keys.

Running the ClickOnce application through the Edge web browser was not an option, because automatizing the execution of Edge gave the user also full and unrestricted access to the web. We decided to create a .NET wrapper program that was able to execute the ClickOnce application as a native Windows executable as a PowerShell script. The code of the wrapper implemented in C# is shown in Fig. 4. The VM was configured to execute the wrapper script at startup every time the remote lab user logs on in the Windows VM.

The TightVNC server was configured to exit the Windows session of the "lab_user" account, once the VNC connection is finished. This ensured that the third-party software is always closed when the reserved time slot ends.

Thus, with these straightforward adaptations we managed to integrate a third-party remote lab application that was OS-dependent, had no integration with a Booking System, while allowing the lab to be fully automatized, without sharing any password with the final users. Figure 5 shows the third-party remote lab control software through the *BridgeServer*.

5 Results and Discussion

Our solution was tested and deployed under different software and hardware configurations in an international network of real-time remote labs, deployed in Latin America, Asia, Africa, and Europe. We used different computers with

```
 9  namespace ExecuteClickOnceApp
10  {
        0 references
11      class Program
12      {
            0 references
13          static int Main(string[] args)
14          {
15              string ApplicationURL = "HERE THE REMOTE LAB URL";
16              string dllbin = "rundll32.exe";
17              string dllargs = " dfshim.dll,ShOpenVerbApplication ";
18
19              var process = new Process{
20                  StartInfo = new ProcessStartInfo
21                  {
22                      WindowStyle = ProcessWindowStyle.Hidden,
23                      FileName = "cmd.exe",
24                      Arguments = "/c " + dllbin +  dllargs + ApplicationURL
25                  }
26              };
27              process.Start();
28              process.WaitForExit();
29              return 0; //exitCode;
30          }
31      }
32  }
33
```

Fig. 4. C# wrapper allowing to execute the ClickOnce program in a standalone script.

Fig. 5. A real-world use of the *BridgeServer* with an isolated host connected to an external ClickOnce service.

diverse Operating Systems (e.g., Linux and Windows) to deploy the *BridgeServer*, using a real machine connected to the lab equipment, or a Virtual Machine when using ClickOnce remote applications. All the software infrastructure was securely deployed using SSL certificates, and the credentials to get access to the VNC server were securely stored *BridgeServer* host, and were not shared with the students. Successful tests were performed with students in different academic programs. The *BridgeServer* code is available as open-source[12] to allow others to use and extend the system, thus enhancing its accessibility and collaborative potential within the global research and educational communities.

We are working on an extension to support RDP (in addition to VNC), similar to the feature of Guacamole. This will simplify even more the access to the Windows session, avoiding the use of the "lab_user" account with no password, which will enforce the security of the system.

6 Conclusions

In conclusion, our solution enhances accessibility and security for real-time remote labs via a secure web platform, eliminating the need for third-party software or password sharing. It includes a tiny web server with features like session validation, web-based remote desktop connection, session time control, live video streaming, and secure file downloads. The solution was successfully tested within an international remote lab network, especially with Windows-based labs, ensuring multi-platform support and security. Finally, our solution is available as open-source software, to promote global collaboration in research and education providing a secure and user-friendly platform for remote learning and research.

Acknowledgements. This work was partially funded by the Erasmus+ Project "EUBBC-Digital" (No. 618925-EPP-1-2020-1-BR-EPPKA2-CBHE-JP). The authors want to thank Alberto Hernandez Neto from University of São Paulo, Brazil, for testing the *BridgeServer* in the use case remote lab.

References

1. Alkhaldi, T., Pranata, I., Athauda, R.I.: A review of contemporary virtual and remote laboratory implementations: observations and findings. J. Comput. Educ. **3**(3), 329–351 (2016). https://doi.org/10.1007/s40692-016-0068-z
2. Achuthan, K., Raghavan, D., Shankar, B., Francis, S.P., Kolil, V.K.: Impact of remote experimentation, interactivity and platform effectiveness on laboratory learning outcomes. Int. J. Educ. Technol. High. Educ. **18**(1), 38 (2021). https://doi.org/10.1186/s41239-021-00272-z
3. Garcia-Zubía, J.: Remote laboratories: empowering STEM education with technology. World Scientific Publishing Company (2021). https://doi.org/10.1142/q0277
4. K. C., N., Orduna, P., Rodríguez-Gil, L., G. C., B., Susheen Srivatsa, C.N., Mulamuttal, K.: Analog electronic experiments in ultra-concurrent laboratory. In: Auer, M.E., May, D. (eds.) REV 2020. AISC, vol. 1231, pp. 37–45. Springer, Cham (2021). https://doi.org/10.1007/978-3-030-52575-0_3
5. Tawfik, M., Sancristobal, E., Martin, S., Diaz, G., Castro, M.: State-of-the-art remote laboratories for industrial electronics applications. In: 2012 Technologies Applied to Electronics Teaching (TAEE), pp. 359–364. IEEE, Vigo, Spain, June 2012. https://doi.org/10.1109/TAEE.2012.6235465

[12] https://github.com/eubbc-digital/bridgeserver.

6. Burd, S.D., Seazzu, A.F., Conway, C.: Virtual computing laboratories: a case study with comparisons to physical computing laboratories. J. Inf. Technol. Educ. Innov. Pract. **8**, 055–078 (2009). https://doi.org/10.28945/173
7. Richardson, T., Stafford-Fraser, Q., Wood, K.R., Hopper, A.: Virtual network computing. IEEE Internet Comput. **2**(1), 33–38 (1998). https://doi.org/10.1109/4236.656066
8. Spice Protocol (2009). https://www.spice-space.org/spice-protocol.html. Accessed 17 Oct 2023
9. Microsoft Corp. Remote Desktop Protocol (2022). https://learn.microsoft.com/en-us/openspecs/windows_protocols/ms-rdpbcgr. Accessed 20 Oct 2023
10. UIT. T.120 Data Protocols for Multimedia Conferencing (2008). https://www.itu.int/rec/T-REC-T.120/. Accessed 25 Oct 2023
11. Kalyan Ram, B., Arun Kumar, S., Prathap, S., Mahesh, B., Mallikarjuna Sarma, B.: Remote laboratories: for real time access to experiment setups with online session booking, utilizing a database and online interface with live streaming. In: Auer, M.E., Zutin, D.G. (eds.) Online Engineering & Internet of Things. LNNS, vol. 22, pp. 190–204. Springer, Cham (2018). https://doi.org/10.1007/978-3-319-64352-6_19
12. Kamruzzaman, M.M., Wang, M., Jiang, H., He, W., Liu, X.: A web-based remote laboratory for the college of optoelectronic engineering of online universities. In,: Optoelectronics Global Conference (OGC), pp. 1–6, Shenzhen, China, August 2015. IEEE (2015). https://doi.org/10.1109/OGC.2015.7336830
13. Sell, R.: Remote laboratory portal for robotic and embedded system experiments. Int. J. Online Biomed. Eng. (iJOE) **9**(S8), 23 (2013). https://doi.org/10.3991/ijoe.v9iS8.3370
14. Villazon, A., Ormachea, O., Orellana, A., Zenteno, A., Fransson, T.: Work in progress: a booking system for remote laboratories – the EXPLORE energy digital academy (EEDA) case study. In: Auer, M.E., Langmann, R., Tsiatsos, T. (eds.) Open Science in Engineering. REV 2023. LNNS, vol. 763, pp. 341–348. Springer, Cham (2023). https://doi.org/10.1007/978-3-031-42467-0_31
15. Schulzrinne, H., Rao, A., Lanphier, R., Westerlund, M., Stiemerling, M.: Real-Time Streaming Protocol Version 2.0, December 2016. https://www.rfc-editor.org/info/rfc7826
16. Neto, A.H., Alves, M.A.L., de Almeida, S.C.A., Rodrigues, C.S.C., Branco, D.C.: The use of remote laboratories for teaching concepts of energy conversion systems. In: Auer, M.E., Langmann, R., Tsiatsos, T. (eds.) Open Science in Engineering. REV 2023. LNNS, vol. 763, pp. 415–426. Springer, Cham (2023). https://doi.org/10.1007/978-3-031-42467-0_39
17. Microsoft Corp. ClickOnce Security and Deployment (2023). https://learn.microsoft.com/en-us/visualstudio/deployment/clickonce-security-and-deployment. Accessed 29 Oct 2023

It's a Marathon, Not a Sprint: Challenges Yet to Overcome for Digital Laboratories in Education

Marcus Soll[4](✉), Louis Kobras[4], Ines Aubel[1], Sebastian Zug[1], Claudius Terkowsky[2], Konrad Boettcher[2], Tobias R. Ortelt[2], Nils Kaufhold[2], Marcel Schade[2], Rajeenthan Sritharan[2], Jan Steinert[2], Uwe Wilkesmann[2], Pierre Helbing[3], Johannes Nau[3], Detlef Streitferdt[3], Antonia Baum[4], Annette Bock[4], Jan Haase[4], Franziska Herrmann[4], Bernhard Meussen[4], and Daniel Versick[4]

[1] TU Bergakademie Freiberg, Akademiestraße 6, 09599 Freiberg, Germany
Ines.Aubel@chemie.tu-freiberg.de, sebastian.zug@informatik.tu-freiberg.de
[2] TU Dortmund University, August-Schmidt-Straße 1, 44227 Dortmund, Germany
{claudius.terkowsky,konrad.boettcher,tobias.ortelt,nils.kaufhold, marcel.schade,rajeenthan.sritharan,jan.steinert, uwe.wilkesmann}@tu-dortmund.de
[3] Technische Universität Ilmenau, Ehrenbergstraße 29, 98693 Ilmenau, Germany
{pierre.helbing,johannes.nau,detlef.streitferdt}@tu-ilmenau.de
[4] NORDAKADEMIE gAG Hochschule der Wirtschaft, Köllner Chaussee 11, 25337 Elmshorn, Germany
{marcus.soll,louis.kobras,antonia.baum,annette.bock,jan.haase, franziska.herrmann,bernhard.meussen,daniel.versick}@nordakademie.de

Abstract. Laboratories play an important role in university education. Digital laboratories and Cross Reality laboratories, such as remote laboratories, provide many benefits to learners compared to traditional laboratories, e.g., access at any hour. However, a number of challenges arise while building and deploying these digital laboratories. This paper has the goal of collecting and presenting (organisational/technical/pedagogical) challenges, both from literature as well as from Crosslab, a large digital laboratory development project. The hope is that by systematically presenting these challenges, new solutions can be found which can be taken into consideration when developing new or evolving existing laboratories.

Keywords: cross reality laboratories · digital laboratories · expert interviews · engineering education research · instructional laboratory

1 What Are Digital Labs?

A persisting trend in the development of higher education is the offer of digital or online laboratories (also known as Cross Reality labs [18]). Zapata Rivera and Larondo Petrie [36] differentiate two kinds of online laboratories: virtual

M. E. Auer et al. (Eds.): STE 2024, LNNS 1028, pp. 220–231, 2024.
https://doi.org/10.1007/978-3-031-61905-2_22

labs, such as simulations, or remote laboratories, where physical hardware can be accessed over a network connection. Over the last years, the focus on research was put mostly on remote laboratories, while this paper aims to include all kinds of (educational) digital laboratories.

Faulconer and Gruss [7] discuss several advantages of online laboratories. For example, online laboratories allow for lower operating and maintenance costs, or allow greater accessibility by allowing students to access the laboratory according to their own schedule. Ortelt et al. [24] argue that remote laboratories enhance explorative laboratory education by illustrating knowledge through realistic processes, facilitating autonomous action and research, and promoting the development of "Work 4.0" competences. Al-Zoubi, Castro, Shahroury, and Sancristobal [1], too, argue that remote laboratories prepare for "Work 4.0".

A plethora of remote laboratory systems already exist, see for example GOLDi [13], LabsLand [23], WebLab-Deusto [21], or VISIR [10]. These systems have in common that their hardware offers little in terms of configuration flexibility – oftentimes only preconfigured experiments are available or the configurability of hardware is strongly limited.

This paper presents the challenges one might encounter while implementing and providing digital laboratories. For that, we include challenges found in literature (Sect. 2) as well as prior experiences from different institutes that were gathered by the CrossLab project group [2] (Sect. 3), with the results being presented in Table 1. We hence pose the question: Are the challenges easy to overcome so we as a community just need a short sprint, or are we in for a marathon?

2 Challenges Found in Literature

Numerous challenges for digital laboratories are already found in literature. This Section aims to summarise these challenges, sorted by organisational, technical, as well as pedagogical challenges.

2.1 Organisational Challenges

According to May et al. [17] organisational challenges start with the acquisition of a remote laboratory. Necessary technology is stated to be an **expensive investment** at times and usually could not be purchased from the funds generally available to educators and departments in a university context. Thus, acquisitions usually were made by projects funded by a third party. Furthermore, **funding of long-term costs** such as maintenance and upkeep after said projects have ended were usually not guaranteed[1]. Providing a laboratory **with**

[1] As Haug and Wedekind [11] show, a majority of projects are not maintained after funding ends. Although admittedly this is an older publication, the situation has changed little since. Mechanisms need to be put in place to combat this institutional oblivion which need to facilitate sustainable operation independently of a singular person or institution.

24/7 availability marks a constraints that often would be unachievable for institutes of higher education. In addition, most remote laboratories would be **tailor-made** by enthusiastic scientific personnel, however for a successful development process different competences between **multiple disciplines** were often required, such as laboratory technology, automisation on a hardware level, software development and pedagogy for higher education.

One important topic for digital laboratories and associated learning management systems is **data privacy**, especially considering the General Data Protection Regulation (GDPR) of the European Union (EU) [25]. This is especially important once laboratories are shared across institutions [33].

Furthermore, **involvement of industrial partners** oftentimes proves difficult. An unpublished survey conducted by the CrossLab project involving several industry representatives found that oftentimes concerns exist regarding data protection, data security, and general IT security issues. Therefore, many companies are hesitant to grant external access to their experimental and laboratory setups.

2.2 Technical Challenges

The technical side, too, brings with it several hurdles for the operation of experiments over multiple institutes. For laboratories to be flexibly available such that different educators can use the experiments in their courses concurrently and **customised to their needs/use cases** the experiments themselves need to be highly configurable. However, as Gomes and Bogosyan [9] show, this is not always the case since most laboratories are developed stand-alone **without standardisation** and lack configuration options. In addition, Soll, Haase, Helbing, and Nau [28] show that an established protocol supporting this **flexibility is still missing**. In particular, specific devices and controllers are hardwired, meaning the devices cannot easily be reused for different experiments or laboratories. Thus, new technical solutions are needed.

Sufficient scaling of the remote laboratories poses yet another challenge. Depending on the experiment, conducting it may take anything between a few milliseconds and up to and more than half an hour. To mitigate bottlenecks in the parallel and simultaneous access by students during class, a sufficiently large number of duplicate experiment instances is needed. This allows for routing students to available instances. As an example, the LabsLand electronics laboratory [35] uses a cloud solution to compensate for temporary failure of an instance by routing to an available one. A different take on solving the scalability issue can be found in so-called Ultra-Concurrent Laboratories[14]. These consist of a (sufficiently large) set of interactive video recordings of experiments. These pre-produced recordings are made available on a server. Users then have the possibility to configure the experiment with a set of pre-defined variables which determines the recordings they are able to view, which then correspond to the experimental procedure and result for the given input. This approach allows for great scalability and is especially feasible for experiments with a long execution time [14] with few degrees of freedom. Another option would be the use of simulation results with a digital twin [4].

One challenge mentioned by Gomes and Bogosyan [9] is **accessibility**. They see two aspects of accessibility which should be considered: Accessibility of laboratories for people with disabilities on the one hand and access to (digital) laboratories for people from developing countries on the other hand.

Nau and Soll [20] define five requirements for the architecture for a flexible laboratory. This includes making it easy to **add new institutes** by both making it **easy to integrate devices** as well as making it easy to integrate the architecture into the existing IT infrastructure. At the same time, the architecture should be partition tolerant so that the system still works if other institutions stop running/supporting it. In addition, adaptability and scalability play an important role. Such a system not only allows for experiments to be shared between institutes – it furthermore offers the technical basis for new forms of learning like hybrid Take-Home Labs [12].

As Uckelmann et al [32] show in their survey, only a few papers consider the **safety/security** of remote laboratories. One example for a safety guideline for federated remote laboratories can be found in [33].

2.3 Pedagogical Challenges

The pedagogical perspective of laboratory education brings with it yet another set of challenges. Many developers of laboratories give little to **no thought to the pedagogical aspects** of a laboratory. Soll [27] shows that about half of all educational laboratories in Computer Science lack a pedagogical concept. Tekkaya et al. [29], too, argue that the field of Laboratory Pedagogy is still lacking – as such, a theory for developing laboratories with a solid pedagogical foundation as well as training opportunities would be missing [29, p. 15]. Terkowsky et al. [30] give an overview over the current status of laboratory development in the context of STEM education with regards to the increasing digital transformation and outcome orientation of a student-centred laboratory education.

Another aspect of the issue is shown by Terkowsky et al. [31]. In their study the authors investigate the learning objectives of already existing laboratories. They found that most laboratories aim for the solution of simple tasks and fall short of the possibilities a student-centered outcome-oriented education offers. Among other tools the authors use a revised version of Bloom's Taxonomy [15] which describes the following cognitive layers in a hiearchical order: Remember, Understand, Apply, Analyse, Evaluate, Create. Based on this taxonomy they were able to show that most learning outcomes of the laboratories tend to fit into the lower levels of Remember, Understand, and Apply. Consequently, **there is a lack of higher pedagogical level** which leaves room for improvement.

In addition, it is important to ensure students have the **required knowledge** to actually complete the laboratory tasks successfully. For example, Uzunidis and Pagiatakis [34] observed in their experiments that students did not have all required knowledge (e.g. mathematics), which made it difficult for the students to complete tasks correctly.

An established canon of high level, **superordinate laboratory learning outcomes**, which has been expanded on, revised and reworked in several directions since, has been described by Feisel and Rosa [8]. The original work lists 13 learning outcomes students should achieve over the course of their Bachelor's degree. While some – like *Creativity* or *Analyse Data* – can be integrated into digital laboratories quite smoothly, others – like *Psychomotor* (operation of devices) or *Sensory Awareness* (perception of, e.g., smell or sound) – cannot be easily implemented. It follows, then, that remote laboratories can only ever be a supplement for physical laboratories, not a replacement. Still different learning outcomes, e.g., *Instrumentation* (selection of fitting sensors or devices) require flexible laboratories as described in Sect. 2.2. For an in-depth discussion of the Feisel and Rosa outcomes in a comparison of hands-on and remote laboratories, see Restivo and Alves [26].

On a positive note, Brinson [5] shows that students achieve a learning effectiveness of equivalent, or even greater, level in remote laboratories in comparison to traditional hands-on laboratories. This is also supported by the results of Biel and Brame [3] who showed in a meta study that well-designed online courses in Biology likely show no difference or even a positive effect size when compared to traditional in-person classes.

3 Experiences from the CrossLab Project

While literature gives a good overview of the challenges of digital laboratories, it is beneficial to combine it with the experiences from creators of digital laboratories to get a complete view, e.g., to find challenges not yet described in literature.

To find new challenges, a total of ten semi-structured interviews [19, p. 34] were conducted. All interviewees were digital laboratory creators from the CrossLab project [2], and interviewees came from a total of four different institutes across Germany. All interviews were conducted in German and all participants gave their informed consent.

All interviews were transcribed and coded both deductively and inductively following Mayring [19]. For deductive coding, the types of challenges found in literature are used. New types found in the interview are added inductively to the already existing types.

Besides confirming some of the challenge types found in literature, the following sections will focus mostly on the new, inductively found challenges in the interviews. The results can be found in Table 1.

3.1 Organisational Challenges

As additional organisational challenges, participants mentioned that **hardware is sometimes fixed in position** and any attempt to move the hardware (e.g., to work from home) requires high amounts of effort. In addition, the **communication in larger projects** that deal with the development of laboratories was often complicated and required hard work. Participants mentioned that for communication, in-person meeting were especially valuable. Another aspect is a **lack of staff**, which made the expensive investment mentioned in Sect. 2.1 even worse. Further problems might involve a **'not invented here'**-effect, where solutions are developed completely new internally instead of adapting pre-existing, well-tested solutions. Further challenges lie with **authorisation, authentication, and access** to experiments for students, especially if the experiments belong to different institutes than the students. Additionally, the degree of access to experiments for students of other institutes needs to be deliberated on an organisational level. Another challenge is posed by institutional **IT regulations** that may vary vastly between different institutes and pose a constraint on making the laboratory available.

It is of equal importance to **involve faculty**. Use of, care for, and embedding into the education of remote laboratories involve a different set of skills from physical laboratories. While the maintenance of remote laboratories can be outsourced to expert contractors, other aspects like pedagogical planning of the course or supporting the students with their use of the laboratory remain with the educators. Without this involvement, laboratories built might not even be useful to educators.

One big factor might be **institutional resistance** to new ideas like digital laboratories. This might occur at different institutional levels, e.g.,

- management might not see the advantages of digital laboratories
- the academic office/examination office might prefer the use of traditional labs
- instructors might not want to include new technology into their teaching
- students might prefer known, traditional labs above unknown, new labs

In total, university structures might prevent the introduction of digital laboratories in university curricula. This was experienced first hand by one of the interviewees.

3.2 Technical Challenges

Participants reported some technical challenges not yet mentioned in literature. They started when **existing hardware was extended** for remote laboratories, where the interfaces of the existing hardware were either not compatible to the rest of the system or were not easy to adapt to remote setting. **Existing buildings may also have bad infrastructure**, including bad internet connection. **Missing or wrong documentation** – for hardware, software or cloud providers – made implementation hard and required an trial-and-error approach. In addition, participants reported that they found both **faulty software and faulty hardware** for which workarounds needed to be developed, if possible at all. Finally, it was reported that by building digital laboratories, such as remote laboratories, a **new dependency on technology** was developed which meant that this new technology needed to be running and supported.

3.3 Pedagogical Challenges

The participants mentioned some new pedagogical challenges in the interviews:

- students have **different pre-existing knowledge**, which must be handled (e.g., by allowing students to have different learning speeds)
- **not enough time for the course or laboratory**, so compromises are needed for the intended learning outcomes
- using digital laboratories instead of physical laboratories require new **pedagogical/teaching concepts**
- digital laboratories should provide the same degrees of **freedom for experiments** as physical laboratories and students need to be able to handle the degrees of freedom
- **tools used in digital labs have knowledge requirements themselves** which need to be taught but do not contribute to the desired learning outcomes of the lab
- digital labs (especially remote labs) should convey a sense of a **real work scenario** and not be experienced as a movie with minor to no interaction
- since digital laboratories allow for unimportant but time-intensive tasks to be skipped (e.g., cleaning of instruments) and therefore allow for ***pedagogical reduction***[16]**, this free time could be filled** with new (pedagogically relevant) tasks
- one difficult problem is the **handling of errors**, especially when they are not in control of the students (e.g. the laboratory hardware lost connection to the network). This might lead to frustration, which even might make students drop the course

3.4 Other Results

Some other aspects were mentioned which do not fit in any of the above categories. Still, those aspects are worth mentioning.

First, many participants mentioned that it is really important to **set goals for the laboratories first**. These goals could either be organisational (e.g., allow students continuous access), technical (e.g., try out if some new technology could be used for laboratories) or pedagogical (e.g., define learning outcomes). When this is not done properly, there is the risk that laboratories are developed without a use case or technology is used just for technology's sake. In the same vein, depending on the goals, **not every laboratory needs to be digital; sometimes a physical laboratory is better suited for the task**.

One open question is how digital laboratories can be **used for official examination**. There are open legal/juridical, technological and organisational questions, e.g., how can one assure access to the laboratory during examination time and what are the juridical consequences on failures of technology?

One participant mentioned that **physical and digital laboratories complement each other** well, with both offering access to different possible learning outcomes. This is especially true for the integration of superordinate learning outcomes, where physical laboratories work better for some (e.g., *Sensory*

Table 1. Types of challenges combined from literature (Lit.) and interviews (Int.). The number in the interview column indicates how often the category was mentioned, no number indicates that the category is not mentioned in interviews.

Category	Challenges	Lit.	Int.	Section
Organisational	authorisation and access		4	3.1
	communication in larger projects		7	3.1
	continuous availability	X		2.1
	continuous funding	X	1	2.1
	data privacy	X		2.1
	expensive investment	X	3	2.1
	hardware fixed in position		1	3.1
	industrial involvement	X		2.1
	institutional resistance		2	3.1
	involvement of faculty/educators		1	3.1
	IT regulations		6	3.1
	lack of multidisciplinarity	X	4	2.1
	lack of staff		3	3.1
	not invented here		2	3.1
	tailor-made solutions	X	2	2.1
Technical	accessibility	X	2	2.2
	addition of new institutes	X	2	2.2
	bad infrastructure of buildings		1	3.2
	customisation for different use cases	X	3	2.2
	ease of integration of new labs/devices	X	2	2.2
	extending existing hardware		1	3.2
	faulty software/hardware		6	3.2
	flexible combinations of experiments	X	2	2.2
	missing/wrong documentation		1	3.2
	new dependency on technology		1	3.2
	no standardisation	X	1	2.2
	safety/security	X		2.2
	scalability	X	1	2.2
Pedagogical	convey real world scenario		3	3.3
	different pre-existing knowledge		2	3.3
	handling errors		1	3.3
	integration of superordinate learning outcomes	X	3	2.3
	knowledge requirement of tools		3	3.3
	lack of higher pedagogical levels	X	1	2.3
	new teaching concepts		3	3.3
	new timeframe through pedagogical reduction		2	3.3
	no pedagogical concept for laboratories	X	9	2.3
	not enough course/lab time		2	3.3
	provide freedom for experiments		2	3.3
	students have required konwledge for laboratory	X		2.3
Other results	complementation of physical and digital		1	3.4
	how to use for official examination?		1	3.4
	just do it		1	3.4
	not every laboratory is better digital		2	3.4
	professionalisation		3	3.4
	set goals for (digital) laboratory first		5	3.4

Awareness), while others might be easier to convey in digital form (e.g., *Communication* through digital collaboration).

One important step might be to increase the **professionalisation** of digital laboratories at universities, for example by standardisation of hardware, by building networks between different institutions, or by building public-private-partnerships. This would allow for cheaper and more robust hardware by bundling development and manufacturing power of participating institutes. In addition, this might make it easier to acquire continuous funding (and therefore covering upkeep) through selling access or professional support. One initiative to tackle the described phenomenon is the establishment of LabsLand[2] at University of Deusto. The goal of LabsLand is to ensure a continuous platform for the use and operation of remote laboratories in teaching. LabsLand serves two purposes here: Not only do they develop laboratories, they also function as a broker between lab offers and lab users and offer the technical infrastructure for these exchanges [22]. Persons that offer laboratories also have the possibility for additional income to secure funding beyond the original runtime of the project. Such a public-private-partnership is also recommended by the experts surveyed by May et al. [17].

Finally, one participant mentioned that it is important to **just do it**, since sometimes the advantages, disadvantages, and unforeseen side-effects of digital laboratories for specific tasks (especially compared to a traditional laboratory) can only be experienced by actually doing it.

4 Combining the Results

The combining results of both the literature and the interviews can be found in Table 1. In total, 40 organisational, technical, and pedagogical challenges for digital laboratories were identified, as well as six other results. Taking everything into consideration, many challenges yet remain unsolved; we are in for the long run, we are in for the marathon.

While this study already presents a comprehensive overview, there is still much room left to incorporate even more perspectives. For example, interviews were only conducted with members of the CrossLab project. Widening the audience to other creators of remote laboratories would be useful. In addition, other focus groups like students or IT administrators are missing completely. It would be beneficial to include those perspective to get an even wider overview over the challenges of remote laboratories.

Acknowledgement. This research was part of the project *Flexibel kombinierbare Cross-Reality Labore in der Hochschullehre: zukunftsfähige Kompetenzentwicklung für ein Lernen und Arbeiten 4.0 (CrossLab)*, which is funded by the Stiftung *Innovation in der Hochschullehre*, Germany. The qualitative analysis of the interviews was done using the QualCoder software [6].

[2] https://labsland.com/eu.

References

1. Al-Zoubi, A., Castro, M., Shahroury, F.R., Sancristobal, E.: Impact of remote labs in preparing students for Work 4.0. In: 2023 IEEE Global Engineering Education Conference (EDUCON), pp. 1–8. IEEE, Kuwait, Kuwait (2023). https://ieeexplore.ieee.org/document/10125216/
2. Aubel, I., et al.: Adaptable digital labs - motivation and vision of the CrossLab project. In: 2022 IEEE German Education Conference (GeCon), pp. 1–6. IEEE, Berlin, Germany (2022). https://ieeexplore.ieee.org/document/9942759/
3. Biel, R., Brame, C.J.: Traditional versus online biology courses: connecting course design and student learning in an online setting. J. Microbiol. Biol. Educ. **17**(3), 417–422 (2016). https://journals.asm.org/doi/10.1128/jmbe.v17i3.1157
4. Boettcher, K., Terkowsky, C., Schade, M., Brandner, D., Grünendahl, S., Pasaliu, B.: Developing a real-world scenario to foster learning and working 4.0 - on using a digital twin of a jet pump experiment in process engineering laboratory education. Eur. J. Eng. Educ. **48**(5), 949–971 (2023). https://www.tandfonline.com/doi/full/10.1080/03043797.2023.2182184
5. Brinson, J.R.: Learning outcome achievement in non-traditional (virtual and remote) versus traditional (hands-on) laboratories: a review of the empirical research. Comput. Educ. **87**, 218–237 (2015). https://linkinghub.elsevier.com/retrieve/pii/S0360131515300087
6. Curtain, C.: QualCoder version 3.4 [computer software] (2023). https://github.com/ccbogel/QualCoder/releases/tag/3.4
7. Faulconer, E.K., Gruss, A.B.: A review to weigh the pros and cons of online, remote, and distance science laboratory experiences. Int. Rev. Res. Open Distrib. Learn. **19**(2) (2018)
8. Feisel, L.D., Rosa, A.J.: The role of the laboratory in undergraduate engineering education. J. Eng. Educ. **94**(1), 121–130 (2005). https://onlinelibrary.wiley.com/doi/10.1002/j.2168-9830.2005.tb00833.x
9. Gomes, L., Bogosyan, S.: Current trends in remote laboratories. IEEE Trans. Industr. Electron. **56**(12), 4744–4756 (2009)
10. Gustavsson, I., Zackrisson, J., Håkansson, L., Claesson, I.: The VISIR project - an open source software initiative for distributed online laboratories. In: International Conference on Remote Engineering and Virtual Instrumentation (REV) (2007)
11. Haug, S., Wedekind, J.: ,,Adresse nicht gefunden" - Auf den digitalen Spuren der E-Teaching-Förderprojekte. E-Learning: Eine Zwischenbilanz. Kritischer Rückblick als Basis eines Aufbruchs 50, 19–37 (2009)
12. Henke, K., Nau, J., Streitferdt, D.: Hybrid take-home labs for the STEM education of the future. In: Uskov, V.L., Howlett, R.J., Jain, L.C. (eds.) SEEL-22 2022. SIST, vol. 305, pp. 17–26. Springer, Singapore (2022). https://doi.org/10.1007/978-981-19-3112-3_2
13. Henke, K., Vietzke, T., Hutschenreuter, R., Wuttke, H.D.: The remote lab cloud "GOLDi-labs.net". In: 2016 13th International Conference on Remote Engineering and Virtual Instrumentation (REV), pp. 37–42. IEEE, Madrid (2016). http://ieeexplore.ieee.org/document/7444437/
14. K. C., N., Orduña, P., Rodríguez-Gil, L., G. C., B., Susheen Srivatsa, C.N., Mulamuttal, K.: Analog electronic experiments in ultra-concurrent laboratory. In: Auer, M.E., May, D. (eds.) REV 2020. AISC, vol. 1231, pp. 37–45. Springer, Cham (2021). https://doi.org/10.1007/978-3-030-52575-0_3

15. Krathwohl, D.R.: A revision of Bloom's taxonomy: an overview. Theory Into Pract. **41**(4), 212–218 (2002). https://www.tandfonline.com/doi/full/10.1207/s15430421tip4104_2
16. Lewin, D.: Toward a theory of pedagogical reduction: Selection, simplification, and generalization in an age of critical education. Educ. Theory **68**(4-5), 495–512 (2018). https://onlinelibrary.wiley.com/doi/abs/10.1111/edth.12326
17. May, D., et al.: Ausblick: Welche Rolle spielen Online-Labore für die Zukunft der Laborlehre? - Antworten einer internationalen Expert*innenbefragung zur fortschreitenden Digitalisierung des Lehrens und Lernens in und mit Laboren. In: Haertel, T., Heix, S., Terkowsky, C., Frye, S., Ortelt, T.R., Lensing, K., May, D. (eds.) Labore in der Hochschullehre: Didaktik, Digitalisierung, Organisation, pp. 283–297. wbv Publikation, Bielefeld (2020). https://www.wbv.de/artikel/6004804w
18. May, D., Terkowsky, C., Varney, V., Boehringer, D.: Between hands-on experiments and cross reality learning environments - contemporary educational approaches in instructional laboratories. Eur. J. Eng. Educ. **48**(5), 783–801 (2023). https://doi.org/10.1080/03043797.2023.2248819
19. Mayring, P.: Qualitative Content Analysis: A Step-by-Step Guide. SAGE Publication Ltd, London (2022)
20. Nau, J., Soll, M.: An extendable microservice architecture for remotely coupled online laboratories. In: Auer, M.E., Langmann, R., Tsiatsos, T. (eds.) REV 2023. LNNS, pp. 97–109. Springer, Cham (2023). https://doi.org/10.1007/978-3-031-42467-0_9
21. Orduña, P., et al.: The WebLab-Deusto remote laboratory management system architecture: achieving scalability, interoperability, and federation of remote experimentation. In: Auer, M.E., Azad, A.K.M., Edwards, A., de Jong, T. (eds.) Cyber-Physical Laboratories in Engineering and Science Education, pp. 17–42. Springer, Cham (2018). https://doi.org/10.1007/978-3-319-76935-6_2
22. Orduña, P., et al.: weblablib: Ein neuer Ansatz zur Einrichtung von Remote-Laboren. In: Haertel, T., Heix, S., Terkowsky, C., Frye, S., Ortelt, T.R., Lensing, K., May, D. (eds.) Labore in der Hochschullehre: Didaktik, Digitalisierung, Organisation, pp. 249–262. wbv Publikation, Bielefeld (2020). https://www.wbv.de/artikel/6004804w
23. Orduña, P., Rodriguez-Gil, L., Garcia-Zubia, J., Angulo, I., Hernandez, U., Azcuenaga, E.: LabsLand: a sharing economy platform to promote educational remote laboratories maintainability, sustainability and adoption. In: 2016 IEEE Frontiers in Education Conference (FIE), pp. 1–6. IEEE, Erie, PA, USA (2016). http://ieeexplore.ieee.org/document/7757579/
24. Ortelt, T.R., et al.: Die digitale Zukunft des Lernens und Lehrens mit remote-Laboren. In: Digitalisierung in Studium und Lehre gemeinsam gestalten, pp. 553–575. Springer, Wiesbaden (2021). https://doi.org/10.1007/978-3-658-32849-8_31
25. Pena-Molina, A., Larrondo-Petrie, M.M.: Privacy considerations in online laboratories management systems. In: Auer, M.E., Langmann, R., Tsiatsos, T. (eds.) REV 2023. LNNS, pp. 123–134. Springer, Cham (2023)
26. Restivo, M.T., Alves, G.R.: Acquisition of higher-order experimental skills through remote and virtual laboratories. In: Dziabenko, O., Javier Garcßía-Zubía (eds.) IT innovative practices in secondary schools: Remote experiments. University of Deusto., Bilbao (2013)
27. Soll, M.: What exactly is a laboratory in computer science? In: 2023 IEEE Global Engineering Education Conference (EDUCON), pp. 1–9. IEEE, Kuwait, Kuwait (2023). https://ieeexplore.ieee.org/document/10125259/

28. Soll, M., Haase, J., Helbing, P., Nau, J.: What are we missing for effective remote laboratories? In: 2022 IEEE German Education Conference (GeCon). pp. 1–6. IEEE, Berlin, Germany (2022). https://ieeexplore.ieee.org/document/9942771/
29. Tekkaya, A.E., Wilkesmann, U., Terkowsky, C., Pleul, C., Radtke, M., Maevus, F., Deutsche Akademie der Technikwissenschaften (eds.): Das Labor in der ingenieurwissenschaftlichen Ausbildung: zukunftsorientierte Ansätze aus dem Projekt IngLab: acatech Studie. acatech Studie, Herbert Utz Verlag GmbH, München (2016)
30. Terkowsky, C., May, D., Frye, S., Haertel, T., Ortelt, T.R., Heix, S., Lensing, K. (eds.): Labore in der Hochschullehre: Didaktik, Digitalisierung, Organisation. wbv Publikation, Bielefeld (2020). https://www.wbv.de/artikel/6004804w
31. Terkowsky, C., Schade, M., Boettcher, K.E.R., Ortelt, T.R.: Once the child has fallen into the well, it is usually too late using content analysis to evaluate instructional laboratory manuals and practices. In: Auer, M.E., Langmann, R., Tsiatsos, T. (eds.) REV 2023. LNNS, pp. 11–23. Springer, Cham (2023). https://doi.org/10.1007/978-3-031-42467-0_2
32. Uckelmann, D., Mezzogori, D., Esposito, G., Neroni, M., Reverberi, D., Ustenko, M.: Safety and security in federated remote labs - a requirement analysis. In: Auer, M.E., May, D. (eds.) REV 2020. AISC, pp. 21–36. Springer, Cham (2021). https://doi.org/10.1007/978-3-030-52575-0_2
33. Uckelmann, D., et al.: Guideline to safety and security in federated remote labs. Int. J. Online Biomed. Eng. (iJOE) **17**(04), 39–62 (2021). https://online-journals.org/index.php/i-joe/article/view/18937
34. Uzunidis, D., Pagiatakis, G.: Design and implementation of a virtual on-line lab on optical communications. Eur. J. Eng. Educ. **48**(5), 913–928 (2023). https://doi.org/10.1080/03043797.2023.2173558
35. Villar-Martinez, A., et al.: LabsLand electronics laboratory: distributed, scalable and reliable remote laboratory for teaching electronics. In: Auer, M.E., Langmann, R., Tsiatsos, T. (eds.) Open Science in Engineering. LNNS, pp. 261–272. Springer, Cham (2023). https://doi.org/10.1007/978-3-031-42467-0_24
36. Zapata Rivera, L.F., Larrondo Petrie, M.M.: Models of collaborative remote laboratories and integration with learning environments. Int. J. Online Eng. (iJOE) **12**(09), 14 (2016). http://online-journals.org/index.php/i-joe/article/view/6129

Towards the Next Generation of Online Laboratory Authoring Tools

Jose C. Baca-Bustillo, Andrea E. Pena-Molina, and Maria M. Larrondo-Petrie(✉)

Florida Atlantic University, Boca Raton, FL 33461, USA
{josebacabustillo,a.e.molina}@ieee.org, petrie@fau.edu

Abstract. This paper explores features that should be included in the next generation of authoring tools to scale the use of online laboratories, and the viability of their implementation using a prototype. Existing online laboratory management systems (OLMS) do not include a comprehensive authoring tool to compose laboratory experiments that connect seamlessly to online laboratory stations. Online laboratory experiments should give seamless access to online laboratory stations, which can be virtual (software simulations), remote physical stations, mobile physical stations, or hybrid stations. Currently, online laboratory authoring tools only give professors access to embed simulations in their online laboratory experiments. Some current OLMSs provide access to a remote laboratory station through fixed laboratory experiments but do not give the professor the flexibility to change or add to the experiment. This limits the use, scalability, and flexibility of online laboratory stations. This paper discusses enhancements to the authoring module of the prototype Smart Adaptive Remote Laboratory (SARL) OLMS. The module includes the implementation of desired features not currently available in OLMSs, gathered by Lab in a Window in interviews of engineering professors and laboratory managers during a National Science Foundation (NSF) Innovation Corps (I-Corps™) training program.

SARL is an initiative of the Latin American and Caribbean Consortium of Engineering Institutions (LACCEI), the Organization of American States' Center of Excellence of Engineering for the Americas. SARL is designed to comply with IEEE 1876-2019 Standard and other standards currently under development related to online laboratories and online education.

Keywords: Online Laboratory Experiment Authoring Tool · Online Laboratory Management System · Engineering Education · Online Labs

1 Introduction

Laboratories are crucial in Science, Technology, Engineering, and Mathematics (STEM) education. ABET accreditation of engineering programs requires laboratory experimentation. During the pandemic, the need to provide access online

M. E. Auer et al. (Eds.): STE 2024, LNNS 1028, pp. 232–244, 2024.
https://doi.org/10.1007/978-3-031-61905-2_23

to laboratory experimentation was a worldwide problem due to school closures. Online laboratories, are classified as virtual (software simulations), remote (physical) laboratory stations, mobile stations, and hybrid stations [1]. To provide access to more than one online laboratory, an Online Laboratory Management System (OLMS) needs to be developed. Most OLMSs currently offer limited functionality, primarily facilitating online access to simulations or to remote laboratory stations through predefined experiments or interfaces. This limits the flexibility and use of existing remote laboratory stations. The next generation of OLMS needs to include a comprehensive authoring tool providing a more dynamic, interactive, educator-centric, and learner-centric approach.

The Institute of Electrical and Electronic Engineering (IEEE) has created the IEEE Standard 1876-2019 for Networked Smart Learning Objects for Online Laboratories and other standards currently under development related to online laboratories and online learning [2]. These standards are designed to enable interoperability and facilitate scaling the use of online laboratories. Additionally, participation in the U.S. National Science Foundation's Innovation Corps (NSF I-Corps™) program resulted in interviews of engineering professors and laboratory managers that yielded a list of desired features for the next generation of OLMSs, focusing on usability and scalability. To this end, the Smart Adaptive Remote Laboratory (SARL) OLMS prototype, initially developed to comply with the IEEE 1876-2019 Standard, has been integrated into an Online Lab Gallery Management System called Lab in a Window, and the SARL authoring module has been greatly enhanced to include the desired features gathered in the NSF I-Corps interviews.

The following section analyzes prior research on authoring tools for e-learning and, in particular, for online laboratories. Next, we describe the architecture of SARL's authoring tool, consisting of two interacting modules. Section 3 defines the module to compose online laboratory stations, which makes new online stations accessible within the SARL laboratory gallery. Section 4 shifts to the module to compose online laboratory experiments, which gives the professor choices of types of activities to include in the experiment, the option to automatically grade an activity, and choices of types of online laboratory stations to connect to the experiment. Section 5 describes the Graphical User Interface (GUI) student view of the online laboratory galleries and experiments and the automatically generated laboratory report. Finally, Sect. 6 concludes the paper with a summary of the findings, reflections on the research, and a discussion on potential avenues for future work.

2 Background

An e-Learning authoring tool is software designed for course creation that provides several features and functionalities to enable the development of courses without requiring programming expertise. These tools are crucial in creating content, organizing the course, and maintaining a standardized appearance for educational institutions [2]. The generated content is compatible with Learning

Management Systems (LMSs) and can be saved in multiple formats. Typically, e-Learning authoring tools come with pre-programmed elements, offering a user-friendly interface with templates, media resources, tools, interactive components, and assessments that professors can readily arrange and customize [3].

Various types of authoring tools are available in the market [4], offering different preferences and addressing different needs. Cloud-based tools offer flexibility by allowing collaborative work on digital learning content in real-time without installation. Desktop-based tools are potent platforms that run locally but need downloading. Open-source tools provide the source code freely, offering flexibility for experienced developers. Authoring tools integrated into LMSs are simple but limited in output, as they are often bundled with LMS packages. Free tools, while cost-free, may compromise on quality, support and scalability.

E-Learning authoring tools [5] are gaining popularity with educational institutions and business organizations creating instructional content. These tools enable trainers to integrate diverse media types and reuse existing course materials. Selecting an appropriate authoring tool is essential in developing e-learning content and can be categorized based on complexity, cost, and intended purpose. Some examples of popular authoring tools are Articulate, Adobe Presenter/Captivate, GLO Maker (Generative Learning Object Maker), E-learning Authoring Tool, ObjectJ Authoring Tool, and Advanced E-Learning Builder. These tools provide a range of features, including interactivity, narration, animations, quizzes, and multimedia integration. Additionally, they adhere to eLearning standards, e.g. SCORM, to ensure compatibility, interoperability, and content reusability.

Easy Java Simulations (EJS), a freeware and open-source software, was one of the first tools employed to create online laboratory experiments with access to simulations [6,7] or remote laboratories [7]. Farias et al. presented an EJS architecture that aligns with the model-view-control (MVC) paradigm, consisting of three integral parts: the Model, which delineates the process in terms of variables and their relationships expressed through algorithms; the Control, specifying user actions within the simulation; and the View, presenting a graphical representation of the process states [6]. In 2011, Fabregas et al. [7] employed Simulink and EJS to build remote laboratories characterized by high interactivity and visualization. Application development in EJS involves three key steps: constructing the model using the EJS built-in simulation mechanism, connecting to a Java Information Manager (JIM), and creating the view to display the model state and reactions to user-induced changes using MathLab's Simulink. The Fabregas methodology enables instructors to manipulate the laboratory experiment through a GUI, offering the students a secure and interactive learning environment [7], but does not have an authoring tool for creating new laboratory experiments that access their lab stations.

In 2015, Broisin et al. discussed the need for an authority tool to create and customize remote laboratories, and incorporated a micro authoring tool in Lab4CE (a Remote laboratory for Computer Education) that enables instructors to design experiments at two levels of granularity in a single computer or a network of computers. In addition, this tool can be used in a virtual or physical laboratory. The authoring tool allows the teacher to enter the metadata of the laboratory experiment (e.g.. the name, description, objective, and availability), describe the laboratory activities, and create the laboratory assignment [8].

Ververidis et al. [9] 2019 introduced an authoring tool that facilitates the development of a virtual laboratory, leveraging Unity3D and WordPress. The implementation of game analytics tracking functions is integrated during game development, with the game data transmitted to an analytics server for visualization. Remarkably, the backend is open source, while the analytics server operates commercially.

Multimedia Educational Resource for Learning and Online Teaching (MERLOT) is a repository of virtual laboratories. This resource enables users to engage in online, interactive simulations and experiments across diverse academic disciplines [10]. It includes the MERLOT Content Builder [11], a free and open-access website development tool seamlessly integrated with the MERLOT platform. Content Builder simplifies web page creation by using a variety of template designs, e.g. e-portfolio structures, lesson plans, pedagogical analyses, student reflections, online courses, tutorials, presentations, and community websites.

In 2020, Zapata-Rivera et al. [12] introduced a prototype of the first online laboratory experiment authoring tool, called the Online Laboratory Experiment Authoring Tool, which could access seamlessly virtual, remote, mobile and hybrid stations. This authoring tool integrates into the SARL OLMS to allow professors to create customized instances of laboratory experiments utilizing the resources available in the platform. The authoring tool facilitates the incorporation of new assets such as online laboratory stations, laboratory experiments, and external resources. Its design includes learning activities, virtual laboratories, local and remote station access, assessment definition, and supplementary learning content. Once a laboratory experiment is composed or modified, it can be stored in the SARL OLMS Laboratory Resource Manager, making it readily accessible to students. A similar authoring tool was presented in 2022 by Hosny et al. [13], called Laboratory Learning System (LLS), which additionally incorporates other modules: a Grader (manually- or automatically-graded), a Virtual Assistant, Cheating Detection and a Session Collaborator.

This paper continues the work done by [12] within the SARL OLMS, and an online laboratory gallery management platform, called Lab In A Window. The following sections describe the redesign and enhancements since 2020 to the SARL online laboratory authoring tool, which now include two separate modules: The Online Laboratory Station Composer registers laboratory stations in the platform and specifies access and interaction methods; and the Online Laboratory Experiment Composer creates or edits existing laboratory experiments that can access and interact with the online stations and external resources.

3 Online Laboratory Station Composer

Included in the desired features in the next generation of OLMSs, gathered during the NSF I-Corps interviews conducted by Lab In A Window, was the ability to easily search for, add to and use online laboratory stations available within SARL, without requiring programming skill. To enable this, laboratory managers or professors should be able to transform traditional, physical laboratory stations into remote laboratory stations, and add them to the SARL platform to make them accessible online. The original SARL authoring tool utilized WebSockets to communicate with the laboratory station. This required the institution hosting remote laboratory stations to use a dedicated Internet Protocol (IP) address for each station. For security purposes, academic Information Technology (IT) officers are reluctant to issue dedicated IP addresses within their institutional network. This greatly limited scaling the number of remote laboratories an institution could host. To eliminate the dedicated IP limitation and facilitate interaction with the registered laboratory stations, a common structured station model was designed. To accomplish this, Internet of Things (IoT) protocols were analyzed, and the Message Queue Telemetry Transport (MQTT) was selected because MQTT facilitates efficient message handling between the system and the laboratory stations without requiring a dedicated IP [14].

A station's functionality within SARL now encompasses message management to regulate the interaction flow with the laboratory station. The *create station* process initiates the generation of a new Smart Laboratory Learning Object (SLLO). The user populates the SLLO with metadata (see Table 1).

Table 1. Metadata for Online Laboratory Station

Field	Details
Discipline	The specific academic discipline or subject area
Lab Type	The nature or category of the laboratory
Location	The physical location of the laboratory station
Institution	The name of the educational or research institution
Manager Information	Details about the laboratory station manager
Sensors/Actuators	List of sensors and actuators integrated into the laboratory station
Cameras	Details about the cameras installed at the laboratory station
MQTT Topic	MQTT Topic for the laboratory station
Streaming URL for Camera	URLs for live streaming from the laboratory's cameras

The SLLO metadata includes fields that enable searching for laboratory stations: *Discipline* indicates the academic disciplines that use the laboratory station, e.g. Physics and Electrical Engineering. *Lab Type* specifies categorizing its primary function, the types of experiments conducted, or the type of equipment, e.g. balance table or oscilloscope, The metadata also includes fields that facilitate asset management and problem handling when equipment is distributed or federated: *Location* details the exact physical location of the laboratory. *Institution* refers to the academic or commercial institution hosting the laboratory. *Manager Information* provides contact details of the laboratory manager responsible for the physical equipment. Finally, the following metadata fields enable control of a remote laboratory using MQTT: The *Sensors/Actuators* field lists the sensors and actuators added to the physical laboratory to permit remote control, including their specifications and default settings, which are vital for the laboratory's interactive capabilities. The *Cameras* field now also includes the type of streaming protocol used for each camera, e.g. RTSP, HLS, or MPEG-DASH, offering insights into the laboratory's visual monitoring and streaming capabilities. The *MQTT Topic* field is a series of words or values delimited by forward slash (for example, sensor/1/temperature); it is used for communication and messaging between remote laboratory equipment and OLMS, a key aspect of an IoT-enabled laboratory. Lastly, the *Streaming URL for Camera* provides direct links for live streaming, enabling remote observation and interaction, essential for remote and hybrid laboratory experiences. Each field collectively contributes to a comprehensive understanding of the online laboratory station, enabling discovery, management, monitoring and interactivity.

4 The Online Laboratory Experiment Composer

Other desired features in next generation of OLMS gathered during the interviews included ability to scale the use of a remote laboratory station by allowing the professor to create new or adapt existing online lab experiments that could easily find and access an online laboratory station available within SARL. The original SARL authoring tool [12] required programming experience, hard-coding the access to and interaction with the associated online lab station, using the station's IP address. This allowed different laboratory experiments to be designed that used the same online laboratory station, access was managed by a scheduler. If the online station's IP changed, the experiment activity using the remote lab would not function, and required a programmer to correct.

The redesigned SARL Online Laboratory Experiment Composer empowers educators to develop diverse and adaptable online laboratory experiments that interface with any of the online laboratory stations registered through the Online Laboratory Station Composer. The new structured creation process begins with initiating an experiment instance and soliciting essential information to create the SLLO instance for the laboratory experiment that includes the metadata described in Table 2. Next, laboratory activities are added to the experiment. Activities are composed of different types of predefined interactions, e.g. incomplete tables or short-answer questions, as well as more complex predefined interactions. Currently implemented complex interactions, include a drawing interaction, which inserts a full-function drawing tool into an activity with the ability to include a partial drawing or image that a student can complete by drawing or annotating. Another complex interaction implemented is a programming interaction, which gives the professor the ability to create a programming template, specifying the programming language (C, C++, C#, Java, or Python) with the option to provide partial code for the student, and sample program input; giving the students a simplified environment to edit or complete the program, and compile and run their program within SARL. Particular engineering disciplines can provide customized interactions. For example, an electrical engineering professor can select truth table interaction that allows the student to specify the output for a combinational circuit, and sequential table interaction to facilitate the student's design of sequential circuits. This variety of interactions ensures that experiments can be designed to cater to a broad spectrum of disciplines, learning objectives and student capabilities. Additionally, an external URL interaction allows the professor to insert a video, audio, or even access to another online laboratory management platform.

Table 2. Metadata for Online Laboratory Experiment Authoring Tool

Field	Description
Experiment Instance Creation	Process to initiate the creation of a new experiment
Title	Title of the experiment
Description	Detailed description of the experiment
Thumbnail Image	Image cover for the experiment
Report Generation Option	Option to enable automatic report generation
Activity Addition	Process to add various types of activities
Activity Type	Includes options like tables, short answers, image templates, sequential tables, truth tables, external links, coding exercises
Activity Description	Comprehensive description for each activity
Difficulty Level	Difficulty level of the activity
Student Content	Guide content provided to students for completing the activity
Lab Station Association	Option to associate an activity with a laboratory station, with customization options for the station's interface
Summary Generation	Process of compiling a summary of the experiment
Publication to Database	Step to publish the experiment template to the database
Access in Creator Landing Page	Location where the experiment template is accessible for publishing to the course gallery

Greatly enriching the composer capability is integrating an online laboratory station with an activity, enhancing the experiment's interactivity and compatibility with specific laboratory requirements. This optional feature provides significant value in creating a more engaging and realistic laboratory experiments that access and control a remote laboratory station. Educators can search for the stations or resources available within SARL and associate a type of station with the experiment, customize these online laboratory stations, adjusting elements, e.g. providing a customized interface enabling or disabling button functionalities and labels to fulfill the experiment requirements while taking into consideration the knowledge level of the student and perhaps modifications required to provide accessibility for students with disabilities. The graphical user interface to interact with the station can also be adaptive by allowing the professor to utilize the image of different brands of laboratory equipment and simplified by presetting some values and limiting the controls that the student can utilize to perform the experiment, for example, using the image of a multimeter but implemented by an online oscilloscope laboratory station.

Once the experiment setup is completed, the experiment template is saved in the professor's individual gallery where the professor can try the experiment using the student view, view, edit, publish, duplicate, or delete the experiment. To create different versions of the experiment the professor can duplicate the experiment template, rename it and edit equations or other activities. When the professor wants to use a laboratory experiment, they create a course gallery and place the experiment template in the appropriate course gallery. The systems generates an access code for each course gallery that the professor distributes to the students in that class to access the experiments in the gallery. The students sign in and add the course gallery using the access code. The next section describes the Graphical User Interface (GUI) for the student view.

5 The Student Graphical User Interface System

Other requested functionalities for the next-generation OLMS included the capability to offer a self-contained laboratory experiment with seamless access to the online laboratory station and other resources, the ability to auto-grade portions of the laboratory activities, and the ability to reduce cheating. These were addressed through the redesign of the GUI described in this section.

When the professor publishes a course gallery and distributes the access code to the students enrolled, the student can create an account, sign in to SARL, and use the access code to view and use a particular course gallery. Figure 1 shows an example of the student view of an Introduction to Logic Design course gallery of a professor, showing 4 of 7 online laboratory experiments. Thumbnails of the experiments, displayed in the order they should be completed, give the student access to conduct each experiment via the browser.

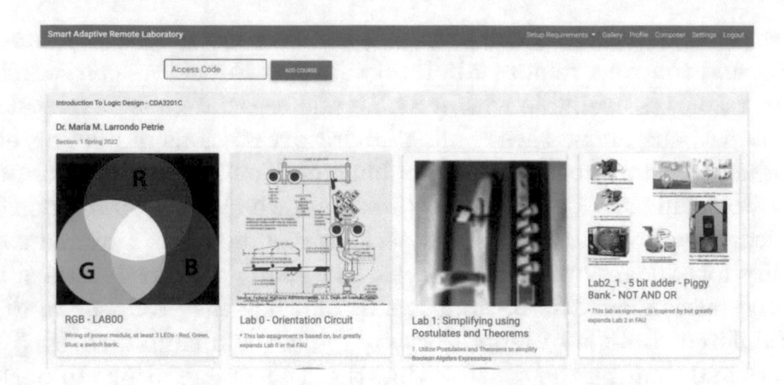

Fig. 1. Student Interface

Figure 2 illustrates the contents of the experiment, displaying all the activities to be completed in a tabular format. Each activity tab displays the instructions on the left side of the screen and the corresponding interactive component for student engagement on the right side. A feature of the system is its ability to support automatic grading, as shown in Fig. 2a where a student completed the output of the partially provided truth table and submitted it for evaluation. The score shows the student missed one of the outputs but does not show which was missed because the student still has a chance to correct the output and re-check it before continuing to activity 2. SARL captures every student interaction using Experience API (xAPI) [15] that will be used in the future to generate learning analytics, such as determining how much time a student spends on an activity, how many tries the student attempted before finalizing submission on an activity, and summary of course result. Activity 2 in Fig. 2b demonstrates the use of a drawing activity where the professor provided a template of NOT, AND, OR gates, and the student drew their circuit design, labeling and connecting the gates they needed. This type of activity demonstrates one that needs manual grading. Figure 3a shows an activity where the students cut and pasted an image of their final wired circuit into the drawing tool. The professor could also have requested a URL of a video where the students demonstrate the workings of their circuit, stepping through all or some of the possible inputs. A significant amount of teaching assistant time is usually devoted to grading each students working circuit. The professor can specify the answer to some of the activities to enable SARL to automatically grade the activity, liberating a large block of time that can be used to help the students who are having trouble doing the experiment instead of those who have completed it. Illustrated in Fig. 3a, a mobile laboratory with a pre-programmed Arduino Nano to connect to SARL was issued to each student. The physical mobile laboratory circuit can be graded remotely by SARL by instructing the student to change the input and output wires to connect to the specified Arduino pins, as shown in Fig. 3b. The Arduino is connected via USB to the laptop. SARL automatically detects the connection and captures the pre-programmed Device ID and matches it to the one issued

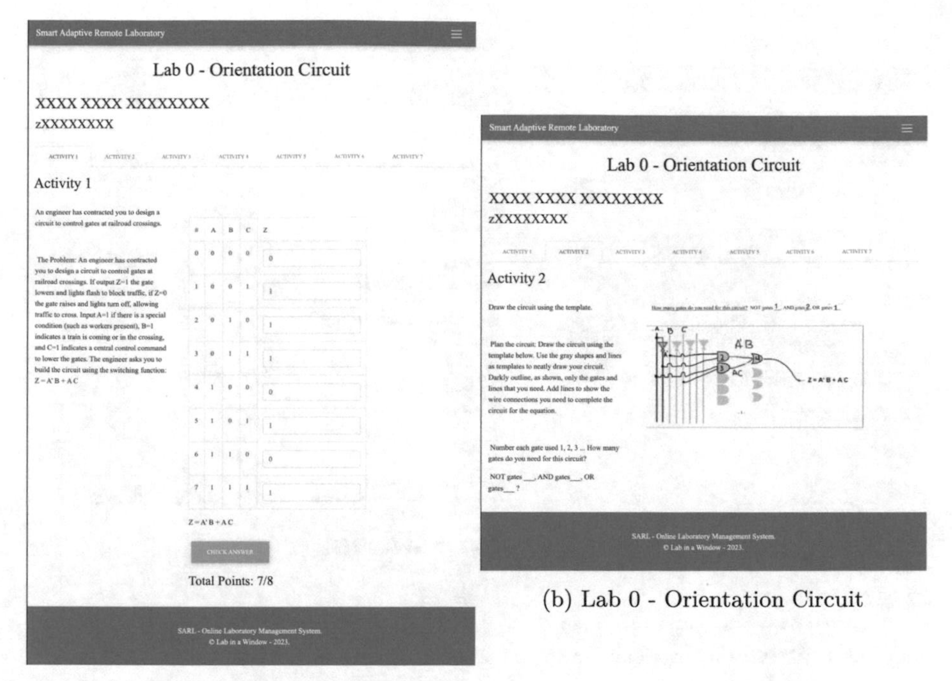

(a) Lab 0 - Orientation Circuit

(b) Lab 0 - Orientation Circuit

Fig. 2. Course Gallery

to the student. SARL is now able to drive the circuit remotely via the Arduino, displaying a partially completed truth table. As it steps through each possible combination of inputs, SARL displays the current value of each assigned pin, enters the output in the truth table with a check mark or X to indicate right or wrong, and tallies the total points scored. If it shows an error, the student has a chance to correct it and autograde it again. When ready, the student submits and moves to the next activity, in some experiments it automatically connects to what we have termed a context laboratory station, that allows the student to deploy their circuit to control a remote station. One laboratory context station depicts a miniature railroad crossing station that uses Virtual Reality (VR) to incorporate in the background the sights and sounds captured by a video of a train crossing. The student's circuit controls the actuators that control the flashing of the warning lights, the audio alert and the lowering or raising of the bar that blocks the crossing while the train travels through the intersection. The activity of deploying the circuit in a context laboratory station permits adding a reflection activity on the impact of failure or success of their design within the context deployed. Submission of the experiments automatically generates a pdf file with protected authentication measures that captures their final results and can be submitted directly the LMS to reduce cheating.

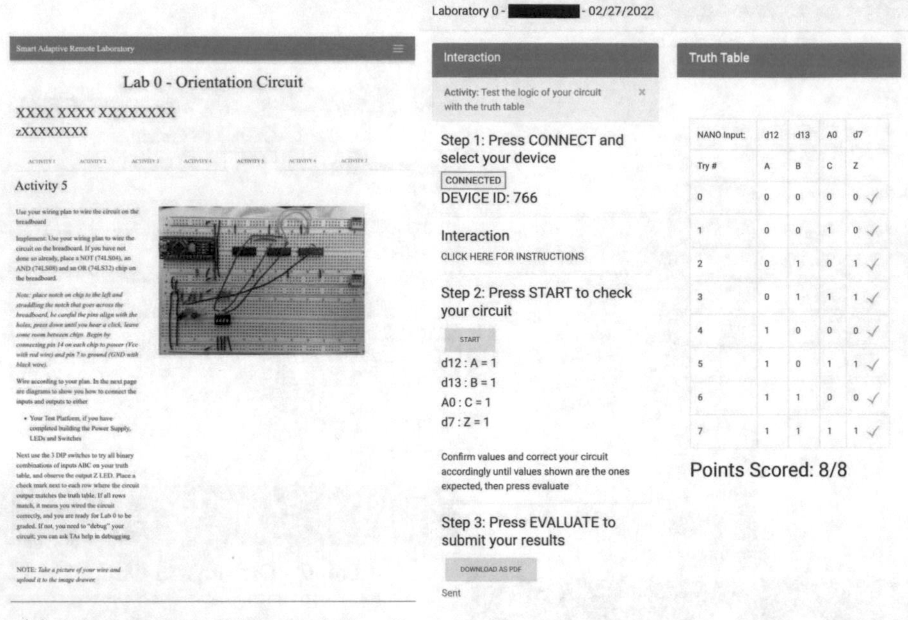

(a) Lab 0 Activity 5 Physical Circuit

(b) Lab 0 Activity 6 Hybrid Station Test

Fig. 3. Student Interface

6 Conclusions

This paper describes desirable features for the next generation of OLMSs identified through interviews conducted through participation in NSF I-Corps program and their implementations in the SARL OLMS and Lab in a Window online laboratory gallery management platform. These include the ability to scale the use of online laboratories (virtual, remote, mobile, or hybrid) through: (1) implementation of a station composer that gives the ability to easily search for, add and access online laboratory stations registered within a gallery. (2) implementation of a laboratory experiment composer that enables professors to search or create online laboratory experiments or adapt existing ones registered within a gallery. (3) implementation of a platform that provides gallery management to enable access to self-contained laboratory experiments including seamless access to the online laboratory stations and other resources, auto-grading of the laboratory activities, automatically generates the laboratory report with measures to reduce cheating.

The implementation of these features moves toward the next generation of online laboratory management systems. Previous to these improvements, the laboratory experiments were hard-coded to run on a specific, dedicate online laboratory station. This next generation of OLMS allows a laboratory station to be used by different, customized experiments, thus providing a more effective, flexible and efficient use of each station. These improvements to the SARL prototype OLMS were alpha-tested successfully with 323 mobile laboratory kits distributed to students in multiple sections of one course (Introduction to Logic Design) taught by one professor over three semesters before and during the pandemic. Future work planned includes incorporation into the OLMS an Intelligent Tutoring System and adaptive interfaces that can be personalized based on the student's knowledge and preferences.

Acknowledgment. This research was supported by the Latin American and Caribbean Consortium of Engineering Institutions (LACCEI), the Organization of American States' Center of Excellence of Engineering for the Americas. The authors deeply appreciate the opportunity to participate and collaborate with the IEEE Education Society, IEEE Standards Association's (IEEE-SA) Industry Connections (IC) program's IC Industry Consortium on Learning Engineering (ICICLE), and IEEE Standards Development Working Groups, including IEEE 1876-2019 and IEEE P2834 standards and others under development related to online laboratories and online education. Participation by Lab in a Window in the NSF Innovation Corps (NSF I-Corps) program at Florida Atlantic University yielded desired features for next-generation OLMSs.

References

1. Zapata-Rivera, L.F., Larrondo-Petrie, M.M.: Models of collaborative remote laboratories and integration with learning environments. Int. J. Online Eng. (iJOE) **12**, 14–21 (2016). https://doi.org/10.3991/ijoe.v12i09.6129
2. Pena-Molina, A.E., Larrondo-Petrie, M.M., Zapata-Rivera, L.F.: The need for E-learning standards for online laboratory management systems. In: 2022 IEEE Learning with MOOCS (LWMOOCS), Antigua Guatemala, Guatemala, pp. 240–245 (2022). https://doi.org/10.1109/LWMOOCS53067.2022.9927880
3. Andreev, I,: What is an Authoring tool?. https://www.valamis.com/hub/authoring-tool. Accessed 13 Nov 2023
4. Penfold, S.: The best eLearning authoring tools, platforms & software. https://www.elucidat.com/blog/elearning-authoring-tools/. Accessed 13 Nov 2023
5. Khademi, M., Haghshenas, M., Kabir, H.: E-learning and authoring tools: at a glance. Int. J. Res. Rev. Appl. Sci. **10**(2), 259–263 (2012)
6. Farias, G., Gomez-Estern, F, De la Torre, L., Muñoz de la Peña, D., Sánchez, C., Dormido, S.: Enhancing virtual and remote labs to perform automatic evaluation. In: IFAC Proceedings, vol. 45, no. 11, pp. 276–281 (2012). https://doi.org/10.3182/20120619-3-RU-2024.00035
7. Fabregas, E., Farias, G., Dormido-Canto, S., Esquembre, F.: Developing a remote laboratory for engineering education. Comput. Educ. **57**(2), 1686–1697 (2011). https://doi.org/10.1016/j.compedu.2011.02.015
8. Broisin, J., Venant, R., Vidal, P.: Lab4CE: a remote laboratory for computer education. Int. J. Artif. Intell. Educ. **25**(4), 154–180 (2017)
9. Ververidis, D., et al.: An authoring tool for educators to make virtual labs. In: Auer, M.E., Tsiatsos, T. (eds.) ICL 2018. AISC, vol. 917, pp. 653–666. Springer, Cham (2019). https://doi.org/10.1007/978-3-030-11935-5_62
10. Virtual Labs MERLOT. https://virtuallabs.merlot.org. Accessed 13 NOv 2023

11. MERLOT Content Builder. https://info.merlot.org/merlothelp/Content_Builder_Welcome.htm. Accessed 13 Nov 2023
12. Zapata-Rivera, L.F., Sanchez-Viloria, J., Aranzazu-Suescun, C., Larrondo-Petrie, M.M.: Design of an online laboratory authoring tool. In: 2020 IEEE Frontiers in Education Conference (FIE), pp. 1-8, IEEE, Uppsala, Sweden (2020). https://doi.org/10.1109/FIE44824.2020.9274018
13. Hosny, H.A., Elmosilhy, N.A., Elmesalawy, M.M., Abd El-Haleem, A.M.: Generic laboratory authoring tool for virtual and remote-controlled laboratories. In: 26th International Computer Science and Engineering Conference (ICSEC), pp. 319–324, Sakon Nakhon, Thailand (2022). https://doi.org/10.1109/ICSEC56337.2022.10049312
14. Sanchez-Viloria, J.A., Zapata-Rivera, L.F., Aranzazu-Suescun, C., Molina-Pena A.E., Larrondo-Petrie M.M.: Online laboratory communication using MQTT IoT standard. In: 2021 World Engineering Education Forum/Global Engineering Deans Council (WEEF/GEDC), pp. 550–555 (2021). https://doi.org/10.1109/WEEF/GEDC53299.2021.9657292.
15. Zapata-Rivera, L.F., Larrondo-Petrie, M.M.: xAPI-based model for tracking online laboratory applications. In: 2018 IEEE Frontiers in Education Conference (FIE), pp. 1–9 (2018). https://doi.org/10.1109/FIE.2018.8658869

Capturing Decision Making in eCommerce Metaverse

Apitchaka Singjai(✉), Noppon Wongta, Keattikorn Samarnggoon, Supara Grudpan, Siraprapa Wattanakul, and Chommaphat Malang

College of Art Media and Technology, Chiang Mai University, Chiang Mai, Thailand
{apitchaka.s,noppon.w,keattikorn.s,supara.g,siraprapa.w, kanokwan.ma}@cmu.ac.th

Abstract. CAMT MetaEd is a metaverse of the College of Art Media and Technology (CAMT). The development team intends to expand the metaverse in many directions. In version 2, the team created a concert where the audience can participate in a live concert and listen to live music in the CAMT metaverse. The challenge appears when the requirement of selling the product has arrived. The upcoming version of CAMT metaverse will feature eCommerce in the metaverse. Rather than traditional eCommerce, the eCommerce Metaverse offers users different virtual shopping experiences.

Goal: The purpose of this study is to gather potential eCommerce Metaverse decision-making from practitioners' perspectives.

Data and Methods: Data is derived from grey literature. The Ground Truth (GT) theory was the fundamental methodology we employed. We encoded our findings using UML-based modeling. The modeling is then validated by a domain expert.

Results: UML reflects the ground truth of decision-making associated with eCommerce metaverse, including decisions, decision options, and their relationships.

Conclusion: We reviewed a large amount of grey literature and introduced four major challenges to the eCommerce metaverse, which are virtual product design, immersive experience, secure transaction, and trust and safety.

Keywords: Decision Making · eCommerce · Grey literature · Ground Truth Theory · Metaverse

1 Introduction

In the digital transformation era, many organizations seek new opportunities to improve themselves. The College of Arts Media and Technology (CAMT) also agrees with this aspect. CAMT Metaverse tends to support the educational purpose at first. After realizing the possibility of breakthrough technology and the potential of the development team, CAMT has extended the use of the Metaverse application to various points of view. The eCommerce Metaverse is

M. E. Auer et al. (Eds.): STE 2024, LNNS 1028, pp. 245–256, 2024.
https://doi.org/10.1007/978-3-031-61905-2_24

an upcoming feature that we are going to explore, which is why it comes into the picture.

The metaverse concept is a virtual realm where people communicate, socialize, work, and play. The metaverse, a term that academics and technology users find fascinating, is a digital environment set to revolutionize our relationship with the virtual and physical worlds. However, the metaverse's growth and appearance are closely linked to rapid technological advances, especially virtual reality (VR) [1–3], augmented reality (AR), and experience in immersive experiences [4,5], which have had a big impact on its growth. This multidimensional concept has far-reaching implications, going beyond simply entertainment into fields as diverse as education, commerce, and social interaction. As a result, the metaverse is one of many topics worthy of investigation. Experts agree that the metaverse needs to offer three-dimensional, immersive virtual environments, but they disagree on how realistic or artificial the environments should be [6,7]. For the digital asset, some favor decentralized approaches [8], while others endorse centralized architectures. Regarding the debates related to the metaverse the topic of the eCommerce Metaverse is supposed to be an interesting area.

The majority of metaverse study literature focuses on customer views, clarifying user drivers in the use and spending of the e-commerce metaverse. When doing an in-depth analysis of metaverse e-commerce development, it is crucial to prioritize the perspective of practitioners. Since the practitioner's expertise is typically found on the internet, we made the decision to gather the data using gray literature. To answer the research question "What are the decisions we should consider in the eCommerce Metaverse?", we used research methodologies such as grey literature-based grounded theory.

This article is structured as follows: First, we discuss the literature Reviews in Sect. 2. Next, we explain our research method in Sect. 3. In Sect. 4, we present the detailed results of our grounded theory and discuss the implications for the research questions, which includes an in-depth comparison of our results to the Metaverse. In Sect. 5, we draw conclusions.

2 Literature Reviews

2.1 Metaverse and eCommerce in General

E-commerce has been a business focus as an impact of the COVID-19 pandemic. E-commerce technologies were promoted and studied, especially in artificial intelligence (AI). Since 2019, the literature on e-commerce has developed studies mainly in machine learning, e-commerce platform development, and service quality in online shopping.

In the meantime, the adoptions of metaverse, AR, and APIs for e-commerce systems were introduced. Metaverse and AR were offered to improve the customer experience in the reality of online shopping. Possibilities and challenges in adopting the metaverse for e-commerce businesses were studied [9].

The major area of metaverse research is exploring the perspectives of prospective consumers and their acceptance of metaverse technology [10]. Studies investigate social, behavioral, and psychological factors influencing how the e-commerce metaverse should be adopted [11]. AR, which is the environment customers experience during shopping, was studied. AR-related factors and their influences on customer confidence and purchase decisions were discussed [12,13]. AR scene spacing, which locates graphic asset positioning, was experimented in order to evaluate the UI design. The study proposed interface design guidelines to manipulate customer decisions on product selection [14]. Customer immersive experience is another research area that identifies tools and techniques to immerse customers into the metaverse shopping experience. [15] proposed a digital twin of retail to enhance shopping similar to the real world. [16] proposed a business model with live-streaming commerce that extended communication between sellers and customers. [17,18] applied AR for interactive cloth fitting. Customers can virtually try on their clothing sizes before purchase. However, few studies in the metaverse were conducted, as reviewed in the next section.

2.2 Metaverse and eCommerce in Decision Making Perspective

From the viewpoint of system design for development, design factors were defined at the application level. Functions such as user customization in product visualization, product searchability, and recommendation were specified as significant factors for customer purchase in metaverse e-commerce [19]. System components and techniques for intelligent marketing and customer engagement were identified [9]. However, the architecture design for the e-commerce metaverse has not been found in this literature area. Some studies have been conducted on the topics of consumer purchasing through application design [19] and the intelligence of ubiquitous semantic metaverse [9].

On the metaverse platform, its system architecture is already complex due to the real-time interaction from several user connections. By integrating e-commerce into the metaverse platform, additional application functions and transaction data transfers are extended.

2.3 Architectural Design Decision

Making decisions regarding the new software requirements and different stakeholders' concerns are the most challenging tasks. Most businesses spend a lot of effort making decisions because different decisions impact different final results. Generally, architecture design decisions (ADDs) refer to the important decisions that influence the selection of hardware and software and the structure of systems and components [20].

ADDs is a process of collaboratively selecting options that provide a set of potential solutions or design rationale to support decision making. As mentioned by [21], software design gets a high advantage from ADDs with a clear and consistent approach to making decisions, allowing new technology adoption with low risk.

ADDs have been developed and implemented to evaluate all candidate solutions in various contexts [22–24].

For this reason, the ADDs are brought in to determine how the metaverse and its extension will continue. Employing ADDs would help to ensure that the metaverse reaches the key objectives and meets all the requirements associated with customer needs [25].

2.4 Practitioner Perspective

In recent years, the practitioner perspective could refer to things that contribute to knowledge [26]. It provides key services to analyze factors or elements that influence decision-making.

To determine the potential factors related to the development and extension of eCommerce Metaverse, we conducted practitioner perspective-based gray literature to derive ground truth theory and UML-based modeling for support decisions. The detailed process and methodology used are described in the next section.

3 Research Methodology

Based on the existing practical knowledge and practices in the field of decision-making, we need a precise approach to acquiring the knowledge we want. We conducted a grey literature-based Ground Thoery (GT) study in which we precisely encoded our findings using UML-based modeling. In this study, as shown in Fig. 1, grey literature is used as a data collection approach. After the first open coding, formal UML-based axial and selective coding modeling is utilized instead of the typically used text-based coding technique. We used a field notes technique and stored our collected information for ongoing comparison in the GT coding process. As a result, we developed UML figures to describe what we discovered in the theoretical samplings.

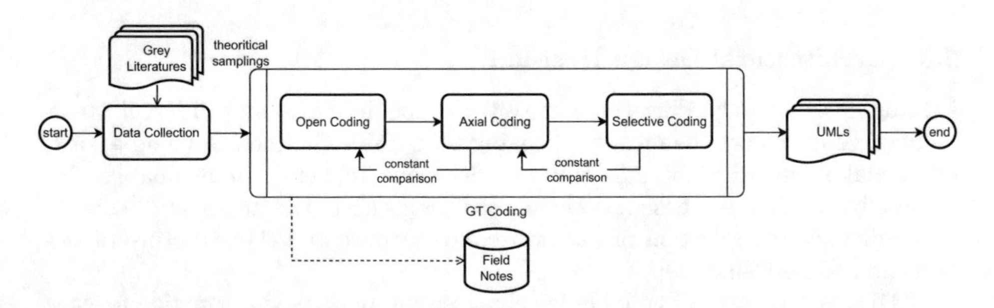

Fig. 1. Research Methodology

4 Results and Discussions

As a result, we developed four major decision-making perspectives: *product virtual design*, *immersive experience*, *secure transaction*, and *trust and safety*. In terms of practice, we can realize the *product virtual design* as the 3D infrastructure and virtual reality avatar (Fig. 2).

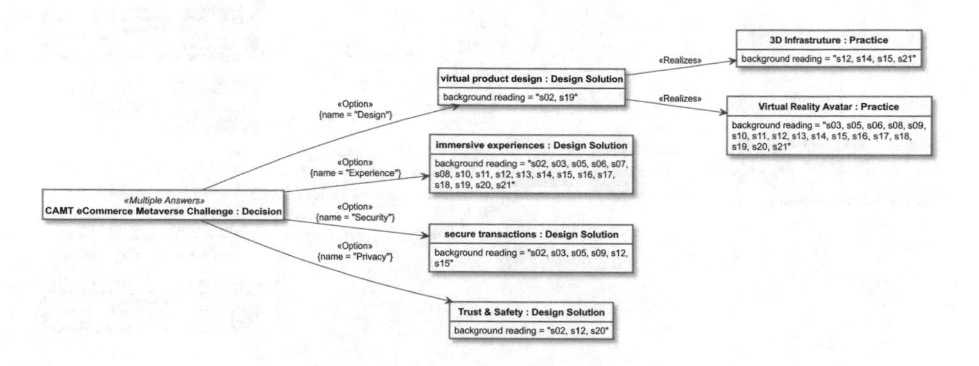

Fig. 2. Main Challenges

4.1 Virtual Product Design

The virtual product design includes considering how to create the 3D infrastructure in the same manner as the virtual reality avatar. It is expected to be one of the most difficult challenges in the eCommerce metaverse because it is tied to the user interface, where the user gets the initial impression of how it seems.

4.2 Immersive Experience

It is not only a difficulty for the metaverse, but also a challenge for eCommerce when it comes to immersive experiences (iCommerce). While we already know that the immersive experience is a unique perspective of the metaverse, eCommerce is also attempting to provide users with an experience distinct from traditional eCommerce. As a result, we divide our inquiries into two parts: 1) how to accommodate iCommerce's users with disabilities or access inequality, and 2) how customers can experience products in a way that is not possible with traditional e-commerce platforms.

For part 1, it is a human concern to deal with iCommerce users with disabilities or access inequality. The eCommerce metaverse should not be restricted to privileged people with internet access. Even with an internet connection, disabled people cannot experience what ordinary people can. To enhance inclusivity, the eCommerce development team should investigate options, including assistive

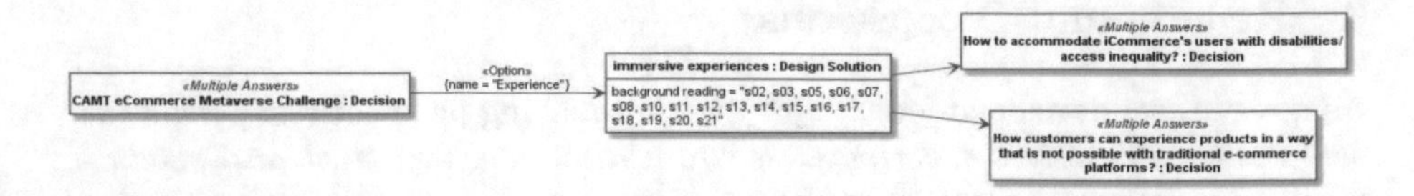

Fig. 3. Immersive Experiences: Design Solution

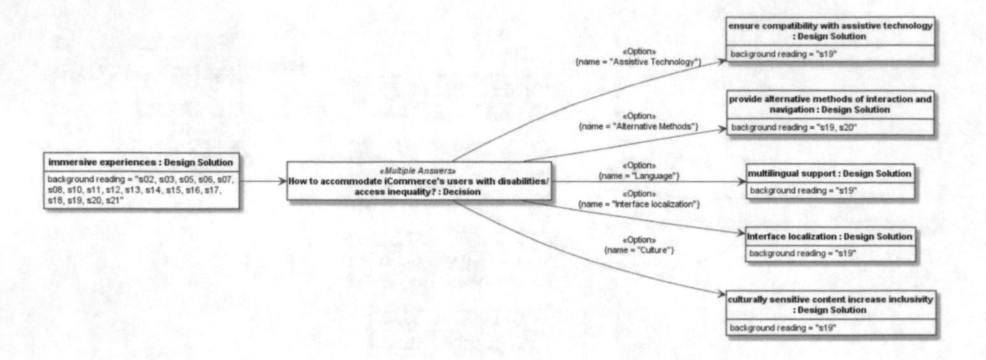

Fig. 4. Immersive Experiences Part1

technology compatibility, alternate interaction and navigation techniques, multilingual support, interface localization, and culturally sensitive content (Figs. 3 and 4).

Conventional online stores are unable to accommodate virtual product design alternatives such as 3D infrastructure and virtual reality avatars (Fig. 5).

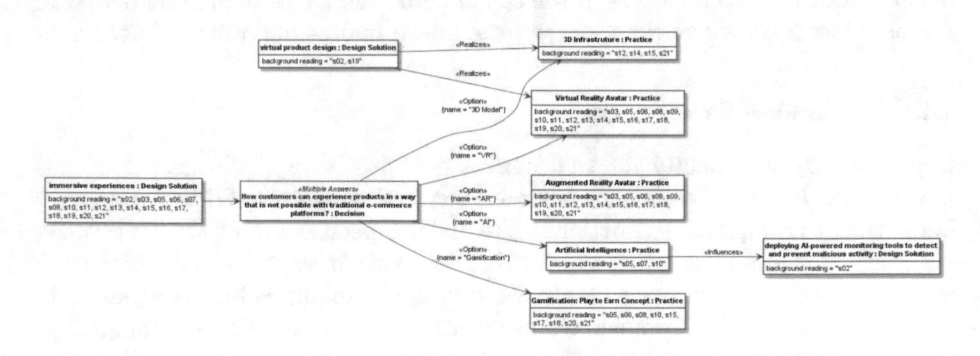

Fig. 5. Immersive Experiences Part2

For part 2, this relationship also demonstrated that while we care about the immersive experience, we cannot overlook the relevance of virtual product design. Furthermore, the eCommerce metaverse should include three additional options, which are augmented reality (AR) avatar, artificial intelligence (AI), and gamification, as shown in Fig. We have to admit that artificial intelligence has an impact on the option in the trust and safety decisions.

Our metaverse serves primarily for education, and features such as sensitive content are fully implemented to filter toxic speech and verbal aggression. The development team creates a 3D infrastructure for cultural information, allowing students to learn about their cultures. Furthermore, the development team also tends to explore other options in detail.

4.3 Secure Transaction

When it comes to secure transaction challenges, we should decide two parts: 1) what are the secure methods for conducting transactions within the metaverse, and 2) digital assets. In terms of exploring products, purchasing products, and checking out products, the eCommerce Metaverse is similar to traditional eCommerce. eCommerce Metaverse offers something unique, particularly digital assets. In any case, if the eCommerce Metaverse development team has not yet considered the digital asset, it is recommended.

Part 1 depicted in Fig. 6 there are three main options (in-house development, integration, and third-party) for implementing a secure transaction in software development in general.

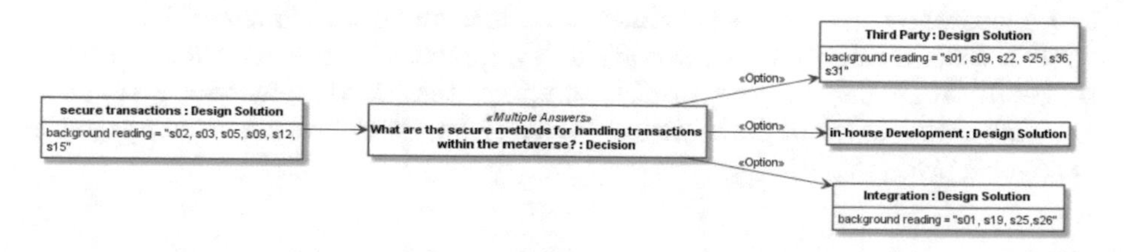

Fig. 6. Secure Transactions - Handing Transactions Part

To elaborate on the options, Fig. 7, illustrates these options in the form of a graphical representation of three stakeholders: the eCommerce Metaverse development team, third-party vendors, and eCommerce Metaverse end users. The first option (a) is internal development. The eCommerce Metaverse team should be able to respond to user requests independently. It indicates that the team is going to develop the transaction service from scratch. The second option (b) is integration development. It means the team integrates some service from a third party to accommodate some part of the transaction service they are unfamiliar with. It is the superior option for controlling development risk. The third option (c) is intended to integrate the transaction service from the third party completely. It implies the development team has a duty only to get a response from the user and transfer the request to the third party. Nonetheless, the third party must submit the response as transaction confirmation.

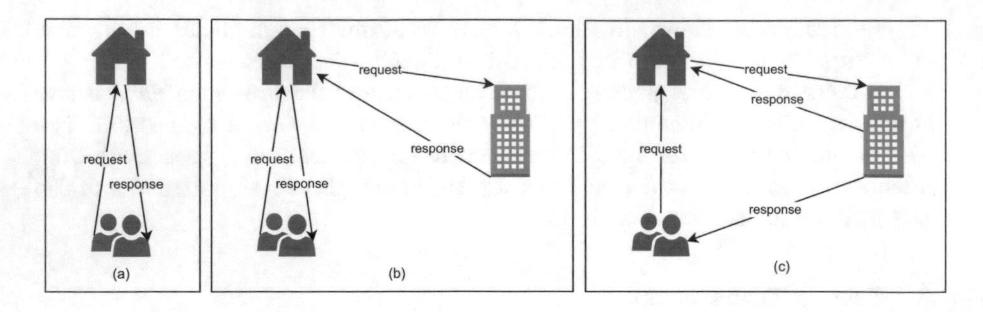

Fig. 7. Graphical Representation of Transaction Service Options

For part 2, we can separate the digital asset decision into three categories: NFTs (non-fungible tokens), cryptocurrency, and digital currency. Please note that cryptocurrency and NFTs use blockchain technology. Meanwhile, digital currency can be anything other than blockchain technology. Despite this, most NFTs are part of the Ethereum blockchain, which is the cryptocurrency. We also distinguish between these two terms (cryptocurrency and NFTs). Blockchain-based cryptocurrency does not require financial institutions to validate transactions. NFTs are blockchain-based, unique digital assets that cannot be reproduced. CAMT Metaverse, depending on the features in our metaverse, we used two approaches: in-house development and integrated development. The development team primarily uses integration development for the eCommerce sector. In the future, it may also be possible to integrate a third-party transaction service fully. We employ our own digital currency for the digital asset to engage the metaverse user (Fig. 8).

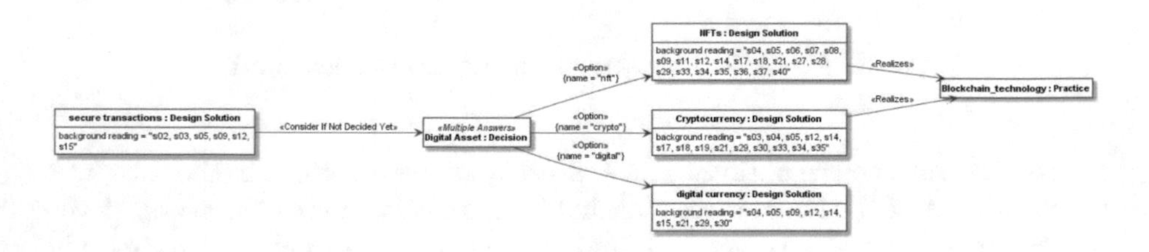

Fig. 8. Secure Transaction - Digital Asset Part

4.4 Trust and Safety

For the trust and safety challenge, four additional judgments are required: security and privacy, youth user awareness, VR-related cybercrime, and legal difficulties. To our understanding, how to cope with regulatory obstacles as technology evolves, and thorny legal concerns are myths that need further exploration from another angle (Fig. 9).

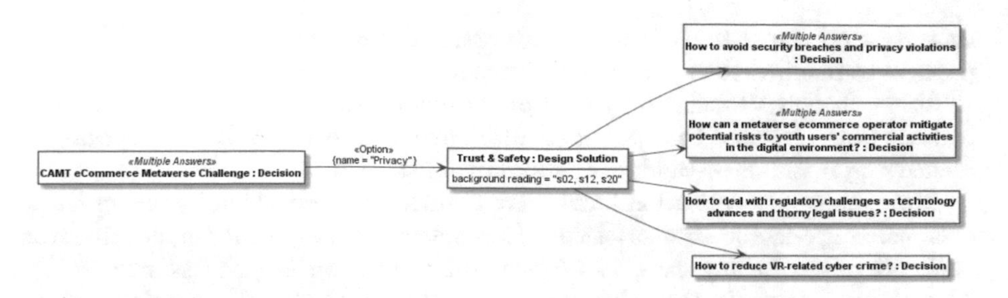

Fig. 9. Trust & Safety Challenge

To avoid breaches of security and privacy violations, decision is possible to select multiple solutions from procedures, encryption, authentication protocol, data protection, guidelines, and monitoring. For the procedures, we talk about establishing effective incident response procedures. For encryption, the key is to put strong data encryption algorithms into effect, which is also possible with blockchain technology. For authentication protocol, implementing a strong authentication protocol is recommended. Two more decisions have remained for this authentication protocol: what are the authentication factors we need? And how many authentication factors do we need? For data protection, implementing strong data protection measures is supposed to be the solution. For guidelines, we should create explicit policies for content control and standards of appropriate behavior. For monitoring, we can use artificial intelligence (AI) in monitoring by implementing AI-powered technologies to identify and stop unwanted activities (Fig. 10).

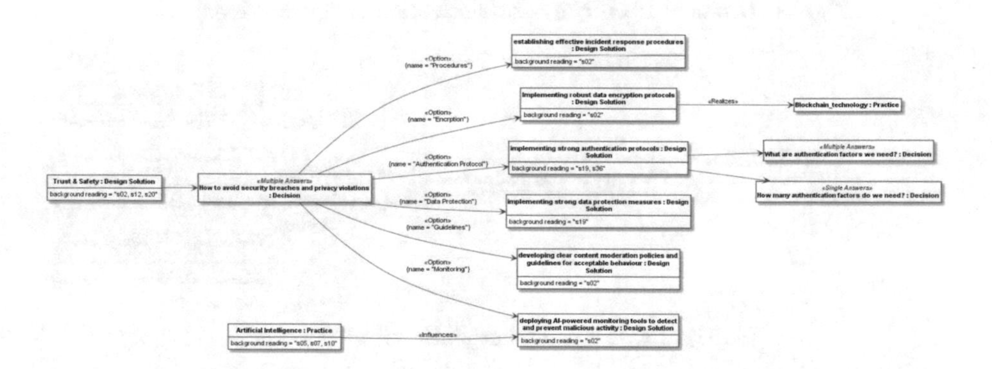

Fig. 10. Decision: How to avoid breach of security and privacy violations?

The most relevant topic for young user awareness is how a metaverse operator limits potential threats to the youth users' commercial activities in the digital environment. There are two outstanding options, which are verification and tracking. The eCommerse Metaverse should provide an age verification system that restricts access to content through more robust privacy concerns. Allowing parents to monitor their children online is another possibility.

Even though VR-related cybercrime is not on the spot for discussion, how to reduce it remains. One potential alternative is to focus on verification by identifying methods to validate IDs and avoid deepfakes.

How many authentication factors are required is determined by one of three options: single-factor authentication, two-factor authentication, or multi-factor authentication. In any case, two-factor authentication is also assumed to be multi-factor authentication. Once the required number of authentication factors has been determined, the selection of specific factors to be employed can be made. These authentication factors can be classified into three distinct groups: password, something the human owns (a token, key, or smartcard), and something the human is (fingerprint, face, or retinal scan). Regarding the authentication topic, our current and future eCommerce is going to employ multifactor authentication because the university can supply authentication services to the development team. As a result, we have the right to access the metaverse user type (staff, lecturer, student, or alumni) that allows the development team to govern the content. In other words, we are capable of handling not only authentication but also authorization (Figs. 11 and 12).

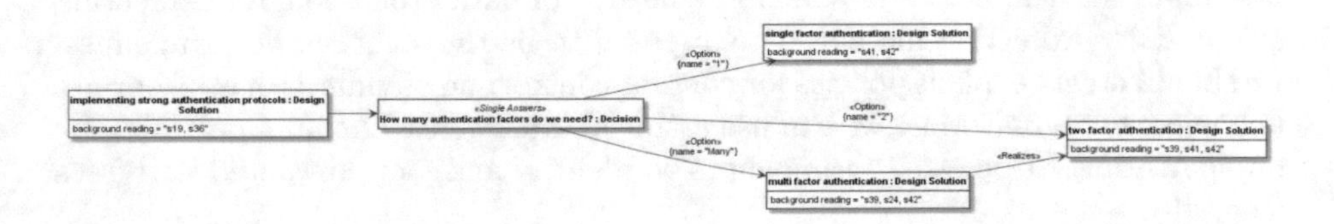

Fig. 11. Decision: How many authentication factors are required?

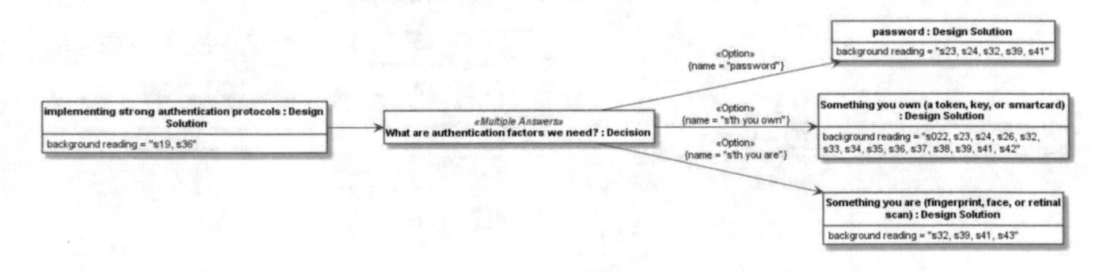

Fig. 12. Decision: Authentication Factor

5 Conclusion

The task of managing the eCommerce Metaverse is considered a crucial responsibility within the team. With regard to the research inquiry, our preferred focus is to investigate the various decision-making factors that deserve consideration. The research approach employed in our study is gray literature-based grounded theory, which was utilized to get insights into the decision-making processes and available options for our forthcoming eCommerce phase. Consequently, we face four primary issues related to virtual product design: immersive experience, secure transactions, trust, and safety. In the future, it is possible to enhance our decision-making processes using other research methodologies. The specific domain in this study that is relevant to decisions and decision options can also be expanded. Furthermore, based on an analysis of the potential decision-making process, it is doable to compare the future eCommerce Metaverse (referred to as the 'to-be' metaverse) with the current state of the eCommerce Metaverse (referred to as the 'as-is' metaverse) that is being developed. It is possible to add the decision driver and promote decision-making at another level, like Architectural Design Decisions (ADDs). In terms of APIs as components in e-commerce architecture, such as advertising APIs, natural language APIs, data and information APIs, and headless APIs are interesting areas to investigate.

References

1. Meta quest VR headsets | accessories & equipment | meta quest | meta store. https://www.meta.com/quest/
2. Pico virtual reality | official website | pico global. https://www.picoxr.com/global
3. PS VR2 features | PS VR2 UI, play area, entertainment & more | playstation (Thailand). https://www.playstation.com/en-th/ps-vr2/ps-vr2-features/
4. Microsoft hololens, mixed reality technology for business. https://www.microsoft.com/en-us/hololens
5. Magic leap 2 | the most immersive enterprise AR device. https://www.magicleap.com/en-us/
6. Neo, J.R.J., Won, A.S., Shepley, M.M.C.: Designing immersive virtual environments for human behavior research. Front. Virtual Reality **2**, 603750 (2021)
7. Jose Rubio-Tamayo, F.G.G., Gertrudix Barrio, M.: Immersive environments and virtual reality: systematic review and advances in communication, interaction and simulation. Multimodal Technol. Interact. **1**, 21 (2017). http://www.mdpi.com/2414-4088/1/4/21
8. Kyun, S., Yi, J., Jang, J.: A decentralized approach to education powered by blockchain technology. Asia-Pac. J. Convergent Res. Interchange **7**, 131–141 (2021). http://fucos.or.kr/journal/APJCRI/Articles/v7n7/13.html
9. Li, K., Lau, B.P.L., Yuan, X., Ni, W., Guizani, M., Yuen, C.: Towards ubiquitous semantic metaverse: challenges, approaches, and opportunities. IEEE Internet Things J. 1 (2023)
10. Mohamed Jahir Hussain, M.S.F.B., Bee Hwa, E.T.: Customer e-satisfaction towards online grocery sites in the metaverse world (2023)
11. Keikhosrokiani, P.: Handbook of Research on Consumer Behavioral Analytics in Metaverse and the Adoption of a Virtual World (2023)
12. Riar, M., Xi, N., Korbel, J.J., Zarnekow, R., Hamari, J.: Using augmented reality for shopping: a framework for AR induced consumer behavior, literature review and future agenda. Internet Res. **33**(1), 242–279 (2023)
13. Barta, S., Gurrea, R., Flavián, C.: How augmented reality increases engagement through its impact on risk and the decision process. Cyberpsychol. Behav. Soc. Netw. **26**(3), 177–187 (2023)

14. De Haas, E., Yiming, H., Bermejo, C., Lin, Z., Hui, P., Lee, L.-H.: Towards trustworthy augmented reality in the metaverse era: probing manipulative designs in virtual-physical commercial platforms. In: Conference Paper, pp. 779–780 (2023)
15. Ahmed, E., Darwish, A., Hassanien, A.E.: A framework for shopping based on digital twinning in the metaverse world. Stud. Big Data **123**, 155–168 (2023)
16. Jeong, H., Yi, Y., Kim, D.: An innovative e-commerce platform incorporating metaverse to live commerce. Int. J. Innov. Comput. Inf. Control **18**(1), 221–229 (2022)
17. Idrees, S., Vignali, G., Gill, S.: Interactive marketing with virtual commerce tools: purchasing right size and fitted garment in fashion metaverse. In: Wang, C.L. (ed.) The Palgrave Handbook of Interactive Marketing, pp. 329–351. Springer, Cham (2023). https://doi.org/10.1007/978-3-031-14961-0_15
18. Shams, M.Y., Elzeki, O.M., Marie, H.S.: Towards 3D virtual dressing room based user-friendly metaverse strategy. Stud. Big Data **123**, 27–42 (2023)
19. Shen, B., Tan, W., Guo, J., Zhao, L., Qin, P.: How to promote user purchase in metaverse? A systematic literature review on consumer behavior research and virtual commerce application design. Appl. Sci. (Switz.) **11**(23), 11087 (2021)
20. Enstrom, D.W.: Guideline: architectural decision (2018). https://www.unified-am.com/UAM/UAM/guidances/guidelines/uam_architectural_decision_BAE7AFA2.html
21. Gonchar, G.: A simple framework for architectural decisions (2023). https://www.infoq.com/articles/framework-architectural-decisions/
22. Singjai, A., Zdun, U., Zimmermann, O.: Practitioner views on the interrelation of microservice APIs and domain-driven design: a grey literature study based on grounded theory. In: 2021 IEEE 18th International Conference on Software Architecture (ICSA), pp. 25–35 (2021)
23. Mohamed, A., Zulkernine, M.: Architectural design decisions for achieving reliable software systems. In: Giese, H. (ed.) ISARCS 2010. LNCS, vol. 6150, pp. 19–32. Springer, Heidelberg (2010). https://doi.org/10.1007/978-3-642-13556-9_2
24. Van der Ven, J.S., Jansen, A.G., Nijhuis, J.A., Bosch, J.: Design decisions: the bridge between rationale and architecture. Rationale Manag. Softw. Eng. 329–348 (2006)
25. Sami, M.: Architectural design decisions (2023). https://melsatar.blog/2017/04/29/architectural-design-decisions/
26. Davison, R.M., Marabelli, M., Tim, Y., Beath, C.: The practitioner perspective. Inf. Syst. J. (2023). https://onlinelibrary.wiley.com/doi/10.1111/isj.12461

Codesys and TIA Portal as a Platform for the Virtual Commissioning of a Multivariable Control System

Patricia Pasmay(✉) and Douglas Plaza

Escuela Superior Politécnica del Litoral, Guayaquil EC090112, Ecuador
{ppasmay,douplaza}@espol.edu.ec

Abstract. The purpose of this work is to develop a control test verification system for a process before its virtual start-up, this process is called virtual commissioning.

Currently, industrial platforms offer the programming of their devices using several languages.

Through this work, the control of a multivariable system is designed, and a virtual start-up is implemented in both TIA Portal and Codesys, analyzing the advantages and disadvantages of both work environments.

The first chapter explains reasons for developing this type of systems, the importance at the level of automation projects and the way in which the problem will be faced.

The second chapter identifies the system model, details the development methodology, and studied techniques used for the control system to be implemented.

The third chapter presents the results of the simulations developed in Matlab Simulink and TIA Portal, Codesys, verifying the control of the system using PID blocks and the implementation of a feedforward controller. Additionally, multivariable control is verified using the library tools LSim in TIA Portal and OSCAT in Codesys, the data is analyzed with an interface in LabVIEW through communication with the OPC server and TIA Portal.

Keywords: Virtual commissioning · Control system · PLC control simulation

1 Introduction

1.1 Commissioning

In the operation of industrial plants such as steam plants and in automation projects, one of the tasks to be carried out is the start-up of systems where the necessary adjustments are made for the correct use of the control programming and the devices involved. Any mismatch in its parameters can cause delays in the implementation of the system, equipment problems or even incidents when verifying the response of the implemented control logic. In commissioning activities, little time is generally left for their execution since they are in the final phase of the project and since they depend on assembly activities, design and control tests cannot be carried out before commissioning. of the system.

M. E. Auer et al. (Eds.): STE 2024, LNNS 1028, pp. 257–266, 2024.
https://doi.org/10.1007/978-3-031-61905-2_25

Systems integrators and manufacturing companies are looking for ways to reduce the costs and times of automation projects. According to (Reinhart and Wunsch, 2007), control software engineering is responsible for more than 50% of the functionalities of automated production equipment. Operating production machines are generally PLC controlled, so it is essential to verify loaded programs offline to reduce the commissioning time of a production system.

The testing and verification of the system control logic before a plant start-up is carried out in the last phase of the projects when all the equipment and the machine are already installed.

With virtual commissioning, the aim is to verify the systems by performing a simulation with the virtual model of the process with a real controller or the simulation with the virtual model of the process with a simulated controller. The fundamental aspect of virtual commissioning is the digitization of all plant devices so that the virtual model works like a real one.

The following benefits of virtual commissioning are named [1].

- A more robust system is created by testing processes while the system is being built, leaving more time for testing.
- It analyzes the performance of the system and identifies possible bottlenecks.
- Reduction of time on site for engineers in charge of carrying out commissioning tests.
- Operator training can be carried out without interrupting the system.

2 Methodology

There are different types of design methodologies to achieve the objectives initially set for the development of the steam boiler system. The following stages are proposed below:

- Mathematical model: The mathematical model of the system is identified by applying the prediction and identification methods studied.
- Parameter configuration: Analyze and identify the variables to control the system to tune the control parameters, through the MATLAB - Simulink software and make the connection between the virtual model and the PLC.
- Programming: Program the system controller using software tools, such as: Codesys, TIA Portal and PLCSIM Advanced, to perform operation tests applying validation criteria of the studied models.
- Simulation: Validate boiler commissioning using Codesys, TIA Portal and MATLAB Simulink.

2.1 Boiler Model

The water introduced into the boiler is converted into steam by transferring heat through the metal of the tubes. The air and fuel are mixed and burned in the hearth, which is usually made up of walls of water tubes that receive the radiant heat of the flame and are where maximum heat transfer occurs. The combustion gases, because of this heat loss, cool and leave the home.

At the operating point, the responses to sudden changes in each one of the inputs have been analyzed, reaching the following conclusions [2].

- The air flow only influences the excess oxygen in the gases.
- Excess oxygen is only affected by fuel and air flow rates and their proportion.
- The vapor pressure presents a stable behavior for the two flows (fuel and water) that affect it and for the steam demand.
- The level in the boiler has an integrating character for the two mentioned flows and for the steam demand. To which is added a non-minimum phase behavior for the fuel flow and for the steam demand (typical of the swelling and contraction phenomena). This non-minimum phase behavior would also have to occur, if the model were more realistic, for the water flow.

This analysis allows us to postulate a linear model at the operating point described by the following matrix equation, between the transfer functions of the inputs and outputs.

$$\begin{bmatrix} Y_1(s) \\ Y_2(s) \\ Y_3(s) \end{bmatrix} = \begin{bmatrix} G_{11}(s) & 0 & G_{13}(s) \\ G_{21}(s) & G_{22}(s) & 0 \\ G_{31}(s) & 0 & G_{33}(s) \end{bmatrix} * \begin{bmatrix} U_1(s) \\ U_2(s) \\ U_3(s) \end{bmatrix} + \begin{bmatrix} G_{1d}(s) \\ 0 \\ G_{3d}(s) \end{bmatrix} * D(s) \quad (1)$$

Y_1 (s) is the vapor pressure, Y_2 (s) is the excess oxygen and Y_3 (s) is the water level, U_1 (s) is the fuel input, U_2 (s) the air input, and U_3 (s) is the water input. G_ij (s) is the transfer function that relates the input i to the input j, D(s) is the system perturbation and G_id (s) is the transfer function that is produced at the input G_ij (s) y the demand of the system [3].

According to the Benchmark [3] the relationship of the input variables with the output is already pre-established, being as follows:

$$Y_1(s) \rightarrow U_1(s) \quad (2)$$

$$Y_2(s) \rightarrow U_2(s) \quad (3)$$

$$Y_3(s) \rightarrow U_3(s) \quad (4)$$

The boiler model on which this design is based, represented as a MIMO block, has three input variables that can be manipulated in the range of 0% to 100% to modify the fuel, air and water flow rates of the boiler. Feeding respectively. Only in the air flow there is also a limitation in the speed of change; the current limitation is 1%. But all of them are affected by certain unknown delays. The model provides, through its three output variables, candidates for controlled variables, information about: the steam pressure in the boiler, the percentage of excess oxygen in the gases from combustion and the water level in the boiler. All outputs are provided as a percentage of an instrumentation range and are affected by noise to the extent to simulate conditions like the industrial plant [2] (Fig. 1).

The identification of the model was carried out by applying System Identification to the system inputs.

2.2 Model Identification

As a previous step to the control system, the mathematical model of the multivariable system was developed, in order to obtain different transfer functions and use them for

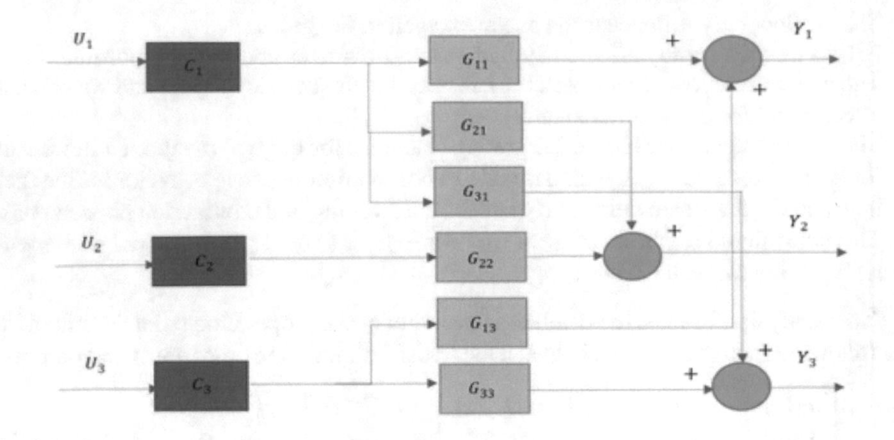

Fig. 1. Mimo control system

the implementation of control in programs for programmable logic controllers. For this, the identification of the model was carried out by applying System Identification to the system inputs.

According to the linearization of the boiler system, the following transfer functions of the model were obtained

$$G_{11}(s) = \frac{0{,}001778}{s + 0{,}003437} \tag{5}$$

$$G_{21}(s) = \frac{-5{,}4453}{10{,}488s + 1} e^{-8{,}57s} \tag{6}$$

$$G_{31}(s) = \frac{0{,}8132s - 0{,}002761}{0.3533s^2 + 1} e^{-10s} \tag{7}$$

$$G_{22}(s) = \frac{3{,}3407}{12{,}761s + 1} e^{-7s} \tag{8}$$

$$G_{13}(s) = \frac{-0{,}094414}{402{,}14s + 1} e^{-10s} \tag{9}$$

$$G_{33}(s) = \frac{0{,}0037578}{s} e^{-15s} \tag{10}$$

The transfer functions of the input (demand) dynamics are presented below:

$$G_{1d}(s) = \frac{-1{,}0235}{364{,}19s + 1} e^{-30s} \tag{11}$$

$$G_{3d}(s) = \frac{0{,}4283s - 0{,}005006}{1{,}34s^2 + s} e^{-10{,}6s} \tag{12}$$

The transfer functions obtained were identified to comply with the dynamics of the process to be controlled.

2.3 Implementation of a MIMO System in Programming Software

For the development of an adequate control system, the boiler must provide a continuous flow of steam at the desired pressure and temperature conditions and operate continuously at the lowest fuel cost while maintaining a high level of safety.

The program to use is the Siemens TIA Portal, a program used for programming medium-scale programmable logic controllers (PLC). At this point it is required to use a free access library such as LSIM.

The PID Compact continuously acquires the process value in a control loop and compares it with the required setpoint. From the result of the comparison, the PID Compact instruction calculates the output value, this output depends on the proportional, integral and derivative variables.

These values were modified according to what was done in the loop tuning carried out in the previous section.

Programming is carried out in Ladder (staircase format) in Cycle Interrupt according to the image of the block diagram made in Simulink.

In the development environment for programming controllers in accordance with the industrial standard IEC-61131–3, CODESYS, the free access licenses, Control Loop and OSCAT, are used.

3 Results

To carry out the PID control, the FRtool tool was used for each coupling at the input and output, the values of the proportional, integral and derivative constants of each system were obtained.

The values obtained from the controller are presented (Figs. 2, 3 and 4):

The programming was developed in the TIA Portal program, for the tuning of a control of a multivariable system of a control boiler, the LSim library (Library for controlled system simulation) was used, as mentioned in Sect. 2, this library allows you to simulate control systems on an S7 CPU.

To enter the transfer functions into the LSim Function Blocks, the type of polynomial order was used; if not equal, it was transformed to be able to use the block.

The FB "LSim_PT1" simulates a PT1 element. The PT1 element is a proportional transfer element with first order delay.

To configure the PID controller blocks, the PID Compact object was used, which offers a control loop optimization and tuning system [4].

The PID Compact continuously acquires the process value in a control loop and compares it with the required setpoint. From the result of the comparison, the PID Compact instruction calculates the output value, this output depends on the proportional, integral, and derivative variables, these values were modified as performed in the loop tuning performed in the previous section.

The tests of the systems were carried out as one input and one output, and the graphs for each system are presented, shown in the following graphs (Figs. 5 and 6),

Programming was carried out in Codesys using the OSCAT libraries, which are used to enter the transfer function according to its polynomial order.

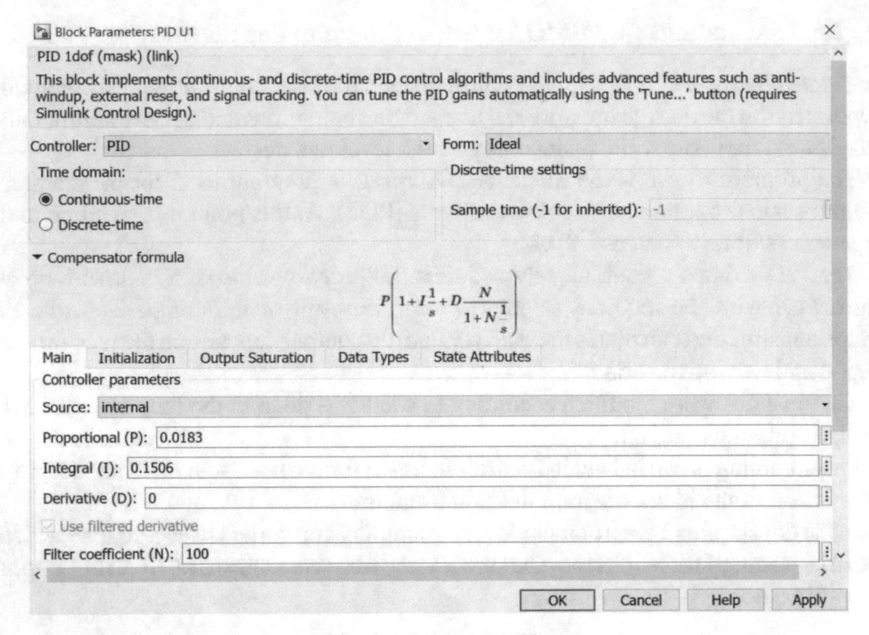

Fig. 2. Control for G11

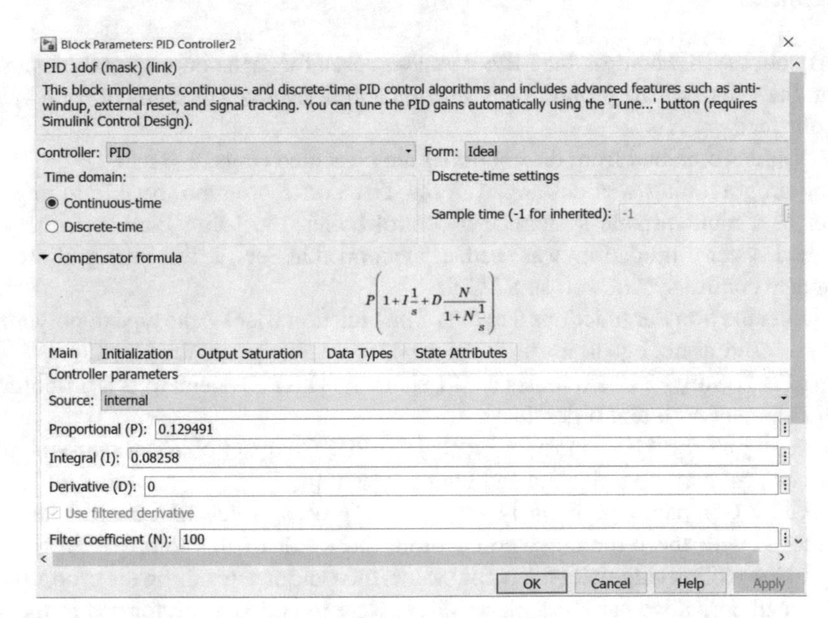

Fig. 3. Control for G22

The connections are made based on blocks in the same way as it was done in Matlab simulink, the tuning tests of the drivers found are carried out.

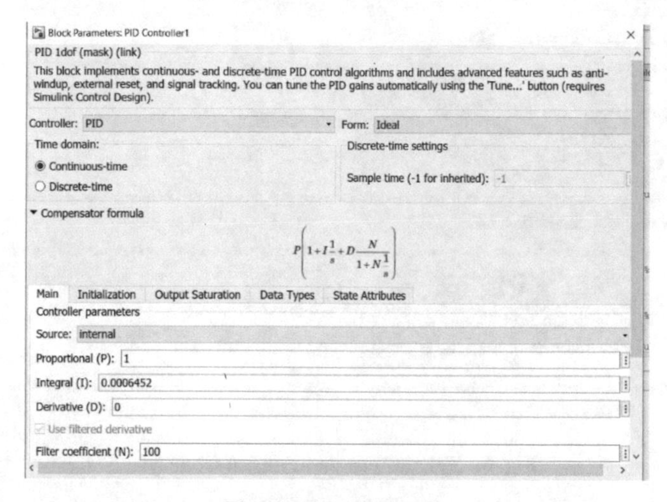

Fig. 4. Control for G33

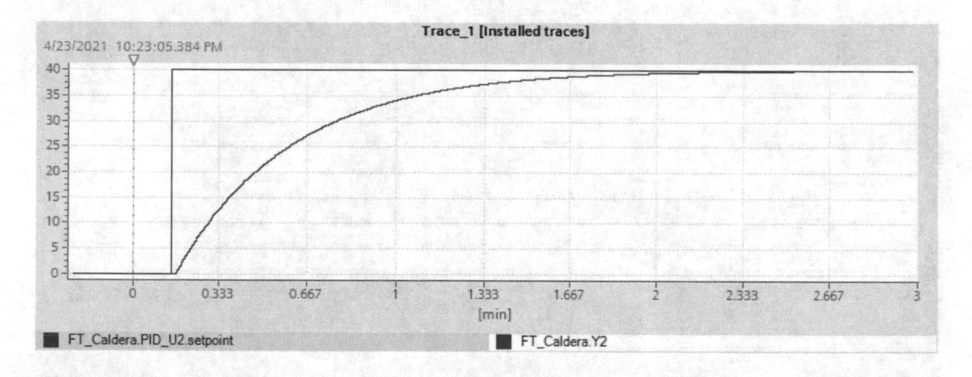

Fig. 5. Air function with PID

For both Tia Portal and Codesys, the PID values are configured, according to the tuning performed. With these values, the system response is verified and modified with trial and error if the desired response is not possible (Figs. 7 and 6).

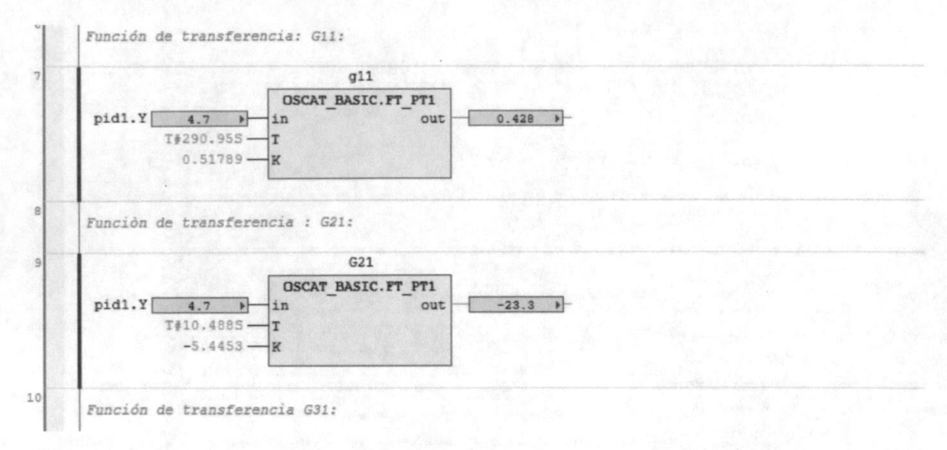

Fig. 6. Systems programming at Codesys

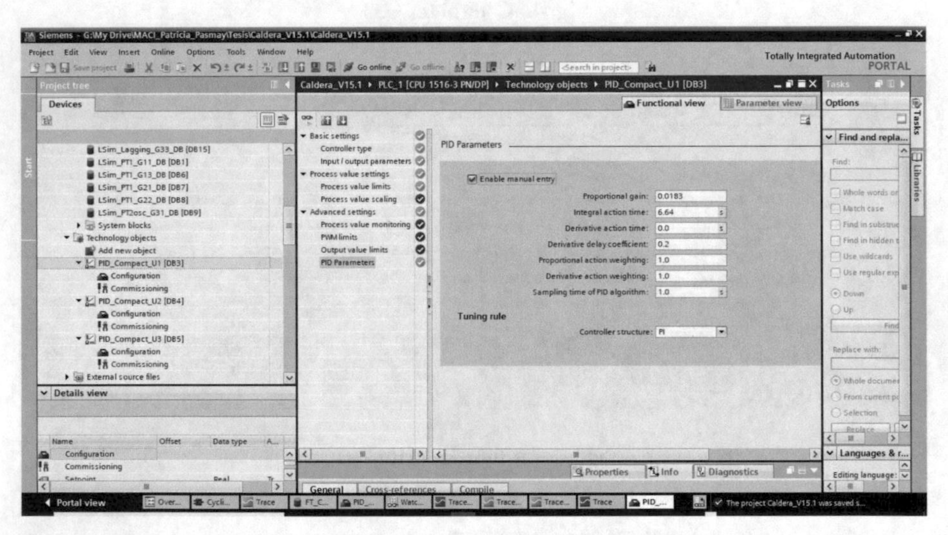

Fig. 7. TIA Portal configuration PID Values

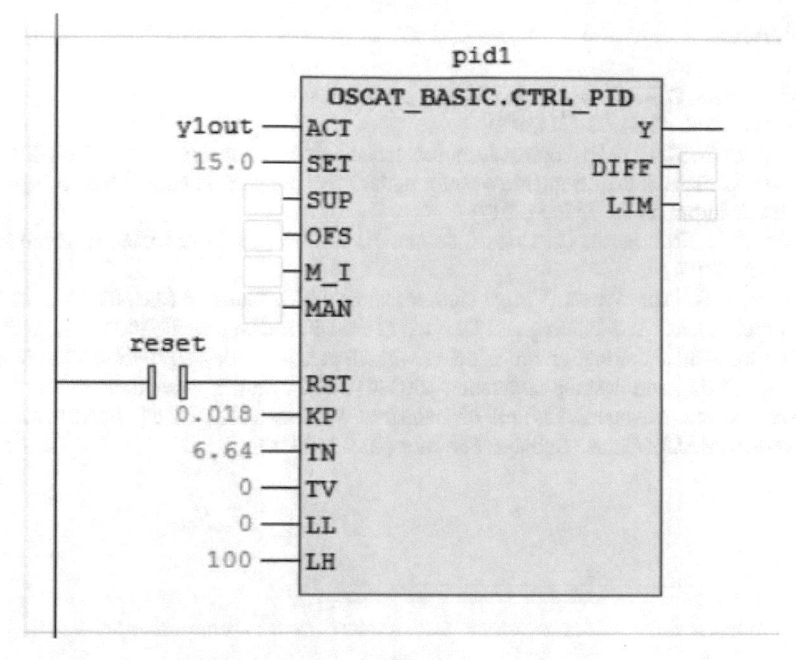

Fig. 8. Codesys, PID values configuration

4 Conclusions

- The control of the decentralized multivariable system was carried out by analyzing the behavior of one input with one output (SISO), prior to the control obtained by applying tuning techniques for a PI controller, where the control efficiency of the dynamic system was evaluated with the MIMO system.
- The control of a multivariable system was designed using default libraries (LSim) in TIA Portal and OSCAT in Codesys. Thus, carrying out commissioning tests and analyzing the behavior of each input variable with respect to the outputs, checking the behavior similar to that of Simulink.
- The digital model of the plant was implemented with a multivariable control structure and the virtual controller using the TIA Portal and Codesys industrial platform, to regulate the parameters of the steam boiler system and its functions using the functional blocks in ladder programming.
- Results closer to the Matlab simulation were obtained using the control blocks of the LSim library in TIA Portal, thus performing simpler configuration in the blocks in the same way as it is done in Simulink.
- The block configuration with the OSCAT library in Codesys did not achieve values as close to the simulation carried out in Simulink, although the simulation and processing of the system is carried out in a simpler way and does not use many machine resources as it does not have to use a program. A separate part for the simulation of the PLC, as in the case of TIA Portal, is the PLC Sim.

References

1. Jain, A., Vera, D.A., Harrison, R.: Virtual commissioning of modular automation systems. IFAC Proc. Vol. **43**(4), 72–77 (2010)
2. Rodríguez, C., Morilla, F.: Control de una caldera de vapor. Concurso en Ingeniería de Control (CIC2017). España: Departamento de Informática y Automática. Escuela Técnica Superior de Ingeniería Informática. UNED (2017)
3. Fernández, I., Rodríguez, C.: Control de una caldera. Grupo de Automática, Electrónica y Robótica (2010)
4. Siemens, A.G.: MathWorks. Virtual Commissioning with Siemens SIMATIC Target 1500S for Simulink, Part 1: Modeling and Desktop Simulation. Obtenido de (2021). https://la.mathworks.com/videos/virtual-commissioning-with-siemens-simatic-target-1500s-for-simulink-part-1-modeling-and-desktop-simulation-1603441648822.html?s_tid=srchtitle
5. Ionescu, C.M.: Advanced Control of Industrial Processes. En C. M. IONESCU. Ghent University, MACI8 Course Syllabus Feb-Mar (2021) (2021)

Work in Progress: Advancing Online Engineering Education: Developing CHM 110 Environmental Chemistry as a Case Study

Rebekah Faber-Starr[1,2] and R. F. A. Hamilton[3](✉)

[1] Northwest State Community College, Archbold, OH, USA
[2] Northern Starr Educational, Patan, Nepal
[3] Lakeland Community College, Kirtland, OH, USA
hamiltrf@gmail.com

Abstract. T Engineering education requires some fundamental courses in the natural sciences. Among those is chemistry. However online courses in chemistry are often lacking in the development of lab techniques and skills. An online chemistry course was developed that could be useful to engineering students and included online and or remote lab pieces. The course was also developed to focus on the principles of sustainability. This focus is in line with many of the Grand Challenges of Engineering [1]. Course content was the same as in an introductory chemistry course covering the same topics but focusing on the sustainability concepts related to them. Titled Chemistry 110 Environmental Chemistry, the course developed to be delivered online. Development of the course in an online space meant changing the approach to course design. Lab work was designed and built around remote teaching rather than trying to take existing labs and make them remote. An emphasis was placed on creating lab exercises that were physically manipulated. Given the remote nature, materials were selected that were sustainable and environmentally friendly.

Keywords: Remote Lab · Green Chemistry · Engineering Education

1 Background

The work is being carried out at Northwest State Community College in Archbold Ohio USA. Northwest State (NSCC) is a small rural public community college that serves a wide geographic area. The area is approximately 2500 square miles or slightly over 5100 square kilometers. This is comparable to the US State of Delaware. NSCC is the only public institution of higher education in the service area, and one of only two institutions of higher education. The campus is non-residential. All students drive to campus. Given the large geographic area the college serves for students the commute may take them more than an hour of driving time. As such the opportunity for distance education is important.

M. E. Auer et al. (Eds.): STE 2024, LNNS 1028, pp. 267–272, 2024.
https://doi.org/10.1007/978-3-031-61905-2_26

2 Introduction

2.1 Rationale for Work

Online engineering education is gaining prominence, driven by the desire to enhance accessibility and democratize learning. Fundamental courses in natural sciences, such as chemistry, are integral to engineering programs, including disciplines like electrical engineering, material science, and mechanical engineering. Most engineering courses in the US require some type of fundamental chemistry course. The transition to online education not only addresses accessibility but also democratizes engineering education. Breaking free from physical constraints opens doors for individuals who might otherwise be unable to pursue engineering due to geographical or logistical limitations [2, 3].

For clarification, for the duration of this work distance chemistry labs will be classified as either virtual or remote. Virtual labs will mean activities that are completely software based and require no physical manipulation. Remote labs are labs that use physical manipulation of materials.

During the COVID pandemic at many institutions all instruction was rapidly moved online. Even as the impacts of the pandemic have subsided, there has still been a demand among students for online learning. Subsequent paragraphs, however, are indented. Chemistry labs have been particularly vexing challenge for convert-ing to remote formats. Work on creating labs that could occur outside of the lab had begun before covid. Elio et. Al. [4] produced a summary of the work in this area prior to COVID. Much of the work at that time was focused on the use of virtual equipment and testing. The lab activities were based on software simulations. During COVID the use of virtual labs accelerated as did several other approaches which included very little hands-on work in the lab setting [5]. Other approaches have emerged that attempt to create remote labs that closely replicate the in-person experience by modifying lab exercises that students do in during in-person lab sessions [6]. This work set about creating activities that would be new and designed specifically for a remote environment.

Moving engineering courses online has the potential to reach a broader audience. By eliminating physical barriers, online education becomes a tool that allows individuals from diverse backgrounds to access quality engineering instruction. The shift towards online engineering education has become pivotal, with an emphasis on making fundamental courses accessible to a wider audience. This paper explores the development of an online chemistry course, Chemistry 110 Environmental Chemistry, tailored for engineering programs. The course not only provides a foundational understanding of chemistry but also integrates online and remote lab components. This initiative aligns with the broader goal of democratizing engineering education by removing physical boundaries.

2.2 Addition of Sustainability

It was also decided to give the course a focus on sustainability, often referred to as green chemistry. This was done to thematically link the chemistry course to engineering. Quite often introductory courses are taught by instructors with training who have built careers in that specific area. Typically, this leads to students taking courses they feel are only marginally connected or completely disconnected from one another.

Learning communities and the linking of classes [7–9] have shown that such initiatives increase student persistence and completion rates.

The sustainability focus points directly at several of the grand challenges of engineering. As they were first identified in 2008 [10] they are:

- Make solar energy economical.
- Provides energy from fusion.
- Develop carbon sequestration.
- Manage the nitrogen cycle.
- Provide access to clean water.
- Improve urban infrastructure.
- Advance health informatics.
- Engineer better medicines.
- Reverse-engineer the brain.
- Prevent nuclear terror.
- Secure cyberspace.
- Enhance virtual reality.
- Advance personalized learning.
- Engineer the tools of scientific discovery.

Students in introductory chemistry courses can quickly see the direct connection between the study of chemistry and the impact the principles of chemistry have on reaching many of these grand challenges.

The use of remote labs also introduces a more practical and immediate need for the inclusion of sustainability. Working with real chemicals and physically manipulating objects results in waste. How to handle the waste stream is an important consideration. Selecting chemicals that can be handled in regular waste stream, and are environmentally friendly is imperative. For example, single displacement reactions are usually demonstrated using silver nitrate. The resulting waste material cannot be safely disposed of in regular waste streams. All the chemicals used in the Chemistry 110 class are specifically selected based upon environmental impact.

3 Course Overview

3.1 Course Description Chemistry 110 Environmental Chemistry

This online course is meticulously crafted to meet the educational needs of engineering students. The course not only serves as an introduction to the chemical intricacies shaping Earth's environment but also integrates green chemistry principles, contributing to the broader goals of sustainability. Some of the questions addressed are:

How can chemistry help prevent, diagnose, and manage environmental problems? How do scientists think about sustainability as it relates to the environment?

How do technical, political, and social issues impact the effectiveness of our environmental stewardship? Laboratory exercises will be conducted weekly throughout the duration of the course to reinforce the Environmental principles and develop analytical thinking and sound laboratory technique and practices.

3.2 Learning Outcomes

Throughout the course, participants will:

1. Apply the scientific method and adhere to essential science protocols in a virtual, remote lab setting.
2. Demonstrate proficiency in employing standard chemistry measurement and conversion methods.
3. Articulate the composition of matter, considering its chemical, elemental, and molecular aspects.
4. Define chemical processes and elucidate their implications for environmental systems.
5. Identify potential environmental impacts through the application of chemical models.
6. Utilize quantitative concepts to assess the environmental impact of substances.
7. Describe chemical techniques used to quantify substance distribution, emphasizing real-world applications.
8. Grasp the intricate balance of competing interests related to sustainability, employing the "triple bottom line" approach.
9. Evaluate the effects of chemicals and chemical processes on the environment, fostering a comprehensive perspective.
10. Apply ethical reasoning to articulate informed positions on pressing environmental issues.

4 Future Work

As we advance into the future of online engineering education, the development and refinement of Chemistry 110 Environmental Chemistry pave the way for continuous improvement and expansion. The integration of green chemistry principles and sustainability in this online course sets the stage for ongoing innovations and enhancements. Several key areas will be the focus of future work:

4.1 Continuous Curriculum Enhancement

The curriculum of Chemistry 110 will undergo iterative reviews and enhancements to ensure its alignment with emerging advancements in both chemistry and engineering. Continuous collaboration with industry experts, incorporating feedback from learners, and staying abreast of evolving sustainability practices will be integral to refining the course content.

4.2 Technological Integration

Leveraging emerging technologies for an enriched online learning experience will be a cornerstone of future work. Exploring virtual reality (VR) and augmented reality (AR) applications, interactive simulations, and advanced online tools will further elevate the course's engagement and effectiveness.

4.3 Research Initiative

Future work will encompass research initiatives to evaluate the effectiveness of the online course in meeting educational objectives. Data analytics and feedback mechanisms will be employed to assess learner performance, engagement, and the impact of sustainability-focused content on their understanding and application of engineering principles.

4.4 Professional Development

Recognizing the dynamic nature of engineering disciplines, future work will explore avenues for professional development embedded within the course. Offering learners access to updated industry insights, certifications, and opportunities for hands-on experiences, even in virtual environments, will enhance their readiness for real-world engineering challenges.

5 Discussion

The incorporation of green chemistry principles into an online format necessitated a paradigm shift in pedagogical approaches. By focusing on sustainability, the course not only provides a foundational understanding of chemistry but also cultivates an awareness of the environmental impact of chemical processes. Exploring the effectiveness of this approach fosters a holistic understanding of engineering and environmental responsibility among learners.

The transition from traditional, in-person labs to remote labs poses unique challenges. Challenges encountered during the design and implementation of remote lab exercises, causes a critical examination of the strategies employed to ensure that the remote labs not only replicate but enhance the learning experience, addressing concerns related to physical manipulation, waste management, and the selection of sustainable materials.

An integral aspect of online education is learner engagement. Maintaining high levels of engagement in an online chemistry course requires interactive elements, realworld applications, and the integration of emerging technologies in sustaining learner interest. Additionally, learner satisfaction surveys and feedback mechanisms are considered in order to assess the overall effectiveness of the course design.

By democratizing access to quality engineering instruction, particularly in the realm of lab-based courses, the course contributes to the ongoing global effort to make education more accessible. The potential for scalability, adaptation to diverse learner needs, and the establishment of partnerships with other educational institutions are vast.

The need for ongoing research initiatives to assess the long-term impact of the course on learners' academic and professional trajectories is critical to success. Additionally, identifying areas for further exploration, such as the integration of emerging technologies, cross-cultural adaptability, and continuous improvement based on learner feedback is key.

By explicitly addressing sustainability and connecting chemistry principles to, Chemistry 110 contributes to shaping a new generation of engineers equipped to tackle complex, real-world problems. The importance of such courses in preparing learners to address the multifaceted challenges of the engineering profession reiterates the course's alignment with the Grand Challenges of Engineering.

References

1. Board, O.S.: National Academies of Sciences, Engineering, and Medicine. Environmental engineering for the 21st century: Addressing grand challenges. National Academies Press v
2. Smith, A., et al.: Online laboratories: a bridge to future engineering education. In: Proceedings of the International Conference on Engineering Education (2022)
3. GOLC. Strategic Plan: Enhancing Global Collaboration in Online Laboratory Education (2023)
4. Elio, S.C., et al.: From a hands-on chemistry lab to a remote chemistry lab: challenges and constrains. In: Auer, M.E., Zutin, D.G. (eds.) Online Engineering & Internet of Things. LNNS, vol. 22, pp. 125–131. Springer, Cham (2018). https://doi.org/10.1007/978-3-319-64352-6_12
5. Forster, J., Nedungadi, S., Mosher, M.: Moving to remote instruction in organic chemistry II laboratories. J. Chem. Educ. **97**(9), 3251–3255 (2020)
6. Bruce, M.R., et al.: Designing a remote, synchronous, hands-on general chemistry lab course. J. Chem. Educ. **98**(10), 3131–3142 (2021)
7. Doolen, T.L., Biddlecombe, E.: The impact of a cohort model learning community on first-year engineering student success. Am. J. Eng. Educ. (AJEE) **5**(1), 27–40 (2014)
8. Johnson, K.E.: Learning communities and the completion agenda. Learn. Commun. Res. Pract. **1**(3), 3 (2013)
9. Leigh Smith, B., MacGregor, J.: Learning communities and the quest for quality. Qual. Assur. Educ. **17**(2), 118–139 (2009)
10. Mote, C.D., Dowling, D.A., Zhou, J.: The power of an idea: the international impacts of the grand challenges for engineering. Engineering **2**(1), 4–7 (2016)
11. Author, F.: Contribution title. In: 9th International Proceedings on Proceedings, pp. 1–2. Publisher, Location (2010)
12. LNCS. http://www.springer.com/lncs. Accessed 21 Nov 2011

Combined Multicomponent Interventions for Older Adults in a Smart Home Living Lab

Sofia Segkouli[1(✉)], Lampros Mpaltadoros[1], Athanasios T. Patenidis[1], Vasilis Alepopoulos[1], Aikaterini Skoumbourdi[1], Margarita Grammatikopoulou[1], Ilias Kalamaras[1], Ioulietta Lazarou[1], Panagiotis Bamidis[2], Despoina Petsani[2], Evdokimos Konstantinidis[2,3], Teemu Santonen[1,2,3,4], Spiros Nikolopoulos[1], Ioannis Kompatsiaris[1], Konstantinos Votis[1], and Dimitrios Tzovaras[1]

[1] Centre for Research and Technology Hellas (CERTH), Information Technologies Institute (ITI), Thessaloniki, Greece
{sofia,lamprosmpalt,apatenidis,vasilisalepopoulos,skouma, marggram,kalamar,iouliettalazarou,nikolopo,ikom,kvotis, tzovaras}@iti.gr, Teemu.Santonen@laurea.fi

[2] Medical Physics and Digital Innovation Laboratory, School of Medicine, Aristotle University of Thessaloniki, Thessaloniki, Greece
bamidis@med.auth.gr, petsanid@auth.gr

[3] European Network of Living Labs, Brussels, Belgium

[4] Laurea University of Applied Sciences, Vantaa, Finland

Abstract. Research indicates the additive effects and interrelation of physical exercise and cognitive training towards cognition maintenance and physical performance in older adults. Combined physical and cognitive interventions enabled by eHealth technologies, web applications and wearable sensors hold a great potential for the improvement of physical and cognitive status of older adults. Living laboratory strategies and mechanisms are considered effective spaces for the implementation of interventions for cognitive and/or physical training. In order to holistically capture the opportunities of the living lab environment, a feasibility study entitled "Use of nZEB Smart Home services as a controlled lab environment to guide real life case studies of older adults", has been conducted in the rapid prototyping and novel technologies' demonstration infrastructure of nZEB smart home of Information Technologies Institute (ITI). The cognitive performance and physical function of 18 users ranging from 55 years old to 70 healthy adults has been assessed in interrelation to measure the significant impact of a holistic approach in lab-based interventions. The study design methods included a) physiological monitoring during exercise, b) balance distribution assessment, c) EEG measurements during Meditation task and d) activities of daily living execution using a Banking transaction and a Virtual Supermarket web application. The feasibility study was successful in incorporating and combining various hardware and software solutions in exercise and meditation tasks as well as in simulations of everyday activities of the participants. Processes and significant insights are presented in a preliminary analysis of the collected data.

Keywords: Living labs · eHealth · Feasibility study · Wearables · Insoles

M. E. Auer et al. (Eds.): STE 2024, LNNS 1028, pp. 273–284, 2024.
https://doi.org/10.1007/978-3-031-61905-2_27

1 Introduction

Living laboratories (LLs) are becoming an effective environment in the well-being and rehabilitation context and interaction spaces for co-design activities and global market benefits [1–3]. LLs equipped with sensors and other technological devices, integrate research and innovation processes in real-life communities and settings for the rehabilitation, transitional care and everyday living environments for older adults [4]. Nonetheless, it is restrictive to conceptualize smart homes as single spaces [5] whereas could incubate joined and functional activities. Previous reviews advocate combined physical and cognitive training bring significant health benefits [6–9]. Recent research investigate the potential link between meditation and cognition especially on older adults due to its perceived therapeutic effects on mind and body balance [10, 11].

LLs provide opportunities for testing and optimizing innovations in respect to human-centric solutions [12], but could be also the incubators of multiple stimuli well combined and performed to achieve physical and cognitive improvement for older adults. In addition, LLs can support clinical assessment in rehabilitation settings and maintenance of a healthy lifestyle [13].

This study places emphasis on a holistic approach of social care of older adults by conceptualizing older adults' needs and preferences through non-pharmaceutical combined interventions. Also sophisticated intervention methods, yoga exercises have been included in the study protocol. Yoga exercises indicate positive effects for older adults health and cognition, joining physical postures with breathing exercises, meditation along with balance training, flexibility, strength, increased muscle tone and improved mood [14]. Mindfulness-Based Stress Reduction (MBSR) and breathing practices cultivate positive thoughts, relax the mind, and enhance the mind-body connection [15, 16].

Several studies have shed light that meditation, specifically MBSR, can induce structural changes in the human brain, by revealing increased brain cell volume in the hippocampus (linked to learning, memory and emotion regulation) and the temporo-parietal junction (linked to empathy and compassion) [17–19]. In electroencephalography (EEG) as an assessment tool, theta and alpha wave synchronization, indicative of internally directed attention, are observed during various meditation practices, including MBSR. Different meditation types exhibit unique EEG patterns, with MBSR linked to increased alpha power. Portable Muse EEG use is also reported for mindfulness evaluation support or to compare meditation techniques in LLs [20–23].

2 Study Aim and Design

The study has been conducted in the context of the Horizon 2020 Project VITALISE (Virtual Health and Wellbeing Living Lab Infrastructure) that aims to enable effective transnational and virtual access to key European health and well-being research infrastructures (RIs), governed by Living Labs. The main objective was to test and train cognitive and physical tasks in a fully operational Smart Home Living Lab infrastructure. A multi-component intervention was built upon a range of technologies (sensors for physiological monitoring, cognitive virtual games, eHealth apps and video assisted exercises of yoga sessions) for a specific target group of healthy older adults.

2.1 Participants

In the feasibility study of the nZEB smart home, we present the testing and evaluation of 18 healthy older adults. Gender balance plays a vital role in research and innovation; to this end an equal number of women and men (10 women and 8 men). Most of the participants were at retirement age, indicating familiarity with the apps and the technologies provided. The study received an ethical approval of the research center committee. A written consent was a prerequisite for the participation in the study.

2.2 Study Protocol

CERTH/ITI nZEB Smart Home (https://smarthome.iti.gr/), services and smart technologies were used to assess the benefits of a multi-component intervention, built upon a range of technologies. Health-related IoT and sensor technologies such as Empatica E4 (https://www.empatica.com/research/e4/) wearable sensors and Muse EEG as well as web applications were incorporated in order to measure physical and brain activity and enable the extraction of valuable data such as patterns and biometric attributes.

The study was conducted on four phases: a. Recruitment pre-planning, participants' enrolment, b. Participant's recruitment: users' engagement c. Session design: incorporating experts' opinion and older adults' preferences, d: Outcome assessment: collection of evidence on predefined Key Performance Indicators (KPIs).

Within sessions, psychological and physiological parameters were assessed such as balance and stability, flexibility and joint, health, regulation of blood pressure, mindfulness, stress and intensity. Ten action steps were required for the participant to complete the protocol, in a specific course of action (Table 1). The full duration per participant was 65–70 min.

Table 1. Protocol Sessions - Steps of action and respective duration

	Session	Duration		Session	Duration
1	Empatica & Insole Set-up	5 min	6	Muse Session Resting	1 min
2	Insole measurement	1 min	7	Muse Session Meditation	10 min
3	Yoga Session	15 min	8	Muse Session Resting	1 min
4	Insole measurement	1 min	9	Banking App	5 min
5	Muse Set-up	10 min	10	Virtual Supermarket	15 min

3 Methods

3.1 Physiological Monitoring During Exercise

During the exercise intervention, every participant followed the instructions as provided with a wearable device for physiological monitoring of their status during the task. The Empatica E4 device was used for data collection of physiological data during the exercise intervention. The E4 is equipped with sensors combining Electrodermal Activity (EDA) also known as Galvanic Skin Response (GSR), Blood Volume Pulse (BVP), Acceleration, Heart Rate (HR), and Temperature data, simultaneously enabling the measurement of sympathetic nervous system activity and heart rate. Each participant was assisted to wear the band before the exercise protocol initiation. Data was collected for the whole duration of the exercise and the band was turned off right after the completion of this task. Data were imported via USB and transferred securely to the Empatica cloud platform.

3.2 Balance Distribution Assessment

Participants wore sensor-equipped shoes and underwent measurements at the beginning and end of a 0.5-h visit for insoles data collection. Activities like yoga were included to assess their impact on standing posture. Using Moticon ReGo insoles, each 16-part smart insole collected pressure data from sixteen sensors on each foot approximately every 40 ms for one minute per session. Using the Pressure Distribution Index (PDI) and the Pressure Symmetry Index (PSI), we evaluated pressure distribution and symmetry both before and after the activity. The PDI measures even pressure distribution across foot areas by utilizing the standard deviation of area pressures:

$$PDI = \frac{\sqrt{\sum (x_i - \mu^2)}}{\mathrm{N}} \tag{1}$$

where x_i is the average pressure on area i during the session, μ is the mean pressure over the whole foot at the same period, and N is the total number of foot areas. Higher PDI values indicate uneven pressure distribution, while values close to zero signify even distribution. PSI measures the symmetry between the two feet, calculated as:

$$PSI = \left| \frac{\mu_{left} - \mu_{right}}{\mu_{left} + \mu_{right}} \right| \tag{2}$$

where μ_{left}, mean left pressure and μ_{right}, mean right pressure. PSI provides information about pressure symmetry in the two feet is. Higher PSI values indicate greater left-right foot pressure asymmetry, whereas values near zero indicate better balance.

We assume optimal values for both indexes are zero in our context, representing ideal state of even pressure distribution. This implies even pressure among the different foot areas for PDI and between left and right feet for PSI. To assess an activity's (e.g. yoga) impact on standing posture, differences in PDI and PSI were calculated before and after the activity.

3.3 EEG Measurements During Meditation Task

During the Meditation task, EEG data were collected to assess the cognitive status of participants in different sessions. EEG measurements of participants were recorded to monitor brain activation in three sessions: **Session 1** - Resting State Baseline: One-minute resting EEG activity (Baseline) before the MBSR task. Participants were instructed to relax, breathe normally, keep their eyes closed, sit still, minimize mouth movements, and allow their minds to wander, while a researcher monitored the procedure. **Session 2** - Mindfulness-Based Stress Reduction (MBSR) - Witness the Breath: Participants engaged in guided meditation, focusing on the natural breath flow. The meditation progressed from concentrating on chest movements to the nose and nostrils and finally to a point just below the nostrils, where the out-breath brushes the upper lip. **Session 3** - RS Follow-Up: Similar to Session 1, one-minute resting EEG activity was recorded after the meditation task, serving as a follow-up to assess any changes.

EEG data were obtained using the MUSE 2 device from InterAxon Inc.. Electrodes were positioned following the international 10–20 system in TP9, AF7, AF8, TP10, and FPz (as a reference), with a sampling frequency of 220 Hz. The device features communication via Bluetooth for streaming, visualizing, and recording EEG data from Muse. Real-time data capture allowed a researcher to manually initiate data streaming and monitor signals, enabling early detection of artifacts.

EEG data analysis was conducted using MNE-Python [24]. Artifacts were firstly removed via manual inspection and discarding of abnormally high amplitude peaks in the signal. Noise and artifacts were further removed via a Finite Impulse Response (FIR) filter (1 Hz–45 Hz). Power Spectral Density values were calculated using the Welch method for each frequency band: 4–8 (Theta), 8–13 Hz (Alpha), 13–30 Hz (Beta). Average PSD values for each channel were used for statistical analysis.

3.4 Activities of Daily Living Execution

Banking App. For the daily living activities and specifically for the assessment of participant execution of a transaction task the Banking app was used [25, 26]. Managing finances is an Instrumental Activity of Daily Living (IADL), usually measured via traditional pen-paper methods and interviews. To simulate this IADL, the 'Banking App' was introduced. In the app, users enter a PIN, an amount to withdraw and confirm all inputs, using a numpad on a tablet's touchscreen that resembles an ATM (Fig. 1, left). Each attempt to complete the task includes entering a PIN, an amount and confirming inputs or starting a new attempt to complete the steps from the beginning.

Bank transaction steps: 1) Start a session, 2) Provide your PIN, 3) Provide amount to transfer, 4) Check your answers and submit. Metrics acquired for inspection include: Count of attempts for each step (attempt as a whole, PIN, Amount, Confirmation), Duration of attempts in msec (attempt as a whole, PIN, Amount, Confirmation), Durations for: a) Providing PIN, b) providing amount to transfer, c) confirm answers.

The purpose of this task is to collect data regarding the count of attempts and durations for each part of the task and inspect data for insightful information regarding a person's behavior towards their daily living tasks, by simulating a bank transaction.

Virtual Supermarket. This study uses data from the Virtual Supermarket game (Fig. 1, right), where participants navigate a virtual supermarket, selecting items based on provided lists. Participants had multiple sessions lasting about 20 min, recording parameters like intended purchases, actual purchases, session duration, and sector time.

To analyze user behavior, sub-scores on a 0–1 scale were calculated. The **Total Playing Time** score measures average session time, normalized across users:

$$s_{tot} = 1 - \frac{t_{tot} - \min(t_{tot}, j)}{\max(t_{tot}, j) - \min(t_{tot}, j)} \tag{3}$$

Using similar formulas, the **Time Per Supermarket Sector** score evaluates average time per sector, and the **Time Per Item** score represents the average time spent on a specific item across all sessions. The **Bought Quantities** score is the normalized ratio of average quantities bought to requested quantities:

$$s_{quan} = \frac{r_q}{\max(t_{item}, j)} \tag{4}$$

and similarly, the **Money Paid** score is the normalized ratio of average money paid to the average cost of items bought. By averaging these sub-scores, an overall score is calculated, reflecting the overall user performance and engagement with the game.

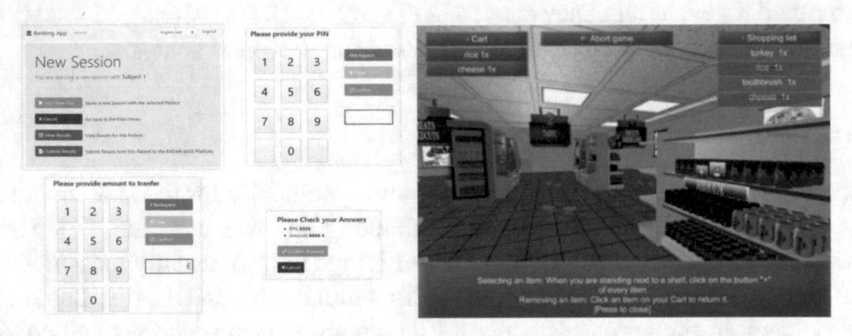

Fig. 1. Left: The Banking App user interface. Right: The Virtual Supermarket user interface.

4 Results

4.1 Physiological Monitoring During Exercise

E4 Empatica offers a secure cloud platform solution, E4 connect, where researchers /clinicians can inspect all available data, in order to extract valuable insights regarding the participant's physiological response to the exercise task. In Fig. 2 we present a specific participant's exercise tasks through the e4 Connect dashboard, where the researcher or clinician can scroll and inspect EDA, BVP, Accelerometer, HR and Temperature data through the whole duration of the exercise. Participants indicated positive feedback to wearable device use. All available data are also stored in.csv form for further analysis in future work.

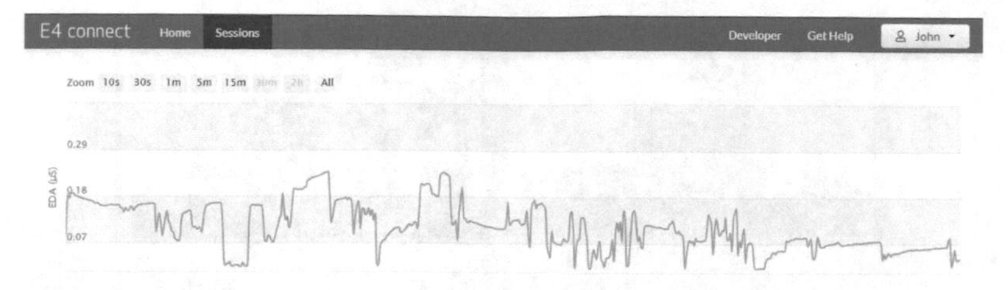

Fig. 2. E4 Connect visualizations: EDA (shown), BVP, Accelerometer, HR, Temperature

4.2 Balance Distribution Assessment

After analyzing PDI and PSI indices along with their differences pre- and post-activity for all participants, we generated histograms illustrating the disparities in PDI for the left and right foot (Fig. 3, top **Error! Reference source not found.**) and PSI differences (Fig. 3, bottom left). Positive differences indicate post-activity values surpassing pre-activity values. PDI histograms reveal a consistent trend, with most users exhibiting values equal to or smaller than zero, signifying improved standing posture. Conversely, PSI differences display an even distribution among users, indicating no clear trend. Figure 3 (bottom right), depicting foot pressure distribution via color scale, highlights that higher-pressure areas are primarily concentrated at the heels, emphasizing their role in foot support and stability. Lastly, Wilcoxon Signed-Rank Tests were performed on the three metrics pre- and post-activity. While the results lacked statistical significance, they offer valuable early indicators of potential system impact.

4.3 EEG Measurements During Meditation Task

There was a statistically significant difference in the three Sessions across the different brain waves, $\chi 2(35) = 333.056, p = 0.001$ according to Friedman test. Post hoc analysis with Wilcoxon signed-rank tests was conducted with a Bonferroni correction applied, resulting in a significance level set at $p < 0.017$. Median (IQR) for the Session 1, Session 2 and Session 3 were ranging across the different channels in the three different brain waves (alpha, theta and beta -Table 2).There were no significant differences between the Session 2-MBSR session and Session 3- RS follow-up or between the Session 1-RS and Session 3-RS follow-up, despite an overall reduction theta waves in AF8 in the Session 1- RS vs Session 2- MBSR ($Z = \text{-}2.025, p = 0.043$).

4.4 Activities of Daily Living Execution

Banking App. Results of the transaction details regarding duration of attempts (attempt as a whole, PIN, Amount, Confirmation) appear in Fig. 4. From the analysis, one participant (Participant 11) was excluded due to extreme duration value occurrence (1563.72 s) – due to banking app response time during the specific task and participant. Average PIN Duration result is 24.42 s, average Amount Duration is 11.54 s and Confirm Duration is 4.99 s. That constitutes a total of 40.95 s for the average of the entire Attempt Duration.

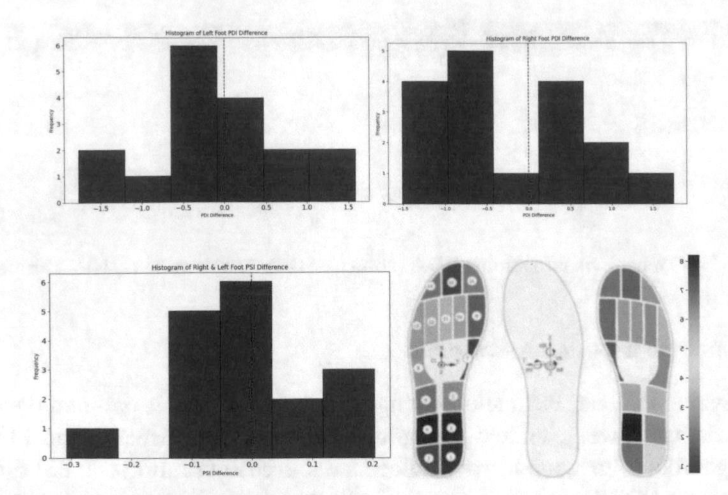

Fig. 3. Histograms of PDI difference between before and after activity, for the left (top left) and the right foot (top right), of the PSI difference between before and after activity measurements (bottom left). Heatmap for the Mean Left and Right Foot Pressure Areas (bottom right).

Table 2. Mean, Median and SD values of the three Muse-EEG sessions in three brain waves (alpha, beta and theta) over the four electrodes (TP9, TP10, AF7 and AF8)

		Theta			Alpha			Beta		
		Mean	SD	Median	Mean	SD	Median	Mean	SD	Median
Session 1	TP9	25.93	28.02	12.44	4.72	2.56	4.22	2.61	1.85	1.85
	AF7	4.16	2.90	2.99	1.58	1.15	1.25	2.45	1.06	1.06
	AF8	3.04	2.84	2.58	1.21	0.84	1.10	1.80	1.45	1.45
	TP10	22.54	30.89	11.68	3.69	1.82	3.34	1.09	1.18	1.18
Session 2	TP9	16.90	24.76	8.76	3.69	1.85	3.06	2.15	1.38	1.38
	AF7	14.74	34.07	1.51	7.49	21.49	0.91	86.96	1.49	1.49
	AF8	1.83	1.19	1.56	1.06	0.68	0.82	2.68	1.36	1.36
	TP10	15.84	24.07	8.20	3.44	1.97	2.67	2.08	1.32	1.32
Session 3	TP9	16.10	15.10	13.66	3.66	1.94	2.85	1.80	1.47	1.47
	AF7	51.13	202.00	1.54	28.60	114.91	0.90	163.90	1.07	1.07
	AF8	1.97	1.53	1.35	3.31	1.93	2.67	2.26	1.00	1.00
	TP10	15.41	15.21	13.58	1.02	0.73	0.73	1.53	1.24	1.24

Participants appeared to show positive feedback to the task, and future work aims to collect data from different groups of participants towards differentiating between groups based on the banking app metrics, e.g. the duration of an attempt in different age groups, or between a healthy participant and a participant living with dementia.

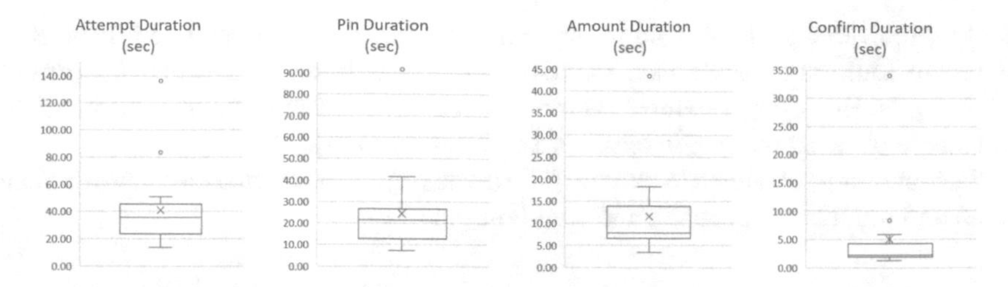

Fig. 4. Banking App boxplots - Attempt, PIN, Amount, and Confirm durations

Virtual Supermarket. Based on the information shown in Fig. 5 (left), the average amount of items people actually bought differs from what was included in the provided shopping list. Figure 5 (middle) adds more detail, showing an equivalence between the average cost of purchased items and their expected values. This highlights that the mismatch between what was acquired and what was expected seems to be due to participants forgetting items, rather than providing an incorrect amount of money.

Figure 5 (right) demonstrates the overall score (Sect. 1.6) for a sample user with several sessions over a period of months. The chart presents a positive trend, implying that engagement with the Virtual Supermarket could potentially have enhanced user perception and attention. The Virtual Supermarket overall score provides a tool to monitor an individuals' development, which could be beneficial for cognitive improvement.

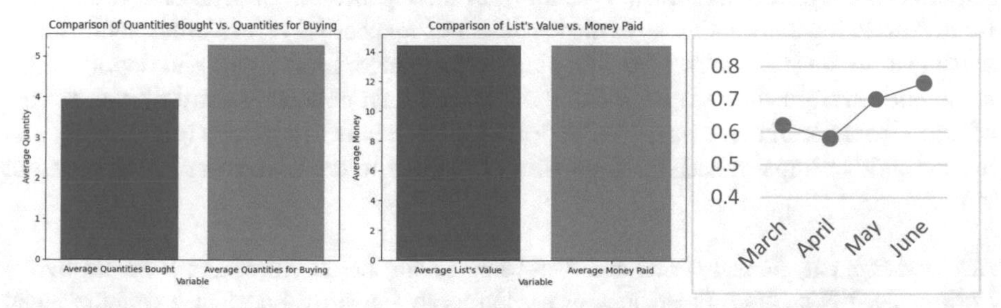

Fig. 5. Left: Supermarket Items Bought vs Items for Buying. Middle: Shopping List's Total Value vs Money Paid. Right: Indicative User's Overall Supermarket Score over time.

5 Discussion

Aligned with the Greek medicine and philosophy notion body and spirit, has always been one of the most puzzling questions in the history of human experience and self-reflection [27]. Despite the promising role of smart homes and LL environment to observe human behaviour while monitoring and controlling the indoor environment and diverse groups of individuals' interactions [28]could also exploit joined physical and cognitive activities and responses as a valid alternative due to their resemblance to real-world settings. In real life, individuals are engaged in physical activities through cognitive functions and vice versa. Multimodal assessment of motor function, physiology and cognition provides rich insights. However, combined training can achieve higher training gains in balance than single physical training in terms of LLs activities [29].

The present work aimed to explore the effects of joined activities to improve mental health and provide better health and well-being.

6 Conclusions

For cognitive and physical outcomes in the LL context a set of devices and apps were used to evaluate older adults' physiological status jointly with participants' cognitive function in terms of daily activities execution. Study results stressed the strong interconnection between physical and cognitive activities and the validity of the multicomponent intervention in real settings. According to study results and more specifically the physiological monitoring, a secure cloud platform offered by E4 Empatica was an effective environment for valuable insights regarding the participant's physiological response to the exercise task. In insoles balance distribution assessment, beneficial insights have been gained that may serve as early indicators of potential impact of the platform. Given that mindfulness meditation is associated with various benefits [30], EEG measurements during the specific task was able to assess participant's cognitive response to the task and observe differences between resting and meditation sessions.

Daily living activities in virtual environment through the Banking app and the Virtual Supermarket applications attracted participants' strong interest. Both tasks were successful in simulating daily activities of the participants, respectively a bank transaction and a daily visit to the supermarket for shopping performance. Limitations and opportunities to be addressed in future work include additional data collection for different groups of participants, will make possible the differentiation between groups (e.g. healthy and mild cognitive impairment) based on similar metrics to those discussed in the present work.

Acknowledgment. Virtual Health and Well-being Living Lab Infrastructure is funded by the Horizon 2020 Framework Programme of the European Union for Research Innovation (grant 101007990).

References

1. Leminen, S., Westerlund, M., Nyström, A-G.: Living Labs as Open-Innovation Networks. Technol. Innov. Manag. Rev. **2**, 6–11 (2012). https://doi.org/10.22215/timreview/602
2. Feurstein, K., Hesmer, A., Hribernik, K., Thoben, K-D., Schumacher, J.: Living labs – a new development strategy, pp 1–14 (2008)
3. Kareborn, B.B., Stahlbrost, A.: Living Lab: an open and citizen-centric approach for innovation. Int. J. Innov. Reg. Dev. **1**, 356 (2009). https://doi.org/10.1504/IJIRD.2009.022727
4. Santonen, T., et al.: Cocreating a harmonized living lab for big data-driven hybrid persona development: protocol for cocreating, testing, and seeking consensus. JMIR Res. Protoc. **11**, e34567 (2022). https://doi.org/10.2196/34567

5. Rodríguez-Gallego, C., Díez-Muñoz, F., Martín-Ruiz, M.-L., Gabaldón, A.-M., Dolón-Poza, M., Pau, I.: A collaborative semantic framework based on activities for the development of applications in smart home living labs. Futur. Gener. Comput. Syst. **140**, 450–465 (2023). https://doi.org/10.1016/j.future.2022.10.027
6. Fullen, M.C., Smith, J.L., Clarke, P.B., Westcott, J.B., McCoy, R., Tomlin, C.C.: Holistic wellness coaching for older adults: preliminary evidence for a novel wellness intervention in senior living communities. J. Appl. Gerontol. **42**, 427–437 (2023). https://doi.org/10.1177/07334648221135582
7. Gheysen, F., et al.: Physical activity to improve cognition in older adults: can physical activity programs enriched with cognitive challenges enhance the effects? A systematic review and meta-analysis. Int. J. Behav. Nutr. Phys. Act. **15**, 63 (2018). https://doi.org/10.1186/s12966-018-0697-x
8. Law, L.L.F., Barnett, F., Yau, M.K., Gray, M.A.: Effects of combined cognitive and exercise interventions on cognition in older adults with and without cognitive impairment: a systematic review. Ageing Res. Rev. **15**, 61–75 (2014). https://doi.org/10.1016/j.arr.2014.02.008
9. Zhu, X., Yin, S., Lang, M., He, R., Li, J.: The more the better? A meta-analysis on effects of combined cognitive and physical intervention on cognition in healthy older adults. Ageing Res. Rev. **31**, 67–79 (2016). https://doi.org/10.1016/j.arr.2016.07.003
10. Tang, Y.-Y., Hölzel, B.K., Posner, M.I.: The neuroscience of mindfulness meditation. Nat. Rev. Neurosci. **16**, 213–225 (2015). https://doi.org/10.1038/nrn3916
11. Vago, D.R., Silbersweig, D.A.: Self-awareness, self-regulation, and self-transcendence (S-ART): a framework for understanding the neurobiological mechanisms of mindfulness. Front. Hum. Neurosci. **6**, 296 (2012). https://doi.org/10.3389/fnhum.2012.00296
12. Ghafurian, M., Wang, K., Dhode, I., Kapoor, M., Morita, P.P., Dautenhahn, K.: Smart home devices for supporting older adults: a systematic review. IEEE Access **11**, 47137–47158 (2023). https://doi.org/10.1109/ACCESS.2023.3266647
13. Korman, M., Weiss, P.L., Kizony, R.: Living labs: overview of ecological approaches for health promotion and rehabilitation. Disabil. Rehabil. **38**, 613–619 (2016). https://doi.org/10.3109/09638288.2015.1059494
14. Hoy, S., Östh, J., Pascoe, M., Kandola, A., Hallgren, M.: Effects of yoga-based interventions on cognitive function in healthy older adults: a systematic review of randomized controlled trials. Complement. Ther. Med. **58**, 102690 (2021). https://doi.org/10.1016/j.ctim.2021.102690
15. Lomas, T., Ivtzan, I., Fu, C.H.Y.: A systematic review of the neurophysiology of mindfulness on EEG oscillations. Neurosci. Biobehav. Rev. **57**, 401–410 (2015). https://doi.org/10.1016/j.neubiorev.2015.09.018
16. Bhardwaj, P., Pathania, N., Pathania, M., Rathaur, V.: Evidence-based yoga and ayurveda lifestyle practices for the geriatric population during Coronavirus disease 2019 pandemic: a narrative. J Prim Care Spec **2**, 38 (2021). https://doi.org/10.4103/jopcs.jopcs_4_21
17. Innes, K.E., et al.: Effects of meditation and music-listening on blood biomarkers of cellular aging and Alzheimer's disease in adults with subjective cognitive decline: an exploratory randomized clinical trial. J. Alzheimer's Dis. **66**, 947–970 (2018). https://doi.org/10.3233/JAD-180164
18. Innes, K.E., Selfe, T.K., Khalsa, D.S., Kandati, S.: Meditation and music improve memory and cognitive function in adults with subjective cognitive decline: a pilot randomized controlled trial. J Alzheimer's Dis **56**, 899–916 (2017). https://doi.org/10.3233/JAD-160867
19. Maddock, A., Blair, C.: How do mindfulness-based programmes improve anxiety, depression and psychological distress? A systematic review. Curr. Psychol. **42**, 10200–10222 (2023). https://doi.org/10.1007/s12144-021-02082-y

20. Hunkin, H., King, D.L., Zajac, I.T.: Evaluating the feasibility of a consumer-grade wearable EEG headband to aid assessment of state and trait mindfulness. J. Clin. Psychol. **77**, 2559–2575 (2021). https://doi.org/10.1002/jclp.23189
21. Sharma, K., Wernicke, A.G., Rahman, H., Potters, L., Sharma, G., Parashar, B.: A retrospective analysis of three focused attention meditation techniques: mantra, breath, and external-point meditation. Cureus (2022).https://doi.org/10.7759/cureus.23589
22. Hawley, L.L., Rector, N.A., DaSilva, A., Laposa, J.M., Richter, M.A.: Technology supported mindfulness for obsessive compulsive disorder: self-reported mindfulness and EEG correlates of mind wandering. Behav. Res. Ther. **136**, 103757 (2021). https://doi.org/10.1016/j.brat.2020.103757
23. Lazarou, I., et al.: Eliciting brain waves of people with cognitive impairment during meditation exercises using portable electroencephalography in a smart-home environment: a pilot study. Front Aging Neurosci **15**, 1167410 (2023). https://doi.org/10.3389/fnagi.2023.1167410
24. Gramfort, A.: MEG and EEG data analysis with MNE-Python. Front. Neurosci. **7**, 71033 (2013). https://doi.org/10.3389/fnins.2013.00267
25. Karakostas, A., et al.: A French-Greek cross-site comparison study of the use of automatic video analyses for the assessment of autonomy in dementia patients. Biosensors **10**, 103 (2020). https://doi.org/10.3390/bios10090103
26. Stavropoulos, T.G., et al.: An app to measure functional decline in managing finances in Alzheimer's disease: preliminary results of the RADAR-AD study. Alzheimer's Dement **17**, e053656 (2021). https://doi.org/10.1002/alz.053645
27. van der Eijk, P.: Body and Spirit in Greek medicine and philosophy (2007)
28. Cureau, R.J., et al.: Bridging the gap from test rooms to field-tests for human indoor comfort studies: A critical review of the sustainability potential of living laboratories. Energy Res. Soc. Sci. **92**, 102778 (2022). https://doi.org/10.1016/j.erss.2022.102778
29. Rieker, J.A., Reales, J.M., Muiños, M., Ballesteros, S.: The effects of combined cognitive-physical interventions on cognitive functioning in healthy older adults: a systematic review and multilevel meta-analysis. Front. Hum. Neurosci.Neurosci. **16**, 838968 (2022). https://doi.org/10.3389/fnhum.2022.838968
30. McQueen, B., Murphy, O., Fitzgerald, P., Bailey, N.: The mindful brain at rest: neural oscillations and aperiodic activity in experienced meditators (2023)

Advances and Challenges in Applied Artificial Intelligence

Process Optimization in the Healthcare Sector Through the Use of AI-Based ECG Analyzis

Scott Meinhardt[1(✉)], Tim Neumann[1], Linus Teich[2], Daniel Franke[1], Laura Schladitz[1], Sven Leonhardt[1], and Sebastian Junghans[1]

[1] Westsächsische Hochschule Zwickau (University of Applied Sciences Zwickau), Kornmarkt 1, 08056 Zwickau, Germany
scott.meinhardt.j57@fh-zwickau.de

[2] Heinrich-Braun-Klinikum (Heinrich Braun Hospital), Karl-Keil-Straße 35, 08060 Zwickau, Germany

Abstract. The e-prescription continues to gain ground in Germany. Since July 1, 2023, the use of the e-prescription has become even more versatile and insured persons are able to redeem the prescription using their electronic health card. The consequences are more convenience for patients and fewer trips to the doctor's office. This eliminates the need for manual signatures, and follow-up prescriptions can be issued without the need for another visit to the doctor's office. This also optimizes and improves the day-to-day work in pharmacies. The e-prescription is just one example of the increasing digitization in healthcare. Artificial intelligence (AI) is improving patient care, particularly in the areas of cancer diagnosis and early detection of heart disease. Digitization can also be understood as a way of optimizing processes. It helps to improve processes in a company (or in this case in the healthcare system or hospital) in such a way that workflows are optimized, quality is increased, and errors are reduced.

In the following, we will explain the basics of a process and process management as well as digitalization as a possibility for process improvement. There will always be a link to healthcare and the healthcare industry. Artificial intelligence will play a major role in this context by demonstrating that AI is already being used in healthcare for imaging and patient self-monitoring. In particular, the integrative development of an AI-based ECG analysis will be presented, illustrating the added value of such an innovation in cardiology and how it can concretely improve processes.

Keywords: Process · Process management · Digitalization · Process optimization · Artificial Intelligence · AI · Cardiology · e-health · ECG · smart technology

1 Introduction

According to an international study by the Bertelsmann Foundation, which analyses and compares the digitalization strategies of 17 countries, Germany is only in 16th place when it comes to shaping the digital transformation in the healthcare sector. The study

M. E. Auer et al. (Eds.): STE 2024, LNNS 1028, pp. 287–298, 2024.
https://doi.org/10.1007/978-3-031-61905-2_28

compares and evaluates the status, intensity of use and key success criteria for effective digitalization strategies. Germany scored 30.0 points on the Digital Health Index, while its neighbor the Netherlands scored 66.1 points [1].

This puts Germany well behind the other countries in the study. One of the main reasons for this is that outdated, analogue processes still exist in the healthcare system. This problem was highlighted by the Corona pandemic. In particular, Germany is lagging behind in terms of digitalization [2]. The electronic patient record, which has become the standard in many countries (including the Netherlands), is still in the implementation phase in Germany and is not expected to be mandatory for everyone until the end of 2024 [3]. Other factors complicating the transformation include heavy bureaucratization, demographic trends, and an increasing shortage of skilled workers [4].

The German Federal Ministry of Health (BMG) [5], the German Medical Association [6] and various scientific publications [7–9] all show that digitalization in the healthcare sector is already a very topical issue and will continue to gain in importance in the future.

On the other hand, it is also important to look at process management and processes in the healthcare sector. Process change and digitalization tend to occur in parallel. One causes the other. Current developments in the introduction of electronic prescriptions in doctors' surgeries and pharmacies show the profound process changes brought about by such a technological innovation [10].

Despite all the current problems and challenges associated with the digitalization of the German healthcare system and the associated implementation of innovative technologies, there is great potential for doctors, hospital staff, health insurers and patients. Improving medical care, reducing the burden on nursing staff, saving time and money, and addressing staff shortages are just a few of the benefits.

In summary, the aim of the work is to show how process optimization in the healthcare sector can be designed through increasing digitalization and artificial intelligence.

2 Process Management

The Gabler Wirtschaftslexikon defines a business process as a *"sequence of value-creating activities [...] with one or more inputs and an output that creates customer value"* [11]. Thus, the term encompasses all of a company's processes. Process management (synonymous with business process management), on the other hand, *"refers to the design of a company's business processes [...]. [Business process management includes the collection, design, documentation and implementation of processes"* [12].

In the healthcare sector, the terms "clinical process management" or "medical process management" are sometimes used [12]. A distinction is always made between medical and non-medical processes. For example, the Caritas hospital St. Josef in Regensburg defines process management as follows *"Process management includes all clinical processes that contribute to the management of patient flow"* [13].

It is also important to think and act in a cross-functional way. Processes in industrial companies are primarily standardized and planned [12]. A comparison of processes in the healthcare and industrial sectors reveals both differences and similarities [11]. In both cases, a customer-to-customer or patient-to-patient orientation and value chains can be identified. In particular, standard treatments can be well modelled in healthcare and are therefore also referred to as clinical pathways. Special cases (e.g., emergencies), on the other hand, are difficult or impossible to model due to their high degree of individuality.

Clinical pathways are not uniformly defined in the literature. ROEDER define the clinical pathway as a consensus reached by the treatment team across professional groups and institutions for the best possible implementation of an entire inpatient treatment. The defined quality of care and the necessary and available resources should always be considered [14].

2.1 Types of Processes

SEIDELMEIER has made a classification of business processes [15]. In his grouping, the proximity of the process to the company's core business is important. For this reason, he divides processes into control processes, core business processes and support processes. Control processes include all processes in a company *"that serve the purpose of setting goals, pursuing goals, planning and strategic orientation of the company and the processes contained in the company"*. [15]. Control processes provide the conceptual framework for the design of core processes. An example of a governance process is strategy development.

The so-called core business processes are based on the specifications of the control processes. These serve the actual value creation of the company. They include, among other things, the production of products or services and all processes from the identification of needs to the satisfaction of customer needs. Examples include customer service, customer support, marketing, and sales processes.

The third type of process defined by SEIDELMEIER are support processes. These are designed to support the core business processes and to build and maintain all the resources needed to deliver the service. Maintenance processes, human resources or cost accounting are some examples of support processes.

The main distinction in healthcare is between medical and business processes. Medical processes, also known as treatment processes, form the core of healthcare. These include, in particular, examinations, treatments, care, rehabilitation, etc., for which there are individual methods of documentation and description (e.g., surgical plans, vaccination plans). Business processes in health care, on the other hand, deal with tasks that are typical for the sector, such as management, human resources, controlling, accounting, logistics or facility management [12].

2.2 Process Optimization

In some definitions, the term process improvement is often used as a synonym. For example, GADATSCH describes process optimization as a *"step-by-step [...] and sustainable [...] improvement"* [17]. It should be noted that process optimization has gained in importance in recent years. Therefore, it can increasingly be assigned to the strategic level of process management. For this reason, process optimization can be used as a basis for business decisions. The primary goal of process optimization is to implement effective and efficient processes by improving existing processes in terms of cost, time, quality, flexibility and use of capital. To achieve this goal, it is important to identify improvement opportunities, evaluate solution approaches and find an optimal solution

[18]. In addition, the principles of good processes [16] should be taken into account to achieve the goals of process improvement. This improvement process follows the classic sequence of the Deming or Plan-Do-Check-Act cycle. The cycle supports the impact of process improvement methods and can be used at all levels of an organization [16].

In health care, it is important not only to know the process, but also to include the medical aspect, e.g., in the sense of treatment specifications that stand for appropriate treatment success. The benefits of process improvement in healthcare can be considered from both a medical and an economic perspective. In terms of medical benefits, there are improvements in the quality of treatment, a basis for discussion for the practitioner, and the assurance of treatment decisions. On the economic side, there is a reduction in length of stay, a basis for cost management, a basis for process interface definition, a basis for risk management and support for quality management.

The literature is full of principles, concepts, methods, and tools for process improvement. However, the distinction between the terms is fluid and not sharply defined. Nevertheless, the goal of process improvement is to improve performance. The path to process improvement is always a defined sequence of changes and rules. The exact design depends on the chosen methods, principles, and tools. Looking at all the concepts and tools, it is striking that the term "digitalization" is used in almost every context. For this reason, digitization can be considered as a possible method for process optimization [19].

3 Digitalization

There are many definitions of digitization. This illustrates the diversity of the term. From a technical point of view, for example, digitization is defined as the preparation of information for processing and storage in a digital system. It can therefore be concluded that digitization is made possible by advances and developments in information and communication technology. Digitization is therefore primarily about the process of converting analogue media into digitally usable signals. In addition, digitization describes a process of change that is currently taking place in many sectors due to rapid technological developments [20], including the health sector.

The digital health publisher mediorbis describes digitalization in the health sector as *"the networking of all actors in the health sector in communication and documentation, including the transmission of health data and information. In addition, health services are to be provided electronically"* [21]. The term digital health is often used in this context. The US Food and Drug Administration (FDA) defines digital health as follows *"The broad field of digital health includes categories such as mobile health (mHealth), Health information technology (IT), wearable devices, telehealth and telemedicine, and personalized medicine. These technologies can empower consumers to make more informed decisions about their own health and provide new opportunities to facilitate prevention, early detection of life-threatening diseases, and management of chronic conditions outside of traditional care settings"* [22].

For healthcare to continue to evolve successfully, digitalization must be further advanced. It offers many opportunities, such as faster communication, more efficient administrative processes and the availability of patient data at any time. Mobile applications ensure that patients play a more self-determined role in the treatment process and increase their health literacy. Digital technologies can help to provide better care for more elderly and chronically ill patients, to pay for expensive medical developments, and to provide good medical care in structurally weak areas [23].

This digital transformation of the healthcare system has been increasingly supported by the Federal Ministry of Health of Germany in recent years. The focus is on secure networking through the telematics infrastructure, the introduction of the electronic health card, the electronic patient record and the electronic prescription as well as the digital health and care applications. The exact framework and goals of the digital transformation in Germany are set out in the digitalization strategy developed by the Federal Ministry of Health [24, 25].

3.1 Drivers of Digitalization

The digital transformation mentioned above has a number of foundations that make this development possible. These foundations are discussed below. In this context, SAP SE has published various IT trends that can be used to identify the initial drivers of digitalization. Increasing digitalization is driven by technological progress in many different areas, such as energy supply [26], the housing of the future [27] or healthcare [8]. NEUMANN identified a total of six drivers of digitalization that are of increasing interest for his work. These are the drivers cloud, mobile, social media, big data, IoT and AI [16]. Some of these will also be of importance in the following.

In the health sector in particular, several key drivers have been identified in the literature, which are interdependent and mutually reinforcing.

The first driver is the rapid technological development in data processing and use. Cloud computing plays a role here by extending the utility perspective of digital technologies. Users now have access to almost unlimited computing power and storage capacity. In addition, mobile devices such as smartphones and tablets enable ubiquitous access to data in the cloud.

A better understanding of the biological basis of human life is seen in the literature as the second driving force in health care. This means that as the biological basis of human life is unraveled, individual differences will become increasingly important for medical diagnosis and therapy. Conversely, this circumstance requires more individualized therapies. This creates many millions of differentiating variables that require support in computing and data storage.

The third driver in healthcare is growing patient empowerment based on increased transparency. The Corona pandemic brought many changes, including in terms of patient sovereignty and transparency. Things like making appointments digitally have become standard, and people are becoming more sovereign but also more demanding when it comes to managing their health as an asset [28].

In summary, the three drivers of healthcare digitalization mentioned above will fundamentally change medicine in the future with the help of digital innovations. This development will require comprehensive regulatory adjustments. A balance must be struck between medical benefits and data security and the protection of patients' privacy.

4 Artificial Intelligence

Artificial Intelligence (AI) *"describes the ability of machines to perform tasks autonomously based on algorithms and to react adaptively to unknown situations. Their behaviour is similar to that of humans: not only do they perform repetitive tasks, but they also learn from their successes and failures and adapt their behaviour accordingly. In the future, machines with artificial intelligence should also be able to think and communicate like humans"* [29]. There are two main types of AI: weak and strong AI. Weak AI is limited to specific areas of application. Strong AI, on the other hand, is designed to match or exceed human intellectual abilities.

AI can also be divided into different sub-areas. Machine learning is used to enable a computer programme to improve its activity in a particular area using its own data. Deep learning is a subset of machine learning because it is based on neural networks and allows the machines with AI to make predictions and challenge them. Artificial neural networks are modelled on the activity of the human brain. The final subfield of Artificial Intelligence is Knowledge Representation. This area enables machines with AI to think logically, draw conclusions and develop arguments [29].

It is clear that AI is very diverse and has enormous potential. For this reason, the industry association Bitkom e.V. has predicted that the retail and healthcare sectors will be the main investors in artificial intelligence in the future [30].

In the healthcare sector, AI is primarily used to create algorithms for the use of medical data. The use of medical data is the basis of digital transformation in healthcare. Two forms of use can be identified: on the one hand, the care of individual patients through the availability of all relevant data and, on the other hand, the availability of data as a basis for research and the development of automated algorithms, as mentioned above. Some literature describes AI as the *"crowning glory of a modern, digitized healthcare system"*. Some working AI applications can already be found in healthcare today. In medical or nursing documentation, text suggestions or notes help to avoid potential errors. In radiology, AI is used to interpret X-rays and can identify images with a high degree of certainty in certain areas, such as breast cancer screening [31].

Of course, the field of cardiology in healthcare is not unaffected by this digital transformation. There are a number of AI-based technologies that are already being used successfully in cardiology today. The focus of this article is on the specific area of cardiology.

4.1 Artificial Intelligence in Cardiology

The interest in AI-based ECG analysis is enormous. This is reflected in the number of scientific articles published on PubMed in recent years. This number has increased exponentially since 2017 [32]. It is no longer uncommon for AI to be successfully applied in cardiology. However, the areas of application are also very diverse, e.g., it is used in image pre-processing and segmentation, in diagnostics or also in clinical decision support. Particularly in image reconstruction, preprocessing and segmentation, AI methods are already widely used and can therefore also be found in device technology products and software [33].

In particular, the AI-based analysis of ECGs will play a predominant role in this work. For example, ECG analysis can be performed using neural networks. This enables the processing of large and high-dimensional data, and deep learning can be used to implement automatic image recognition and natural speech recognition. Most importantly, the goal is to further optimize traditional ECG diagnostics and thus improve processes.

5 AI-Based ECG Assessment Method

The development of this artificial intelligence is now described in this chapter. It is the first known algorithm for AI-based development in children. The aim now is to expand the existing possibilities and extend current use to the pediatric field using appropriate software [34].

5.1 Cardiological Basics

Typically, ECGs are only performed in a clinical setting or in outpatient clinics. Smart watches now make it possible to record and analyze ECGs automatically. This means that such examinations can be integrated into the domestic environment within a smart home. This creates a link between health/medicine and information technology on the one hand, and between health and housing on the other. The function of recording ECGs on the smartwatch has been available on Apple models since the Apple Watch Series 4. The continuous derivation of the potential results in a curve. On this curve, similarly, configured complexes consisting of the P-wave, QRS complex and T-wave can be seen at regular intervals, representing the different stages of the heart's excitation propagation. Ideally, this curve results in a sinus rhythm, the physiological rhythm of the heart. If there is no sinus rhythm, this is an important indication, and the problem can be clarified on an outpatient basis. However, evaluation of the recorded ECG must always be done in a hospital due to the lack of analysis capabilities. Apple has developed an ECG analysis via the Apple Watch, although it is limited to a few diagnoses. Nevertheless, the software was approved by the Food and Drug Administration (FDA) in 2018 [34].

5.2 Technical Implementation

Basically, application software that aims to make AI models available to the user consists of three basic components. A backend, which is responsible for all data processing within the software, a frontend, which handles the visualization of the data and interaction with the user, and the AI model, which is an algorithmic system that processes data and recognizes patterns to perform tasks based on its training and programming. The backend acts as an interface to the AI model and processes the necessary inputs and outputs for the AI model. Outputs are prepared for the frontend and made available to the user.

In order to train an AI model, extensive training data is required, which in this case is available as.csv files generated by an Apple Watch. These files contain digitized readings of the analogue ECG signal at a digitization rate of 512Hz. A 30-s ECG recording therefore contains 15360 readings, expressed in microvolts. However, to be suitable for processing by a CNN (Convolutional Neural Network), the data must be available as two-dimensional images instead of one-dimensional. Using the above parameters and appropriate Python libraries, the data can be converted into appropriate images. The goal of the CNNs used here is to classify individual beats in the ECG, so the available ECGs were divided into individual beats. A segmentation procedure is used to do this. The result is images of individual beats, which represent the training data. Unlike human observers, the images used to train the AI do not have a temporal and electrical reference grid. This is not needed here.

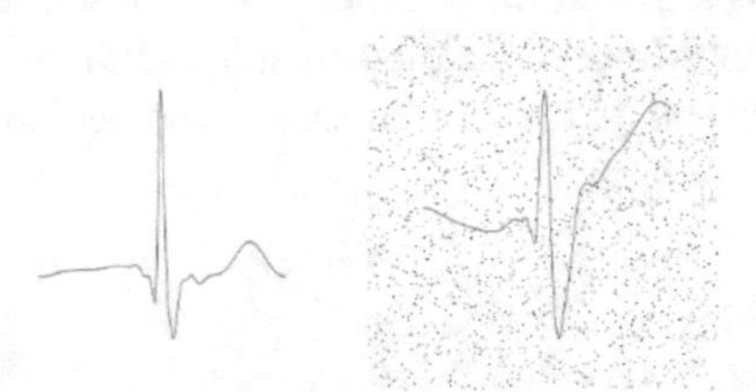

Fig. 1. Training picture with and without noise

Proven model architectures for image recognition are used to train the models. To generate a model, training is performed using training data that has previously been divided into classes and the model architecture. Depending on the type of images to be classified, different architectures may produce different results. The pre-processing of the data or images also has a significant impact on the accuracy of the model. It is often impossible to say at the outset which approach will produce the best results, so different approaches need to be tested and compared. To optimize the training data to increase the accuracy of the models, data augmentation is used. In this process, the data set is expanded by artificial manipulation. Techniques such as cropping, rotating, flipping, scaling, adding noise or changing color intensity are used to create new variations of the original images (Fig. 1). An example for the ECG images is the addition of salt-and-pepper noise, where random pixels of the image are colored [34].

5.3 Contributing to Healthcare Process Improvement Through AI-Based ECG Interpretation

The use of AI technology can assist both medical staff and patients. AI can perform basic tasks that, while not replacing the doctor or nurse, can take a lot of the workload off their shoulders and thus lead to significant time savings in medical treatment pathways. For example, data can be directly stored and recorded electronically, documentation can be reduced, and entire processes can be restructured. As a result, process quality is significantly improved. There is more time for more individualized, patient-focused treatments that benefit the patient. This applies to the clinical, inpatient sector as well as to discharge management and subsequent outpatient follow-up by general practitioners or specialists. This point is directly linked to the principle of "outpatient before inpatient", which is laid down in §39 of the German Social Code, Book V. The smartwatch supports this approach by allowing ECGs to be recorded and, in some cases, analyzed. As the

AI takes over routine evaluations, time and human resources are saved both in the hospital and in the subsequent outpatient care and can be used elsewhere. This makes the process much more patient-centered. In general, the software improves the entire care situation, which can currently be classified as precarious, minimizes the length of stay and morbidity, and increases the success of treatment.

The use of AI in this area enables upstream analysis of ECG data, which reduces the burden on hospital structures and staff. This results in a clearly measurable time saving, which relieves the care situation in terms of scarce resources and staff shortages, as well as the overload of medical staff. As a result, existing capacities can be better controlled, and their use optimized. With the help of the smartwatch, ECGs can be recorded more frequently and thus cover a wider time window.

The support provided by the smartwatch enables the patient to actively participate in his or her treatment, even outside of medical therapy. Patients have the opportunity to be proactive at all times and can achieve a greater degree of autonomy than is possible in the traditional patient-doctor relationship. Increased patient participation via the smartwatch also has positive effects in terms of saving time and resources.

Furthermore, the smartwatch examination can be seen as a preventive measure, as it can also be used in moments of risk and suspicion. The patient is thus empowered to take an active part in his or her treatment and to take targeted and independent preventive measures, which, given the seriousness of the disease and its prevalence, offer the possibility of good protection against or mitigation of such cardiological events.

The smartwatch is simple and intuitive to use. This makes it more effective across all age groups than a 12-lead ECG. In addition, wearable ECGs are significantly shorter and the timing of the recording can be adjusted to suit the circumstances.

There are also benefits for statutory health insurers. Through improved and more structured cardiological care, deaths can be minimized, more patients can be reached and medical outcomes can be improved. This leads to a sustainable reduction in costs for the statutory health insurance funds. Given the severity and prevalence of heart disease, the cost savings would be enormous.

Overall, it can be said that all the points mentioned in this chapter can fulfil many aspects of process improvement goals.

6 Conclusion

It is clear that digitalization plays an important role in process optimization. Processes can be significantly improved through the development and implementation of new innovations. Digitalization is based on many different drivers. One of these drivers is artificial intelligence, which has gained enormously in importance in recent years and has reached the middle of society at the latest with ChatGPT. Digitalization also plays an important role in the healthcare sector. However, as the study by the Bertelsmann Foundation [1] presented at the beginning of this article shows, Germany lags far behind the rest of the world when it comes to digitalization in the healthcare sector. Artificial intelligence, as a driver of digitalization, is seen as the key technology of the future in healthcare. However, it is important to note that there are a number of challenges that need to be identified and addressed. Regulatory issues such as data protection need to be addressed. Acceptance of AI by patients and doctors also seems to be a challenge, as some studies have shown [35–37].

In the future, the technology and software developed will be tested and evaluated using appropriate research methods. In addition, the AI-based ECG evaluation will be used in a real laboratory under real conditions as part of the Jenergiereal research project [38]. In addition, the software will be evaluated in a field study in a different context. It will be tested in a clinical setting as well as in an outpatient setting to investigate the use of the smartwatch at home in everyday life.

Overall, this work shows that AI-driven ECG interpretation is a promising tool for the diagnosis of heart disease. As described in the paper and shown in a concrete example, AI clearly has the potential to improve healthcare processes in a sustainable way. It is inevitable that AI will play an increasingly important role in cardiology and healthcare as a whole. In order to reap the full benefits of this technology, it is important to ensure that the development of AI applications is guided by ethical and moral principles and that the needs of patients and society are paramount.

References

1. #SmartHealthSystems (2021, 10. November). https://www.bertelsmann-stiftung.de/de/publikationen/publikation/did/smarthealthsystems. Accessed 09 Nov 2023
2. Sievers, U.: Digitales Gesundheitswesen: kaum am Start, schon veraltet. VDI nachrichten - Das Nachrichtenportal für Ingenieure (2023, 13. Juni). https://www.vdi-nachrichten.com/technik/gesundheit/digitales-gesundheitswesen-kaum-am-start-schon-veraltet/. Accessed 02 Sep 2023
3. Lauterbach: Elektr. Patientenakte ab Ende 2024 für alle verbindlich. (o. D.). App Title. https://www.bundesgesundheitsministerium.de/presse/interviews/interview/fas-030324-elektronische-patientenakte.html. Accessed 09 Feb 2023
4. Online, F. (2023, 7. Februar). Ärzte, Schwestern und Pfleger schlagen Alarm: „Das muss sich JETZT ändern!" FOCUS online. https://www.focus.de/gesundheit/unsere-kliniken-sind-krank-aerzte-schwestern-und-pfleger-schlagen-alarm-das-muss-sich-jetzt-aendern_id_184030389.html. Accessed 07 Sep 2023
5. Digitalisierung im Gesundheitswesen. (o. D.). App Title. https://www.bundesgesundheitsministerium.de/themen/digitalisierung/digitalisierung-im-gesundheitswesen.html. Accessed 21 Sep 2023
6. Bundesärztekammer. (o. D.). Digitalisierung in der Gesundheitsversorgung. https://www.bundesaerztekammer.de/themen/aerzte/digitalisierung. Accessed 17 Sep 2023
7. Jorzig, A., Sarangi, F.: Digitalisierung im Gesundheitswesen. In: Springer eBooks (2020).https://doi.org/10.1007/978-3-662-58306-7
8. Stachwitz, P., Debatin, J.F.: Digitalisierung im Gesundheitswesen: heute und in Zukunft. Bundesgesundheitsblatt - Gesundheitsforschung - Gesundheitsschutz **66**(2), 105–113 (2023).https://doi.org/10.1007/s00103-022-03642-8
9. Kretschmer, C., Wdr, S.Z.: Digitalisierung im Gesundheitswesen: Praxistest fürs E-Rezept. tagesschau.de (2023, 28. August). https://www.tagesschau.de/inland/gesellschaft/e-rezept-start-100.html. Accessed 12 Sep 2023
10. Gabler Wirtschaftslexikon. In: Springer eBooks (2019). https://doi.org/10.1007/978-3-658-19571-7
11. Gadatsch, A.: Geschäftsprozessmanagement im Gesundheitswesen. In Springer eBooks (S. 5–28) (2013). https://doi.org/10.1007/978-3-658-01166-6_2

12. Home. (2023, 11. Juli). http://www.caritasstjosef.de/content/node_11888.html
13. Kleemann, T.: Die dritte Generation von Krankenhausinformationssystemen – Workflowunterstützung und Prozessmanagement. In: Schlegel, H. (eds.) Steuerung der IT im Klinikmanagement. Vieweg+Teubner (2010). https://doi.org/10.1007/978-3-8348-9393-2_15
14. Bergmann, K. (2007). Klinische Behandlungspfade: mit Standards erfolgreicher arbeiten ; mit 4 Tabellen. Deutscher Ärzteverlag
15. Seidlmeier, H.: Prozessmodellierung mit ARIS®. In: Springer eBooks (2015). https://doi.org/10.1007/978-3-658-03905-9
16. Neumann, T.: Prozessverbesserung durch die Digitalisierung von Infrastrukturen [Dissertation] (2021)
17. Gadatsch, A.: Geschäftsprozesse analysieren und optimieren. In: Essentials (2015). https://doi.org/10.1007/978-3-658-09110-1
18. Hellmann, W.: Klinische pfade: Konzepte, Umsetzung, Erfahrungen (2002)
19. Nissen, V., Stelzer, D., Straßburger, S. Fischer, D.: Multikonferenz Wirtschaftsinformatik (MKWI) 2016: Technische Universität Ilmenau, 09. - 11. März 2016; Band III. Multikonferenz Wirtschaftsinformatik (2016). https://www.db-thueringen.de/receive/dbt_mods_00027211
20. Petry, T.: Digital leadership: Erfolgreiches Führen in Zeiten der Digital Economy (2019)
21. Ross-Büttgen, M.: Digitalisierung im Gesundheitswesen. mediorbis (2023, 12. Juni). https://mediorbis.de/ratgeber/ehealth/digitalisierung-im-gesundheitswesen#page
22. Center for Devices and Radiological Health. What is digital health? U.S. Food and Drug Administration (2020, 22. September). https://www.fda.gov/medical-devices/digital-health-center-excellence/what-digital-health. Accessed 02 Sep 2023
23. Digitalisierung im Gesundheitswesen. (o. D.-b). App Title. https://www.bundesgesundheitsministerium.de/themen/digitalisierung/digitalisierung-im-gesundheitswesen.html. Accessed 05 Sep 2023
24. Digitalisierungsstrategie. (o. D.). App Title. https://www.bundesgesundheitsministerium.de/themen/digitalisierung/digitalisierungsstrategie.html. Accessed 02 Oct 2023
25. Kempf, M., Pihl, C.: Closed loop medication management im Krankenhaus. In: Leonhardt, S., Neumann, T., Kretz, D., Teich, T., Bodach, M. (eds.) Innovation und Kooperation auf dem Weg zur All Electric Society. Springer Gabler, Wiesbaden (2022). https://doi.org/10.1007/978-3-658-38706-8_17
26. Doleski, O.D.: Utility 4.0: Transformation vom Versorgungs- zum digitalen Energiedienstleistungsunternehmen. Springer-Verlag (2015)
27. Schubert, J., Leonhardt, S., Schneider, M., Neumann, T., Gill, B.T.: Tobias: Smarte Quartiere 2050– flexibel, resilient und intelligent. Forschungsagenda für umwelt- und sozialgerechtes technisch assistiertes Wohnen. In: Weidner, Robert (Hrsg.): Zweite transdisziplinäre Konferenz Technische Unterstützungssysteme, die die Menschen wirklich wollen : Konferenzband. Hamburg: Helmut-Schmidt-Universität. S. 129-138 (2016)
28. Mindsquare. Künstliche Intelligenz. mindsquare (2023, 24. Oktober). https://mindsquare.de/knowhow/kuenstliche-intelligenz/. Accessed 24 Oct 2023
29. EV, B.: Europäischer KI-Markt verfünffacht sich binnen fünf Jahren. Bitkom (2022, 8. September). https://www.bitkom.org/Presse/Presseinformation/Europaeischer-KI-Markt-verfuenffacht-sich-binnen-fuenf-Jahren. Accessed 24 Oct 2023
30. Kassel, K., Pfannstiel, M.A.: Einsatzgebiete künstlicher Intelligenz bei chronischen Erkrankungen – Ein erster Überblick im Diagnostik- und Therapiebereich. In: Pfannstiel, M.A. (eds) Künstliche Intelligenz im Gesundheitswesen. Springer Gabler, Wiesbaden (2022). https://doi.org/10.1007/978-3-658-33597-7_19
31. Hennemuth, A., Hüllebrand, M., Doeblin, P., Krüger, N., Kelle, S.: Anwendungen von künstlicher Intelligenz in der diagnostischen kardialen Bildanalyse. Der Kardiologe **16**(2), 72–81 (2022). https://doi.org/10.1007/s12181-022-00548-2

32. Friedrich, S., et al.: Applications of artificial intelligence/machine learning approaches in cardiovascular medicine: a systematic review with recommendations. Eur. Heart J. **2**(3), 424–436 (2021). https://doi.org/10.1093/ehjdh/ztab054
33. Fotaki, A., Puyol-Antón, E., Chiribiri, A., Botnar, R.M., Pushparajah, K., Prieto, C.: Artificial intelligence in cardiac MRI: Is clinical adoption forthcoming? Front. Cardiovasc. Med. **8**, 818765 (2022). https://doi.org/10.3389/fcvm.2021.818765
34. Paech, C., Teich, L., Franke, D.: Agile Entwicklung einer KI-basierten EKG-Auswertung im Kindesalter. In: Leonhardt, S., Neumann, T., Kretz, D., Teich, T., Bodach, M. (eds.) Innovation und Kooperation auf dem Weg zur All Electric Society. Springer Gabler, Wiesbaden (2022). https://doi.org/10.1007/978-3-658-38706-8_18
35. Chen, M., et al.: Acceptance of clinical artificial intelligence among physicians and medical students: a systematic review with cross-sectional survey. Front. Med. **9**, 990604 (2022). https://doi.org/10.3389/fmed.2022.990604
36. Fritsch, S., et al.: Attitudes and perception of artificial intelligence in healthcare: a cross-sectional survey among patients. Digit. Health **8**, 205520762211167 (2022). https://doi.org/10.1177/20552076221116772
37. Reffien, M.A.M., et al.: Physicians´ attitude towards artificial intelligence in medicine, their expectations and concerns: an online mobile survey. Malays. J. Publ. Health Med. **21**(1), 181–189 (2021). https://doi.org/10.37268/mjphm/vol.21/no.1/art.742
38. JenErgieReal - Stadtwerke Jena. (o. D.). https://www.stadtwerke-jena.de/nachhaltigkeit/energiewende/jenergiereal.html. Accessed 25 Oct 2023

Sentiment Analysis for Sarcasm of Video Gamers

Zhen Li[(✉)], Leonardo Espinosa-Leal, and Kaj-Mikael Björk

Graduate School and Research, Arcada University of Applied Sciences, Jan-Magnus Janssons plats 1, 00560 Helsinki, Finland
{zhen.li,leonardo.espinosaleal,kaj-mikael.bjork}@arcada.fi

Abstract. The video game industry has been growing rapidly over the years, and video gamers have started to form a subculture where sarcasm is highly appreciated. Despite the important influence of the video gaming industry, the research focusing on the video game community is still very rare. In addition, the emerging large language models based on transformer architecture provide new tooling and perspective to study culture-related topics. Therefore, in this work, we would like to focus on studying the sarcasm culture among video gamers with the help of state-of-the-art large language models and massive game reviews. We show that a general large language model can be fine-tuned to specialize in capturing the gamer's sarcastic sentiment in Steam game reviews. The fine-tuned model has a 16% point improvement in accuracy compared to general sentiment models.

Keywords: NLP · Sentiment Analysis · Video Games · Sarcasm

1 Introduction

In October 2023, Microsoft finalized its acquisition of video game publisher Activation Blizzard with 69 billion US dollars [18] The purchase was delayed by lengthy regulatory reviews, and the regulatory interference demonstrated the growing influence of the gaming industry. According to Statista, the revenue of the gaming market was about 347 billion US dollars in 2022 [8] In the meanwhile, the game streaming industry is also growing rapidly. As of 2023, the game live-streaming market is believed to reach about 11 billion US dollars. The annual growth rate is around %10 every year, which will lead to a 17 billion market volume by 2027. By 2027, the user population is forecasted to reach 1.6 billion [2].

With such a big revenue and community, video gamers start to form a specific video game culture [19]. In video-game culture, sarcasm is heavily used and appreciated. Sarcasm is believed to be a high level language phenomenon because it expresses the opposite sentiment compared to plain text. In order to understand sarcasm, a certain amount of common sense or common understanding, tonal stress, and gesture clues are crucial. For language models, the lack of access to tonal stress and gesture clues makes detecting and classifying sarcasm very

M. E. Auer et al. (Eds.): STE 2024, LNNS 1028, pp. 299–308, 2024.
https://doi.org/10.1007/978-3-031-61905-2_29

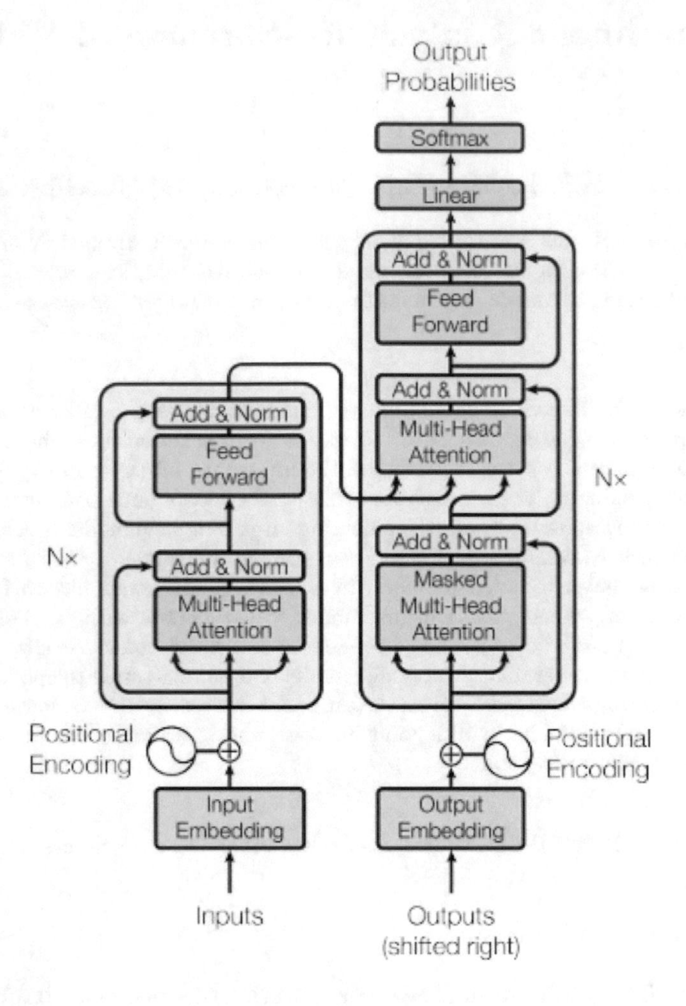

Fig. 1. Transformer architecture. The picture is reprinted from [21]

difficult and challenging. Such difficulty draws the attention of language model researchers. There are many efforts to detect sarcasm on X (formerly known as Twitter) and Reddit posts [9,10,12,14,15]. Some researchers [22] have also tried to study sarcasm in game reviews using traditional language models, such as Stanford CoreNLP [17], NLTK [11], and SentiStrength [20].

However, there is not much work focusing on a large number of game reviews using state-of-the-art transformer-based language models yet. We would like to fine-tune a state-of-the-art transformer language model with a large amount of Steam reviews to understand the sarcasm of video gamers and analyse their sentiment. Such sentimental analysis helps the gaming industry to capture the demand and interest of gamers, so the game developers can react to the market more accurately.

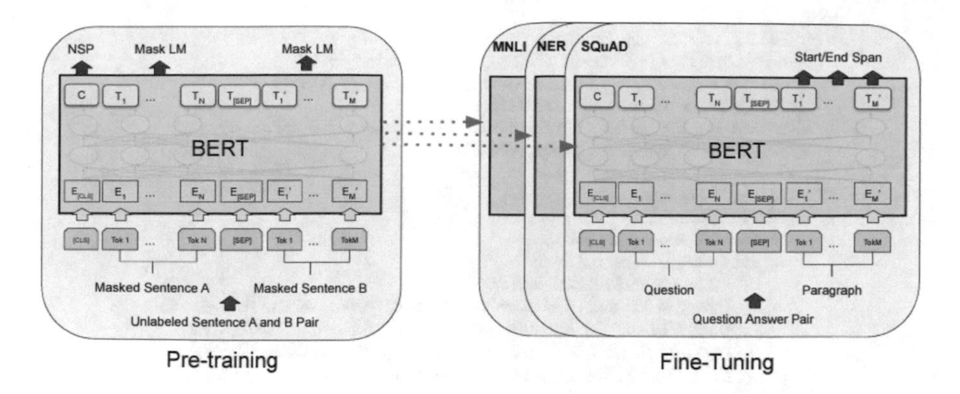

Fig. 2. BERT overall illustration for pre-training and fine-tuning. The picture is reprinted from [13]

2 Background

Transformer architecture was first proposed in 2017 [21] as shown in Fig. 1, reprinted from [21]. Since then, transformer models and their variations have been the dominant option for large language models. The most eye-catching ChatGPT [4] is also transformer based model, as indicated in its full name - Chat Generative Pre-trained Transformer. The attention mechanism, shown in Eq. 1 (where Q is the query, K is key, V is value, d_k is the dimension of the key), in a transformer is concise and effective in capturing temporal relations, which makes it very suitable for language modelling.

$$\text{Attention}(Q, K, V) = \text{softmax}\left(\frac{QK^T}{\sqrt{d_k}}\right)V \tag{1}$$

ChatGPT and GPT-4 are commercial models with massive success, but they are no longer open-source. However, there are many transformer based open source models on Hugging Face Co [3]. BERT [13] based models are popular for sentiment analysis and produce decent performance. Its variation, roBERTa [16] is also a suitable alternative. BERT stands for Bidirectional Encoder Representations from Transformers. The structure is shown in Fig. 2. The masked language model objective that BERT uses enables the transformer to be bidirectional, which boosts the performance in 11 NLP tasks defined in GLUE [23]. Furthermore, researchers found that BERT is under-trained in the original BERT work, and proposed improvement to unleash the potential of BERT, which is then called roBERTa. Normally, most of the times, fine tuning involves some substantial task-specific architecture modifications; however, both BERT and roBERTa are naturally fine-tuning friendly and don't need much of architecture changes as a result of the bidirectional structure.

```
(roberta): RobertaModel(
  (embeddings): RobertaEmbeddings(
    (word_embeddings): Embedding(50265, 768, padding_idx=1)
    (position_embeddings): Embedding(514, 768, padding_idx=1)
    (token_type_embeddings): Embedding(1, 768)
    (LayerNorm): LayerNorm((768,), eps=1e-05, elementwise_affine=True)
    (dropout): Dropout(p=0.1, inplace=False)
  )
  (encoder): RobertaEncoder(
    (layer): ModuleList(
      (0-11): 12 x RobertaLayer(
        (attention): RobertaAttention(
          (self): RobertaSelfAttention(
            (query): Linear(in_features=768, out_features=768, bias=True)
            (key): Linear(in_features=768, out_features=768, bias=True)
            (value): Linear(in_features=768, out_features=768, bias=True)
            (dropout): Dropout(p=0.1, inplace=False)
          )
          (output): RobertaSelfOutput(
            (dense): Linear(in_features=768, out_features=768, bias=True)
            (LayerNorm): LayerNorm((768,), eps=1e-05, elementwise_affine=True)
            (dropout): Dropout(p=0.1, inplace=False)
          )
        )
        (intermediate): RobertaIntermediate(
          (dense): Linear(in_features=768, out_features=3072, bias=True)
          (intermediate_act_fn): GELUActivation()
        )
        (output): RobertaOutput(
          (dense): Linear(in_features=3072, out_features=768, bias=True)
          (LayerNorm): LayerNorm((768,), eps=1e-05, elementwise_affine=True)
          (dropout): Dropout(p=0.1, inplace=False)
        )
      )
    )
  )
)
(classifier): RobertaClassificationHead(
  (dense): Linear(in_features=768, out_features=768, bias=True)
  (dropout): Dropout(p=0.1, inplace=False)
  (out_proj): Linear(in_features=768, out_features=2, bias=True)
)
```

Fig. 3. roBERTa structure from twitter-roberta-base-sentiment [1]

3 Method

Fine tuning deep pre-trained models is a common technique for developing models specialized on specific pre-defined tasks on top of existing deep and large pre-trained models. For example, ChatGPT is a fine-tuned model on top of GPT-3.5 or GPT-4 that specialises in conversations. With fine-tuning, general purpose models trained on massive texts can focus on particular subjects. In the meanwhile, no training from scratch also saves much computing power and resources. Most of modern-day state-of-the-art pre-trained models require millions of dollars to train. Fine tuning allows developers to produce decent performance models with limited resources.

In this work, we start with two fine-tuned pre-trained BERT based models specializing in general sentiment analysis. The two starting models are bert-base-multilingual-uncased-sentiment [6] (referred as BERT-sentiment) and twitter-roberta-base-sentiment [1] (referred as roBERTa-sentiment, structure is shown in Fig. 3), which can be found on Hugging Face. Both models are popular pre-trained models, with downloads being 1.6 million and 10.1 million respectively.

We first conduct sentiment analysis using the two sentiment analysis models on Steam Reviews [7] from Kaggle [5] to estimate the initial sentiment analysis performance without fine tuning for game review data, which contains many gamer touches of sarcasm. We then fine tune the roBERTa based model on part of the Steam reviews. Then, evaluate the fine tuned roBERTa model on reserved steam reviews that are not used for fine-tuning.

4 Experimental Setup

In order to compare results between different models, the sentiment analysis has been downgraded to a binary case with only two sentiments, i.e. positive and negative. Because both BERT and roBERTa base sentiment models have neutral sentiment, and the fine tuned model doesn't have a neutral option as there's no neutral option in the Steam review dataset; therefore, reviews classified to be neutral by either BERT-sentiment or roBERTa-sentiment model are removed for final performance analysis.

For BERT-sentiment model, as the model has 5 classes indicated by stars, this means that 1 and 2 stars mean negative, 3 stars are neutral, and 4 and 5 stars refer to the positive. As for roBERTa-sentiment model, 0 means negative, 1 means neutral, and 2 points to positive, as described in the model card. As mentioned before, eventually neutral ones are moved to maintain equality in comparison.

For fine tuning of roBERTa, we maintained the structure from roBERTa-sentiment, which is shown in Fig. 3. For fine tuning data, 63% of the total data is used for training and 7% of the data is used for validation, and the rest 30% is used for testing, which is unknown to the fine-tuned model. The classic binary classification metrics are used to evaluate the model performance on the reserved test dataset. In addition to the removal of neutral records, this means fewer reviews such as "Early Access Review" are removed as well.

The fine tuning is conducted on NVIDIA Volta V100 GPU for three days (72 h), which means it's trained for 2.24 epochs on about 4 million examples. The starting learning rate is set at $2e-5$, and the batch size is set to 32. PyTorch version is version 2.0. For the rest of the hyper parameters, default would be good. The key for fine-tuning process is normally on the fine-tuning data, the higher the data quality and quantity, the better the results.

5 Results

The fine tuning loss and learning rate changes, along with epoch numbers, can be found in Fig. 4. From the figure, we can see that there are some stages during the fine tuning process. At the end of the fine tuning, the model reaches the stable stage. Maybe future training will improve performance as well, but due to the limitation of training resources, we decided to stop at 2.24 epochs.

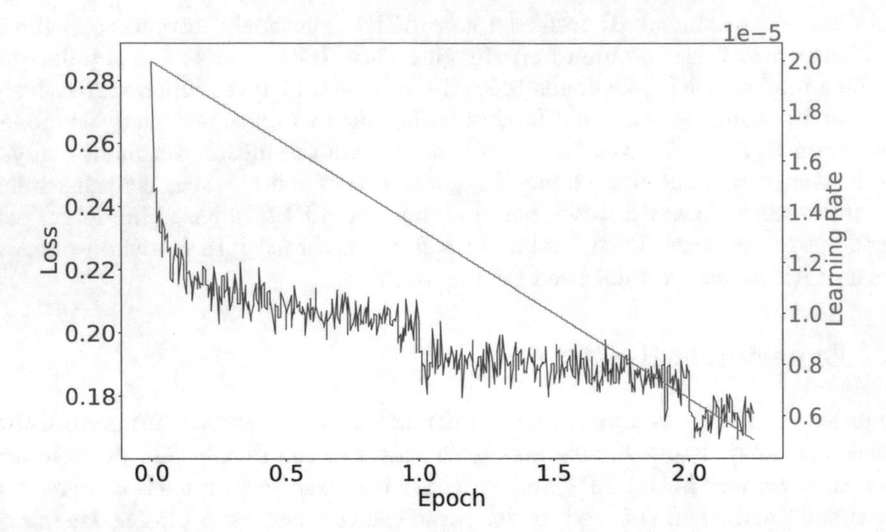

Fig. 4. Loss changes and learning rate decaying during fine tuning process

Table 1. Confusion matrices in percentage for different sentiment models

	BERT-sentiment	roBERTa-sentiment	Fine-tuned roBERTa
True Negative	14.8	14.5	**14.6**
True Positive	72.2	72.4	**81.3**
False Negative	11.2	10.9	**1.99**
False Positive	**1.88**	2.19	2.07

The confusion matrices of the BERT-sentiment, roBERTa-sentiment, and fine-tuned roBERTa are shown in Table 1. As we can see from the table, the fine tuned model outperforms the general sentiment BERT and roBERTa model on True Positive by about nine percentage points and on False Negative by about nine percentage points. This means that the fine tuned model captures the gamers' subculture and gets used to how gamers express emotions. In the meanwhile, the performance of BERT and roBERTa indicates that the wording and emotion expression approaches for gamers are different from Twitter or Redditers because the BERT-sentiment and roBERTa-sentiment were trained on posts from those platforms.

In Tables 2 and 3, one can find some examples of which fine tuned models correctly classified, while BERT-sentiment and roBERTa-sentiment didn't. There seem to be some typical sarcastic patterns that gamers prefer, such as joking about spending too much time on the game to reflect its quality, however still with a negative tone. In the meanwhile, some negative sentiment in general daily life could mean positive compliments on comments for horror games as well. In the example, someone commented that "i **** my pants". If it's a comment for

Table 2. Some sarcastic examples that are correctly classified by fine tuned model, but not others. The review should reflect POSITIVE sentiment.

Review Example
Bricks, They'‘re so heavy and make me so sad
Hi buddies, is there any cheat engine for this game? cauz it's hard as f.* -*-k...[a]
the biggest horrer experience ive ever had i almost peed myself
10/10 Would lose all friends via trust issues again
10/10 would jump off Maze Bank again
a light broke while i was walking and i ****[b] my pants
VERY depressing and atmospheric
gameplay made me crap myself
Ga...Ma...N...! Is he Romeo, or is he Lambo? I guess I can 'tolerate' this game
This game is so horribly pointless, it's amazing. I've been playing this the way I used to play GTA as a kid, I'd just run around and kill people. Goat Simulator takes that to the next level and gives you pointless achievements! Pointless Trading cards! Pointless milestones! I'm being completely serious. This is the most fun I've had by not accomplishing anything in my life. 10/10, will goat again
Starts to ruin my life. 10/10
At the end of a mission, i got kicked by host. 11/10 would play again
Those fools who this game bad ratings give pity I
You will loose your social life, your girlfriend, jobs and whatever else you had. You will not touch any other game. It'll be only you and V. V and you. No game should be THIS good. Next time I hope Rockstar makes a worse game than this so I don't loose my life
buy it or live to regret it

[a] Hard language is censored for publication
[b] Reviewer used symbol instead of hard language

horror game, it definitely means positive attitude towards the game. Therefore, in addition to the sarcasm that the fine-tuned model is able to capture, domain knowledge and overall gaming context are learned by fine tuning as well.

Similar to all the fine tune models, the fine-tuning of a model relies heavily on the fine tuning data quality. The Steam reviews could have noises as there could be misleading comments and false recommendation. One could try to use some other rules to further clean the dataset such as using the "recommendation usefulness" indicator. In Steam game reviews, other players and thumb up the review to indicate the usefulness of the comment. Further cleaned data could help with the performance improvement.

Table 3. Some sarcastic examples that are correctly classified by fine tuned model, but not others. The review should reflect NEGATIVE sentiment.

Review Example
Keeping it short... Loved the game in the past. Now...? It's easily P2W, Steer Clear of it!
This is a good game! Is what i thought until it removed text to speech, now it's just a pile of poopoo waiting to be flushed :)
pille of ****[a] [0.001 / 100]
Fun for the first 15 s
If you like to play games where you Queue for hours without finding a game with friends then this is the game for you. good concept, a lot of bugs I wish they would fix before worrying about adding to the game please fix whats all ready there
Nice slideshow simulator i5 3570K GTX 780 8gb Ram Game runs smooth 20–30 fps
I like the concept, the idea, the point of it. It is just done in a way that makes me want to sit in school to learn rather than play this. If there would be puzzles or reasons for me to look at items, that would turn the game into a game, i would not be borred after the 10 first words. Also the controlls are unnatural to me. Hope some update will come around to improve it from 'looking at yellow stickers to an object', to an interesting way learning a language
Decided to change it from being on Mac and WIndows to ust Windows. Thanks for taking my money you ****[a]
Payday : Global Offensive 2,the best way to name this game now
Beautiful game. Have to make an account with a 3rd party company to play. Had issues with save games lost mysave over 20 times making me start over. Was very fun. I wouldnt recogmend due to Never getting a reply to any single help thread for the fix
This game is amazing, it's a ton of fun. That being said, I can only finish about half of my games without the server failing and having to close the window. Other players confirm this, do not buy at this time. When they fix their connection issues this game will be fantastic. EDIT: This game still crashes and freezes all the time. You can play for free on their website, this is just a money-grab on Steam
great game but needs to be fixed and taken away from windows live but everything else about it is great the game story and modding community is just amazing also its a bitxh to get working on win 8 so i wouldn't recommend but if they patch it so theres no windows live login bullshit then i might change my mind

[a] Reviewer used symbol instead of hard language

6 Conclusion

In this work, we showcase that fine-tuned models outperform general purpose models on specific tasks. The improvement in accuracy can be as high as 16% points. The fine-tuned model both improves on true positive and false negative. In addition, the results verify that the gamer subculture is full of sarcasm and is quite different from that of Twitter and Reddit users.

This type of fine-tuned sentiment detection model can help the domain experts to better analyse the demand and feedback of the specific target audience to improve and satisfy users. Moreover, it's possible to fine tune a model to generate gamer style sarcasm to mimic gamers' personalities so that generative models can impersonate a certain type of people. A chatbot with such a generative model can adapt more to gamers' subcultures and is likely to be more popular than general chatbots. Reviews are very short, and it's well known that context length is very crucial to language models; therefore, more work on longer conversations for such fine tuning models still needs to be done in order to thoroughly evaluate the potential of building a gamer chatbot with gamer humour.

References

1. cardiffnlp/twitter-roberta-base-sentiment · Hugging Face — huggingface.co. https://huggingface.co/cardiffnlp/twitter-roberta-base-sentiment. Accessed 25 Dec 2023
2. Games Live Streaming - Worldwide — Statista Market Forecast — statista.com. https://www.statista.com/outlook/amo/media/games/games-live-streaming/worldwide. Accessed 25 Dec 2023
3. Hugging Face - The AI community building the future. — huggingface.co. https://huggingface.co/. Accessed 25 Dec 2023
4. Introducing ChatGPT — openai.com. https://openai.com/blog/chatgpt. Accessed 25 Dec 2023
5. Kaggle: Your Machine Learning and Data Science Community — kaggle.com. https://www.kaggle.com/. Accessed 25 Dec 2023
6. nlptown/bert-base-multilingual-uncased-sentiment · Hugging Face — huggingface.co. https://huggingface.co/nlptown/bert-base-multilingual-uncased-sentiment. Accessed 25 Dec 2023
7. Steam Reviews — kaggle.com. https://www.kaggle.com/datasets/andrewmvd/steam-reviews. Accessed 25 Dec 2023
8. Topic: Video game industry — statista.com. https://www.statista.com/topics/868/video-games/#topicOverview. Accessed 25 Dec 2023
9. Bamman, D., Smith, N.: Contextualized sarcasm detection on Twitter. In: Proceedings of the International AAAI Conference on Web and Social Media, vol. 9, pp. 574–577 (2015)
10. Barbieri, F., Saggion, H., Ronzano, F.: Modelling sarcasm in twitter, a novel approach. In: proceedings of the 5th Workshop on Computational Approaches to Subjectivity, Sentiment and Social Media Analysis, pp. 50–58 (2014)
11. Bird, S.: NLTK: the natural language toolkit. In: Proceedings of the COLING/ACL 2006 Interactive Presentation Sessions, pp. 69–72 (2006)
12. Bouazizi, M., Ohtsuki, T.O.: A pattern-based approach for sarcasm detection on Twitter. IEEE Access **4**, 5477–5488 (2016)
13. Devlin, J., Chang, M.W., Lee, K., Toutanova, K.: BERT: pre-training of deep bidirectional transformers for language understanding. arXiv preprint: arXiv:1810.04805 (2018)
14. González-Ibánez, R., Muresan, S., Wacholder, N.: Identifying sarcasm in twitter: a closer look. In: Proceedings of the 49th Annual Meeting of the Association for Computational Linguistics: Human Language Technologies, pp. 581–586 (2011)

15. Kumar, A., Anand, V.: Transformers on sarcasm detection with context. In: Proceedings of the Second Workshop on Figurative Language Processing, pp. 88–92 (2020)
16. Liu, Y., et al.: RoBERTa: a robustly optimized BERT pretraining approach. arXiv preprint: arXiv:1907.11692 (2019)
17. Manning, C.D., Surdeanu, M., Bauer, J., Finkel, J.R., Bethard, S., McClosky, D.: The stanford coreNLP natural language processing toolkit. In: Proceedings of 52nd Annual Meeting of the Association for Computational Linguistics: System Demonstrations, pp. 55–60 (2014)
18. Novet, J.: Microsoft closes $69 billion acquisition of Activision Blizzard after lengthy regulatory review — cnbc.com. https://www.cnbc.com/2023/10/13/microsoft-closes-activision-blizzard-deal-after-regulatory-review.html. Accessed 25 Dec 2023
19. Shaw, A.: What is video game culture? Cultural studies and game studies. Games Cult. **5**(4), 403–424 (2010). https://doi.org/10.1177/1555412009360414
20. Thelwall, M., Buckley, K., Paltoglou, G., Cai, D., Kappas, A.: Sentiment strength detection in short informal text. J. Am. Soc. Inform. Sci. Technol. **61**(12), 2544–2558 (2010)
21. Vaswani, A., et al.: Attention is all you need. In: Advances in Neural Information Processing Systems, vol. 30 (2017)
22. Viggiato, M., Lin, D., Hindle, A., Bezemer, C.P.: What causes wrong sentiment classifications of game reviews? IEEE Trans. Games **14**(3), 350–363 (2021)
23. Wang, A., Singh, A., Michael, J., Hill, F., Levy, O., Bowman, S.R.: GLUE: a multi-task benchmark and analysis platform for natural language understanding. arXiv preprint: arXiv:1804.07461 (2018)

Automatic Speech Recognition of Finnish-Swedish Dialects: A Comparison of Three Cutting-Edge Technologies

Leonardo Espinosa-Leal[1(✉)], Kristoffer Kuvaja Adolfsson[1], and Andrey Shcherbakov[2]

[1] Graduate School and Research, Arcada University of Applied Sciences, Jan-Magnus Janssons plats 1, 00560 Helsinki, Finland
{leonardo.espinosaleal,kristoffer.kuvajaadolfsson}@arcada.fi

[2] School of Engineering, Culture and Wellbeing, Arcada University of Applied Sciences, Jan-Magnus Janssons plats 1, 00560 Helsinki, Finland
andrey.shcherbakov@arcada.fi

Abstract. This paper explores the performance of two different automatic speech recognition models for the Finnish-Swedish language. The first model, Whisper V1 released by OpenAI and the second, the KBLab model trained using a large dataset by the National Library of Sweden. These models were trained initially using data from the Swedish language from Sweden, and the results were compared with previous work trained using a dataset of Finnish-Swedish audio. Our results indicate that general models perform at the same level, opening up the possibility of using these in Finland for the inclusion of the Finnish-Swedish minority.

Keywords: Speech recognition · Finnish-Swedish · Dialects

1 Introduction

Finland is a Nordic country with two official languages: Finnish and Swedish. The Swedish language is an important element in the history and culture of Finland, as the country was part of the Swedish empire for more than six centuries. The Swedish dialects of Finland belong to the East Swedish family of dialects. Their origins trace back to Old Swedish, which spread to Finland from Central Sweden with Swedish settlers from the 12th century onwards. Swedish dialects are spoken in four regions of Finland: Ostrobothnia, the autonomous island province of Åland, Åboland and Nyland (Uusimaa) [9] (see Fig. 1). Finland is a country with a high level of digitalization of services [5]. Therefore, it is of paramount importance that research has a strong focus on studying and developing artificial intelligence algorithms that contain no bias for the inclusion of these minority groups (e.g. healthcare [3]).

This study encompasses a comparison between new AI tools, specifically Open AI Whisper [11], a cutting-edge model, against the popular scientific tool

M. E. Auer et al. (Eds.): STE 2024, LNNS 1028, pp. 309–315, 2024.
https://doi.org/10.1007/978-3-031-61905-2_30

Kaldi in automatic speech recognition [10] on the Finnish-Swedish spoken language and a Swedish model created by the national library in Stockholm [6].

1.1 Related Work

Kaldi was released in 2011 and has been the de facto standard in scientific research on speech recognition for years [10]. The project is still getting more robust and has integrated the algorithms developed recently. However, in 2020, Meta released their Wav2vec 2.0 model, introducing a new trend in speech recognition, a significant boost in communities, and the development of automatic speech recognition (ASR). Since then, OpenAI has joined the ranks, releasing Whisper in 2022, streamlining the deployment and making the field even more accessible. As these newer models gain momentum, the gap between human and machine speech recognition is closing, and for Finnish-Swedish to stay relevant, it must keep up with the evolving field and catch on to the latest trends as society gets increasingly digitalized.

Regarding the Finnish-Swedish language, previous projects have focused on the collection and creation of datasets for training speech recognition algorithms [2,8]. Unfortunately, these efforts have been limited: the size of the dataset is small, or the audio doesn't contain the full audio transcriptions. Therefore, finding alternative models is an open challenge.

2 Research Methodology

We used the openly available dataset from the Finnish Parliament, pre-extracted by the Aalto Speech Recognition group. The dataset contains a total of 6 h of individual parliament speeches in the Finland-Swedish dialect. A total of 44 individuals [1]. Our work complements previous research using the same dataset but using a Deep Neural Network (DNN) Hidden Markov Model (HMM) [12].

Table 1. Example of three obtained transcriptions for the KBLab and Whisper V1 models on the parliament's dataset.

Reference	KBLAb	Whisper V1
på juridisk nivå men också i	å juridisk nivå men också	på juridisk nivå men också i
praktiken garantera att människor får möjligheter att tala sitt eget språk inför domstol det är kanske	praktiken garantera att människor får möjligheter att tala sitt eget språk inför domstol det kanske	praktiken garantera att människor får möjligheter att tala sitt eget språk inför domstol det kanske är
av de tillfällen i livet där det faktiskt är väldigt avgörande för en människas	na de tillfällen i livet där det faktiskt är väldigt avgörande för en människas	av de tillfällen i livet där det faktiskt är väldigt avgörande för människans

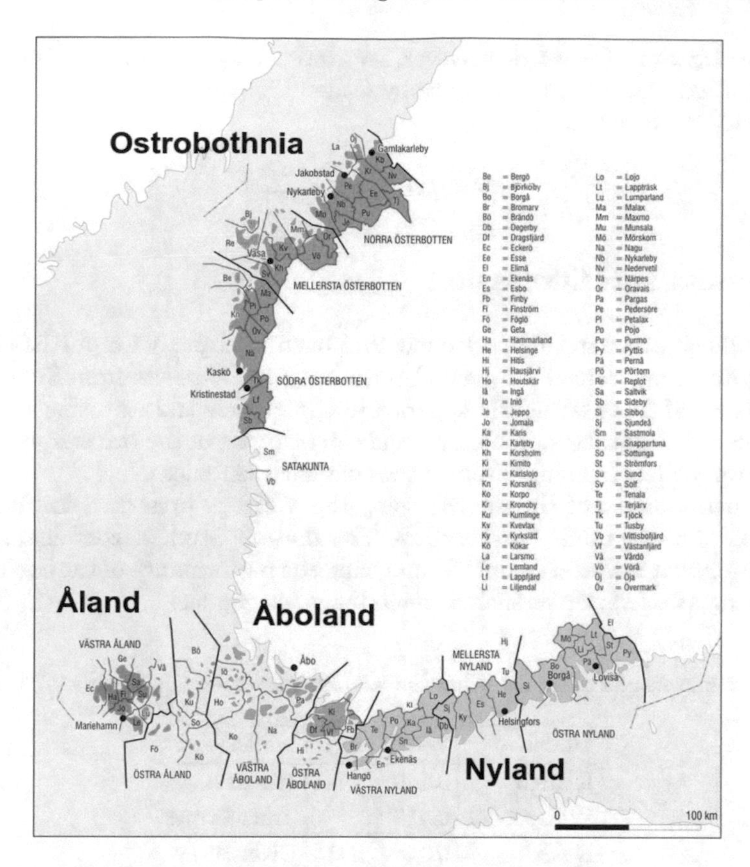

Fig. 1. Map of the southwest area of Finland, showing the distribution of the four main dialectal forms of the Finnish-Swedish languages. North-west: Ostrobothnia, south-west: Åland, south-centre: Åboland and south-east: Nyland (figure modified from [4]).

We used Whisper V1 (Released 20230314) [11] and the first version of the KBLAB model (Wav2vec 2.0 large VoxRex), an autoregressive model trained by the dataset of the Library of Sweden [6,7] available in the HuggingFace NLP library [14].

Both experiments were performed locally using a PC (Windows 11 and Python 3) with Intel 12Gen and NVIDIA 3060 GPU with 12 GB of Virtual Memory. Transcription time was done. Using the full GPU, Whisper's calculations took around 40 mins in total. KBLab took around 6 h, but the pipeline supported only the CPU model and had to be reloaded for each file.

Speech recognition can commonly be evaluated using Word Error Rate (WER), which is also used throughout our comparison. By adding the number of substitutions (S), deletions (D) and correct words (C) done by the model and dividing it by the number of words (N) in the reference you can calculate the WER. Another common evaluation is Character Error Rate, using per character instead of a word, however, to keep up with the consistency of the previous Kaldi model and work done by Otto-Ville [13] at Aalto only WER was performed. WER formula:

$$WER = \frac{S + D + C}{N} \tag{1}$$

3 Results and Discussion

The results obtained in this work show that both Whisper V1 and KBLab models, despite being trained for general purposes with datasets from Sweden, the performance of these is slightly superior to the models trained using data from Finnish-Swedish speakers. Table 1 presents an example of the transcriptions done by the two models in comparison to the reference sentences.

The performance of the models using the WER is presented in Table 2 on the whole dataset used. Furthermore, The dataset studied contains valuable information that makes it useful for analyzing the performance of the two models, considering other variables such as region, age, or gender.

Table 2. Main results obtained using the KBLab and OpenAI's Whisper V1 models.

Model	WER	Model
KBLab	0.180	Wav2Vec
Whisper V1	0.213	Transformer
Kaldi	0.145 (test)	DNN HMM
Kaldi	0.162 (train)	DNN HMM

3.1 Distribution by Region

In Table 3, the results of the models analyzed by the Finnish-Swedish dialectal regions. The dataset is integrated by Finnish politicians; therefore, as public figures, information is available on the internet. Wikipedia was used to find the birth region of each politician. The results indicate that both models perform similarly by dialectal region. However, each model performs slightly over the other in specific cases.

Table 3. Results by Dialectal Region

Dialectal Region	Birth Region	Count	KBLab	Whisper V1
Ostrobothnia	Pohjois-Pohjanmaan	1	0.149	0.149
	Keski-Pohjanmaan	1	0.311	0.200
	Pohjanmaan	7	0.156	0.157
	Etelä-Pohjanmaan	1	0.143	0.176
	Lapin	3	0.111	0.143
Åland	Ahvenanmaan	1	0.103	0.154
Åboland	Varsinais-Suomen	5	0.147	0.190
Nyland	Uudenmaan	18	0.214	0.200
others	Etelä-Karjalan	1	0.273	0.227
	Etelä-Savon	2	0.143	0.174
	Pirkanmaan	1	0.129	0.120
	Pohjois-Savon	3	0.147	0.124

3.2 Distribution by Age

The distribution per age group is presented in Table 4. The distribution of group ages shows that most of the people in the dataset are in the range of 60–50 and 70–60. However, the exact range of age is difficult to estimate, as the dataset was collected by using recorded videos from the Finnish parliament in a range of 5 years, still, it is important to consider that the dataset is obtained from adults therefore, the assessing of the accuracy of the speech recognition models should be biased.

3.3 Distribution by Gender

The results of the models categorized by gender are consigned in Table 5. The gender consigned in the table was the one that appears on the Wikipedia page of each member of the Finnish parliament. The male population is larger than the female population, which also implies a bias in the dataset, however, the performance of the model for both genders and both models is good in comparison to the Kaldi model.

Table 4. Result by Age group

Age Group	Count	KBLab	Whisper V1
80–70	2	0.233	0.076
70–60	13	0.160	0.169
60–50	15	0.161	0.164
50–40	10	0.240	0.252
40–30	4	0.122	0.122

Table 5. Result by Gender

Age Group	Count	KBLab	Whisper V1
F	17	0.172	0.167
M	27	0.182	0.185

In general, our results indicate that both models perform well in comparison to the baseline model using Kaldi-DNN HMM. The results also show a positive performance when geography, gender, and age are considered. These results show the robustness of the model, although some limitations are expected due to the distribution of the tested data.

4 Conclusion and Future Research

After testing the two general automatic speech recognition models trained on the Swedish language pronunciation used in Sweden, we found that they improved the state-of-the-art word error rate score when tested on a dataset containing audio from the Finnish-Swedish speakers. This result opens up the possibility of creating better ASR models combining data from different sources. Our results are, however, limited by the size of the dataset tested and the lower variety in the data, e.g. in the age distribution of the speakers.

While cutting-edge general multilingual ASR models continue to improve, we found that they perform slightly worse than models trained specifically on the Swedish language. A closer examination of ASR performance on languages with a smaller user base further confirms that training on national or even local datasets is crucial for achieving the most robust performance in automatic speech recognition.

For Finland Swedish to remain attractive, it is of paramount importance that we create the right tools. The number one lacking resource for creating better and more approachable technical solutions in automatic speech recognition is the creation of substantial open and available datasets. Moreover, technology should offer a path for the inclusion of minority populations, such as Finnish-Swedish speakers in Finland. Further research directions plan to explore the new Whisper models released by OpenAI and the last update of the KBLab model. Moreover, we aim to test these with a larger dataset of Finnish-Swedish dialects. Another direction is to fine-tune pre-trained large models using collected data from Finland.

References

1. Aalto University, Department of Signal Processing and Acoustics: Aalto Finland Swedish Parliament ASR Corpus 2015-2020. http://urn.fi/urn:nbn:fi:lb-2022052004
2. Espinosa-Leal, L., et al.: Tafidiai: Taligenkänning för finlandssvenska dialekter genom artificiell intelligens (2022). https://doi.org/10.5281/zenodo.7495136

3. Hägglund, S., et al.: Stakeholders' experiences of and expectations for robot accents in a dental care simulation: a Finland-Swedish case study. In: Social Robots in Social Institutions, pp. 125–134. IOS Press (2023)
4. Kotus: Swedish dialects in Finland - Institute for the Languages of Finland — kotus.fi. https://www.kotus.fi/en/on_language/dialects/swedish_dialects_in_finland_7542. Accessed 03 Dec 2023
5. Leal, L.E., Chapman, A., Westerlund, M.: Autonomous industrial management via reinforcement learning. J. Intell. Fuzzy Syst. **39**(6), 8427–8439 (2020)
6. Malmsten, M., Börjeson, L., Haffenden, C.: Playing with words at the national library of Sweden–making a Swedish BERT. arXiv preprint: arXiv:2007.01658 (2020)
7. Malmsten, M., Haffenden, C., Börjeson, L.: Hearing voices at the national library–a speech corpus and acoustic model for the Swedish language. arXiv preprint: arXiv:2205.03026 (2022)
8. Moisio, A., et al.: Lahjoita Puhetta: a large-scale corpus of spoken Finnish with some benchmarks. Lang. Resour. Eval. **57**(3), 1295–1327 (2023)
9. Östman, J.O., Mattfolk, L.: Ideologies of Standardisation: Finland Swedish and Swedish-language Finland. Stan. Lang. Lang. Stan. Changing Eur. **1**, 75–82 (2011)
10. Povey, D., et al.: The Kaldi speech recognition toolkit. In: IEEE 2011 Workshop on Automatic Speech Recognition and Understanding. IEEE Signal Processing Society (2011)
11. Radford, A., Kim, J.W., Xu, T., Brockman, G., McLeavey, C., Sutskever, I.: Robust speech recognition via large-scale weak supervision. In: International Conference on Machine Learning, pp. 28492–28518. PMLR (2023)
12. Raitolahti, O.V.: Finland Swedish automatic speech recognition. Master Thesis (2022)
13. Raitolahti, O.V.: Finland Swedish automatic speech recognition (2022). https://urn.fi/URN:NBN:fi:aalto-202203272601
14. Wolf, T., et al.: HuggingFace's transformers: state-of-the-art natural language processing. arXiv preprint: arXiv:1910.03771 (2019)

Predicting the Duration of User Stories in Agile Project Management

Asif Raza[1,2] and Leonardo Espinosa-Leal[2](✉)

[1] VizTrend OY, 02620 Espoo, Finland
asif.raza@yodiz.com
[2] Graduate School and Research, Arcada University of Applied Sciences, Jan-Magnus Janssons plats 1, 00560 Helsinki, Finland
leonardo.espinosaleal@arcada.fi

Abstract. Effective effort estimation in agile project planning is vital because it helps organizations build product plans that they can stick to, have shorter turn-around time, and have better cost discipline. Machine learning can play an essential role in planning and estimating the project schedule. In this paper, a series of supervised machine learning models were studied, analyzed, and implemented to solve the problem of predicting effort estimates in *Agile Scrum*. The obtained results are compared with similar previous studies. We performed experiments using different Natural Language Processing (NLP) methods such as Term Frequency-inverse document frequency (TF-IDF), fastText, and different Neural Networks, including Recurrent Neural Networks (RNN), Long Short Term Memory (LSTM), and Bidirectional Encoder Representations from Transformers (BERT). The trained models were fitted with three publicly available datasets. Our findings show that fastText (with a pre-trained model) significantly performed better in predicting the story-points of user-stories. The second-best performing model was bidirectional LSTM. Moreover, distilBERT performs poorly among all the models analyzed. This study can pave the way for organizations to benefit from these machine learning models and accurately predict project deadlines and schedules.

Keywords: Agile Scrum · Effort Estimation · NLP · Project Management · fastText · TF-IDF · BERT · distilBERT · bi-LSTM

1 Introduction

Business leaders realize that the pace of competition is increasing, so they need to find ways to respond to change and deliver value faster to their customers. Organizations are adopting *Agile* project management methodology in response to these challenging environments to keep pace with customer demand and develop efficient, high-quality products faster. Agile is becoming a popular project management approach because of its emphasis on efficiency and quality [22].

M. E. Auer et al. (Eds.): STE 2024, LNNS 1028, pp. 316–328, 2024.
https://doi.org/10.1007/978-3-031-61905-2_31

Agile Scrum methodology has now become a de facto standard in the Information Technology (IT) industry [18,25]. Agile Scrum [28] deals with the estimate and schedule of project planning by answering questions like, how considerable is the effort? When will it be done? Moreover, when will it be delivered? These questions circle during the life cycle of a project, but due to the iterative nature of Scrum, these questions are raised often, and the team adapts to change.

Agile Scrum first decomposes a project into smaller requirements to understand the bigger picture of architecture, user experience (UX), quality assurance (QA), development, and deployment. This helps to understand a project's risk, duration, schedule, and cost. The smaller steps are taken to achieve the big picture. In agile, this work is done using a series of repeated cycles (sprint) and a short timebox iteration (the ideal duration of a sprint is 2–4 weeks). The project is subdivided into small subprojects, which are represented by sprints. In each sprint, a potentially shippable product increment is delivered, which consists of design, implementation, testing, and deployment stages. Thus, in every sprint, new features represented by user-stories are added to the product, which results in gradual project growth. The user-story selected in each sprint is based on business priority to ensure that the essential user-stories are developed first.

Determining effort estimates using story-point is a common practice in Agile Scrum. Story-point is a relative measure of the size of a user-story. Story-point is purely an estimate of the time it will take to finish a user-story with a quality product. For example, a user-story with ten story-point is twice as big, risky, and complex as a user-story estimated as five story-point.

Research conducted by McKinsey & Company [1] in collaboration with the University of Oxford suggested that, on average, large IT projects run 45% over budget and 7% over time while delivering 56% less value than predicted. Another study [2] indicated that, while analyzing more than 1,100 software projects, it was found that only 30% met their original delivery deadline, with an average overrun of around 25%. Delay in launching or delivering projects means lost sales, an advantage for the competitor to get ahead, and potentially long-lasting damage to reputation and unhappy customers.

The motivation of this work is to help organizations be effective and precise in project planning when it comes to effort estimation. Our findings improve the state-of-the-art and show that machine learning can play an essential role in planning and estimating the project schedule.

2 Dataset

The data used in this study is from open-source projects and used in earlier similar studies [5,21] respective git repository [4], additionally some more data collected from another public git repository [17]. Here, this study has 25 different projects fetched from 3 stated sources. The format of the data is CSV or Excel.

These are user-stories of functional and non-functional requirements. All the user-stories contain a title, description, and commit-logs. The attributes of user-stories are described in Table 1. All the user-stories have a story-point, which

Table 1. Attributes in User-stories

User-Story Attribute	Explanation	Data Type
Title	1–3 line of text describing the title	String
Description	Describing requirement or enhancement in detail	String
Commit-Logs	SW code commit	String
Story-Point	Fibonacci series [0,1,2,3,5,8,...], which describes the effort estimate of work	Integer

is the label. At the top of Fig. 1 shows the data as Pandas [27] data-frame; the three columns represent the title, description and story-point. The length of the text in user-stories is essential; as longer the text, the requirement will grow. The bottom-left of Fig. 1 shows the word count in user stories. The story-point in the dataset is the actual effort estimated by the team regarding a user-story, so they are the actual duration of a user-story completed in a sprint. The bottom-right of Fig. 1 shows the user-story view from a commercially available Agile Scrum tool[1].

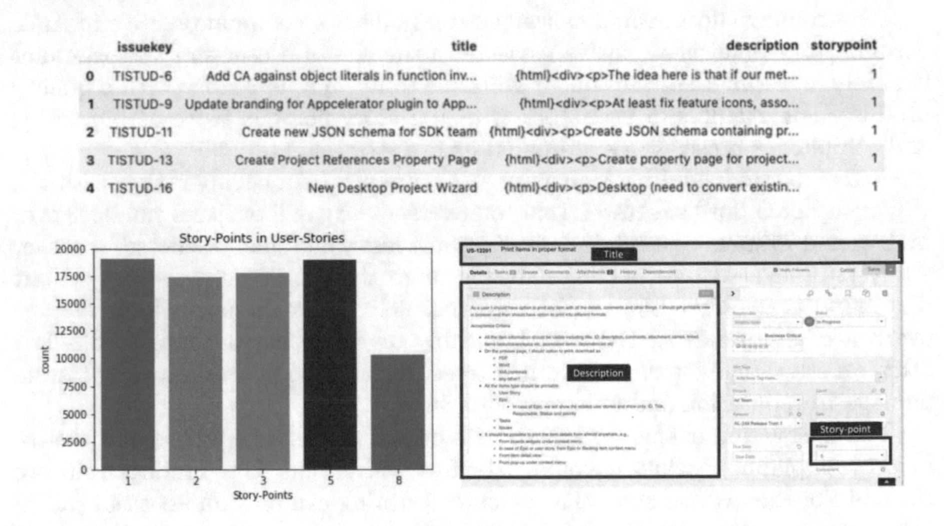

	issuekey	title	description	storypoint
0	TISTUD-6	Add CA against object literals in function inv...	{html}<div><p>The idea here is that if our met...	1
1	TISTUD-9	Update branding for Appcelerator plugin to App...	{html}<div><p>At least fix feature icons, asso...	1
2	TISTUD-11	Create new JSON schema for SDK team	{html}<div><p>Create JSON schema containing pr...	1
3	TISTUD-13	Create Project References Property Page	{html}<div><p>Create property page for project...	1
4	TISTUD-16	New Desktop Project Wizard	{html}<div><p>Desktop (need to convert existin...	1

Fig. 1. Top: User-story's title, description and story-point in the corpus. Bottom-left: User-stories and story-points distribution in the dataset, and bottom-right: User-Story view from Agile-Scrum tool.

2.1 Data Cleanup

By analyzing the data, it was realized that several items in the data needed to be removed. These items consisted of *NaN* values, special characters, stop-words, single characters, numbers, URLs, HTML and JSON tags, punctuation,

[1] https://www.yodiz.com/.

and others. After cleaning all of these items, data was converted to lower cases. "Title" and "Description" of user-stories were combined to obtain a single input as a feature input.

Another observation regarding the dataset was that it has story-points from a 1–100 scale. Occasionally, the team puts a large number at the beginning regarding the effort estimations. Story-points in the data are grouped along with the values 1, 2, 3, 5 and 8. This effort estimate represents the classes the model needs to predict in our data.

The dataset consisted of 90000 datapoints, and it was split randomly with a ratio of 83.3% (75000 data points) for the training set and 16.7% test set (15000 datapoints). In total, there are five target classes. The training dataset is used for model training, and the test dataset is used for model validation.

3 Related Work

Artificial intelligence has become an essential element for the industrial optimization of processes [7]. For example, machine learning research in project management and product design has come a long way; some approaches and related work are available.

In the study by Porru et al. [21], the authors proposed the Term Frequency - inverse document frequency (TF-IDF) based model for estimating the story-point in the Agile Scrum project for issue reports. Scott et al. [26] further extended Porru's study [21] by investigating if more features can be utilized along with TF-IDF feature. Choetkiertikul et al. [5] stated that combining two deep learning architectures, long short-term memory (LSTM) and recurrent highway network (RHN), was very effective in predicting story-points for user-stories.

Panda et al. [20] used various neural networks to enhance the prediction accuracy of agile software effort estimates. Different authors used different types of neural networks (General Regression Neural Network (GRNN), Probabilistic Neural Network (PNN), Group Method of Data Handling (GMDH), Polynomial Neural Network, and Cascade-Correlation Neural Network). The performance of the models is usually assessed using MSE, R^2, MMRE, or PRED metrics.

The authors used a unique approach for predicting the effort estimate in the study by Gultekin and Oya Kalipsiz [9]. They stated that SW development is done in different phases. Each of the phases has its own iterations, so first, the effort is estimated on each iteration. Aggregation is done on the sum of iterations in the SW development phase, and then the cost and effort estimate for the phase is calculated using the Neural Networks, Support Vector Machine (SVM), Random Forest (RF), Logistic Regressing (LR), and Gradient Boosting (GB). The model is analyzed using model accuracy and MEA. Other authors also analyzed their model against other similar studies [5,11,20,21,24]. Based on the empirical evaluation, the authors stated that the error rate of their story point-based estimation model is better than others.

In another research aiming for text classification [8], the authors compared the Bidirectional Encoder Representations from Transformers (BERT) [6] against the traditional machine learning approach like TF-IDF. Another study proposed the Hierarchical Attention Network (HAN) for document classification [30].

4 Material and Methods

4.1 TF-IDF

Term Frequency - inverse document frequency (TF-IDF) is a widespread algorithm that transforms text into a meaningful representation of numbers in a vector. Porru et al. [21] proposed this method with the aim of predicting the story-point using the "Title" and "Description" provided in text format.

TF-IDF is a simple way to calculate the "score" of the words in a document relative to a corpus. The score is calculated on each term in a user-story across the whole corpus. This will give each word based on its occurrence weighting factors, represented as weight within a corpus. In Eq. 1 the TF-IDF score w is defined as a term t in document d with respect to corpus D. Term frequency is defined as the raw count of term t in the document d, divided by the total number of the term in D in Eq. 2.

$$w(t, d, D) = tf(t, d) \times idf(t, D) \tag{1}$$

$$TF = \left(\frac{\text{No. of repet. of term in a doc.}}{\text{Total number of terms in a document}} \right) \tag{2}$$

In Eq. 3, idf is defined as the logarithm of the total number of the documents in the collection D, divided by the number of documents where t is present.

$$\text{idf}(t, D) = \log \frac{|D|}{1 + |\{d \in D : t \in d\}|} \tag{3}$$

To use the TF-IDF matrix for predicting the story-point in user-stories, two prominent machine learning models: Logistic Regressing (LR) [19] and Support Vector Machine (SVM) [14] are chosen.

4.2 fastText

Two different approaches of fastText [15] as proposed by Joulin et al.[3] are applied in this paper: fastText with hyperparameter and fastText with a pre-trained model.

fastText with Hyperparameters. The first approach is to tune parameter values provided by fastText. Additional values and parameters are used to optimise the model performance. The following parameters and range of values were used for tuning the values: i) learning rate (lr): $[0.05, 0.1, 0.2]$, ii) n-gram: $[1, 2, 3, 5]$, iii) number of epochs: $[5, 10, 15]$, iv) window size (size of the context window): $[5, 10, 15]$, and v) loss: $[ns, hs, softmax]$. These stated parameter values are initialized in fastText by looping through different parameter values. There are a total of 324 ($lr \times 3 \times n - gram \times 4 \times epoch \times 3 \times ws \times 3 \times loss \times 3$) iterations, each provided with different accuracy based on the parameter and values used; then the high performing parameter values are picked on which the model performs best.

fastText with Pre-trained Model. The second approach is to use a pre-trained model. The pre-trained model is trained using the project's data without labels. Data for training the model is not used in the training or validation process. There is also a possibility to use a pre-trained model trained on Wikipedia data or Google's pre-trained model on books. However, since the model trained on domain-specific data generally performs better, that is why it trains its model using fastText. The advantage of having a pre-trained model is to reduce the training time and have a better quality vector [10]. That is why we built pre-train the model with domain-specific data with fastText and used it as input.

4.3 DAN

Deep Averaging Network (DAN), proposed by Iyyer et al.[13], is the continuous bag-of-words (CBoW) or neural bag of words representation, computes a sum or mean of word embeddings over a document.

As described by the author, DAN is a deep unordered model that can obtain near state-of-the-art accuracy on various sentence and document level tasks with just minutes of training time on an average laptop computer. DAN works in three simple steps: 1. Take the vector average of the embeddings associated with an input sequence of tokens 2. Pass the average through one or hidden layers, and 3. Perform (linear) classification on the fully connected layer.

4.4 LSTM

The following choice of the neural network model is Long Short-Term Memory (LSTM) [12]. LSTM works well to capture the long-term dependencies. It introduces a memory cell to remember values over arbitrary time. Input, output, and forget gate regulates the flow of information in and out of the cell. Bidirectional LSTM (bi-LSTM) is also used in this study. In bi-LSTM, forward and backward hidden layers are concatenated and applied to drop out. Parameters used in bi-LSTM: input dimension is the size of the vocabulary of the corpus, the size of the embedding dimension is 100, the size of the hidden dimension is 256, and the output dimension is the same as the number of classes, which is 5. The number of layers is 2.

4.5 BERT

In this paper DistilBERT [23] is used, which is a lighter, smaller and faster version of BERT [6]. DistilBERT is a pre-trained model used in this work and is provided by HuggingFace, which is a library of Transformers Wolf et al. [29]. DistilBERT model is used by interface class **DistilBertForSequenceClassification** by HuggingFace [29], which provide model for text classification. **DistilBertTokenizer** is used for tokenization.

5 Results

5.1 TF-IDF

Table 2 shows LR and SVM models' train and test accuracy. Analyzing the precision and recall (shown in Fig. 2), all classes have a very stable and equal threshold of accuracy except class "1", which could be the reason that data for class 1 is in balance with others or the class threshold boundary from class "1" or other classes are not clearly classified by SVM. For all classes, precision and recall values are very consistent. Class "2", "5", and "8" have a precision of 79% while class precision is slightly less. The average accuracy of the SVM model is 77%.

As described by Joachims et al.[14], SVM is very well suited for text categorization. SVM consistently achieved good performance on text categorization tasks. Furthermore, SVM does not require any parameter tuning since it can find good parameter settings automatically. All this makes SVM a very promising and easy-to-use method for text classification. LR, on the other hand, does not perform so well compared to the SVM. The confusion matrix (normalized and non-normalized) of SVM validation accuracy is shown in Fig. 2.

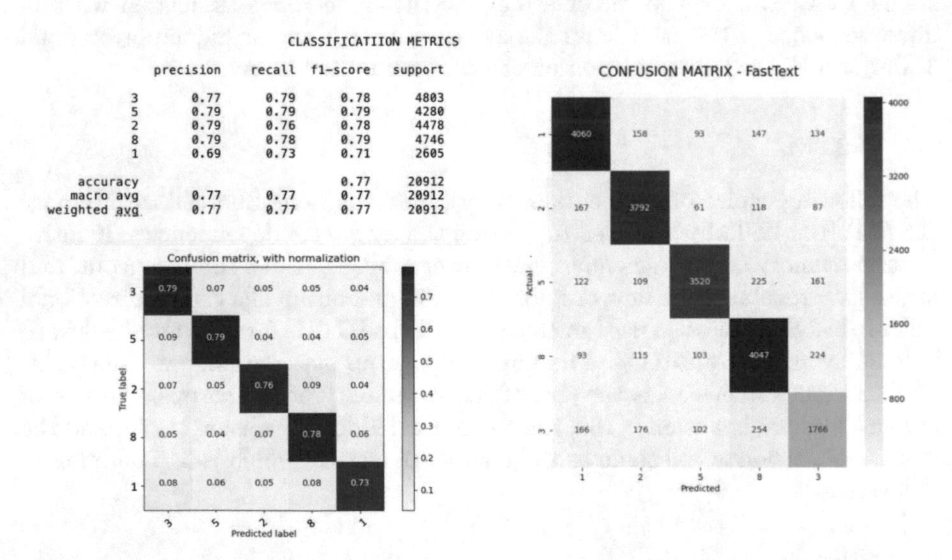

CLASSIFICATIION METRICS

	precision	recall	f1-score	support
3	0.77	0.79	0.78	4803
5	0.79	0.79	0.79	4280
2	0.79	0.76	0.78	4478
8	0.79	0.78	0.79	4746
1	0.69	0.73	0.71	2605
accuracy			0.77	20912
macro avg	0.77	0.77	0.77	20912
weighted avg	0.77	0.77	0.77	20912

Fig. 2. Top-left: Classification report for SVM model with TF-IDF. Bottom-left: SVM model validation confusion matrix and right: fastText test confusion matrix.

5.2 fastText

fastText Result with Hyperparameter Values: In the first approach, the model was trained with different parameter values to identify the best values on which the model performed well. The training and test accuracy (shown in Table 2) with this method was 96.54% and 85.92%. Figure 2 shows the fastText test accuracy confusion matrix.

Result from fastText Pre-trained Model Approach: The fastText model was used, including five-fold cross-validation for training the data. The data is divided into five-folds, and mean accuracy was calculated on each fold. Table 2 shows the 98.32% train and 87.4% test mean accuracy of this approach. A pre-trained model was used in this approach. It helps the accuracy, as the model learns faster and better. This paper applies a similar technique proposed by Joulin et al. [16], and the result obtained is aligned with the stated study. In the stated study, the authors quoted, "fastText is often on par with deep learning classifiers in terms of accuracy, and many orders of magnitude faster for training and evaluation." Also, the authors mentioned that they were able to achieve approximately 1% higher accuracy in their study when using a pre-trained model. In this study, using the pre-trained model approach, the accuracy is increased by 1.48%, compared to the model which does not use a pre-trained model (hyper-parameter approach). This shows that the result achieved is similar to the study mentioned earlier.

5.3 DAN

The technique followed is similar to that proposed by Iyyer et al. [13]. It is a DAN network, which is a special variant of RNN. One hundred epochs are used to train the network, training accuracy, validation accuracy, and loss are calculated after every five epochs, and mean accuracy is calculated after all the epochs. Training mean accuracy 84.18% and validation mean accuracy 69.75%, shown in Table 2. Training the network was fast but did not produce as good results as the fastText or TF-IDF approach. After 50 epochs, the validation loss does not decrease much. Training loss shows a good decrease, but not significantly in the validation set.

5.4 LSTM

This section describes the result of LSTM, as shown in Table 2. The technique followed in this paper is similar to that proposed by Choetkiertikul et al. (2019) [5]. The main difference is, in this study, bi-LSTM is used, which is a special variant of LSTM. Another difference is that no additional RHN (Recurrent High way Network) layer is added on top of bi-LSTM. The concept is similar to the extent that LSTM is used for text classification. bi-LSTM performs significantly well with a training mean accuracy of 93.1% and validation mean accuracy of

83.2%, but is still behind the fastText pre-trained model. The bi-LSTM is trained using a similar method described in DAN (Sect. 5.3). Training loss shows a good decrease, but not significantly in the validation set.

5.5 BERT

This section describes the result of BERT. The BERT model has been creating tremendous popularity. However, it does not produce a good result in this study. Table 2 shows the validation accuracy results. In fact, it is one of the least performing models in our methods. DistilBERT performs worse than the TF-IDF approach. There could be two explanations: a lighter variant of BERT, which this study is using. Another explanation is that NLP problems relating to text classification depend on the dataset's domain in the investigation. BERT has proven excellent results for different kinds of NLP problems, like generic email spam detection, sentiment analysis in the public domain, etc. The study strongly suggested that distilBERT is not suitable for the dataset used in this study. The validation accuracy of distilBert model was 70.7%. Training loss shows a good decrease, but is not as significant in the validation set.

6 Result Comparison

The consolidated results from all the techniques used in this study: TF-IDF, fastText, RNN, bi-LSTM, and BERT are consigned in Table 2.

Table 2. Consolidated results view of all the models

Techniques	Validation Acc. %	Training Acc. %
fastText with pre-trained model	**87.4%**	**98.32%**
fastText without pre-trained model	85.92%	96.54%
bi-LSTM+word embedding	83.2%	93.1%
TF-IDF+SVM+3-gram	77.4%	96.8%
TF-IDF+LR+3-gram	72.1%	83.8%
DistilBERT	70.7%	85.92%
DAN+word embedding	69.7%	84.17%

Finally, when looking at the results of all the methods, it is evident that the fastText (pre-trained model approach) performs best with 87.4% accuracy, followed by the second approach of fastText with 85.92%. So, we can clearly say that both approaches of fastText performed best in our study and is state-of-the-art. Furthermore, in the neural network domain, bi-LSTM was the top-performing model with 83.2% accuracy. 4th in our list is the TF-IDF+SVM model with an accuracy of 77.4%, followed by TF-IDF+LR model with an accuracy of 72.1%.

Table 3. Comparison of proposed models with previous studies

Source	Methods	Results		Data
Panda et al. [24]		Acc. %	MMRE	21 projects
	GRNN	89.9	0.351	
	PNN	87.6	1.57	
	GMDA	89.6	0.15	
	CCNN	94.7	0.14	
Gultekin et al. [9]		Acc.%	MAE	3233 user-stories
	NN	92.5	7.48	
	SVM	97.7	2.27	
	GB	99.8	0.2	
	RF	97.8	2.2	
Porru et al. [21]		Acc. %	MMRE	699 user-stories
	SVM	59	0.50	
	NB	44	0.85	
	KNN	36	0.70	
	DT	23	0.98	
Choetkiertikul et al. [5]		Acc. %	MAE	3233 user-stories
	Deep-SE	69.67–50.29	0.64–5.97	
	LSTM+RF	68.51–17.86	0.66–9.86	
	LSTM+SVM	66.61–44.19	0.70–6.70	
	LSTM+ATLM	66.51–16.92	0.70–9.97	
	LSTM+LR	63.20–16.9	0.77–9.97	
Scott et al. [26]		Acc. %	MAE	3233 user-stories
	SVM	46.33–93.15	0.36–3.45	
This work		Acc. %	MAE	90k user-stories
	FastText pre-trained vector	87.4	0.4342	
	FastText Hyperparameter	85.8	0.4863	
	bi-LSTM + word embeddings	83.2	0.7052	
	TF-IDF+3-Gram+SVM	77.4	0.9680	
	TF-IDF+3-Gram+LR	72.1	0.8618	
	DistilBERT	70.7	1.2133	
	DAN + word embeddings	69.7	0.8891	

The two models at the bottom of the list are distilBERT with an accuracy of 70.7% and list one DAN with an accuracy of 69.7%.

The results in Table 3 demonstrate the comparison of work done in this study and by other authors. This comparison is only for evaluation purposes. Direct comparison is not possible for various reasons, like the dataset used in each of the studies, the size of the datasets, the number of classes used in evaluating the model, the number of user-stories, and issues in the dataset and project domain. The purpose of summarizing this paper's past studies and results is to provide what techniques have been used so far and the best techniques suitable for predicting the effort estimate of agile projects. The MAE is one factor that can be used interdependently in evaluating the individual model and technique.

7 Conclusions

To conclude, the fastText proves to be a much more powerful text classification library, and it is much easier to use and provides much better results than other approaches. The classical approach (TF-IDF) is much faster but still lacks accuracy and speed. On the other hand, neural networks (LSTM) were harder to train as they required longer training time and GPU power but still offered reasonably good performance when compared to other approaches. The fastText with a pre-trained model is state-of-the-art in our model list by providing the best solution to this problem. Its excel is at all levels of performance, speed, and simple to implement.

To summarize, the most crucial factor in problems relating to NLP is the dataset itself. It is challenging to tell which approach works well, depending on the data domain and data characteristics. Some models or techniques may work well with one type of data, but may not work on the same type but in a different domain. In the NLP domain, the context of words and the semantics of the document play an important role. As the morphology of words is different in languages and domains, it belongs. Similarly, the approaches used in this study may or may not work in other projects if they belong to different domains, e.g., health, social science, or marketing.

This study concludes that document presentation and word embedding play a vital role in NLP problems. The straightforward explanation is that fastText works due to the skipgram mechanism, and LSTM works over DAN because it handles the information effectively due to its excellent gating mechanism.

References

1. Ben-Moshe, O., Johnston, S., Pyle, D., Silbey, A.: R&d that's on time and on budget? yes, with predictive analytics (2020). https://www.mckinsey.com/business-functions/operations/our-insights/rd-thats-on-time-and-on-budget-yes-with-predictive-analytics
2. Bloch, M., Blumberg, S., Laartz, J.: Delivering large-scale it projects on time, on budget, and on value. Harv. Bus. Rev. **5**(1), 2–7 (2012)
3. Bojanowski, P., Grave, E., Joulin, A., Mikolov, T.: Enriching word vectors with subword information. Trans. Assoc. Comput. Linguist. **5**, 135–146 (2017)
4. Choetkiertiku, M.: Datasets (2019). https://github.com/morakotch/datasets
5. Choetkiertikul, M., Dam, H.K., Tran, T., Pham, T., Ghose, A., Menzies, T.: A deep learning model for estimating story points. IEEE Trans. Software Eng. **45**(7), 637–656 (2018)
6. Devlin, J., Chang, M.W., Lee, K., Toutanova, K.: BERT: pre-training of deep bidirectional transformers for language understanding. arXiv preprint: arXiv:1810.04805 (2018)
7. Espinosa-Leal, L., Chapman, A., Westerlund, M.: Autonomous industrial management via reinforcement learning. J. Intell. Fuzzy Syst. **39**(6), 8427–8439 (2020)
8. González-Carvajal, S., Garrido-Merchán, E.C.: Comparing BERT against traditional machine learning text classification. arXiv preprint: arXiv:2005.13012 (2020)

9. Gultekin, M., Kalipsiz, O.: Story point-based effort estimation model with machine learning techniques. Int. J. Software Eng. Knowl. Eng. **30**(01), 43–66 (2020)
10. Gutierrez-Bustamante, M., Espinosa-Leal, L.: Natural language processing methods for scoring sustainability reports-a study of Nordic listed companies. Sustainability **14**(15), 9165 (2022)
11. Hamouda, A.E.D.: Using agile story points as an estimation technique in CMMI organizations. In: 2014 Agile Conference, pp. 16–23. IEEE (2014)
12. Hochreiter, S., Schmidhuber, J.: Long short-term memory. Neural Comput. **9**(8), 1735–1780 (1997)
13. Iyyer, M., Manjunatha, V., Boyd-Graber, J., Daumé III, H.: Deep unordered composition rivals syntactic methods for text classification. In: Proceedings of the 53rd Annual Meeting of the Association for Computational Linguistics and the 7th International Joint Conference on Natural Language Processing (volume 1: Long papers), pp. 1681–1691 (2015)
14. Joachims, T.: Text categorization with support vector machines: learning with many relevant features. In: Nedellec, C., Rouveirol, C. (eds.) Machine Learning: ECML-98. Lecture Notes in Computer Science, vol. 1398, pp. 137–142. Springer, Berlin (1998). https://doi.org/10.1007/BFb0026683
15. Joulin, A., Grave, E., Bojanowski, P., Douze, M., Jégou, H., Mikolov, T.: Fasttext.zip: compressing text classification models. arXiv preprint: arXiv:1612.03651 (2016)
16. Joulin, A., Grave, E., Bojanowski, P., Mikolov, T.: Bag of tricks for efficient text classification. In: Proceedings of the 15th Conference of the European Chapter of the Association for Computational Linguistics: Volume 2, Short Papers, pp. 427–431. Association for Computational Linguistics (2017)
17. Koralage, R.: Agile scrum sprint velocity dataset (2020). https://github.com/RandulaKoralage/AgileScrumSprintVelocityDataSet
18. Kumar, G., Bhatia, P.K.: Impact of agile methodology on software development process. Int. J. Comput. Technol. Electr. Eng. (IJCTEE) **2**(4), 46–50 (2012)
19. Menard, S.: Applied Logistic Regression Analysis, vol. 106. Sage, Newcastle upon Tyne (2002)
20. Panda, A., Satapathy, S.M., Rath, S.K.: Empirical validation of neural network models for agile software effort estimation based on story points. Procedia Comput. Sci. **57**, 772–781 (2015)
21. Porru, S., Murgia, A., Demeyer, S., Marchesi, M., Tonelli, R.: Estimating story points from issue reports. In: Proceedings of the The 12th International Conference on Predictive Models and Data Analytics in Software Engineering, pp. 1–10 (2016)
22. Ramin, F., Matthies, C., Teusner, R.: More than code: contributions in scrum software engineering teams. In: Proceedings of the IEEE/ACM 42nd International Conference on Software Engineering Workshops, pp. 137–140 (2020)
23. Sanh, V., Debut, L., Chaumond, J., Wolf, T.: DistilBERT, a distilled version of BERT: smaller, faster, cheaper and lighter. arXiv preprint: arXiv:1910.01108 (2019)
24. Satapathy, S.M., Panda, A., Rath, S.: Story point approach based agile software effort estimation using various SVR kernel methods. In: Proceedings of the International Conference on Software Engineering and Knowledge Engineering, SEKE (2014)
25. Schwaber, K., Sutherland, J.: The scrum guide. Scrum Alliance **21**(19), 1 (2011)
26. Scott, E., Pfahl, D.: Using developers' features to estimate story points. In: Proceedings of the 2018 International Conference on Software and System Process, pp. 106–110 (2018)

27. pandas development team, T.: pandas-dev/pandas: Pandas (2020). https://doi.org/10.5281/zenodo.3509134
28. Vargas, D.A.D., et al.: Implementing scrum to develop a connected robot. arXiv preprint: arXiv:1807.01662 (2018)
29. Wolf, T., et al.: HuggingFace's transformers: state-of-the-art natural language processing. arXiv preprint: arXiv:1910.03771 (2019)
30. Yang, Z., Yang, D., Dyer, C., He, X., Smola, A., Hovy, E.: Hierarchical attention networks for document classification. In: Proceedings of the 2016 Conference of the North American Chapter of the Association for Computational Linguistics: Human Language Technologies, pp. 1480–1489 (2016)

Data Obfuscation Scenarios for Batch ELM in Federated Learning Applications

Anton Akusok(✉), Leonardo Espinosa-Leal, Tamirat Atsemegiorgis, and Kaj-Mikael Björk

Graduate School and Research, Arcada University of Applied Sciences, Jan-Magnus Janssons plats 1, 00560 Helsinki, Finland
{anton.akusok,leonardo.espinosaleal,atsmegit,kaj-mikael.bjork}@arcada.fi

Abstract. The batch formulation of the Extreme Learning Machines (ELM) method fits well with federated learning scenarios. This paper proposes and investigates the strategies for data obfuscation that can be used in combination with ELM to create a secure distributed learning environment. Results show that the model allows for significant levels of added noise with minimal impact on its predictive performance; enabling secure federated learning in tasks that can benefit from it.

Keywords: Extreme Learning Machine · Federated learning · Batch processing

1 Introduction

Federated Learning (FL) is one of the prominent machine learning techniques in which the training data is distributed to the edge devices while safeguarding it from exposure to the central server. The conventional method of ML adopts a different approach, where the server collects data from the edge components and conducts centralized model training using identically distributed training data [13]. Federated learning enables the concurrent collaboration of multiple edge devices for model training, eliminating the necessity of centralizing data storage. This approach has found application in various domains, such as healthcare, manufacturing, telecom, transportation, etc [3,11,13].

High-quality and extensive datasets lead to improved performance in machine-learning models. To address the data isolation challenges encountered by researchers, federated learning presents a viable alternative [3,6]. Figure 1 shows FL, each edge component assumes responsibility for training a model using its local data and subsequently transmits the updated weights to the central server. The server then updates by aggregating the received weights from the various edge components and redistributes the updated weights back to each individual edge component for model retraining with the latest weights. This iterative process continues until specific predefined criteria are met. Fundamentally, Federated Learning is concerned with centralizing model weights while decentralizing the data, thus addressing the pressing concern of data privacy.

M. E. Auer et al. (Eds.): STE 2024, LNNS 1028, pp. 329–338, 2024.
https://doi.org/10.1007/978-3-031-61905-2_32

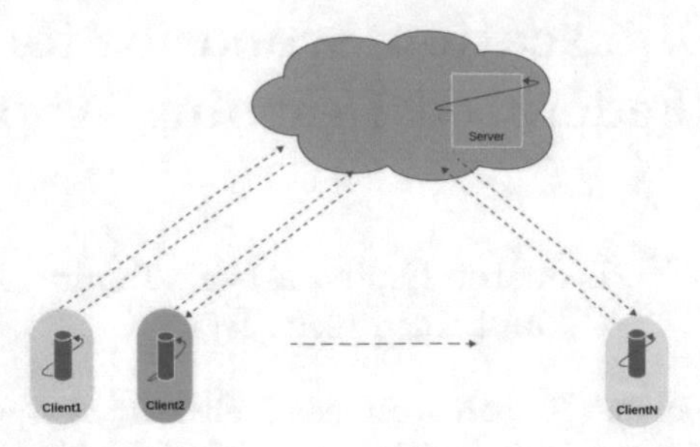

Fig. 1. Basic Federated Learning Model with N Clients Connected to the Cloud Serve

One use case of FL was presented by Google's keyboard called *Gboard*, and the model was trained to predict the next word in a smartphone [6]. The model's training was conducted on millions of smartphone devices without compromising the customers' privacy [3,6]. The use case was found to be with a better recall than centralized training using stochastic gradient descent [6].

Challenges related to data privacy and security have emerged as significant barriers to the centralized collection and processing of data [3,13]. Legal regulations such as the GDPR (General Data Protection Regulation) and similar measures taken by different countries and states demand that personal data be processed transparently with a clearly stated purpose and the data subject's consent. This makes collecting data in a centralized repository increasingly difficult, as a result, pursuing a new data training paradigm is not just a preference; it has become a necessity [3,6,13].

Edge computing [3] complements federated learning well, because it allows the data to stay at users' devices avoiding the risk of transferring sensitive data over the network, and leaving the users in full control over their own data. Extreme Learning Machines (ELM) [7] is a method that can be used for edge computing for its fast non-iterative solution and a distributed solution, as explained below. This paper investigates ways of adding data obfuscation to ELM, that is the key feature of practical federated learning systems.

2 Methodoloty

2.1 Batch Solution for Extreme Learning Machines

Extreme Learning Machine [7,12] is the method of choice in this paper. It has a fast exact solution, fits well to computational architecture, and can be computed very fast [2], and fits well with other feature extraction methods [4,9]. ELM solution can be written in batch form, which fits especially well with distributed systems like federated learning.

In brief, an ELM is a linear model computed over a fixed random projection of input data into a new (usually higher-dimensional) feature space, with a bounded non-linear function applied to the projected features element-wise.

Unlike the basic linear model, ELM can vary the dimensionality of its (randomly projected) feature space controlling the model learning capacity. Non-linear function addresses the inability of a linear model to capture non-linear dependencies. The general formulation of ELM looks like this:

$$\mathbf{H} = f(\mathbf{X}\mathbf{W} + \mathbf{1}b) \tag{1}$$

$$\mathbf{H}\boldsymbol{\beta} = \mathbf{Y} \tag{2}$$

where $\mathbf{W}$ and b are randomly chosen and fixed parameters of the data projection, and $\boldsymbol{\beta}$ is the computed solution of the ELM model that minimizes the error term, usually an L2-regularised mean squared error.

The practical issue of Eq. 2 is its dependency on the data size; a very large dataset makes it expensive to compute. Equation 3 presents a different formulation with its matrix size independent of the number of data points, a beneficial property for large data analysis [5, 10].

$$(\mathbf{H}^T\mathbf{H})\boldsymbol{\beta} = \mathbf{H}^T\mathbf{Y} \tag{3}$$

It's important to note that the matrices $\mathbf{H}^T\mathbf{H}$, $\mathbf{H}^T\mathbf{Y}$ for the full dataset are sums of the corresponding matrices for the same dataset split in batches of data. The random projection part $\mathbf{H} = f(\mathbf{X}\mathbf{W} + \mathbf{1}b)$ is fixed, so it can be done on batches of data independently, and does not introduce any inter-dependencies - as would be the case if the system optimized both matrices $\mathbf{W}$ and $\boldsymbol{\beta}$. The batch solution of ELM looks like this:

$$\begin{aligned}
\mathbf{X} &= \begin{bmatrix} \mathbf{X}_1 \\ \mathbf{X}_2 \\ \dots \\ \mathbf{X}_k \end{bmatrix}, \mathbf{Y} = \begin{bmatrix} \mathbf{Y}_1 \\ \mathbf{Y}_2 \\ \dots \\ \mathbf{Y}_k \end{bmatrix} \\
\mathbf{H}_i &= f(\mathbf{X}_i\mathbf{W} + \mathbf{1}b) \\
\mathbf{H}^T\mathbf{H} &= \sum_{i=1}^{k} (\mathbf{H}_i^T\mathbf{H}_i) \\
\mathbf{H}^T\mathbf{Y} &= \sum_{i=1}^{k} (\mathbf{H}_i^T\mathbf{Y}_i) \\
(\mathbf{H}^T\mathbf{H})\boldsymbol{\beta} &= \mathbf{H}^T\mathbf{Y}
\end{aligned}$$

Note that the ELM solution stays the same as in 3, but the data can be computed and represented in batches as pairs of matrices $(\mathbf{H}_i^T\mathbf{H}_i, \mathbf{H}_i^T\mathbf{Y}_i)$. The final ELM solution adds together all the batches and computes the output weights $\boldsymbol{\beta}$.

The output weights computation has the complexity of $O(l^3)$ and the $(\mathbf{H}_i^T\mathbf{H}_i)$ matrix computation has the complexity of $O(l^2N)$ for l hidden neurons and N data samples. In practice with $N >> l$ the output weights computation is "fast" compared to the whole model training, and can be done repetitively.

Another useful feature of batch ELM solution is fast L2 parameter tuning. ELM requires L2 regularisation to balance between the model learning capacity and overfitting. This regularisation can be ignored while computing batch data $(\mathbf{H}_i^T\mathbf{H}_i, \mathbf{H}_i^T\mathbf{Y}_i)$. Once the data is ready, the model solution with different values α of L2 regularisation can be computed as in Eq. 4. The best value of α is then selected on a validation set.

$$\left(\sum_{i=1}^{k}\mathbf{H}_i^T\mathbf{H}_i + \alpha^2\mathbf{I}\right)\boldsymbol{\beta} = \sum_{i=1}^{k}\mathbf{H}_i^T\mathbf{Y}_i \tag{4}$$

Batches can be computed on individual clients of a federated learning system and exchanged between the clients without sharing the original data. Alternatively, new data can be added to an existing ELM model by computing a batch for the new data, adding it to the previous values of $(\mathbf{H}^T\mathbf{H}, \mathbf{H}^T\mathbf{Y})$, and re-computing the output weights. Moreover, computing a batch of data for some data samples already used in training and then **subtracting** it from the $(\mathbf{H}^T\mathbf{H}, \mathbf{H}^T\mathbf{Y})$ matrices before re-computing the solution allows the ELM to **forget** these data samples as if they never existed [1].

2.2 Data Obfuscation Scenarios

Batch ELM formulation is useful for federated learning because data batches can be computed and exchanged between the clients while keeping the original data private. However, we assume that the ELM configuration, specifically its fixed random projection matrix $\mathbf{W}$, bias b, and the non-linear function $f()$ are publicly known. This creates a risk of reverse-engineering the original data features from the data batches.

We propose the strategy of adding random noise to data before exchanging it, making exact reconstruction of the original data features $\mathbf{X}$ and $\mathbf{Y}$ impossible. The noise level is selected to have a small effect on the model performance, so federated learning remains beneficial over learning from the local data only. Two strategies are proposed for obfuscating input data features $\mathbf{X}$ and data labels $\mathbf{Y}$.

2.3 Input Feature Obfuscation

ELM formulation includes a bounded non-linear function $f()$, often a sigmoid or a hyperbolic tangent function. The bounded function helps to obfuscate the data $\mathbf{H}$ as the difference between the returned function value and the boundary becomes so small that it cannot be represented with the default machine precision, and the precise information needed for reconstruction $\hat{\mathbf{X}} \approx f^{-1}(\mathbf{H})$ is lost.

However, the reconstruction ability can be decreased further by adding a small random noise to $\mathbf{H}$. Because of the inverse function, the small noise in $\mathbf{H}$ would often lead to a very large noise in $\mathbf{X}$ or even make the reconstruction impossible by creating the value in $\mathbf{H}$ that is outside the bounds of a bounded function $f()$. While some reconstruction is technically possible, the obtained values $\hat{\mathbf{X}}$ would be different from the original values $\mathbf{X}$. The experimental section evaluates the scale of noise that can be added to $\mathbf{H}$ without significantly degrading the ELM performance.

2.4 Label Obfuscation with Differential Privacy

Data labels $\mathbf{Y}$ are harder to obfuscate because they are not transformed by a non-linear function. Instead, a differential privacy approach is adopted that replaces some percentage of labels $\mathbf{Y}$ with random values sampled from the same distribution. This means that even if the data labels are reconstructed by another client, it is impossible to tell whether a specific label is a genuine measurement or a random number.

The experimental section evaluates the effect of replacing different percentages of data labels on model performance.

3 Experimental Results

The experiments run on the challenging California housing dataset [8]. The applied ELM model has 200 hidden neurons, and the input features are set to the same scale. The data is split across three virtual "clients" having 100, 1000, and 10000 training samples correspondingly, and 400 client-specific test samples. Model performance is reported as the R2 score for the client-specific test set. The results are averaged over 10 initializations with random data sample assignment to clients.

3.1 Federated Learning

Federated learning helps individual actors achieve better predictive performance of their model by including data of other similar actors, thus enlarging the training data size. The benefits depend on the amount of data an actor has, and the amount of data it can receive from its peers. The model performance on the client's own data, and in a federated learning scenario where one client uses another client's data, is shown in Fig. 2. Client with the least amount of data benefits the most from having access to more training data.

Separating client data geographically (Fig. 3) presents another interesting case. Here the data points are assigned to clients not randomly, but by cutting the dataset into three chunks with latitude = 40 and longitude = −118 lines. Federated learning does not help in this scenario, as more training data from a different location leads to the degradation in predictive performance at a client's own location.

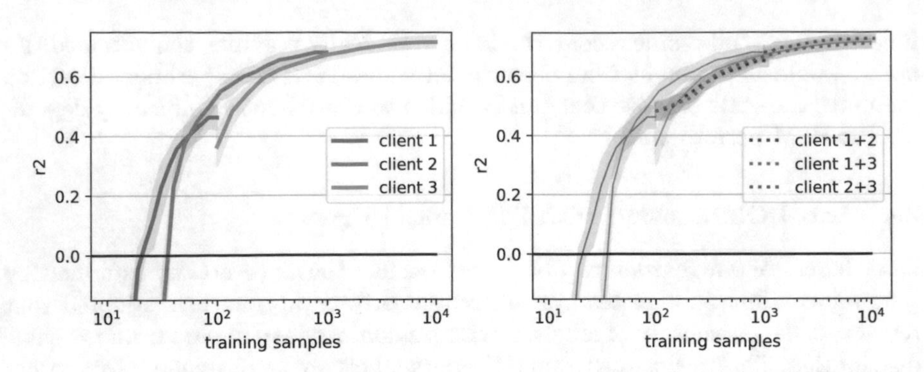

Fig. 2. Model performance on the three clients with 100, 1000, and 10000 samples independently (left) and in federated learning scenario (right)

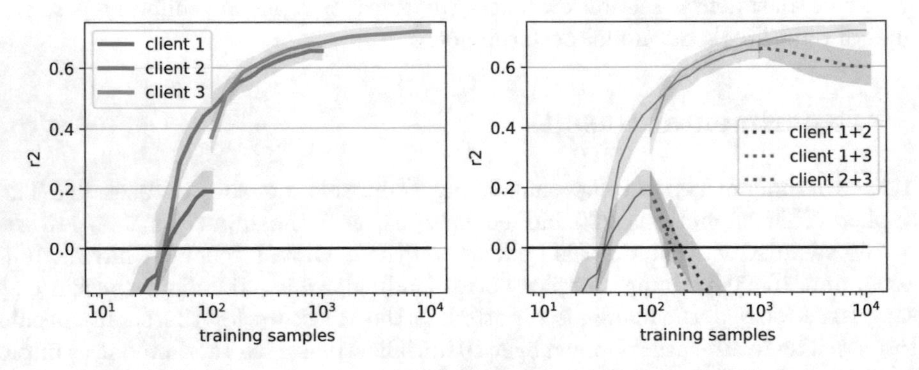

Fig. 3. Model performance on the three clients separated geographically, on their own (left) and in federated learning scenario (right). While per-client results are similar to the random data assignment scenario, including training data from another location decreases the model's performance.

3.2 Obfuscating Input Features

ELM works not with input features directly but with a non-linear transformation of their random projection called $\mathbf{H} = f(\mathbf{XW}+\mathbf{1}b)$. The batches of training data are exchanged in the form of $(\mathbf{H}^T\mathbf{H},\ \mathbf{H}^T\mathbf{Y})$. It is impossible to reconstruct $\mathbf{H}$ precisely back from $\mathbf{H}^T\mathbf{H}$, so this fact gives a strong protection to raw values of data features.

In case a malicious actor could obtain $\mathbf{H}$, the feature values of $\mathbf{X}$ can be protected by adding a random noise to $\mathbf{H}$ by a client before sharing it with peers. The noise is fixed and sample-specific, so repetitive requests for data always return data $\mathbf{H}$ with the same noise, avoiding the possibility of canceling out the noise by repetitively querying the same data and averaging received values.

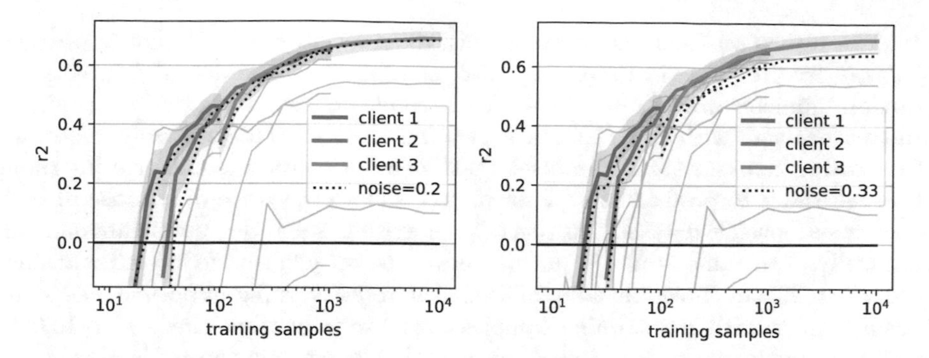

Fig. 4. Model performance with noise added to **H**. Original performance is shown in color, dotted line highlights specific noise levels, gray lines show stronger added noise.

The amount of noise should allow the model to perform close to its best. The noise is modeled as a normally distributed random value with a given standard deviation. In tests shown in Fig. 4, the added noise standard deviation of up to 0.2 does not significantly affect model performance, while 0.33 decreases the results. Higher noise levels make the model unusable. While the exact value is experiment-specific, the tests are done with input data normalized to zero mean and unit variance.

The effects of noise with a standard deviation of 0.2 are shown in Fig. 5. Many values of **H** are changed by significant amounts by noise, and it's impossible to precisely reconstruct **X** while the noisy data remains useful for modeling purposes.

Original H

```
[[-1.     0.517  0.403 -0.994  1.   ]
 [-1.     0.854  0.586 -0.47   0.998]
 [-0.9    0.509  1.    -0.926 -0.484]
 [ 1.    -0.057 -0.798 -0.909  0.997]
 [-0.989 -0.211  0.403 -0.773  0.559]]
```

Noisy H

```
[[-1.187  0.279  0.194 -0.849  1.298]
 [-0.901  1.108  0.434 -0.884  1.081]
 [-0.679  0.542  1.012 -0.981 -0.503]
 [ 0.945  0.349 -1.011 -0.742  0.933]
 [-1.004 -0.442  0.382 -0.982  0.833]]
```

Fig. 5. Effect of noise with standard deviation of 0.2 on **H**. Only part of the data is shown for clarity.

3.3 Obfuscating Output Labels

Output targets can be obfuscated in a differential privacy approach. A percentage of targets **Y** can be replaced by random numbers drawn from a normal distribution with the parameters computed for the original data **Y**. The replacement targets are assigned once and don't change over the client's lifetime.

Let's call the percentage of replaced targets as "noise" for simplicity, although it is not noise because a part of the targets is fully replaced with random numbers, not modified by the addition of some random numbers.

The results on federated learning with different levels of differential privacy "noise" are presented in Fig. 6a-b. Noise, or replacement of actual y targets by random numbers drawn from the same distribution, does not significantly affect the performance with up to 10% targets being replaced. This probably relates to the stochastic nature of the dataset that does not contain exhaustive information necessary to produce the target y (to evaluate the price of a house in this California housing dataset), so the targets exhibit some degree of randomness naturally. The same 10% of "noise" allows for an efficient federated learning too. Even 20% of "noise" enables efficient federated learning in optimal cases, as when a client with 100 training samples learns from another client with 10,000 training samples with added target noise. Only the extreme "noise" levels of over 35% of randomly replaced data targets can render federated learning unhelpful.

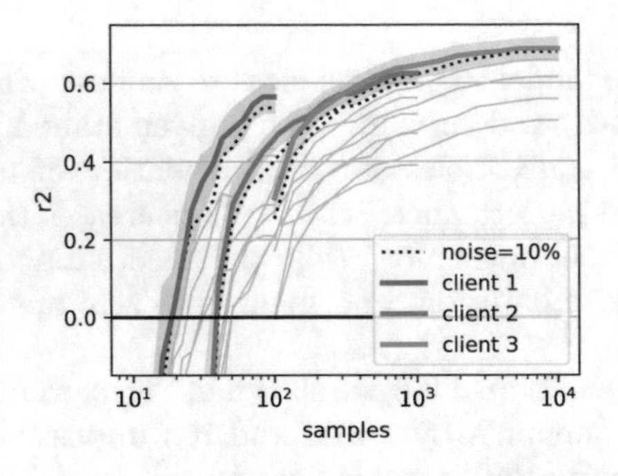

(a) Individual client performance at no noise (color), 10% noise in $\mathbf{Y}$ (dotted), and 20-50% noise in $\mathbf{Y}$ (grey).

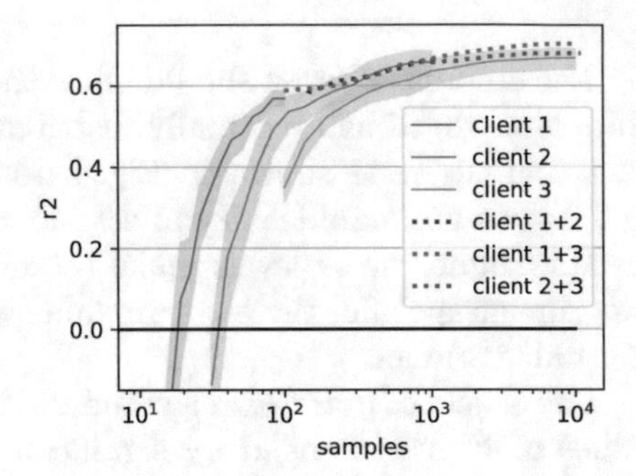

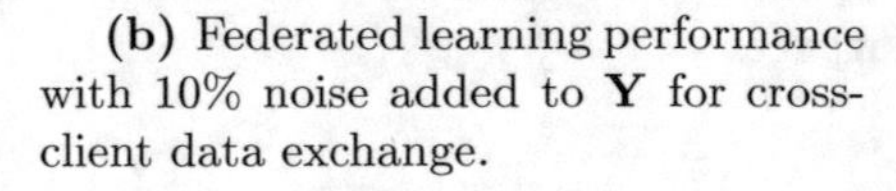

(b) Federated learning performance with 10% noise added to $\mathbf{Y}$ for cross-client data exchange.

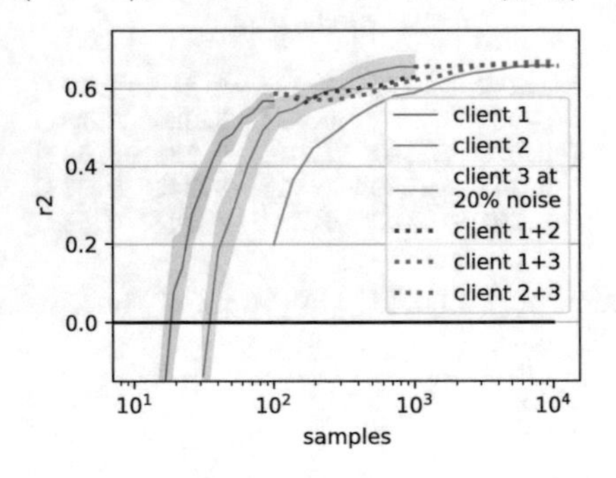

(c) Federated learning performance with 20% noise added to $\mathbf{Y}$ for cross-client data exchange.

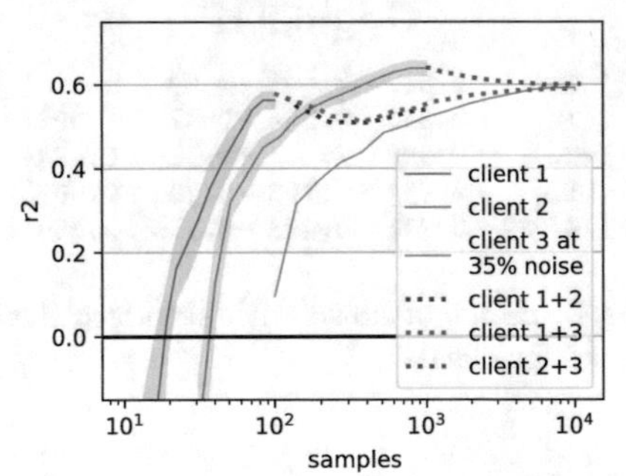

(d) Federated learning performance with 35% noise added to $\mathbf{Y}$ for cross-client data exchange.

Fig. 6. Effect of label noise on training data.

4 Conclusions

Extreme Learning Machine model in its batch solution formulation fits very well to the federated learning use case, allowing for simple information sharing while keeping the original data features private. The exact solution of ELM enables single-pass data sharing, unlike the iterative solutions of other methods. ELM is tolerable to noise, allowing for the addition of significant amounts of noise to the processed data while preserving the model performance: the normally distributed random noise with a standard deviation of up to 0.2 to matrix $\mathbf{H}$ computed from input data normalized to zero mean and unit variance, and up to 10% labels replaced by random values sampled from the same distribution. Such an amount of noise removes the ability to precisely reconstruct the original data values by other clients but keeps the benefits of improved model performance from combining training information in a federated learning setup.

In the case of significant data imbalance, e.g. a client learns from another client with 100× more data, even larger amounts of noise can be used. For example, random noise with a standard deviation of 0.33 in $\mathbf{H}$ and replacing 20% of data labels with random values. This amount of noise decreases the peak model performance, but still improves the model performance of the client with little training data, compared to learning only from the local data. This means that the clients can use noise also to limit the maximum benefit they are willing to share in federated learning scenarios, e.g. keeping their competitive advantage while providing beneficial data for the general public.

The feasibility of federated learning must be validated for each use case, and it cannot be taken for granted. Federated learning may be detrimental to model performance in non-iid data distribution cases. This can be tested safely by sharing data with a large amount of added noise, and validating if the performance stays at a constant level pointing to potential benefits from federated learning with less noisy data, or if the performance degrades beyond random guessing pointing to the specific problem not being suitable for federated learning.

References

1. Akusok, A., Leal, L.E., Björk, K.M., Lendasse, A.: Scikit-ELM: an extreme learning machine toolbox for dynamic and scalable learning. In: Cao, J., Vong, C.M., Miche, Y., Lendasse, A. (eds.) ELM2019. Proceedings in Adaptation, Learning and Optimization, vol. 14, pp. 69–78. Springer, Cham (2021). https://doi.org/10.1007/978-3-030-58989-9_8
2. Akusok, A., Miche, Y., Hegedus, J., Nian, R., Lendasse, A.: A two-stage methodology using K-NN and false-positive minimizing ELM for nominal data classification. Cogn. Comput. **6**(3), 432–445 (2014). https://doi.org/10.1007/s12559-014-9253-4
3. Brecko, A., Kajati, E., Koziorek, J., Zolotova, I.: Federated learning for edge computing: a survey. Appl. Sci. **12**(18), 9124 (2022). https://doi.org/10.3390/app12189124

4. Espinosa-Leal, L., Akusok, A., Lendasse, A., Björk, K.M.: Extreme learning machines for signature verification. In: Cao, J., Vong, C.M., Miche, Y., Lendasse, A. (eds.) ELM2019. Proceedings in Adaptation, Learning and Optimization, vol. 14, pp. 31–40. Springer, Cham (2021). https://doi.org/10.1007/978-3-030-58989-9_4
5. Espinosa-Leal, L., Akusok, A., Lendasse, A., Björk, K.M.: Website classification from webpage renders. In: Cao, J., Vong, C.M., Miche, Y., Lendasse, A. (eds.) ELM2019. Proceedings of ELM2019, vol. 14, pp. 41–50. Springer, Cham (2021). https://doi.org/10.1007/978-3-030-58989-9_5
6. Hard, A., et al.: Federated learning for mobile keyboard prediction (2018). https://arxiv.org/abs/1811.03604
7. Huang, G.B., Zhu, Q.Y., Siew, C.K.: Extreme learning machine: theory and applications. Neurocomputing **70**(1), 489–501 (2006). https://doi.org/10.1016/.neucom.2005.12.126, neural Networks
8. Kelley Pace, R., Barry, R.: Sparse spatial autoregressions. Stat. Probab. Lett. **33**(3), 291–297 (1997). https://doi.org/10.1016/S0167-7152(96)00140-X
9. Lauren, P., Qu, G., Zhang, F., Lendasse, A.: Discriminant document embeddings with an extreme learning machine for classifying clinical narratives. Neurocomputing **277**, 129–138 (2018). https://doi.org/10.1016/j.neucom.2017.01.117, hierarchical Extreme Learning Machines
10. Leal, L.E., Björk, K.M., Lendasse, A., Akusok, A.: A web page classifier library based on random image content analysis using deep learning. In: Proceedings of the 11th PErvasive Technologies Related to Assistive Environments Conference, pp. 13–16. PETRA 2018, Association for Computing Machinery, New York, NY, USA (2018). https://doi.org/10.1145/3197768.3201525
11. Leopold, G.: Federated learning applied to cancer research (2019). https://www.hpcwire.com/2019/10/17/federated-learning-applied-to-cancer-research/
12. Liu, X., Lin, S., Fang, J., Xu, Z.: Is extreme learning machine feasible? a theoretical assessment (part i). IEEE Trans. Neural Netw. Learn. Syst. **26**(1), 7–20 (2015). https://doi.org/10.1109/TNNLS.2014.2335212
13. Spacagna, G.: Federated learning and differential privacy (2019). https://medium.com/data-science-vademecum/federated-learning-and-differential-privacy-cbbec1961c30

Valohai-CSC Integration: A Machine Learning Management Platform for Finnish Academic Institutions

Leonardo Espinosa-Leal[1](✉), Andrey Shcherbakov[2], and Magnus Westerlund[1]

[1] Graduate School and Research, Arcada University of Applied Sciences, Jan-Magnus Janssons plats 1, 00560 Helsinki, Finland
{leonardo.espinosaleal,magnus.westerlund}@arcada.fi

[2] School of Engineering, Culture and Wellbeing, Arcada University of Applied Sciences, Jan-Magnus Janssons plats 1, 00560 Helsinki, Finland
andrey.shcherbakov@arcada.fi

Abstract. We present and review the results of the integration of Valohai (https://valohai.com), a machine learning management platform that supports training, pipeline and version control machine learning, and deep learning network capabilities in combination with the computational resources from the CSC-IT Center for Science (https://www.csc.fi), a Finnish center of expertise in information technology provider of supercomputer capabilities owned by the Finnish state and higher education institutions. We tested this integration against different commercial cloud services by training a YOLOv3 model with the COCO2014 dataset as benchmarking. Our findings reflect the technical advantages of using this infrastructure for research and teaching in higher education institutions in Finland.

Keywords: Machine Learning · Valohai · CSC · MLOps

1 Introduction

Data scientists, students, and even some researchers rave about the Jupyter Notebook: a sandbox-type environment that allows us to create and share documents that contain live code, equations, visualizations, and narrative text [6]. Writing code in a Jupyter Notebook is akin to authoring a book that features executable code. It allows you to compartmentalize code into distinct cells and sections, each accompanied by detailed descriptions explaining the function of each cell, thereby offering intermediate outputs for enhanced clarity and understanding.

Version Control Systems (VCS) are nowadays the core of software development and prototyping [1,9]. The software industry's success and the capacity to adapt fast to any requirement have been partially thanks to the complete integration between the needed computational resources and the VCS.

M. E. Auer et al. (Eds.): STE 2024, LNNS 1028, pp. 339–351, 2024.
https://doi.org/10.1007/978-3-031-61905-2_33

Machine learning and big data's advances are strongly linked to the easy adaptation of software development techniques and strategies. Notebooks are often used for prototyping and exploratory data analysis rather than final deployment in production environments. While Jupyter Notebooks are excellent for certain stages of a project, they are not as effective for managing complex workflows or collaborative development necessary for production environments. This is where MLOps comes into play.

1.1 Machine Learning Operations (MLOps)

MLOps, or Machine Learning Operations, is an emerging discipline that combines Machine Learning, DevOps, and Data Engineering to streamline and automate the end-to-end machine learning lifecycle. As outlined in recent research [7], MLOps aims to foster a collaborative environment for data scientists, developers, and operations teams. This collaboration ensures that machine learning models can be developed, tested, deployed, and maintained efficiently and reliably.

Key aspects of MLOps include:

- **Automated Workflows:** MLOps introduces automation in the ML workflow, enabling the seamless transition of machine learning models from experimentation to production.
- **Version Control and Collaboration:** It emphasizes version control and collaboration, addressing the challenges posed by notebooks in versioning and collaborative development.
- **Continuous Integration/Continuous Deployment (CI/CD):** MLOps integrates CI/CD practices into machine learning, ensuring that models are continually updated, tested, and deployed in a controlled manner.
- **Monitoring and Maintenance:** Continuous monitoring and maintenance of machine learning models in production environments are crucial components of MLOps, ensuring model performance and accuracy over time.

MLOps bridges the gap between the exploratory nature of Jupyter Notebooks and the structured, collaborative environment required for production. It provides the framework and tools needed to manage complex data science projects effectively, ensuring that the transition from prototype to production is smooth and efficient.

2 Motivation

When multiple users need to collaborate in a software development project, the core idea is simple: every time someone wants to contribute to the project with some changes, they bundle them together in a commit and that contribution is added at the end of the project history. Every commit is an atomic unit of contribution, and the history is nothing more than an ordered sequence of commits. It allows to observe the status of the project at any point in time: from the first prototype to the latest production change, it's simply a matter of choosing what commit to look at.

Machine Learning version control is special. Unlike a software project, changing data, hyperparameters, learning rate, algorithms, and architecture, among others, makes a huge difference, and it is difficult to capture in traditional VCS such as `git`. We can identify two different phases:

- Prototyping: build your pipeline to solve the problems, try different parameters, models, even add more data for training and testing.
- Production: Deploy a machine learning model alongside all the relevant machinery required to make it work.

2.1 Enhanced Deep Learning Project Management with Valohai

Since `git` can't help much in a Machine learning project, we performed some benchmark experiments using Valohai[1]. Valohai is a platform that automates Machine Learning Operations (MLOps) [10] and presents a solution for managing deep learning projects, offering a suite of features that streamline the process of model development, data handling, and team collaboration. It allows data scientists to train models faster by automating all cloud machine startup and shutdown, parallel hyperparameter tuning, and record keeping. Basically, Valohai manages the cloud instance from cloud service provider (Google Cloud platform, Microsoft Azure, Amazon Web Services, CSC, among others). Valohai can point to your code and data, launch the workers, run the code through a ML pipeline and, more importantly, shutdown the instances when the jobs stop running. The integration of Valohai into deep learning workflows brings several benefits.

- Valohai's platform automatically versions each run of a model, creating a timeline of the project's progression. This feature is useful for maintaining a record of all modifications and experiments, allowing data scientists to trace the evolution of their models over time. As deep learning projects typically involve numerous iterations, providing an automatic VCS which ensures that no detail of the model's development is lost, thereby enhancing reproducibility and auditability.
- A critical aspect of deep learning is evaluating and comparing the performance of different model iterations. Valohai facilitates this by allowing users to compare metrics across various runs. This functionality is essential for teams to track their progress, identify the most promising models, and make data-driven decisions. By visualising and comparing key performance indicators, teams can discern trends, improvements, or regressions in model accuracy, thereby guiding further development efforts.
- In deep learning, managing and versioning datasets is as crucial as the models themselves. Valohai addresses this need by providing tools for dataset curation and versioning without necessitating data duplication. This approach is beneficial for handling large datasets and ensuring consistency across model training iterations. It allows teams to maintain a clear record of which dataset versions were used with specific model versions, thereby ensuring traceability and consistency in experiments.

[1] https://valohai.com/.

- Deep learning projects often encounter challenges related to data dependencies and collaboration, especially in a distributed team environment. Traditional tools like `git` are not designed to handle large datasets or binary data dependencies effectively, which is a common scenario in machine learning projects. Researchers often find themselves manually managing data, thus providing inefficiencies and potential errors. Valohai's platform offers a structured way to manage data dependencies. It simplifies the process of accessing and sharing datasets, whether they are stored locally or in cloud storage.
- Deep learning models, particularly those involving complex neural networks, often require substantial computational resources for training. Furthermore, these models may be written in various programming languages and depend on multiple frameworks and libraries. Valohai's platform supports managing these dependencies and hardware requirements. It allows developers to specify the necessary computational resources and automatically handles the allocation of these resources, whether on-premises or in the cloud. This capability is useful for optimizing model training times and ensuring the smooth running of experiments across different environments.

3 Overview of the Workload Processing Cloud Frameworks

Here we describe the cloud services from CSC and Valohai in comparison with the most used cloud services available in the market. In Fig. 1 the scheme of the Valohai data pipeline is presented, and for comparison in Fig. 2 the scheme of the AWS data pipeline.

3.1 CSC

CSC - IT Center for Science[2] is a Finnish organization offering a wide range of information technology services. Owned by the Finnish state and higher education institutions, CSC is committed to providing high-quality ICT expertise for various sectors.

CSC plays a pivotal role in supporting higher education institutions and research institutes, offering them the necessary tools and services to facilitate advanced research and education. The center caters to a wide array of fields including culture, public administration, and enterprises, contributing to societal advancement through technology. CSC is known for its high-quality ICT services, providing technology solutions that meet international standards. Center offers a range of cloud services and computing resources tailored to support research, education, and various data-intensive tasks.

- cPouta and ePouta Cloud Services: These are CSC's Infrastructure as a Service cloud platforms. cPouta is suitable for a wide range of workloads, while ePouta is designed for sensitive data requiring a high level of security.

[2] https://www.csc.fi/solutions-for-research.

- Rahti Container Cloud Service, a Kubernetes-based container orchestration service, allowing users to manage and scale containerized applications and workloads.
- Part of the EuroHPC Joint Undertaking, LUMI Supercomputer is one of the world's fastest supercomputers. It offers exceptional computational power for a wide range of research and data analysis tasks.
- Puhti Supercomputer which is designed for a broad range of research needs, offering powerful CPUs and GPUs for various computing tasks including AI and machine learning workloads.
- Mahti Supercomputer: Another high-performance computing resource, which is optimized for large-scale simulations and demanding analytics tasks.
- Allas Object Storage, a scalable storage service, suitable for storing large amounts of data in various formats.
- Kaivos Database Service is dedicated to handling large databases.
- Notebooks and Virtual Desktop Infrastructure (VDI): CSC offers cloud-based notebook and VDI services for interactive data analysis and remote desktop usage, facilitating flexible and accessible computing environments.
- Training and Development Environments: These are dedicated environments for training, development, and testing, offering resources for educational purposes and development projects.

CSC is an important component of Finland's technological infrastructure, especially in the realms of education and research.

3.2 Valohai

Valohai operates as a workload processing framework, designed specifically to facilitate machine learning tasks. Functionally comparable to continuous integration systems, Valohai automates the execution of machine learning pipelines by integrating code, configurations, and data sources from VCS. This integration enables seamless operation of pipelines on an automated infrastructure, complemented by a user-friendly interface.

Project Configuration. Valohai integrates with VCS repositories to access and manage machine learning projects. Central to this process is the use of a YAML configuration file, defined as *valohai.yaml*, which outlines the permissible executions within the project's context. This configuration file is a requisite for every Valohai-enabled `git` repository, detailing the specific runs to be executed and the REST endpoints that can be auto-generated for the project.

Execution Process. The execution process in Valohai delineates specific workload types including data anonymization, data generation, feature extraction, model training, and evaluation. Essential components for execution are:

- **Environment:** The computational environment, including machine type and cloud specifications. For instance, neural network training might require a server with specific GPU, CPU, and RAM specifications. Valohai supports integration with various computational environments, including CSC computational flavours.
- **Docker Image:** A Docker image is necessary to encapsulate all tools, frameworks, libraries, and the Linux system required for the project.
- **Repository Commit Contents:** The code for the project is fetched from the linked VCS repository, typically comprising training scripts and other essential code.
- **Training Dataset:** Valohai automates the retrieval of training datasets from various data storage services such as AWS S3, Google Cloud Bucket, or CSC Allas object storage.

Execution commands are queued and cached by the Valohai master, which then instructs the computational server (e.g., CSC server) to execute the downloaded code from the VCS repository. The server, upon receiving commands, prepares the environment using the Docker image and downloads the necessary dataset for execution. Real-time logs generated during the execution process are relayed back to Valohai and displayed on the Web UI. These logs are also stored for later access.

Data Storage and Management. Valohai projects utilize one or more data stores for secure file management. These data stores are used to download training data and upload trained models. Supported data store types include:

- AWS S3
- Azure Storage
- Google Cloud Storage
- OpenStack Swift (utilized for CSC Allas object storage)

Besides data stores, Valohai also allows downloading files from any public HTTP(S) addresses.

Task Management. In Valohai, "tasks" refer to collections of related executions. A prevalent use case is hyperparameter optimization, where multiple executions of a model are run with varying parameters to ascertain the most effective neural network configuration. This process involves exploring different layouts, weights, and biases to optimize model performance.

3.3 Amazon SageMaker

Amazon SageMaker[3] is an machine learning service that streamlines the entire ML lifecycle. It combines a user-friendly interface with tools for data scientists [5]. Key features of Amazon SageMaker include:

[3] https://aws.amazon.com/sagemaker/mlops.

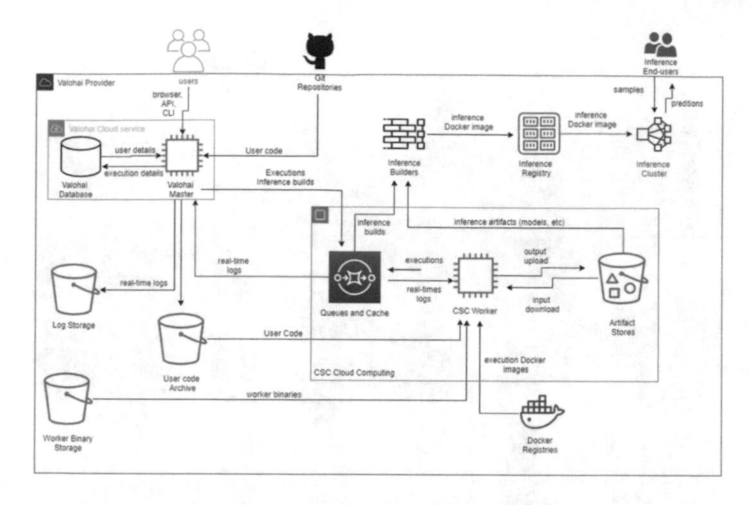

Fig. 1. Valohai Data pipeline [11].

- **Jupyter-based Authoring notebooks:** SageMaker provides Jupyter notebooks as a primary tool for data exploration, analysis, and model building, simplifying the initial stages of the ML workflow.
- **Optimized Algorithms for Distributed Systems:** The platform includes built-in algorithms that are specially optimized for efficiency and performance on large datasets and distributed systems.
- **Comprehensive Model Building and Deployment Tools:** SageMaker offers integrated tools that enable quick and efficient model building, training, and deployment, streamlining these processes for users.
- **MLOps Capabilities:** SageMaker supports MLOps practices, offering features like automated model building, testing, deployment, and monitoring.
- **Scalable Machine Learning Environment:** It provides a scalable environment, allowing users to manage resources effectively based on their project requirements.
- **Integration with AWS Ecosystem:** SageMaker is seamlessly integrated with the broader AWS ecosystem, providing access to a wide range of cloud services and tools.

In comparison to Valohai, which specializes in automated execution and orchestration of machine learning pipelines, Amazon SageMaker offers a more holistic and integrated environment. While Valohai emphasizes automation and efficient pipeline management, SageMaker provides a broader range of tools and features, focusing on usability, integration with the AWS ecosystem, and comprehensive management of the ML lifecycle from data preparation to model deployment and monitoring

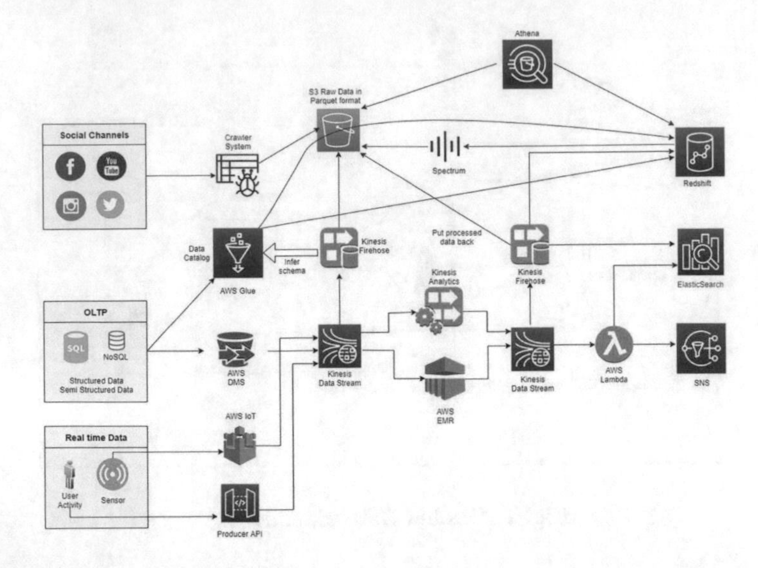

Fig. 2. AWS Data pipeline https://www.linkedin.com/pulse/streaming-data-pipelines-cloud-platforms-aws-gcp-tam-nguyen)

3.4 Google Vertex AI

Google's Vertex AI, as detailed in its training overview[4], offers a platform designed for experienced data scientists, emphasizing flexibility and integration with popular machine learning frameworks. Its key features include:

- **Support for multiple frameworks:** Vertex AI is compatible with TensorFlow, XGBoost, Scikit-learn, and Keras, allowing data scientists to work with their preferred tools and libraries.
- **Emphasis on flexibility:** The platform provides an environment that caters to the needs of experienced users, particularly in terms of cloud infrastructure and computational resources.
- **Comprehensive service model:** Vertex AI offers both infrastructure-as-a-service and platform-as-a-service solutions, providing scalable resources and managed services to streamline the development and deployment of machine learning models.
- **Integrated tooling:** It includes integrated tooling for the entire machine learning workflow, from data preparation and model training to deployment and monitoring.
- **MLOps features:** Vertex AI provides MLOps features, including pipeline automation, model monitoring, and versioning, ensuring efficient management of machine learning lifecycle.
- **AI Platform Unified:** Vertex AI brings together Google Cloud's AI offerings into a unified API, client library, and user interface.

[4] https://cloud.google.com/vertex-ai/docs/training/overview.

- **Pre-built algorithms and custom model training:** Users can leverage pre-built algorithms or train custom models, offering a balance between ease-of-use and customizability.

In contrast to Valohai, which provides an automated workflow management system with a strong focus on orchestration and automation of machine learning pipelines, Google Vertex AI emphasizes flexibility, a broad choice of tools and deeper integration with Google Cloud's infrastructure and services.

3.5 IBM Watson Machine Learning

IBM Watson Machine Learning[5], as detailed on its official documentation, is a platform designed to accommodate a broad spectrum of users, from beginners to experienced data scientists. It combines user-friendly interfaces with automated tools to streamline the machine learning process. Its key features include:

- **Flexibility in Model Building:** Watson Machine Learning provides both automated and manual options for model building, accommodating various project requirements.
- **Advanced AutoAI Capabilities:** The platform's AutoAI feature automates the entire process of data preparation, model building, and deployment, reducing the time and effort required to develop machine learning models.
- **Integration with IBM Cloud Services:** Watson Machine Learning is seamlessly integrated with various IBM Cloud services, providing users with a comprehensive set of tools for data processing, analytics, and model deployment.
- **Collaborative Environment:** The platform encourages collaboration among team members, enabling sharing and management of projects, datasets, and models.
- **Scalable Deployment Options:** Users can deploy models in various environments, including on the cloud, on-premises, or at the edge, offering flexibility in deployment strategies.
- **Robust Security and Compliance:** The platform ensures high standards of security and compliance, making it suitable for enterprise-level applications and sensitive data handling.

In comparison to Valohai, IBM Watson Machine Learning Studio places a stronger emphasis on automated processes and user-friendliness, making it particularly suitable for users who prefer a more guided experience in model development. Valohai, on the other hand, offers more detailed project configuration and task management features, catering to users who seek greater control over automated pipeline execution. While both platforms aim to simplify the machine learning workflow, Watson Machine Learning leans more towards accessibility and integration with IBM's ecosystem, whereas Valohai focuses on streamlining and automating complex ML workflows.

[5] https://dataplatform.cloud.ibm.com/docs/content/svc-welcome/wml.html.

3.6 Microsoft Azure Machine Learning

Microsoft Azure Machine Learning, as detailed on its official documentation[6], is a comprehensive cloud-based service for machine learning that caters to a wide range of users from data scientists to developers. Its features and capabilities are designed to streamline the machine learning process, offering robust support for MLOps and collaboration. Key features of Azure Machine Learning include:

- **Support for Various Roles:** Azure Machine Learning is designed to be accessible for different roles involved in machine learning projects, including data scientists, ML engineers, and software developers. This inclusivity ensures a collaborative environment that can cater to a diverse skill set.
- **Automated MLOps:** The platform provides tools to automate the entire machine learning lifecycle, from data preparation and model training to deployment and management. This automation streamlines the workflow and reduces the manual effort required in managing machine learning projects.
- **Collaboration and Integration Tools:** It offers seamless integration with popular development tools like GitHub and Visual Studio Code (VSCode), enhancing collaboration among team members. These integrations facilitate version control, code sharing, and continuous integration/continuous deployment (CI/CD) practices.
- **Emphasis on Open-Source and Integrations:** Azure Machine Learning heavily emphasizes the use of open-source tools and supports a range of open-source frameworks and libraries. It also integrates with other platforms like Databricks, enabling users to leverage the best tools available in the ecosystem.
- **Advanced Experimentation and Model Management:** The service provides advanced tools for managing and tracking experiments, which is crucial for the iterative nature of machine learning development. Users can track metrics, manage models, and compare different runs to optimize performance.
- **Scalable Training and Deployment Options:** Azure Machine Learning supports scalable model training and deployment options, including on-demand cloud resources and edge computing. This flexibility allows for efficient resource utilization and faster deployment cycles.
- **Integrated Data Science and AI Tools:** The platform offers a suite of integrated tools for data science and artificial intelligence, including Azure Synapse Analytics and Azure Cognitive Services. These tools enable users to build and deploy sophisticated AI solutions.
- **Enterprise-grade Security and Compliance:** Security and compliance are key aspects of Azure Machine Learning, offering enterprise-grade features that ensure the security of data and models, compliance with industry standards, and governance across the machine learning lifecycle.

These features collectively position Microsoft Azure Machine Learning as a robust and versatile platform, ideal for teams seeking an integrated, collaborative, and efficient approach to machine learning and AI development.

---

[6] https://learn.microsoft.com/en-us/azure/machine-learning/.

4 Experiments

The computational capabilities from the CSC have been successfully used for research [3,4] and academic purposes by the industrial and educative institutions. More recently, the integration between Valohai and the CSC resources has been previously successfully used [2] but without comparing with the other available platforms. In this work, we compare the capabilities and advantages of the integration between Valohai and the CSC with other players.

For comparison, we benchmark using a YOLOv3 model [5] trained with a COCO2014 dataset [8] on the Google Cloud platform, Amazon AWS and Valohai with CSC instances. A summary of the instances is as follows:

- **Google Cloud Platform:**
  Machine type: n1-standard-16 (16 vCPUs, 60 GB memory) CPU: Intel Skylake GPU: K80, T4, V100 CUDA with Nvidia FP16/32 HDD: 1 TB SSD.
- **AWS:**
  Machine type: p2.8xlarge, p3.2xlarge and p3.8xlarge (4,8,32 and 64 CPUs1) CPU: Intel Skylake GPU: K80, V100 HDD: 1 Tb SSD.
- **Valohai platform with CSC integration:**
  Machine type: CSC Puhti-AI CPU: Intel Skylake GPU: Nvidia P100 CUDA with Nvidia Apex F16/32 HDD: 1 TB SSD.

In Table 1 the full combination of technical parameters and instances is presented. In Table 2 contains the results in training time and value per hour as well as the number of imagers per second. Here it is easy to see that Valohai-CSC presents the best ratio of computational performance/cost.

Table 1. Valohai vs other Cloud Platform price

	Cloud service	GPU(Nvidia)	CPUs	CPU RAM	GPU RAM	Cost per hour $
Google Colab K80 (free)	GG Colab	K80	1	13 GB	12 GB	
GCP K80	GCP	K80	6	17 GB	12 GB	0.20
GCP P4	GCP	P4	4	26 GB	8 GB	0.33
GCP P100	GCP	P100	6	20 GB	16 GB	0.50
GCP V100	GCP	V100	8	20 GB	16 GB	0.82
GCP V100×2	GCP	V100×2	8	30 GB	64 GB	1.57
AWS EC2 K80	AWS	K80	4	61 GB	12 GB	0.28
AWS (p2.8xlarge)	AWS	K80×8	32	96 GB	488 GB	2.35
AWS (p3.2xlarge)	AWS	V100	8	61 GB	16 GB	1.05
AWS (p3.8xlarge)	AWS	V100×4	6 4	488 GB	128 GB	4.05
Amazon Sagemaker	AWS	K80	4	61 GB	12 GB	1.26
Valohai Cloud	Valohai	P100×2	14	128 GB	32 GB	
Computer RTX2080ti		2080ti	8	32 GB	11 GB	

Table 2. Number of GPU instances, batch size, and cost of the use of different cloud instances.

GPU	Quantity	Batch	img/s	epoch time	epoch cost	$ per hour
GCP K80	1	32	11	175 min	0.58$	0.20$/h
GCP T4	1	32	41	48 min	0.28$	0.35$/h
GCP V100	4	32 64	122178	16 min 11 min	0.23$ 0.314	0.83$/h 0.57$/h
Valohai P100	2	32	120	31 min		
2080ti	1	3264	81140	24 min 14 min		

5 Conclusions

This work shows that the integration between the Valohai platform and the computational resources of the Finnish Supercomputer center CSC is a competitive alternative for training deep learning models. Development of such alliances between academy and private companies is of paramount importance for the digital sovereignty of Europe and education. Furthermore, this kind of national infrastructure is relevant for the use of sensitive data in the creation of machine learning models, respecting the guidelines of the European General Data Protection Regulation (GDPR) guidelines.

References

1. Defaix, F., Doyle, M., Wetmore, R.: Version control system for software development (2010). US Patent 7,680,932
2. Espinosa-Leal, L.A., Björk, K.M., Lendasse, A., Akusok, A.: A web page classifier library based on random image content analysis using deep learning. In: Proceedings of the 11th PErvasive Technologies Related to Assistive Environments Conference, pp. 13–16 (2018). https://doi.org/10.1145/3197768.3201525
3. Espinosa-Leal, L.A., Karpenko, A., Caro, M.: Optimizing a parametrized thomas-fermi-dirac-weizsäcker density functional for atoms. Phys. Chem. Chem. Phys. **17**(47), 31463–31471 (2015). https://doi.org/10.1039/c5cp01211b
4. Giedraityte, Z., et al.: Three-dimensional uracil network with sodium as a linker. J. Phys. Chem. C **120**(46), 26342–26349 (2016). https://doi.org/10.1021/ACS.JPCC.6B08986
5. Joshi, A.V.: Amazon's Machine Learning Toolkit: Sagemaker. In: Machine Learning and Artificial Intelligence, pp. 233–243. Springer, Cham (2020). https://doi.org/10.1007/978-3-030-26622-6_24
6. Kluyver, T., et al.: Jupyter notebooks-a publishing format for reproducible computational workflows. In: ELPUB, pp. 87–90 (2016). https://doi.org/10.3233/978-1-61499-649-1-87
7. Kreuzberger, D., Kühl, N., Hirschl, S.: Machine learning operations (mlops): overview, definition, and architecture. IEEE Access **11**, 31866–31879 (2023). https://doi.org/10.1109/ACCESS.2023.3262138

8. Lin, T., et al.: Microsoft COCO: common objects in context. CoRR **abs/1405. 0312** (2014). http://arxiv.org/abs/1405.0312
9. Loeliger, J., McCullough, M.: Version Control with Git: Powerful Tools and Techniques for Collaborative Software Development. O'Reilly Media Inc, Sebastopol (2012)
10. Mäkinen, S., Skogström, H., Laaksonen, E., Mikkonen, T.: Who needs mlops: What data scientists seek to accomplish and how can mlops help? In: 2021 IEEE/ACM 1st Workshop on AI Engineering-Software Engineering for AI (WAIN), pp. 109–112. IEEE (2021). https://doi.org/10.1109/WAIN52551.2021.00024
11. Valohai: IT Infrastructure for Machine Learning. [White paper] (n.d.). URL https://get.valohai.com/hubfs/Valohai_Whitepaper_A4_IT_Infrastructure_4pages.pdf

From Biased Towards Affirmative Artificial Intelligence Tools in Education

Milena Parland[1(✉)] and Andrey Shcherbakov[2]

[1] Åbo Akademi University, 20500 Turku, Finland
milena.parland@abo.fi
[2] Arcada University of Applied Sciences, 00550 Helsinki, Finland
andrey.shcherbakov@arcada.fi

Abstract. Previous research has highlighted that AI tends to contribute to bias and the marginalisation of minority groups. This paper will focus on Artificial Intelligence Tools in Education (AITED) equality and equity in the context of nation-states in the European Union, with examples from Finland. To avoid reproducing discrimination in AITED is a basic aim enforced by the law both on EU level and on national levels in EU. So far, the discussion focuses mostly on how to decrease bias and discrimination in AI, but in this paper, we aim further and introducing the idea of affirmative measures actively promoting equality and equity. The research question for this paper is: What existing methods could help us to shape more equal and affirmative AITED? We find that promoting equality and equity instead of just avoiding discrimination takes us towards affirmative rights for minorities. In this paper, we find that there is a need to engage representatives for minorities and a minority/non-discrimination expert while shaping the AITED and to create an inclusive milieu for them. We discuss the urge to use a language that specifies each minority, and we look at the difference between non-discrimination and affirmative rights for minorities. Shaping more equal AITED could also be promoted by using the WILPF, Women's International League for Peace and Freedom, tool from Political Economic Analyses and Critical Race Theory. Introducing datasheets that accompany every data set seems also beneficial for more equal AITED.

Keywords: Big Data Bias · Education · Critical Race Theory · Affirmative Rights · National Minorities

1 Introduction

Today, there are expectations that Artificial Intelligence (AI) will bring new, benevolent, flexible tuition models to educational institutions. AI is already being used in educational settings for various applications, including grading, personalised learning, and administrative tasks. While AI has the potential to revolutionise the education sector, there are growing concerns about the quality of the big data on which these systems are trained and about systemic reproducing of bias (Balayn et al., 2021; *EUAFR*, 2022). Previous research has highlighted that AI tends to contribute to the marginalisation of minority

M. E. Auer et al. (Eds.): STE 2024, LNNS 1028, pp. 352–362, 2024.
https://doi.org/10.1007/978-3-031-61905-2_34

groups: "Without concerted efforts, the reinforcement of systemic bias and discrimination will continue to perpetuate through these technology systems that are becoming ubiquitous" (Ravanera & Kaplan, 2021, p. 9). This paper will focus on Artificial Intelligence Tools in Education (AITED) and equality and equity in the context of nation-states in the European Union, with examples from Finland introducing a discussion about affirmative measures to promote equity in AITED.

AI used in primary and middles schools should also promote equality and equity to fulfil the curriculum[1] and the Convention on the Rights of the Child[2], as well as the Framework Convention for the Protection of National Minorities (FCNM)[3]. The convention not only urges the states to promote effective equality and to preserving and develop minority cultures, religions, and languages, but also urges to make education accessible to all and foster intercultural knowledge. Therefor knowledge about the national minorities and their culture should be spread to all students, and AITED used in educational settings should consider this and be part of the implementation processes linked to the FCNM.

2 Background

The question of how to improve equality and equity AITED seems urgent. In this paper, equality and equity are defined according to Article 21 in the EU Charter of Fundamental Rights[4]; "Any discrimination based on any ground such as sex, race, colour, ethnic or social origin, genetic features, language, religion or belief, political or any other opinion, membership of a national minority, property, birth, disability, age or sexual orientation shall be prohibited." Indirect discrimination is also prohibited, and this form of discrimination might often be hidden in AITED. This paper focuses four one ground: ethnicity, membership of a national minority, race and religion. We base our choice on the agreement Framework Convention for the Protection of National Minorities from 1995. In Finland, the national minorities are the Sami, the Roma, the Jews, the Tatars, the Old Russians, the Karelians, and the Swedish-speaking Finns. There is a possibility to include more groups (*CoE*, 2014). The minorities presented are multi-layered. In this paper we are not going further into the discussion about how national minorities in Finland want to define themselves, and how the dominating group define them and how ethnic, racial, and religious categories intersect. Instead, we suggest some concrete solutions for shaping more equal and affirmative AITED. Indeed, the solutions and methods presented in this paper can easily be used in context concerning various minority groups.

Previous research shows that it is demanding to avoid the reproduction of bias and to find solutions for shaping equal AITED (Ravanera & Kaplan, 2021). This means there is a thrilling field for research and inventions.

[1] https://www.oph.fi/en/statistics-and-publications/publications/new-national-core-curriculum-basic-education-focus-school.

[2] https://www.ohchr.org/en/instruments-mechanisms/instruments/convention-rights-child.

[3] https://rm.coe.int/168007cdac.

[4] https://fra.europa.eu/en/eu-charter/article/21-non-discrimination.

3 Research Question

In this paper, we move on from discussing the mere obligation that the law enforces on AI developers, i.e., to create solutions and tools and instruments that are non-discriminative, towards the idea of affirmative AITED. The whole processes of shaping the AITED must be considered while striving towards equity (Fenu et al., 2022).

We will look for solutions and methods to promote the shaping of more equal and affirmative AITED. According to Brunner and Küpper (Brunner & Küpper, 2002), minority rights can be divided into three levels: non-discrimination, affirmative and autonomous. The first level concerning non-discrimination merely guarantees the right to be physically alive and exist in society, but it often allows assimilative policies that lead to the cultural and linguistic extinction of minorities. They need protection that is "not solely antidiscrimination and undifferentiated citizenship, but rather various groupdifferentiated minority rights" (Kymlicka, 2007, p. 91). AITED solutions without affirmative strategy make the minority groups invisible and will contribute to processes of assimilation and oppression because of the convergence of racism, explained by Delgado and Stefanic (Delgado & Stefancic, 2017, p. 9). Ensuring that databases used for AITED are free from abusive statements, hate speech, and threats only meets the most basic standards of equality. It is not sufficient for AITED, when employed by learners of all ages, to simply avoid causing harm to minorities. A more substantial goal is to actively promote positive affirmation and respectful, autonomous representation of minority groups within AITED. This includes ensuring that minority visibility is respectfully differentiated within the technology. Efforts toward this end should proceed alongside initiatives to reduce the prevalence of harmful, hateful, and threatening content in the large datasets that could be replicated within AITED. The research question for this paper is: What existing methods could help us to shape more equal and affirmative AITED?

4 Homogeneity and Missing Data

A problem that often exists is the homogeneity among the groups that take part in the processes to develop the AITED: administrators, pedagogues, IT engineers, and visual designers. Alkhatib states that "designers consolidate and ossify power" creating maps and systems where no friction exist. At the same time the people most affected by the abridgments systems are left with no tools to challenge it (Alkhatib, 2021, p. 2). Mainly, the people, who take part in creating the AITED, belong to the dominant groups in society; if there are exceptions, they might blend in due to group pressure. Educational institutions tend to uphold nationalistic paradigms due to their historical origins, connected to the national state.

> "All Western countries continue to adopt a range of policies to inculcate overarching national identities and loyalties, including the mandatory teaching of the nation's language, history, and institutions in schools, language tests for citizenship, the funding of national media and museums, and the diffusion of national symbols, flags, anthems, and holidays, to name just a few" (Kymlicka, 2007, p. 72).

For example, when visualizing Finland as a geographic entity, it is important to consider whether the visualization incorporates geographical names in Finnish exclusively, or if it also acknowledges the Swedish language, the three Saami languages (North Sámi, Skolt Sámi, and Inari Sámi), and the Karelian language. Taking the task one step further would be to think about how to make the national minorities visible without any clear territorial links as the Finnish Kale Roma, the Tatar, the Jews and the old Russians and reduce assimilation into the dominating group and its discourse. Research shows that minorities experience exclusion in schools and that tuition in their own groups with their own teachers is beneficial and gives better academic results (Parland & Kwazema, 2023; Phinney & Chavira, 1992; Quintana, 2007; Seaton et al., 2006). The risk that AITED and the models they are built on might reproduce experiences of exclusion on the one hand, through ignoring and hiding the fact that minorities exist and, on the other hand, through reproducing bias.

The reason to this is to be find in the heart of coding and shaping of algorithms as these processes include theorizing and simplification and what Alkhatib calls a homogenizing effect that promotes monoculture. Those who don't already fit the existing pattern are erased or oppressed and this happens wherever AI wields dominating power. (Alkhatib, 2021).

In USA and UK, racial data exist and is available for researchers. In EU, such data does not exist. Anyway, EU allows "the collection of data needed to combat racial or ethnic discrimination, especially if this is done with the explicit and informed consent of the persons concerned" (Ringelheim, 2009, p. 54). The lack of regular collection of racial censuses in the EU is linked to European history and the Holocaust. This dark heritage leads to the choice to have no racial data, which is understandable. Anyway the choice also hinders minorities from looking numerous, which serves to make the nation-states look more homogenous than they are. According to the statistical database in Finland, there were 3,802 Muslims aged 0–14 years in Finland, while 13,416 pupils were studying Islam in primary schools (Salmenkivi & Åhs, 2022). The number of Jews in Finland is around 1,100, according to the official statistical database, but this number is based on membership in one of the two Jewish parishes, ignoring all non-members. An additional example concerns the official statistics regarding the Roma population in Finland; these figures do not account for Roma individuals from other parts of Europe who reside in Finland. Currently, the national statistical database[5], which includes census information, is not really accessible. Moreover, it is unclear how decisions regarding the accessibility and visibility of certain data have been made and the rationale behind these choices.

Discussions around non-discrimination, equality, and equity often occur in broad terms without distinguishing between different minority groups. In Finland, the discourse typically contrasts migrants with Finns. However, this simplistic discourse may contribute to othering and exoticizing, and also preserves the in-visualization of minorities. Anti-racism practices and Critical Race Theory (CRT) are American-developed and their application in Europe is not without challenges. Within the European context, there is a tendency to downplay the significance of race as a social differentiator, shifting the focus to 'culture' and creating hierarchies that may deem some cultures superior to

[5] https://pxdata.stat.fi/PxWeb/pxweb/en/StatFin/.

others, as noted by Kóczé in (Kóczé, 2018). CRT originates from American Critical Legal Studies, and careful consideration is needed when integrating its principles into European settings, as the legal traditions differ, a point echoed by Moschel in (Moschel, 2007). It is obvious that when working to adopt both methods to the European context it is essential to be groupdifferentiative and consider the national minorities including the transnational, pan-European group of the Roma people. Officially, Roma does not even exist as a minority or as a racial category in the US. These means most of the existing and popular discourses about how to tackle racism and discrimination is not helpful when it comes to discrimination against Roma. Nevertheless, more that 12 million people of Roma descent live in Europe. Roma people is a crucial minority in the European context, and the non-Roma has always tried to control and define the Roma identity (Hancock, 2007). It is crucial to let minorities define themselves. When or if AITED in Europe and nation-states as Finland choose to ignore the Roma and their presence in the history and in the present life of the continent this means that there is reproduction of oppression and discrimination of the group.

In this paper we choose one concrete idea about how to promote equality and equity and non-discrimination AITED, and we choose to focus on FCNM and the national minorities including Roma. The choice to work from this specific platform makes it easier to offer concrete solutions and methods. If we would refer to a nonspecific variety of othered groups as "migrants" or "racialised" we could risk reproducing othering, and un-willingly be re-enforcing a discourse where "*the other*" exists beside the normal group, i.e. the "original" dominating group. When we are groupdifferentiating we make national minorities visible. Simultaneously we are deconstructing the nationalistic dichotomy used in the mention discourse based on only two groups: the migrants versus the "original" dominating group.

5 Silence

When it comes to minorities, it is also essential to understand that minorities often remain silent and do not want to share their experiences and desires with the dominant group. In the Finnish context, this has been explored both Stenroos (Stenroos, 2018) and Lehtola (Lehtola, 2022). The latter points out how the majoritarian group dominates through a paternalistic way of talking "Sanelivat isännällisesti…" (Lehtola, 2022, pp. 178–179). A member of a religious minority also explains his attempt to talk about his concerns at school: "And I just see people looking at me rolling their eyes… So suddenly what I thought was important and part of my identity was actually something ridiculous" (*Cultura/IF*, 2021). Also, children from ethnic and religious minorities share narratives about wishes to blend in (Parland, 2023).

The challenges faced by minorities in expressing themselves suggest that the dominant group's normative behaviors subconsciously incorporate silencing others. This phenomenon of silence and its implications have been extensively studied (Ahlvik-Harju, 2016; Spivak, 1988). This means that aside from the problem with lacking racial and minority census, there is also a problem with how to develop a social environment where members of the marginalised can speak. Involving representatives from minority communities and giving them a mandate at all stages while developing the AI is essential.

However, it will not automatically be efficient if the dominant groups' representations remain uneducated about their position bias and stereotypes and the historical context surrounding the minority in the specific nation. This can be compared to engineers developing tools without consulting end-users—resulting in dysfunctional solutions[6]. Machine learning algorithms, ranging from traditional linear models to modern deep neural networks, are highly susceptible to the biases inherent in the data sets they are trained on. As a result, these models often perpetuate the very predictions that—as also language through history—have served to "justify social hierarchy and violence" (Jindal, 2022, p. 2). When models are routinely trained on data that has not been critically examined and curated, they reflect and reinforce the prevailing social order, thereby perpetuating discrimination.

6 Representation

It is important to break the pattern of ignoring the voices of minorities when planning and shaping AITED. Indeed, there are many possibilities for teams of engineers, anti-discrimination experts, and representatives of minority communities to develop sustainable methodological solutions if they cooperate. There are challenges for minorities to speak out instead of blending in; therefore, expertise in minority studies and equity is needed.

In the process of eliciting data requirements and utilizing database management systems, the implementation of bias constraints can foster non-discriminative practices and support affirmative equity. However, it is vital to maintain awareness of the necessity for group-differentiated minority affirmation when developing these methods (Kymlicka, 2007). Evidently the ethical and the political discussion about what categories we choose while shaping the data requirements always remains.

7 Tackling Missing Data

The problem with the lack of census data and other missing data has to be acknowledged and taken into consideration. If we get back to Finland, there is a lack of relevant data about national minorities or different migrant groups. There are methods for compensating missing data by imputation, estimation and pooling estimates[7]. However, all these methods can be problematised, and the decision to use them must be agreed upon by the group concerned. Missing data makes minorities invisible and increases the processes of blending in and assimilation.

For providers of AITED, the least they can do is acknowledge the limitations of the raw data and the databases used and openly declare the missing data and what this might mean in terms of equity, equality, and pedagogical quality.

In any case, the first step would preferably be to conduct interviews with data scientists, engineers, representatives for minorities, educators, experts on exclusion/minority studies, and policymakers involved in data collection and curation for educational AI systems.

[6] https://matr.net/news/why-it-projects-fail-lack-of-user-involvement.

[7] https://egap.org/resource/10-things-to-know-about-missing-data/.

8 Political Economy Analysis

There are also practical tools that can be applied while developing policies for AITED and simultaneously educating the persons using them as they become more aware of structural and systemic problems in education that produce and reproduce oppression in the societal context. Political Economy Analysis (PEA) offers tools for such work. Political analysis examines who controls wealth and power, why certain decisions are made, who makes them, and who benefits from them. It also explores the impact of these decisions on society in terms of economics, politics, and social aspects. This field is interdisciplinary, primarily drawing from political science, economics, and sociology, but it also incorporates elements of law, history, and other academic disciplines. A tool that could inspire the development of concrete tools with sets of questions is the Women's International League for Peace and Freedom (WILPF) guide to feminist political economy which helps "to deconstruct seemingly fixed and unchangeable economic, social, and political parameters" (Isakovic, 2018, p. 2) through concrete, fixed questions. Questions from this framework also work for analysing equality and equity related to ethnicity, race and religion if slightly modified.

Research shows that for example classification bias can be reduced in commercial APIs following methods promoting gender equality and equity (Raji & Buolamwini, 2019). Obviously, there could be specific methods and tools developed for marginalized groups of different types, as national minorities in Finland or other European countries, but so far, we start by referring to existing tools.

On the most concrete level, the Swedish Sida's Gender Toolbox[8], which has been engaged with gender strategies for years, recommends always striving to specify who is referred to. It is advised to avoid using unspecified categories such as "people" and instead use specified terms such as women, men, girls, boys, and an inclusive option for those who are non-binary. This same method can be applied when shaping policies and strategies regarding national minorities. The initial step involves an examination of the training dataset to verify its inclusion of distinct groups like the Sami, Roma, Jews, Karelians, etc. Subsequently, an analysis should be conducted to ascertain if the solution or tool utilizes the term "people", potentially standardizing the dominant group as the benchmark.

9 Critical Race Theory

Researchers refer to Critical Race Theory (CRT) and how its lenses could improve AI (Alkhatib, 2021; Delgado & Stefancic, 2017). CRT could be utilised to analyse and interpret how racial biases may be embedded within these data collection and curation processes and to investigate systemic and societal factors influencing these biases. This step involves analysing reports and feedback focusing on data handling practices and their implications. Utilising CRT to analyse and interpret biases in data collection and curation processes might work well as it reveals power structures and blind spots.

[8] https://www.sida.se/en/for-partners/methods-materials/gender-toolbox.

- CRT provides a framework to understand biases within a broader historical, cultural, and social context. This perspective is crucial in identifying biases and their origins and perpetuation mechanisms.
- CRT emphasises the examination of systemic and institutional factors that contribute to inequality. By applying this theory, researchers can uncover deeper systemic issues influencing data biases, which might otherwise be overlooked in a purely technical analysis.
- CRT encourages a holistic approach to examining race and power dynamics. This leads to a more comprehensive analysis of how racial biases are embedded and sustained within AI systems, considering both overt and subtle forms of discrimination.
- Insights gained from a CRT perspective can inform more effective and equitable solutions. Understanding the root causes and systemic nature of biases enables the development of targeted strategies to address and mitigate these issues in AITED development and application.

CRT offers a critical lens to explore and address the complexities of racial biases in data collection and curating, ensuring that solutions are not just technically sound but also socially responsible and equitable. There is a need to examine data availability and assess how practices of collection methods and curation processes for developing educational AI systems contribute to or mitigate racial biases. This involves compiling findings about the impact of data collection and curation on racial bias in AI.

10 Datasheets Accompanying Data Sets

Timnit Gebru et al. (Gebru et al., 2021) propose that every dataset should include a datasheet detailing its purpose, composition, collection methodology, and suggested uses to enhance transparency and accountability, a practice also endorsed by the World Economic Forum to prevent discriminatory outcomes (*WEF*, 2018). Drawing a parallel to the electronics industry, where every component comes with a datasheet that describes its characteristics and other vital information, implementing similar datasheets in AITED could significantly enhance its quality. This is because it would provide clarity on the origins and processing of data.

In educational contexts, the use of datasheets can be likened to scholars in science and humanities citing sources and declaring the basis of their work. Introducing datasheets in the field of education, particularly within official school and university settings linked to state education systems, could align with existing conventions, laws, and curricula to promote equality and equity. As AI increasingly relies on data that is often collected in an unstructured manner, all stakeholders—from developers to end-users—face challenges. Users need to comprehend the development process, foundations, and limitations of AI tools to prevent potential negative impacts on the educational process. This is especially critical when AITED is used with children, where ensuring equality, equity and quality is imperative.

It is evident that creating data sheets takes time and that resources must be allocated to this job. According to Gebru et al. (Gebru et al., 2021) "datasheets provide an opportunity for dataset creators to distinguish themselves as prioritizing transparency and accountability". The rapid advancement of Large Language Models could potentially lower the costs associated with developing AITED. The internet is already populated with educational tools, increasingly incorporating AI.

Standing out in this domain and achieving visibility is a considerable challenge. For EU-based AITED providers, one strategy for differentiation could involve support for equality and equity and adherence to the minority policies of the EU. National educational institutions might find it beneficial to prioritize AITED solutions that align with EU directives and policies. Moreover, the idea of establishing a special certification for AITED that meets EU equality and equity standards could be compelling. This certification would include compliance with the EU's Directives on the Prohibition of Discrimination[9] and the Framework Convention for the Protection of National Minorities (FCNM) at the national level.

11 Conclusion

Several layers of problems can amplify the inequity in AITED: societal practices in the majoritarian power distribution, missing data about minorities, and technocratic focus on the solutions for gathering data that should be solved with the tools from social science to guarantee equality and equity in emerging AITEDs. We also promote the suggestion that every data set should be accompanied by a datasheet that documents its motivation, composition, collection process, recommended uses etc. as this will increase transparency and accountability (Gebru et al., 2021).

Preventing the perpetuation of discrimination within AITED is a fundamental goal mandated by laws at both the EU level and within individual EU member states. However, simply avoiding discrimination is insufficient; actively promoting equality and equity includes affirmative measures with groupdifferentiation promoting minorities and their visibility. This paper concentrates on national minorities, exploring ways to ensure their recognition and representation within AITED.

The paper identifies two key principles essential for developing AITED: 1) Including minority representatives and an expert on minority/non discrimination within the AITED development team is crucial. 2) Shaping an inclusive and affirmative milieu for representants from minorities in the team.

Additionally, the paper recommends using tools from Political Economic Analyses, such as the WILPF tool, and applying Critical Race Theory to develop AITED solutions that not only prevent bias and stereotypes but also promote equality and equity and affirm the rights of minorities.

By addressing the challenges of bias reproduction, silencing, and invisibility of minorities, AITED might contribute to advancing equality and equity within educational contexts. Educational institutions have the potential to lead the way in applying new, methods and solutions promoting equality and equity including affirmative rights for minorities in the realm of applied AI.

[9] https://eur-lex.europa.eu/legal-content/EN/TXT/HTML/?uri=CELEX%3A52014DC0002.

References

Ahlvik-Harju, C.: Resisting indignity: a feminist disability theology. Doctoral thesis. Åbo Akademi University (2016)

Alkhatib, A.: To live in their utopia: why algorithmic systems create absurd outcomes. In: Proceedings of the 2021 CHI Conference on Human Factors in Computing Systems (2021). https://doi.org/10.1145/3411764.3445740

Balayn, A., Lofi, C., Houben, G.-J.: Managing bias and unfairness in data for decision support: a survey of machine learning and data engineering approaches to identify and mitigate bias and unfairness within data management and analytics systems. VLDB J. **30**(5), 739–768 (2021). https://doi.org/10.1007/s00778-021-00671-8

Bias in algorithms—Artificial intelligence and discrimination, p. 103 (2022). European Union Agency for Fundamental Rights. https://doi.org/10.2811/25847

Brunner, G., Küpper, H.: European options of autonomy: a typology of autonomy models of autonomy self-governance. In: Gál, K. (ed.) Minority Governance in Europe, p. 26. LGI/ECMI (2002)

Cultura Archive Transcripts (IF mgt 2021/021-029) (2021). Cultura archive of Åbo Akademi University, Finland; Cultura/IF mgt 2021/021-029

Delgado, R., Stefancic, J.: Critical race theory: An introduction, 3rd edn. NY University Press (2017)

Fenu, G., Galici, R., Marras, M.: Experts' view on challenges and needs for fairness in artificial intelligence for education. In: Rodrigo, M.M., Matsuda, N., Cristea, A.I., Dimitrova, V. (eds.) Artificial Intelligence in Education, pp. 243–255. Springer, Cham (2022). https://doi.org/10.1007/978-3-031-11644-5_20

Fourth Report Submitted by Finland Pursuant to Article 25, Paragraph 2 of the Framework Convention for the Protection of National Minorities, p. 106. (2014). [Country Report]. Council of Europe. https://rm.coe.int/CoERMPublicCommonSearchServices/DisplayDCTMContent?documentId=09000016802f299f

Gebru, T., et al.: Datasheets for datasets. Commun. ACM **64**(12), 86–92 (2021). https://doi.org/10.1145/3458723

Hancock, I.: The struggle for the control of identity. In: hAodha, M.Ó., Hancock, I. (eds.) Migrant and Nomad: European Visual Culture and the Representation of 'Otherness', pp. 41–60. Blaue Eule Verlag (2007)

How to Prevent Discriminatory Outcomes in Machine Learning (p. 30). (2018). World Economic Forum Global Future Council on Human Rights. https://www.weforum.org/whitepapers/how-to-prevent-discriminatory-outcomes-in-machine-learning

Isakovic, N.P.: A WILPF Guide to Feminist Political Economy, 4th edn. Women's International League for Peace and Freedom (2018). https://www.wilpf.org/publications/a-wilpf-guide-to-feminist-political-economy/

Jindal, A.: Misguided artificial intelligence: how racial bias is built into clinical models. Brown Hosp. Med. **2**(1), 6 (2022). https://doi.org/10.56305/001c.38021

Kóczé, A.: Transgressing borders: challenging racist and sexist epistemology. In: Roma Activism: Reimagining Power and Knowledge, pp. 111–129. Berghahn Books (2018)

Kymlicka, W.: Multicultural Odysseys: Navigating the New International Politics of Diversity. OUP, Oxford (2007)

Lehtola, V.-P.: Entiset elävät meissä Saamelaisten historia ja Suomi. Gaudiamus (2022)

Moschel, M.: Color blindness or total blindness-the absence of critical race theory in Europe. Rutgers Race L. Rev. **9**, 57 (2007)

Parland, M.: When they tell you who you are. Approaching Religon **13**(3) (2023). https://doi.org/10.30664/ar.131085

Parland, M., Kwazema, M.: Looking for hidden notebooks: analysing social exclusion experienced by teachers of minority religions in Finnish schools. In: Alemanji, A.A., Meijer, C.M., Kwazema, M., Benyah, F.E.K. (eds.) Contemporary Discourses in Social Exclusion, pp. 91–118. Springer, Cham (2022). https://doi.org/10.1007/978-3-031-18180-1_5

Phinney, J.S., Chavira, V.: Ethnic identity and self-esteem: an exploratory longitudinal study. J. Adolesc. **15**(3), 271–281 (1992). https://doi.org/10.1016/0140-1971(92)90030-9

Quintana, S.M.: Racial and ethnic identity: developmental perspectives and research. J. Couns. Psychol. **54**(3), 259–270 (2007). https://doi.org/10.1037/0022-0167.54.3.259

Raji, I.D., Buolamwini, J.: Actionable auditing: investigating the impact of publicly naming biased performance results of commercial AI products. In: Proceedings of the 2019 AAAI/ACM Conference on AI, Ethics, and Society, pp. 429–435 (2019). https://doi.org/10.1145/3306618.3314244

Ravanera, C., Kaplan, S.: An Equity Lens on Artificial Intelligence. Gender and the Economy (2021). https://www.gendereconomy.org/artificial-intelligence/

Ringelheim, J.: Collecting racial or ethnic data for antidiscrimination policies: a US-Europe comparison. Rutgers Race Law Rev. **10**(1), 39–142 (2009)

Salmenkivi, E., Åhs, V.: Selvitys katsomusaineiden opetuksen nykytilasta ja uudistamistarpeista (13; Opetus- ja kulttuuriministeriön julkaisuja 2022:13, p. 253). Opetus- ja kulttuuriministeriö (2022). https://julkaisut.valtioneuvosto.fi/bitstream/handle/10024/164015/OKM_2022_13.pdf

Seaton, E.K., Scottham, K.M., Sellers, R.M.: The status model of racial identity development in African American adolescents: evidence of structure, trajectories, and well-being. Child Dev. **77**(5), 1416–1426 (2006). https://doi.org/10.1111/j.1467-8624.2006.00944.x

Spivak, G.C.: Can the Subaltern Speak? Macmillan (1988)

Stenroos, M.T.: Power and hierarchy among Finnish Kaale Roma: insights on integration and inclusion processes. Crit. Romani Stud. **1**(2), 6–23 (2018). https://doi.org/10.29098/crs.v1i2.12

Bridging the AI Knowledge Gap with Open Online Education in Europe

Dario Assante[1] (✉), Claudio Fornaro[1], Luigi Laura[1], Daniele Pirrone[1], Ali Gokdemir[2], and Veselina Jecheva[3]

[1] Università Telematica Internazionale Uninettuno, Corso Vittorio Emanuele II 39, 00186 Rome, Italy
d.assante@uninettuno.it
[2] Innomate Ltd., Merkez, 67030 Zonguldak, Zonguldak Province, Turkey
[3] Burgas Free University, San Stefano Street 62, 8001 Burgas, Bulgaria

Abstract. With the outbreak of the Covid-19, many countries had to interrupt in-person trainings and the adoption of online learning accelerated dramatically. Distance education shifted from an opportunity to a need, and the lack of innovative digital content, particularly in the field of artificial intelligence (AI), became apparent. In this context, in this paper we introduce the EU2AI project, funded in the framework of the Erasmus + programme, with the aim to address these challenges and contribute to the development of digital learning materials and tools, specifically focused on AI education.

Keywords: Artificial Intelligence · Distance Learning · Gamification · Skills development · Vocational education

1 Introduction

The outbreak of the Covid-19 pandemic has not only disrupted traditional modes of education but has also accentuated the need for innovative digital content and distance learning opportunities. Among the fields that have witnessed a surge in demand and relevance is artificial intelligence (AI). The rapid advancements in AI technology and its increasing impact across various industries have underscored the importance of providing comprehensive AI education to students and professionals alike [1–6].

The European Union (EU) has been proactive in formulating policies and regulations to govern the development and deployment of artificial intelligence (AI) technologies [7]. The EU's approach to AI focuses on fostering innovation, ensuring ethical and trustworthy AI, and safeguarding fundamental rights and values. The EU's AI policies emphasize the importance of human-centric AI, transparency, and accountability. The EU's regulatory framework includes the General Data Protection Regulation (GDPR) and the proposed AI Act, which aims to establish a comprehensive framework for AI governance, covering areas such as data usage, risk assessment, transparency, and human oversight. The EU seeks to position itself as a global leader in AI while upholding principles of fairness, transparency, and inclusivity, and promoting the responsible and

M. E. Auer et al. (Eds.): STE 2024, LNNS 1028, pp. 363–369, 2024.
https://doi.org/10.1007/978-3-031-61905-2_35

sustainable development and deployment of AI technologies [8–10]. However, it has become apparent that there is a significant knowledge gap in AI education, particularly in terms of accessible and engaging digital learning materials [11–16].

To address these challenges and contribute to the development of AI education, the EU2AI project was launched within the framework of the Erasmus + programme. The primary objective of the EU2AI project is to bridge the knowledge gap in AI education by training students with in-demand qualifications, enhancing the proficiency of teachers and students, fostering collaboration between educational institutions and small and medium-sized enterprises (SMEs), promoting digital skills and distance education, and aligning skills with the labor market. By doing so, the project aims to empower individuals with the skills and knowledge needed to thrive in the rapidly evolving field of AI.

2 The EU2AI Project and Its Outcomes

By addressing the knowledge gap in AI education, fostering collaboration between educational institutions and businesses, and equipping individuals with the necessary skills, the EU2AI project contributes to the growth of the European digital industry and propels Europe towards becoming a frontrunner in the field of AI. The project is coordinated by Burgas Free University (Bulgaria) and includes partners from Italy, Romania, Spain and Turkey. Including both Universities and VET training providers, the partnership brings complementary and multidisciplinary competences to better address the formulated challenges.

The outcomes of the EU2AI project are not only tangible, such as the creation of a gamification interface, curricula for AI education, and an assessment system, but also intangible. These intangible outcomes encompass increased knowledge on measurement and evaluation systems and AI applications, heightened awareness of AI, improved institutional capacity in AI, personal skill development for employment, and enhanced assessment and evaluation systems. Furthermore, the EU2AI project aligns with the strategic plans of the European Commission in becoming a global leader in developing cutting-edge and trustworthy AI.

The project main outcomes are the self-assessment tool and the online gamified training course. These outcomes are complementary, since the first one is used to assess potential lacks of skills and the second one is used to fill them. In addition, the developed virtual platform provides users access to both resources.

2.1 The Self-assessment Tool

The development of a specialized self-assessment tool for AI training is a crucial and innovative aspect of the project. Currently, there is a lack of such tools in existing research. Therefore, the project aims to fill this gap by creating a unique self-assessment system. This system enables trainers and students to assess their AI competencies, identify their strengths and weaknesses, and motivate themselves to improve. The tool fully aligns with the DigComp Competence Framework, which has served as the reference source during its development phase [17]. The self-assessment tool categorizes competencies into three levels (basic, intermediate, and advanced), allowing participants to determine their qualification status. This method enables educators and students to assess their current level and receive targeted education accordingly. The data obtained from the self-assessment tool will serve as a valuable reference for teachers and facilitate students' learning progress. Over 12 types of reports generated by the tool provide comprehensive information about students, enabling educators to identify and prioritize the skills that require further development.

The self-assessment tool not only measures the quality and competence of instructors and teachers in vocational training and AI education but also provides valuable feedback. By determining the knowledge and information gaps of the target group, the self-assessment tool will enhance the clarity and purpose of the training. Personalized teaching plans will be prepared based on the set proficiency levels for educators and students, optimizing efficiency and motivation. Based on these results, students can provide valuable insights for the curriculum, suggesting additional resources that can support their skill development.

The transferable potential of the web-based self-assessment system is significant as it will have a free license adaptable to all courses. This feature enhances its potential for implementation in various educational environments, making it highly versatile and accessible.

2.2 The EU2AI Curriculum

The EU2AI training course is the main project outcome. It provides a comprehensive and practical understanding of the theoretical aspects, software tools, and hands-on applications in the field of AI. Through a structured curriculum, learners will delve into the fundamental concepts of AI, explore machine learning techniques, delve into neural network concepts, and gain insights into the emerging field of deep learning. The course emphasizes the importance of datasets in AI and equips learners with the necessary knowledge to work with them effectively.

By the end of the course, learners will have a solid theoretical foundation in AI, hands-on experience with software tools, and practical knowledge in applying AI techniques to various domains. This comprehensive curriculum equips learners with the necessary skills and knowledge to contribute to the rapidly evolving field of AI and opens up opportunities for them to pursue careers in AI research, development, and implementation.

The training course on AI aims to address several important objectives. Firstly, it focuses on training qualified individuals at the vocational school, associate, and undergraduate levels to lead AI research and contribute to the national digital industry. The course also aims to develop innovative approaches and training methods to empower educators in effectively teaching AI concepts. By providing qualified manpower in the field of artificial intelligence, the course aims to meet the growing demand for AI expertise in various industries. Additionally, the course seeks to promote the use of the E-learning platform, recognizing the necessity of distance learning systems, particularly in light of the Covid-19 pandemic.

The curriculum covers the following topics:

THEORETICAL ASPECTS	**HANDS-ON**
• Introduction to Artificial Intelligence	• Create a Datasets
• Machine Learning	Regression analysis
• Neural network concept	KNN samples & analysis
• Deep Learning concepts	• SVM samples & analysis
• Dataset	• ANN samples & analysis
SOFTWARE	• RNN samples & analysis
• Libraries for Python	• CNN samples & analysis
• Algorithms for machine learning	• AI into the robotics systems
• Analysis	AI into the IoT systems

Another crucial aspect of the training course is ensuring the alignment between supply and demand in the labor market by equipping learners with the necessary basic skills and capabilities sought by employers. The course also emphasizes strengthening the international dimension of education and training by fostering cooperation between vocational education and training (VET), higher education institutions (HEIs), and SMEs. This collaboration aims to support the development of both corporations and individuals in the field of AI.

2.3 Interactive and Gamified Learning Activities

For the practical part of the learning activities, we designed some interactive exercises using the Colaboratory platform from Google, usually called Colab [18]. Colab is a cloud-based notebook environment derived from the open source Jupyter project that "is the original web application for creating and sharing computational documents. It offers a simple, streamlined, document-centric experience" (jupyter.org). Colab notebooks are hosted on Google's servers, so users can access them from any device with a web browser; it allows users to write and execute Python code without having to install any software. This makes Colab a convenient platform for teaching, learning, and research in machine learning, data science, and other fields that use Python.

Colab notebooks combine executable code and rich text in a single document, along with images, HTML, LaTeX, and more. Colab notebooks are also shareable with others, so collaborators can easily work on projects together; we use this feature to allow students to visualize the notebook, whilst they can modify their own personal copy of each notebook in order to complete the exercises it provides.

Thanks to the features of the Colab environment, the exercises are interactive: each notebook is divided into cells, that can either contain text or code to be executed. See for example the notebook shown in the figure above, where there is a textual introduction that describes the goal of the exercise, and a second cell with the code, in this case the declaration of the necessary libraries to be included. All our exercises follow this structure:

1. We begin with a description of the exercise, and the expected learned outcome.
2. Then there is the code for one or more full working examples, that the students need to execute in order to techniques employed.

3. Finally, there are some questions that require the student to modify the code of the working examples in order to reach the goals requested. In some cases, it is depicted the expected results, so the students can check whether its code is correct (Fig. 1).

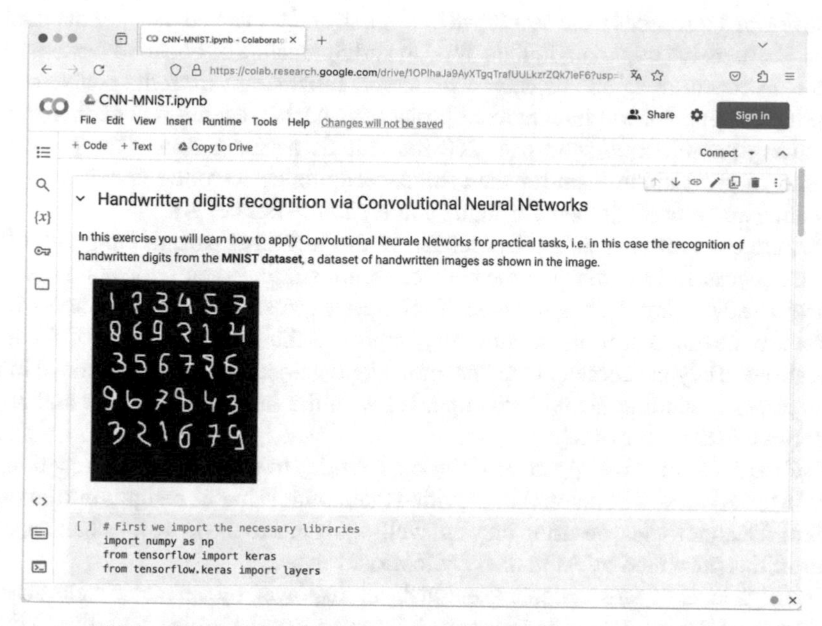

Fig. 1. An example of an interactive exercise in the Colab platform

Currently, in the platform there are exercises that help the student to dive into the practical aspects of:

- The most popular Python libraries for machine learning and data analysis, including keras, tensorflow, scikit-learn, and pandas, just to mention some of them.
- General Artificial Neural Networks for classification, using both Keras, that can be considered a high-level interface for TensorFlow, and TensorFlow alone; as usual, the dataset is split intro training and test sets and the students learn also how to assess the quality of the models built.
- Convolutional Neural Networks (CNN), that is particularly well-suited for analyzing images and other grid-like data. CNNs are inspired by the structure of the human visual cortex, which is the part of the brain that is responsible for processing visual information. The exercise cover tasks of classification (like in the figure shown, where each image needs to be classified into the ten categories corresponding to one of the ten digits) and labeling, where each image is labeled by its (main) content.
- Recurrent Neural Networks (RNN), that is a type of artificial neural network that is designed to recognize patterns in sequences of data. Unlike traditional feedforward neural networks, which process each input independently, RNNs can incorporate information from previous inputs into their current output. This makes them well-suited for tasks such as natural language processing (NLP), speech recognition, and

time series forecasting. The exercises proposed, in particular, cover the tasks of text classification and text generation.

2.4 Piloting

The developed resources were developed in English and translated into 4 of the partners' languages, namely Bulgarian, Italian, Turkish and Spanish. The obtained feedback from students is encouraging and reveals their general satisfaction from the course content. The pilot training for the international project on AI has been a resounding success, with strong participation from over 200 individuals across Europe. The participants have shown great enthusiasm for the course, recognizing its value in enhancing their knowledge and preparing them for future careers in the field of AI.

Despite encountering difficulties related to the technical part, the majority of participants successfully completed the course, demonstrating their dedication and commitment to advancing their skills in AI. They have expressed their satisfaction with the program, believing that it has significantly improved their understanding of AI and its applications. They are confident that the knowledge gained will prove beneficial in their future careers, enabling them to stay updated with the latest technologies and remain competitive in the job market.

The participants have appreciated the opportunity to stay abreast of new developments in the AI field. The course has provided them with valuable insights into emerging trends and technologies, ensuring they are well-equipped to navigate the challenges and opportunities presented by AI in their professional lives.

The high level of satisfaction and enthusiasm displayed by the participants attests to the success of the pilot training program. It underscores the project's ability to deliver quality training and empower individuals with relevant AI skills, ultimately contributing to their personal and professional growth.

3 Conclusions

The EU2AI project is expected to yield significant outcomes in the field of AI education. Through the development of a gamification interface for the e-learning platform, the creation of curricula for AI education, and the implementation of an assessment system, the project aims to enhance digital skills and competence in AI for both educators and students. The collaboration between educational institutions and the development of innovative training materials will contribute to bridging the gap in digital content and improving AI education. Furthermore, the project aligns with the strategic plans of the European Commission to become a global leader in developing cutting-edge and trustworthy AI. By enhancing AI education, promoting awareness and understanding of AI, and empowering individuals with the necessary skills and knowledge, the EU2AI project contributes to the growth of the European digital industry and addresses the challenges in AI education.

Acknowledgement. This work has been co-funded by the European Commission within the project EU2AI – *Increasing the Quality of Vocational Education with the Samples of Artificial Intelligence Technology in Different Fields* (2021-1-BG01-KA220-VET-000034626).

References

1. Borah, S., Kama, C., Rakshit, S., Vajjhala, N.R.: Applications of artificial intelligence in small- and medium-sized enterprises (SMEs). In: Mallick, P.K., Bhoi, A.K., Barsocchi, P., de Albuquerque, V.H.C. (eds.) Cognitive Informatics and Soft Computing. LNNS, vol. 375, pp. 717–726. Springer, Singapore (2022). https://doi.org/10.1007/978-981-16-8763-1_59
2. Baabdullah, A.M., Alalwan, A.A., Slade, E.L., Raman, R., Khatatneh, K.F.: SMEs and artificial intelligence (AI): antecedents and consequences of AI-based B2B practices. Ind. Mark. Manage. **98**, 255–270 (2021)
3. Hansen, E.B., Bøgh, S.: Artificial intelligence and internet of things in small and medium-sized enterprises: a survey. J. Manuf. Syst. **58**, 362–372 (2021)
4. Bhalerao, K., Kumar, A., Kumar, A., Pujari, P.: A study of barriers and benefits of artificial intelligence adoption in small and medium enterprise. Acad. Market. Stud. J. **26**, 1–6 (2022)
5. Wei, R., Pardo, C.: Artificial intelligence and SMEs: how can B2B SMEs leverage AI platforms to integrate AI technologies? Ind. Mark. Manage. **107**, 466–483 (2022)
6. Ingalagi, S.S., Mutkekar, R.R., Kulkarni, P.M.: Artificial intelligence (AI) adaptation: analysis of determinants among small to medium-sized enterprises (SME's). In: IOP Conference Series: Materials Science and Engineering, vol. 1049, no. 1, p. 012017. IOP Publishing (2021)
7. Annoni, A., et al.: Artificial intelligence: a European perspective. Joint Research Centre (2018)
8. Cohen, I.G., Evgeniou, T., Gerke, S., Minssen, T.: The European artificial intelligence strategy: implications and challenges for digital health. Lancet Digit. Health **2**(7), e376–e379 (2020)
9. Cath, C., Wachter, S., Mittelstadt, B., Taddeo, M., Floridi, L.: Artificial intelligence and the 'good society': the US, EU, and UK approach. Sci. Eng. Ethics **24**, 505–528 (2018)
10. Smuha, N.A.: The EU approach to ethics guidelines for trustworthy artificial intelligence. Comput. Law Rev. Int. **20**(4), 97–106 (2019)
11. Assante, D., Caforio, A., Flamini, M., Romano, E.: Smart education in the context of industry 4.0. In: 2019 IEEE Global Engineering Education Conference (EDUCON), pp. 1140–1145 (2019)
12. Chen, L., Chen, P., Lin, Z.: Artificial intelligence in education: a review. IEEE Access **8**, 75264–75278 (2020)
13. Sapci, A.H., Sapci, H.A.: Artificial intelligence education and tools for medical and health informatics students: systematic review. JMIR Med. Educ. **6**(1), e19285 (2020)
14. Fiok, K., Farahani, F.V., Karwowski, W., Ahram, T.: Explainable artificial intelligence for education and training. J. Def. Model. Simul. **19**(2), 133–144 (2022)
15. Richardson, M.L., et al.: Review of artificial intelligence training tools and courses for radiologists. Acad. Radiol. **28**(9), 1238–1252 (2021)
16. Assante D., Flamini M., Romano E.: Open educational resources for industry 4.0: supporting the digital transition in a European dimension. In: 2021 IEEE Global Engineering Education Conference (EDUCON), pp. 1509–1513 (2021)
17. The digital competence wheel. https://digcomp.digital-competence.eu/
18. Nelson, M.J., Hoover, A.K.: Notes on using google Colaboratory in AI education. In: Proceedings of the 2020 ACM conference on Innovation and Technology in Computer Science Education, pp. 533–534 (2020)

Human-Robot Interaction for Sustainable Development

Deploying Humanoid Robots in a Social Environment

Kristoffer Kuvaja Adolfsson[1(✉)], Christa Tigerstedt[1], Dennis Biström[2], and Leonardo Espinosa-Leal[1]

[1] Graduate School and Research, Arcada University of Applied Sciences, Jan-Magnus Janssons Plats 1, 00560 Helsinki, Finland
{kristoffer.kuvajaadolfsson,christa.tigerstedt, leonardo.espinosaleal}@arcada.fi

[2] School of Engineering, Culture and Wellbeing, Arcada University of Applied Sciences, Jan-Magnus Janssons Plats 1, 00560 Helsinki, Finland
dennis.bistrom@arcada.fi

Abstract. From early 2020 Arcada has pursued state-of-the-art research in human-robot-interaction (HRI). Deploying robots in a multitude of use cases and designing applications with experts from their respective fields. However, getting a robot service from the drafting stages in a lab out into a real-world professional situation is a complex endeavour that takes management, communication, and a multitude of talents. The paper aims to suggest a method, divided into three-stages for deploying humanoids in social environments. By participatory research with autoethnographic influences, this method is proposed through the experience, insight and knowledge of the authors. Having worked with field experts and customers to deploy humanoid robots in a multitude of environments over the past five years in different research projects. The three-stage method suggested (discovery, development, and deployment) aims to further other HRI projects.

Keywords: Human Robot Interaction · Humanoid robots · Interaction design

1 Introduction

According to the World Health Organization, the world population is ageing faster than before, and it is estimated that the proportion of people aged over 60 will almost double by 2050 [9]. This puts a larger demographic of the population at risk of needing care, while simultaneously shrinking the age group of the workforce. A suggestion to prepare society for this inevitability is automation, through AI and robotics, enhancing functions and promoting independence and autonomy for the people in need [13].

The interest in humanoid robots, robots that resemble the human body in shape and form, has been reinvigorated with recent technological breakthroughs

M. E. Auer et al. (Eds.): STE 2024, LNNS 1028, pp. 373–380, 2024.
https://doi.org/10.1007/978-3-031-61905-2_36

that brought AI into the spotlight of publicity. As the 6th industrial revolution, with AI and automation at the forefront, progresses, we should see large improvements to the portability and accessibility of said technologies [5]. This should open an opportunity for deploying more robots, and humanoid robots, in society. This paper aims to suggest a method for deploying humanoid robots in social environments using a three-step method; discover, develop and deploy to increase the benefits in holistic research for human-robot interaction of all those involved.

2 Background

Since early 2020, Arcada University of Applied Sciences has pursued state-of-the-art research in human robot interaction (HRI), starting with a report from the AFORA project [3]. Since then, Arcada has concluded several projects with humanoid robots. Facilitating them as informants for dental care, waiters in restaurants, caretakers in elderly care and recently beginning a project in early childhood education. As such, the team at Arcada has years of expertise in the field of deploying humanoids in a multitude of social environments.

Humanoid robots can primarily be identified by their humanoid features, they have shapes and forms that confirm human anatomy, commonly with facial features like eyes and mouths allowing for expressive interactions often combined with voice capabilities and interaction cited as a valuable key feature [19].

Robots are fundamentally built using hardware and implemented using software. Many robots come with preconfigured and proprietary software and are manufactured for specific markets and functions. It is still possible to repurpose many humanoid robots for different tasks in research, as the humanoid robot NAO is being used in research works across the world, among other robot platforms [1].

When developing and deploying software, there are multiple methods to choose from like agile, scrum, lean or kanban. It is, however, noted that robots are more than just software and as such these methods are found lacking for our scope.

The agile software development method, as outlined by the agile manifest [4], puts a stronger emphasis on change and individuals over keeping a to a plan and while the waterfall methodology [12] builds a foundation for any scaled software development it does not encompass idea of bringing the field, and the end-users, closer to the development and the pre-integration phases as Pekkarinen [10] and Tuisku [18] stresses. In other words, the users need to be involved from the beginning and throughout to gain understanding and training to ensure a successful deployment. Hence, UX design and design thinking tools are proposed as methods or approaches to take the HRI deployment to the next level [11].

This is something our experiment has gained from in a different use case [14] and also where ideas of putting effort into the research phase and the end-user has been proven purposeful, sustainable and more ethical [8].

In the model (Fig. 1) by Prati, the UX design flow is visualised. It includes several activities for researching the needs of the end user in the very beginning. For example, focus groups, observations and several tools for mapping insights [11].

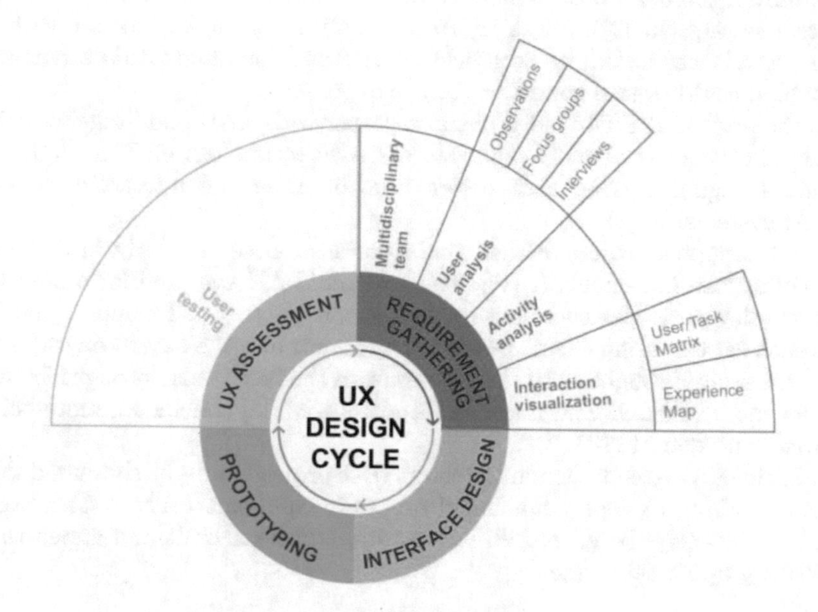

Fig. 1. UX design workflow [11]

3 Method

The method applied is inspired by participatory research with autoethnographic influences, consolidating insights and experiences gathered over time from work in research projects by the authors. The remaining objective throughout the process and the reflective phase is strengthened by the numerous team dialogues over the years in research.

Alf saw use directly in the immediate project after AFORA called MäRi (Människa-robot interaktion, Swedish for HRI). Together with Åbo Akademi University (ÅAU), Arcada studied and collected data in a dental care use-case where customers (patients) interacted with Alf at Arcada or with the humanoid robot Pepper at ÅAU. The study was conducted in controlled environments simulating a dental clinic, where participants got to experience flossing techniques and instructions from the humanoid. The use-case was designed together with dental care experts from Finland and Sweden, and the application on Alf was developed as a bachelor thesis from Arcada students [7]. MäRi culminated in the paper *Stakeholders' Experiences of and Expectations for Robot Accents in a Dental Care Simulation: A Finland-Swedish Case Study* presented at the *Robot Philosophy 2022* conference in Helsinki [13].

Humanoid robot waitress Amy has seen work in restaurant settings in *soCial ASSIstive robOts dePloyEd in the service professIons* (Cassiopeia). Working in student-run Restaurant Prakticum. Designing use-cases with experienced experts from the field, both waitresses and chefs, and serving paying customers in a real setting [16,17]). Tasks for Amy mainly involved serving and dish collection, and it was found, by both field experts and customers, that a humanoid companion could overall contribute to the experience.

In the project TaFiDiAI [6] Arcada, together with ÅAU and StageZero Technologies, identified the benefits and values of ASR technology on Finnish Swedish dialects, bringing a user-centric design from human-robot interaction together with AI systems.

The Cassiopeia project further conducted a use-case in elderly care together with Folkhälsan in Brunakärr [15]. The humanoid Alf was used for a multitude of day-to-day activities such as sing-alongs and quizzes or for more repetitive tasks such as reminding clients to wash their hands and the day-to-day schedule. While the study brought forth clear issues from the field it also brought forth the need for more research, and much like Andtfolk. M. [2] concluded, more holistic research is needed in HRI.

Additionally, Arcada has initiated an HRI project in early childhood development, working to deploy humanoid robots in child-care centres. This project is still in the early design phases but has already been built and added to our autoethnographic catalogue.

4 Results

As each project comes to a conclusion valuable insights and reflections are garnered in the team. With time, a trend for how the projects develop has been identified. The first step is assessing our resources like partners, humanoid robots at hand, in-house expertise and researcher situation. This is then followed up by contacting experts in the field of research, such as dental experts or early childhood education experts, for respective projects. Meeting with said experts out in the field, the social environment where the experts operate, would be preferable, as Pekkarinen put it [10]. Bringing humanoid robots for these meetings is ideal, as for many field experts it can be their first encounter with humanoid robots.

When introducing humanoids for HRI projects, we have also experienced success with keeping discussions open-ended so as not to stifle creative thinking and ideas further into the project. Further, experts out in the field are seldom accustomed to state-of-the-art technologies and will invariably have many questions and thoughts. To stimulate collaboration, the authors found that it is important to keep an open-mind and give answers respectfully and truthfully in these situations.

As an example, it has been suggested that our humanoids utilise popular services for certain tasks, like online services for large language models to chat with clients or image recognition for tracking clients. It is then important for the team to break down the service into feasible alternatives. For example a large language model or image recognition software can be run locally with the correct hardware. However that would also said hardware and a space for it to run locally if the research is to keep with regulations.

In our use case with a restaurant setting it should also noted that more practical limitations and discussions need to be made. The restaurant in questions would from time to time accommodate changes to the layout for specific customer groups. The humanoids used in the use case was not equipped with dynamic mapping and so any changes made would require a remapping. As such staff could adjust and keep changes in the restaurant during the experiments to a minimum and additionally they could inform personnel in good time if any remapping had to be done.

Next stage is defining the groups involved and affected by the research. This is done using user-centric personas. The focus lies on creating a persona for customers, experts and the humanoid robot following templates that outline attributes like challenges, opportunities and needs of the persona in conjunction with the social environment the research is based on. Based on these templates, use-cases are drafted and decided upon, service professionals needs if deemed technically feasible by the research team. This will determine for what purpose the humanoid robots can support the field experts and customers in their day-to-day activities and tasks.

As the use-cases are specified, the research team initiates the development process. Experts on selected humanoid robots and software developers are called to fulfil the specifications. Software development and operations methods like agile, kanban or scrum can be great tools when developing software. However, as HRI research often involves multiple parties and stakeholders and humanoid robots combine both hardware as well as software, the circumstances diverge. For a more holistic research approach with both customers and field experts, it is important that development is ongoing and does not stop when the specifications have been fulfilled. When customers and experts engage with the humanoid out in the field new discoveries often reveal themselves that changes the specification and new features or functions need implementation.

To bring practical examples of this; in the restaurant setting, it was determined that the humanoid Amy should have a very polite way of excusing itself when its collision detection got triggered while working as a waiter. However, after the field experts, I.e., the waiters, experienced this in a real-world situation, they wished for a more concise alert from Amy. This was in due to perceiving customers as not adjusting well to the longer approach.

When working with the elderly, the humanoid robot Alf had predetermined schedules on display for the day activities. But as the weather is not predictable, changes were sometimes made by caretakers from day to day in the morning. Changes developers could implement if available in quick succession, but field experts simply are not equipped to deal with at hand.

Other traditional methodologies, like the waterfall methodology, follow a more rigid and fixed timeline. Traditionally, it leaves little room for the client and product owner (field experts) once specifications have been made, making it unsuitable for the experiences described above. These factors, along with the fact that most common needs for a development process stem from a practical need, rather than a research aim, means field experts are not as goal aligned. Consequently, common managerial aspects must account for, and be observant of, these nuances when leading a research project in HRI.

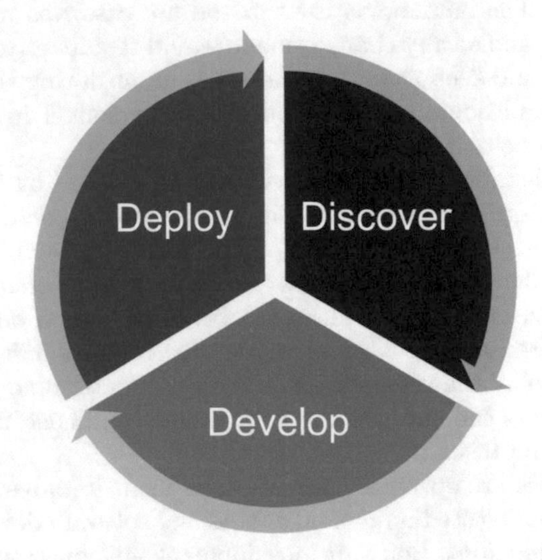

Fig. 2. Three-step-method

With our methods outlined, we would suggest a three-step method for deploying humanoid robots in social environments. The three steps follow an incremental approach of discovering, developing and deploying. Each stage concludes with the initiation of the following. The method can be summarised in Fig. 2 a three-step-method.

Table 1. Three-steps method breakdown

Stage	Actions	Target
Discover	Explore field, humanoid robots, capabilities, opportunities, risks and needs	Personas, use-cases, specifications, human-centric design
Develop	Bring specifications, software and hardware together	Functioning and trustworthy use-cases
Deploy	Implement humanoid robots, integrate experts and customers in usage	Iterate on development, observations and research

5 Conclusion

In this paper, we have suggested a three-stage model for deploying humanoid robots in social environments. The foundation for this method is inspired by human-centred approaches to HRI deployment in a multitude of human robot interaction projects in different roles and scopes. The three stages, discover, develop and deploy encompass a more holistic approach, aiming to best involve experts in different social fields too. Furthermore, the method is well aligned with UX design and design thinking principles [8,11]. Table 1 summarises the different actions and goals described in the result on what the method aims to encompass and achieve.

References

1. Amirova, A., Rakhymbayeva, N., Yadollahi, E., Sandygulova, A., Johal, W.: 10 years of human-nao interaction research: a scoping review. Front. Robot. AI **8** (2021). https://doi.org/10.3389/frobt.2021.744526. https://www.frontiersin.org/articles/10.3389/frobt.2021.744526
2. Andtfolk, M.: The Possibilities for Using Humanoid Robots as a Care Resource, 1st edn. (2022). https://urn.fi/URN:ISBN:978-952-12-4201-4
3. AU of Applied Sciences: Humanoid robots at arcada research fair online 31.3 (2020). https://www.arcada.fi/en/permalink/156
4. Beck, K., et al.: Manifesto for agile software development (2001). https://agilemanifesto.org/
5. Espinosa-Leal, L., Chapman, A., Westerlund, M.: Autonomous industrial management via reinforcement learning. J. Intell. Fuzzy Syst. **39**(6), 8427–8439 (2020)
6. Espinosa-Leal, L., et al.: Tafidiai: Taligenkänning för finlandssvenska dialekter genom artificiell intelligens (2022). https://doi.org/10.5281/zenodo.7495136
7. Kuvaja Adolfsson, K.: Apputveckling för roboten alf : Ett arbete inom märi projektet (2022). https://urn.fi/URN:NBN:fi:amk-2022053013052
8. Lewrick, M., Link, P., Leifer, L.: The Design Thinking Playbook: Mindful Digital Transformation of Teams. The Design Thinking Playbook: Mindful Digital Transformation of Teams, Hoboken, New Jersey (2018)
9. World Health Organization: Ageing and health (2022). https://www.who.int/news-room/fact-sheets/detail/ageing-and-health. Accessed 24 Nov 2023
10. Pekkarinen, S., Melkas, H., Hyypiä, M.: Elderly care and digital services: toward a sustainable sociotechnical transition. In: Toivonen, M., Saari, E. (eds.) Human-Centered Digitalization and Services. TSS, vol. 19, pp. 259–284. Springer, Singapore (2019). https://doi.org/10.1007/978-981-13-7725-9_14
11. Prati, E., Peruzzini, M., Pellicciari, M., Raffacli, R.: How to include user experience in the design of human-robot interaction. Robot. Comput.-Integrated Manuf. **68**, 102072 (2021). https://doi.org/10.1016/j.rcim.2020.102072. https://www.sciencedirect.com/science/article/pii/S0736584520302805
12. Royce, W.W.: Managing the development of large software systems: concepts and techniques. In: Proceedings of the IEEE WESTCON, Los Angeles, pp. 1–9 (1970). http://www.cs.umd.edu/class/spring2003/cmsc838p/Process/waterfall.pdf. Reprinted in Proceedings of the Ninth International Conference on Software Engineering, March 1987, pp. 328–338

13. Susanne, H., et al.: Stakeholders experiences of and expectations for robot accents in a dental care simulation: a Finland-Swedish case study. In: Hakli, R., Mäkelä, P., Seibt, J. (eds.) Social Robots in Social Institutions. Frontiers of Artificial Intelligence and Applications, vol. 366, pp. 125–134. IOS Press (2023). https://doi.org/10.3233/FAIA220611. https://cas.au.dk/en/robophilosophy/conferences/rpc2022, robophilosophy 2022; Conference date: 16-08-2022 Through 19-08-2022
14. Tigerstedt, C., Biström, D.: Teaching and learning with humanoid social and service robots in higher education -learnings from service design and it framework use case development modules (2021). https://doi.org/10.3389/frobt.2021.744526
15. Tigerstedt, C., Dennis Biström, D., Kuvaja Adolfsson, K.: Underhållning och trygghet - inblick i en robots dag i äldre omsorgen (2023). https://www.arcada.fi/
16. Tigerstedt, C., Dennis Biström, D., Kuvaja Adolfsson, K.: Waiter please! the capable service robots amy and alex (2023). https://www.arcada.fi/en/article/blog/2023-09-07/waiter-please-capable-service-robots-amy-and-alex
17. Tigerstedt, C., Fabricius, S.: Unannounced (2023). https://www.arcada.fi/en
18. Tuisku, O., Pekkarinen, S., Hennala, L., Melkas, H.: "Robots do not replace a nurse with a beating heart": the publicity around a robotic innovation in elderly care. Inf. Technol. People **32**, 47–67 (2019). https://doi.org/10.1108/ITP-06-2018-0277
19. Tung, V., Au, N.: Exploring customer experiences with robotics in hospitality. Int. J. Contemp. Hospitality Manag. **30** (2018). https://doi.org/10.1108/IJCHM-06-2017-0322

Open-Sourcing a Humanoid Robot

Ensuring Integrity and Privacy of Robot Sensor Data

Dennis Biström[1(✉)], Kristoffer Kuvaja Adolfsson[2], Christa Tigerstedt[2], and Leonardo Espinosa-Leal[2]

[1] School of Engineering, Culture and Wellbeing, Arcada University of Applied Sciences, Jan-Magnus Janssons Plats 1, 00560 Helsinki, Finland
dennis.bistrom@arcada.fi

[2] Graduate School and Research, Arcada University of Applied Sciences, Jan-Magnus Janssons Plats 1, 00560 Helsinki, Finland
{kristoffer.kuvajaadolfsson,christa.tigerstedt, leonardo.espinosaleal}@arcada.fi

Abstract. This paper outlines the process of open sourcing key components of a commercial humanoid service robot. Humanoid service robots collect data from their environment, and the integrity of these data streams cannot be ensured for proprietary solutions. The integrity of this data needs to be ensured when deploying robots in sensitive use cases.

We have identified which data streams are most delicate and investigated the implementation of the sensor or feature on the robot using basic data- and network analysis tools. A comparison of sensor data was conducted to prioritise the order of modifications. Some features can be modified to achieve data stream integrity, but other sensors and systems need to be replaced. Other open-source robotics projects, commercial and non-commercial, serve as subjects for comparison.

There are few open-source humanoid service robots. The integrity of humanoid robot sensor data can be ensured and verified. All sensor data is not equally sensitive by nature; some sensors were not considered a priority and needed to be modified. Extensive technical knowledge is required for successful modification as well as validation of the data.

Commercial humanoid service robots can be modified to ensure sensor data integrity with the purpose of deployment in sensitive use cases. The applicability of the implementation is yet to be confirmed.

Keywords: open-source · humanoid · robots · privacy

1 Introduction

When working with humanoid robots, several privacy concerns need to be considered. Humanoid service robots collect data from the environment by using sensors like cameras and microphones. This collected data can include personal information. As such, ensuring data integrity and sufficient security on

M. E. Auer et al. (Eds.): STE 2024, LNNS 1028, pp. 381–389, 2024.
https://doi.org/10.1007/978-3-031-61905-2_37

the humanoid robot and its sensors is essential. Under the general data protection regulation, personal data collection must be made with the user's consent [4]. Thus, the user must be informed about what data is being collected and have a chance to opt-out. The user must be informed about what the purpose of the data collection is, in what way, and for how long the data will be stored. The environment in which the robot is active might impact the sensitivity of the data collected. For example, robots in people's homes or workplaces can pose a threat to privacy and anonymity [2].

There is always a threat of data intrusions, and sufficient methods must be undertaken to ensure safe storage of collected data. There is a need for transparency in humanoid robots since users need to understand how robots operate, what data they collect, and how that data is used. There are also several ethical considerations to be made; however, this paper is about the technical component of the privacy and transparency in humanoid service robots [17].

Our starting hypothesis is that humanoid robots with proprietary software do not provide the level of transparency needed for the use of humanoids in sensitive use cases. We argue that the integrity of all data streams a robot utilizes needs to ensure privacy for the humanoid to be suitable for sensitive use cases. The approach of the paper is practical as the work has been conducted on a physical humanoid service robot called Snow, manufactured by CSJBot.

The starting point of the paper is to identify all sensors and associated data streams of a proprietary humanoid service robot and modify or replace hardware or software until the integrity of all data streams have been verified. The sensitivity of each sensors data stream will be assessed, and the most sensitive data streams will be prioritized in the replacement process. The results of the modifications will be summarized in the result chapter and an analysis of the impact of the modifications is presented in the conclusions of the paper.

1.1 Open Source

Free and open-source software (FOSS) has some advantages over its proprietary counterparts. Open-source software is transparent, meaning the code is available for review. This enables a third party to ensure the software is up to date with modern security standards. Auditability is a key feature when it comes to evaluating privacy and security of any piece of software [10].

Both open source and proprietary software is vulnerable to issues with maintenance. If a project does not have active contributors with a long-term commitment, it might not get necessary updates like security patches in time [1].

Public code repositories enable collaborative efforts and independent code verification. Furthermore, FOSS-software provides equal access to anyone in the world, not just the ones who can afford a license. Most open-source licenses also enable derivative works, enabling software integration into larger software systems or services. Interoperability is a core value in many projects and several humanoid robots running the same framework of software can be beneficial.

There are a few commercial open-source humanoid service robots, for example Reachy and Poppy from Pollen Robotics and QTRobot from Lux AI [9,11].

2 Methods

2.1 Network Analysis

A network analysis using Wireshark was done on the router connected to a commercial humanoid robot in our robot lab (Sanbot Elf). Since the robot's software is closed source, efficient analysis methods like code review are not possible. The purpose of the network analysis is to assess how much traffic is being sent and received by a proprietary humanoid robot. By reading the network logs of the local router we can assess which end points the robot is connecting to and how much data is transferred [16].

2.2 Component Mapping

A humanoid robot (CSJBot Snow, a reception robot) was inspected with the purpose of mapping what components are necessary in a commercial humanoid service robot. The contents of the humanoid robot were compared to the contents of a similar humanoid built for a different use case (CSJBot Amy, a waiter robot). The connections of each electronic board to one another defines the possible data streams.

2.3 Priority Assessment

Based on the results of the network analysis and the component mapping, an order of operations can be established according to the potential risk of each sensors data stream on privacy. A live video feed is much more sensitive than a collision detection sensor only reporting a simple boolean to the rest of the robot (eg. true for collision detected, false for no collision) [7].

3 Results

We have identified which data streams are most delicate and investigated the implementation of the sensor or feature on the Snow robot using basic data- and network analysis tools. It's worth mentioning that a total of three humanoid robots were subject to the network analysis and the component mapping, therefore the results are specific to these robots, not generally applicable. A comparison of different sensors data streams was conducted to prioritise the order of modifications to the robot. Some features can be modified to achieve data stream integrity, other sensors and systems need to be replaced. Other open-source robotics projects, commercial (Poppy) and non-commercial (inMoov), serve as subjects for comparison (Fig. 1).

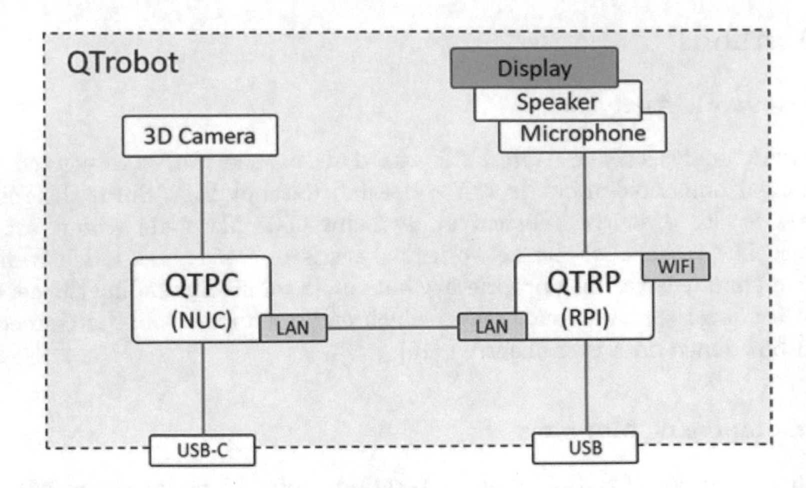

Fig. 1. QTrobot components and sensors

3.1 Network Analysis

(See Fig. 2).

Time	Da	Source	Destination	Protoc	Length	Info
5.080137	…	192.168.2.3	47.91.74.100	TCP	70	37269 → 11001 [PSH, ACK] Seq=10185 Ack=10185
0.000458	…	192.168.2.3	47.91.74.100	TCP	138	37269 → 11001 [PSH, ACK] Seq=10189 Ack=10185
0.040122	…	47.91.74.100	192.168.2.3	TCP	66	11001 → 37269 [ACK] Seq=10185 Ack=10261 Win=2
0.000001	…	47.91.74.100	192.168.2.3	TCP	142	11001 → 37269 [PSH, ACK] Seq=10185 Ack=10261
0.003681	…	192.168.2.3	47.91.74.100	TCP	66	37269 → 11001 [ACK] Seq=10261 Ack=10261 Win=1

Fig. 2. Screencap from Wireshark looking at network traffic from the humanoid robot

The IP 47.91.74.100 can be traced to Frankfurt am Main in Germany, the owner is listed as Alibaba.com LLC, that is an Alibaba Cloud server [8]. While the size of the network package is so small (just 142 bytes) this could not be audio, image nor video data we can concur that continuous communication is present on the humanoid robot to cloud services whenever the robot has an active internet connection. As for text, 140 bytes can fit 160 characters *(140 bytes * 8 bits)/7 bits = 160 characters).*

Even though network traffic is low, without access to the source code of the software we cannot ensure that sensor data is stored locally on the robot. During previous examinations of the humanoid robots (Sanbot Elf) file system we found that clips of video camera footage was saved to the root file system without informing the user. The video footage was related to a home security app, where intruders would be recorded when detected. Since the Wireshark network monitoring trials were only done for a limited amount of time (8 trials at 8 different times during the course of one day), there is no way of ensuring that large amounts of data are not sent over the network at some point.

3.2 Component Mapping

(See Fig. 3).

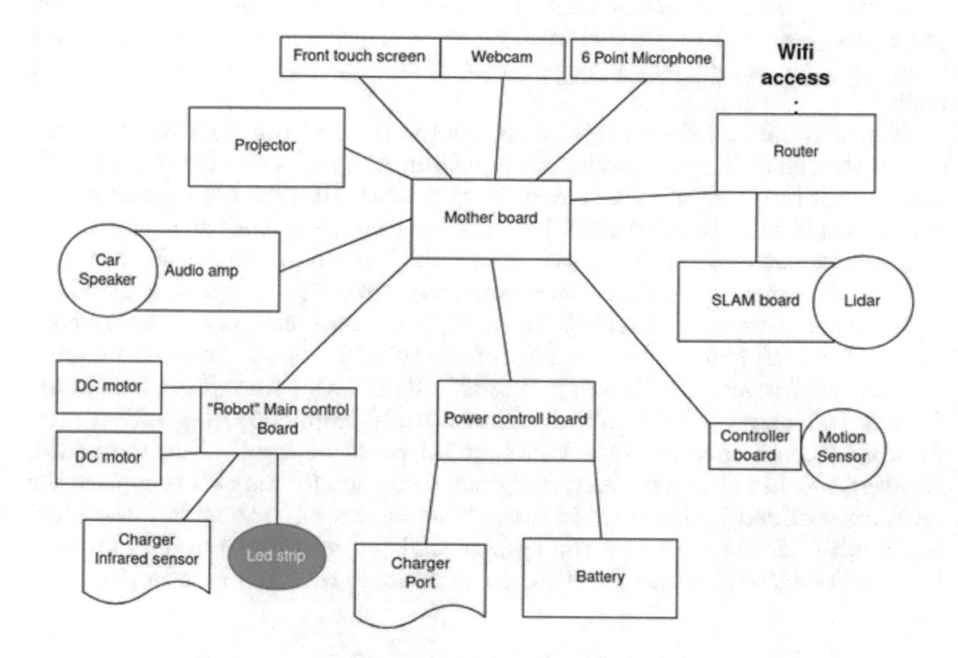

Fig. 3. Component mapping

The following components of the humanoid robot send or receive data streams from sensors:

- Webcam & Microphone
- Lidar and Motion sensor
- Charger infrared sensor
- Touch panel

All data needs to be sent or received through the router as it is the only component of the robot that connects to the outside world. The navigational platform of the robot is fully functional offline.

3.3 Priority Assessment

As outlined by the component mapping, the centre of operations on the CSJBot Snow is an Android based main board. The data streams from the webcam and the microphone as well as the motion sensor and the SLAM radar are all connected to this main board.

The order of which the sensor data integrity was evaluated is in relation to the sensitivity of the produced data streams. A live audio/video stream is by nature more privacy sensitive than e.g. a proximity sensor [5]. The navigation functionality uses the LIDAR sensor to map the surroundings. There have been publications related to possible infringements of privacy by utilizing the data from the lidar sensors [13], however proof of concept papers is hardly proof of malice.

The webcam and the microphone sensors were ruled the most sensitive sensors of the robot. The connectors of the microphone and web camera were both proprietary in the form of a split eDP connector. Reverse engineering of the wirings could have been possible, but the replacement of both the camera and microphones was deemed the easier approach. The proprietary eDP connector used for the camera, audio and touch panel was not compatible with some of the replaced components of the robot, for example the new main board. As a result, the touch screen and display assembly of the robot needed to be swapped out.

The main board is a Rockchip RK3399e IOT device from Shenzhen Smart Device Technology Co.. While the hardware is proprietary, theoretically an Android Open-Source Project (AOSP) ROM could be installed on the board, however, we did not find the necessary documentation for the part to replace the software provided by the manufacturer. Instead, we chose to replace the main board with an Ubuntu based x86 system. This ensures that the data streams from the replaced sensors are not fed to an opaque software platform (Fig. 4).

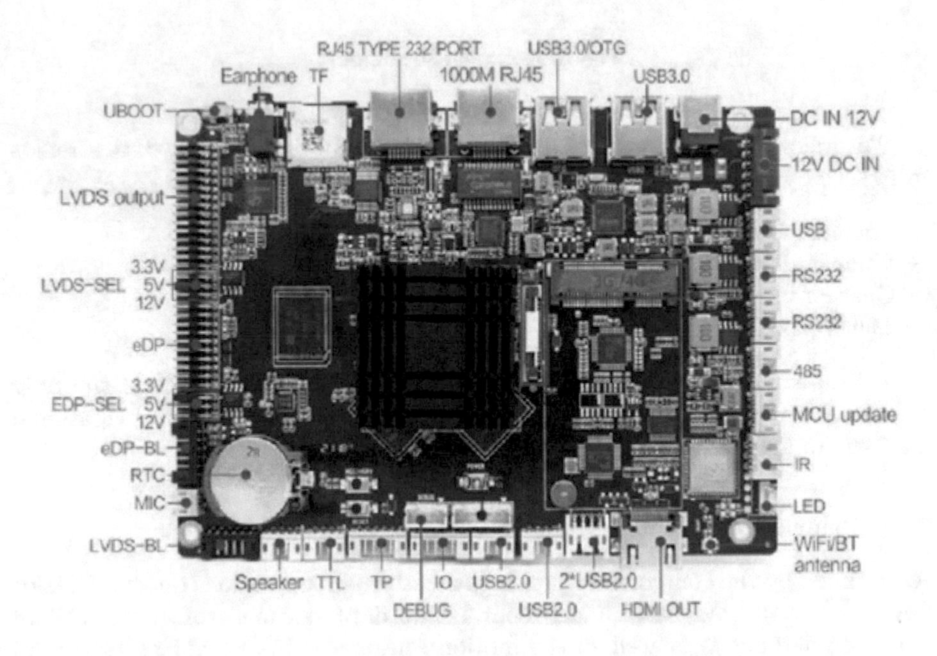

Fig. 4. Proprietary main board

Out of the original components, the motherboard, webcam, microphone and touch interface, was also replaced. The router and SLAM boards were "off the shelf" products and not considered high risk. As of this moment the small bi-directional arms (up, down) are no longer functional. The humanoids facial features, projected on a curved plastic dome were dependant on the Android based motherboard. To achieve modularity like that found in other humanoid service robots [9] the facial features were implemented on a Raspberry Pi 4. New software for the facial features was developed and implemented as part of a bachelor thesis as the original proprietary software used was not available for modification [6] (Fig. 5).

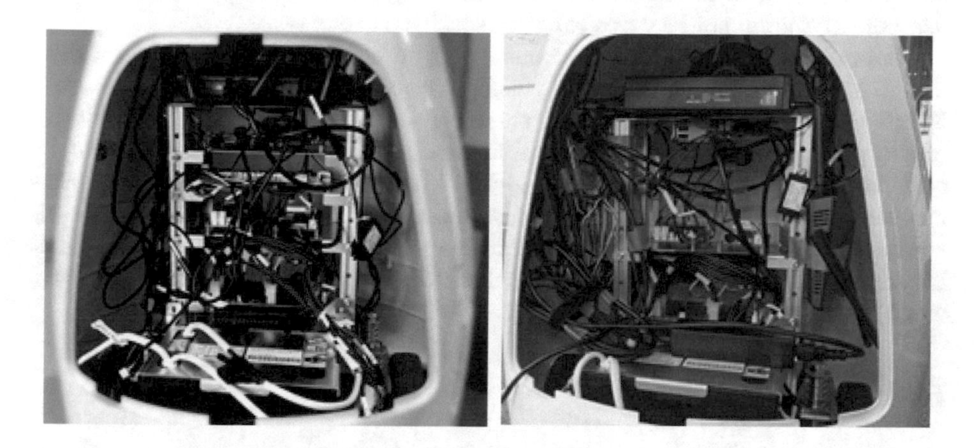

Fig. 5. Snow before (left) & after (right) conversion to open-source hardware

3.4 Data Storage and Integrity

The transmission of data from the robot to servers for processing is problematic from a privacy standpoint. Data can be intercepted, and the endpoints may be insecure. "Data security in the cloud computing is more complicated than data security in the traditional information systems" [14]. A deeper review of the endpoints was not possible since the manufacturer does not grant access to the endpoints or the source code of the software running on them. By hosting services in-house on local servers, we maintain control of the data generated by the sensors.

Several sensors generate data that can be analysed in real time. Real time data analysis or data analysis on streams is less problematic since the data is not stored. One popular example of this is a local real-time object detection algorithm YOLOv7 [15]. Data that is not stored does not need a data management plan or consent from the user.

4 Conclusion

There are established guidelines on how to design robots with privacy in mind, the concept of privacy by design was presented by Ann Cavoukian as early as 2009 [3]. While the concept is simple, critics say that while clear regulations are good, there are immense challenges when it comes to integration of privacy into practice [12]. Within the field of robotics several attempts have been made to provide hands-on guidelines on how to fulfill the privacy by design principles [7] however the discussion of privacy and transparency in closed-source implementations of social humanoids remains slim. This paper delivers concrete examples on which the physical sensors data stream integrity needs to be verified in the use case of a humanoid service robot in sensitive use cases.

Commercial humanoid service robots can be modified to ensure sensor data integrity with the purpose of deployment in sensitive use cases. All sensor data is not equally sensitive by nature, some sensors were not considered priority and were not modified. The applicability of the implementation is yet to be confirmed. Humanoid robots, when locked behind proprietary software, makes evaluation of data integrity and streams unfeasible. Extensive technical knowledge is required for successful modification as well as validation of the data. Replacing hardware, while strenuous when locked behind proprietary software, can be done to guarantee data privacy.

References

1. del Bianco, V., Lavazza, L., Morasca, S., Taibi, D.: Quality of open source software: the QualiPSo trustworthiness model. In: Boldyreff, C., Crowston, K., Lundell, B., Wasserman, A.I. (eds.) OSS 2009. IAICT, vol. 299, pp. 199–212. Springer, Heidelberg (2009). https://doi.org/10.1007/978-3-642-02032-2_18
2. Cardiell, L.: A robot is watching you: humanoid robots and the different impacts on privacy. Masaryk Univ. J. Law Technol. 247–278 (2021). https://doi.org/10.5817/MUJLT2021-2-5
3. Cavoukian, A.: Privacy by design: the 7 foundational principles (2009)
4. European Parliament, Council of the European Union: Regulation (EU) 2016/679 of the European Parliament and of the Council. https://data.europa.eu/eli/reg/2016/679/oj
5. Gray, S.: Always on: privacy implications of microphone-enabled devices. In: Future of Privacy Forum (FPF) Annual Conference (2016). https://fpf.org/wp-content/uploads/2016/04/FPF_Always_On_WP.pdf
6. Hallik, S.: Sociala robotens ansikte: animerade ansiktsuttryck med js och svg. URN:NBN:fi:amk-2023061924165 (2023)
7. Heuer, T., Schiering, I., Gerndt, R.: Privacy framework for context-aware robot development. Paladyn J. Behav. Robot. **12**(1), 468–480 (2021). https://doi.org/10.1515/pjbr-2021-0032
8. KeyCDN: IP Location Finder (2023). https://tools.keycdn.com/geo?host=47.91.74.100
9. LuxAI: QTRobot for research and development (2023). https://luxai.com/humanoid-social-robot-for-research-and-teaching/

10. Peterson, C.: How I coined the term open source (2018). https://opensource.com/article/18/2/coining-term-open-source-software
11. Robotics, P.: Reachy by Pollen Robotics (2023). https://www.pollen-robotics.com/
12. Spiekermann, S.: The challenges of privacy by design. Commun. ACM **55**(7), 38–40 (2012). https://doi.org/10.1145/2209249.2209263
13. Sriram, S., Xu, C., Jena, M.K., Huang, Y.: Spying with your robot vacuum cleaner: eavesdropping via lidar sensors. In: Proceedings of the18th Conference on Embedded Networked Sensor Systems (SenSys 2020), pp. 354–367 (2020). https://doi.org/10.1145/3384419.3430781
14. Sun, Y., Miao, Z., Li, X., Wang, W.: Data security and privacy in cloud computing. Int. J. Distrib. Sens. Netw. (2014). https://doi.org/10.1155/2014/190903
15. Wang, C.Y., Chen, Y., Tsai, H., Lin, W.J., Wang, M.C., Wu, C.H.: YOLOv7: trainable bag-of-freebies sets new state-of-the-art for real-time object detectors. arXiv (2022). https://doi.org/10.48550/ARXIV.2207.02696
16. Wireshark: Introduction (2023). https://www.wireshark.org/docs/wsug_html_chunked/ChapterIntroduction.html
17. Wisniewski, P.J., Page, X.: Privacy theories and frameworks. In: Knijnenburg, B.P., Page, X., Wisniewski, P., Lipford, H.R., Proferes, N., Romano, J. (eds.) Modern Socio-Technical Perspectives on Privacy, pp. 15–41. Springer, Cham (2022). https://doi.org/10.1007/978-3-030-82786-1_2

Rehabilitation with Humanoid Robots: A Feasibility Study of Rehabilitation of Children with Cerebral Palsy (CP) Using a QTRobot

Ira Jeglinsky-Kankainen[1], Thomas Hellstén[2], Jonny Karlsson[2](✉), and Leonardo Espinosa-Leal[2]

[1] Graduate School and Research, Arcada University of Applied Sciences, Jan-Magnus Janssons plats 1, 00560 Helsinki, Finland
ira.jeglinsky-kankainen@arcada.fi

[2] School of Engineering, Culture and Wellbeing, Arcada University of Applied Sciences, Jan-Magnus Janssons plats 1, 00560 Helsinki, Finland
{thomas.hellsten,jonny.karlsson,leonardo.espinosaleal}@arcada.fi

Abstract. The use of humanoid robots, in the field of rehabilitation, has increased over the last years as the technology has become more sophisticated. The aim of this paper is to explore the possibility of using humanoid robots in rehabilitation for children and youth with cerebral palsy (CP). A total of 9 papers were found with inclusion criteria of children with CP, rehabilitation and humanoid robots. In these 9 papers there were 40 children with reported CP diagnosis, age range from 2.11 to 18 years. We found that humanoid robots tested in real rehabilitation scenarios were NAO, ZORA, MARKO and URSUS. These were mostly used in simple rehabilitation interventions as motivators in different exercises for joint mobility or therapeutic exercises in legs or arms. Our findings show that humanoid robots need to be toughly tested following a complete standard process for planning, experimenting and reporting before being used in rehabilitation. In addition, our analysis shows that QTrobot has a huge potential for the rehabilitation of CP children.

Keywords: Cerebral palsy · children · rehabilitation · humanoid robot · human-robot · interaction

1 Introduction

With the prevalence of 1.5–2.7 per 1000 live birth cerebral palsy (CP) is one of the most common lifelong disabilities. It is caused by a non-progressive brain injury occurring before the age of two in the developmental brain. Motor impairments are always present even if there is a great variation in functioning. Challenges in cognition, perception, sensation, behavior and comorbidities such as epilepsy are common as well as secondary conditions such as musculoskeletal problems and pain [1]. CP has traditionally been classified in relation to the pathology and aetiology of the impairment [1], with the subtypes spastic CP (tetraplegic, diplegic, hemiplegic or bilateral, unilateral), dyskinetic CP,

M. E. Auer et al. (Eds.): STE 2024, LNNS 1028, pp. 390–400, 2024.
https://doi.org/10.1007/978-3-031-61905-2_38

ataxic CP and mixed CP. However, for these children, it is not pathology and etiology that determines their everyday life, but function, activity and participation. The difficulties faced by people with CP in their daily lives have led to the development of classifications that reflect their functioning in everyday life [2]. Gross motor function is classified using the Gross Motor Function Classification System (GMFCS), a five-level scale based on everyday functional performance, use of assistive devices and the quality of movement. Level one describes the least and level five the most impaired gross function level [3]. Likewise, hand function is categorized using the Manual Ability Classification System (MACS) [4], while communication is assessed through the Communication Function Classification System (CFCS) [5]. These three classification systems together capture the intricate nature of the condition.

Due to their complex and often heterogeneous clinical presentation, as well as functional status changing over time, many of these children need lifelong care and rehabilitation [6]. Rehabilitation is defined by WHO as "a set of interventions designed to optimize functioning and reduce disability in individuals with health conditions in interaction with their environment" [7]. Rehabilitation is a planned process based on the needs and goals of the rehabilitated person, who with the support of professionals strives to be as independent as possible in everyday activities [8]. Rehabilitation also helps the individual to participate in meaningful activities in daily life. In rehabilitation functioning is usually defined as an individual's physical, mental and social capability to manage meaningful and essential everyday activities at home, at work, at school and during leisure time [8] In a study on evidence for various interventions for children with CP, robotics was found to have weak to moderate evidence for training hand function, passive ankle mobility, and cognitive training [9].

In studies on robotics and rehabilitation for children with CP, various forms of robots are present [10]. Considering that the goal of rehabilitation often focuses on activity and participation, and that children with CP may face multifaceted challenges that make participation challenging, it may be justified to use humanoid robots within rehabilitation.

Rehabilitation e.g. physiotherapy has traditionally been a hands-on profession, however use of technology in the field of rehabilitation has increased due to the COVID-19 pandemic [11] and may be as effective as traditional rehabilitation for certain groups, such as for people with neurological diseases [12], musculoskeletal conditions [13] as well as heart and lung diseases [14]. Technology in healthcare, such as robotics can be implemented by health care professionals at a clinic or independently by rehabilitee at home [15]. Advantages for rehabilitee to use technology independently at home instead of having conventional rehabilitation in a clinic, is that the rehabilitee does not need to travel to clinic, thus saving time and travel costs. Another positive consequence is that the rehabilitee can decide for themselves when to perform their therapeutic exercise and it is easier to implement the exercise into their daily activity [15, 16]. However, there are also barriers that exclude wider use of technology in rehabilitation. These include the professional competence in using technical equipment, resistance to technology in health care and technical investment costs [17].

Thanks to the advances in artificial intelligence and industry 4.0 [18] there has been a revived interest in the use of robotic entities in the field of rehabilitation. One subset of these are the humanoid robots that thanks to their human-like features, make them more trustable by the final users for rehabilitation activities. New AI algorithms with human accuracy level for computer vision or speech recognition and more recently new socioeconomical changes such as the low interest in healthcare professional carriers or others global events like the recent COVID- 19 pandemic have opened the door for the exploration of new technologies for rehabilitation.

The aim of this study was to explore the possibility of using humanoid robots in rehabilitation for children and youth with CP. Our goal was to analyze the technical capabilities of these and compare them with the ones in the QTRobot humanoid robot, that has been shown to be highly efficient in the treatment of autistic children [19]. We will analyze the state of the art in the field. Our research questions were: 1) what kind of rehabilitation is realized with the humanoid robot? 2) what kind of human robots are used and how are they used?

2 Methodology

We used a scoping review, following the JBI methodology for scoping reviews [20] as well as the five-step process outlined by Arksey and O'Malley in 2005 [21]. These steps encompass the formulation of a research question, the search for suitable studies, the selection of pertinent studies, the charting of key data points, and the appropriate presentation of the findings. The final step involves summarizing and articulating the significant findings in the context of the study's objective. The reporting was done using the Preferred Reporting Items for Systematic Reviews and Meta-Analyses extension for Scoping Reviews (PRISMA-ScR) [22] Scoping review was chosen as we wanted to explore the breadth and depth of the topic, identify knowledge gaps and to develop a research agenda [23].

2.1 Search Strategy

In this work we have used a collection of search string agreed among the authors given their expertise in each subfield: Physical therapy, CP, Children rehabilitation and Robotics and artificial intelligence. These keys were combined in one search string and used in five different scientific databases: SpringerLink, ScienceDirect, ACM, IEEE and Google Scholar. The temporal range used was from 2013 onwards. The non relevant papers for this study were discarded by checking the title, abstract and keywords (See Table 1).

2.2 Study Selection

Table 1 and Fig. 1 shows the study selection. Inclusion criteria for the studies were studies published 2013–2023, rehabilitation for children with CP and a humanoid robot involved. Studies where a comprehensive literature search of databases returned 1399 records. These articles were screened for title and abstract. In this phase, 1318 were excluded; the main criteria for exclusion in this phase were articles not specifically targeting children with CP, rehabilitation and humanoid robots. After removing duplicates 81 peer reviewed papers were assessed for full text.

Table 1. Search string and records found in the five different databases scanned. The number of papers discarded was 94% of the total.

Search string	Database					Discarded
	Google Scholar	SpringerLink	Science Direct	ACM	IEEE	
("CHILD-ROBOT INTERACTION" OR "CHILD ROBOT INTERACTION" OR "HUMANOID ROBOT") AND ("CEREBRAL PALSY") AND ("REHABILITATION" OR "PHYSICAL THERAPY")	1100	234	27	32	6	1318 (94%)

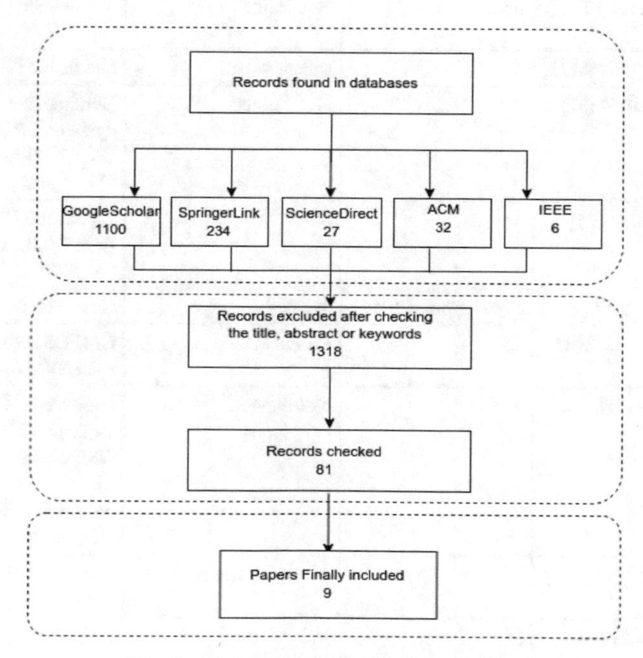

Fig. 1. The search and selection process.

2.3 Data Analysis

Following the guidelines set by Arksey and O'Malley (2005) [21], the results are presented in a two-part format. Initially, fundamental data related to the characteristics of the study were charted. In the subsequent phase, the data were thematically organized to address the aim of the study. Pertinent data were extracted from the articles included in the study and then categorized into themes. Figure 1 shows a diagram of the selection process.

3 Results

Based on the screening of the title and abstract, we left 9 papers for full-text screening. After removing duplicates nine papers were critically assessed and included in our analysis. In these nine papers there were 40 children with reported CP diagnosis, age range from 2.11 to 18. In three studies CP- sub-group was mentioned, and classification in four (Table 2).

Table 2. Characteristics of reported children with CP

Author	Children with CP, n	Age	Sub-group and classifications
Suárez Meijías et al. (2013) [29]	6*	Not defined	Not defined
Kozyavkin et al. (2014) [33]	6	Range 4–6	Not defined
Rahman et al. (2015) [32]	2	Not defined	Hemiplegic CP (n = 1), dystonic CP (n = 1) GMFCS II (n = 1), IV (n = 1)
Buitrago et al. 2017 [25]	1	8	Dyskinetic CP GMFCS III, MACS III, CFCS III
Gnjatovic´ et al. 2017 [27]	5 (of reported 13**)	Mean 8.4 (range 5–13)	Not defined
van den Heuvel et al. 2017 [28]	13	Range 2.11–18	GMFCS II (n = 13), III (n = 2), IV (N = 2)
Chen et al. 2018 [31]	7	Not defined	Hemiplegic CP (n = 5), Diplegic CP (n = 2) GMFCS I (n = 5), III (n = 2) MACS I (n = 1), II (n = 4), III (n = 2)
Martín et al. 2020 [26]	3	Mean 7.6 (range 7–9)	Dystonic CP, hemiplegic CP, nd
Butchart et al. 2021 [24]	3	Range 10–11	GMFCS II (n = 1), III (n = 2) MACS I (n = 3) CFCS I (n = 3)

* In the article it is reported children with CP and Brachialis paresis, however not how many of these were children with CP, ** In the article 13 children with CP and other physical disabilities were mentioned, of which 5 had CP.

3.1 Humanoid Robots Used for Rehabilitation

The findings of this research showed that humanoid robots used in rehabilitation of children with CP were mainly the NAO robot [24–26], MARKO [27], ZORA [28] and URSUS [29]. These robots were used mainly as a motivational companion and the technical analysis was usually performed using external sensors such as cameras.

NAO is a small humanoid designed for human interaction. It is equipped with sensors, including touch sensors, cameras, microphone etc. It is also equipped with several motors and joints making it able to move its head, legs and arms, and thus it is also able to make different types of gestures. NAO can interact with humans through speech and facial expressions. NAO can have several roles during a therapy session. In [24] different roles were defined, including demonstrator, motivator, companion and coach. Due to the advanced motoric capabilities of NAO, it is well suited for physically showing how different types of exercises should be performed. However, NAO is not suitable for imitating all types of physical movements. As stated in [25], NAO has some physical limitations and limited mobility compared to humans and therefore it is not ideal to use NAO for demonstrating to children with CP how walking should be correctly performed. NAO has great capabilities to communicate verbally with children and thus it has been widely tested for providing verbal encouragements and feedback before, during and after therapy sessions, thus making it a good companion and coach.

External camera sensors have also been utilized in some test setups for analyzing the pose of a CP patient. In Martin et al. [26], a therapy setup is presented where the NAO robot first shows a series of predefined poses to the patient including specific positions of upper extremities where the arms are bent in specific angles. Then, NAO verbally asks the children to perform the same upper body posture. The robot then analyses the postures of the children, with the help of an external Kinect sensor, by comparing the joint angles to the predefined angles. If the children have performed the posture correctly, the NAO robot proceeds to the next exercise.

The ZORA humanoid robot is physically the same as NAO but uses unique software tailored for healthcare, education and customer service. In van den Heuvel et al. [28] ZORA was involved in five intervention sessions, including movement exercises, dance exercises, robot control and cognitive exercises.

MARKO is a conversational humanoid robot, developed at the University of Novi Sad, as a tool for assisting treatment of children with developmental disorders [30]. In appearance, MARKO is similar to NAO with several movable joints in the head, arms and legs making it possible to imitate several types of therapeutic exercises. MARKO is though tailor-made for participating in rehabilitation of children, while NAO is a general-purpose humanoid. In addition to imitating therapy-relevant exercises, MARKO can also generate basic emotional facial expression and autonomously engage in natural language dialogues. In the research papers selected for this study, MARKO was used only in one experimental setting with children with CP [27] but due to the sensitivity of the presented study MARKO was strictly controlled by a human operator. However, the autonomous verbal communication feature provides great potential for MARKO to be included in treatment of children with CP without strict involvement of human therapists in the future.

URSUS can be considered the most advanced platform for the rehabilitation of children with CP. Its design is humanoid-like but the appearance is of a teddy bear. Its technical capabilities are to some degree superior to the previous described humanoid robots because it can not only act as a motivational companion, but also can be used to analyze the movements performed during the therapy session using its integrated technologies such as RGB-D camera and Speech recognition. The main disadvantage of URSUS is that it is not a commercial robot, therefore the potential use in other cases is limited.

3.2 Rehabilitation with Humanoid Robots

Exercise of body function was mostly used in the articles included in this study. Either focused on lower extremities or upper extremities. Exercises in lower extremities included hip and knee extension exercises or hip abduction exercises [24, 28]. Exercises in upper extremity or head included exercises in shoulder, elbow, hand [26, 27, 29, 31] and head [27, 30]. In one study, upper extremity training was performed through a game projected via a webcam, and the robot was used as a supporter, giving directives and cheering [31]. In one study cognitive exercises were included in the form of a card game [30]. Activity was practiced through different exercises. Walking was practiced by repeating what the robot did [25], or by a goal-directed protocol where the child started from a sitting position then walked to the robot where he touched the robots head, who then gave feedback and asked the boy to walk back to the chair [25]. Sit-to stand was also practiced by standing up and touching the head of the robot, which was standing on a table. Balance was practiced by imitating the robot, lifting one leg for 10 s and kicking a ball [32]. One study mentioned nine exercise scenarios but did not specify what these were. The robot was used as a starter of the training, by introducing itself and telling what kind of training is coming up [33].

Only two of the included study assessment methods were reported [29, 31]. To assess finger dexterity a Nine-Hole Peg Test (9HPT) was used with the URSUS humanoid robot [29], shoulder and elbow range of motion, movement time and number of movements was assessed with computer vision combined with a humanoid robot [30]. In all studies, verbal instructions were used to facilitate, motivate and encourage the child to do the exercise during the rehabilitation session [24–27, 29–33]. In some of the studies the therapist assisted the children in their exercise contact with the robot.

4 Discussion

Our study showed that integration of robots of different kinds in rehabilitation of children with CP has been considered in a wide range of research papers, but only a limited number of studies is reporting on experiments with specifically humanoid robots in real rehabilitation test setups. The dominant humanoid in these test setups has been NAO and its version with healthcare specific software, ZORA, but also other humanoids have been used, including MARKO and URSUS.

The robots were mainly used to do different exercises that trained joint mobility or certain functions in the legs or arms. More functional movements were trained by rising from a chair, balance or walking. Most of the children in the studies were GMFCS levels I–III, i.e. children with relatively good function, moving either unaided or with walking aids. The selected movements are well suited to these children but given the multifaceted difficulties a child with CP may have, it was surprising that there were not more children with severe disabilities. A shortcoming of the studies was that there were few children, and the description of the children was rather incomplete, which makes it difficult to relate the function of the children in relation to the training performed.

In most cases, the robot had an important role in encouraging children to exercise, showing examples of movements or standing by to cheer or motivate. In view of that humanoid robots seem to be suitable for rehabilitation. They can perform versatile movements that children can easily imitate, and they have expressive faces, so they easily get the attention from children, and thus children may find it easy to communicate with them.

In only one specific case, NAO was used in combination with an external Kinect sensor for analyzing joint angles of the upper extremities to examine whether the children performed the rehabilitation exercises in the correct way. This is an interesting way of integrating humanoids in rehabilitation as it enables them to not only encourage children and instruct them what to do but also to potentially provide feedback without the intervention of a human therapist. However, the camera sensor and computational features of humanoid robots have so far not been widely used for these kinds of purposes. Thus, an interesting field of future research is to analyze how e.g. computer vision and machine learning can be effectively used for providing the capability of a humanoid to understand the exercises a child needs to perform and to supervise children in a more autonomous manner.

4.1 The QTrobot

Recently, LuxAI (www.luxai.com) a European company, has launched on the market a new kind of humanoid robot specially designed for targeting children with special needs. QTrobot is an expressive humanoid social robot flexible, open source, with cutting-edge technology that can potentially be used in the rehabilitation of children with CP. The review analysis of the scientific literature done in this work, shows that humanoid robots are used mainly as motivators during the therapy sessions, and the analysis of the movement via, e.g., computer vision is done via external technical setups. QTrobot can be used in both cases because it has integrated an RGB-D camera for pose estimation analysis and speech recognition capabilities similar to the URSUS robot, however as mentioned before, URSUS is not available commercially, so this limits the use in different scenarios. Furthermore, the main advantage is that applications can bc developed easily as QTrobot runs on the Ubuntu open-source operating system.

5 Conclusion

Implementing humanoid robots in rehabilitation of children with CP in real rehabilitation interventions is scarce in the scientific literature. Our findings show that most used humanoid robots tested in real scenarios are the NAO (healthcare specific software),

ZORA, MARKO and URSUS humanoid robots. More specifically, humanoid robots are used in simple rehabilitation interventions as motivators in different exercises for joint mobility or therapeutic exercises in legs or arms. Only in few studies more functional movements were trained as rising from a chair, balance or walking. Our findings in this paper are twofold. First, before humanoid robots are implemented in rehabilitation these need to be toughly tested, and more importantly, it is highly important to stablish a complete standard process for planning, experimenting and reporting that integrates the three core elements: CP specialists, physiotherapists and technical support. This will ensure scientific reproducibility. These results may help organizations (healthcare and technology) in developing, planning and implementing humanoid robots effectively in everyday work. Secondly, our technical analysis shows that QTrobot has a strong potential for the rehabilitation of children with CP due to its technical and social capabilities. Future research directions include the technical development of physiotherapy exercises and testing on children with CP using one QTRobot.

References

1. Baxter, P., et al.: The definition and classification of cerebral palsy. Dev. Med. Child Neurol. **49**, 1–44 (2007)
2. Rosenbaum, P., et al.: A report: the definition and classification of cerebral palsy April 2006. Dev. Med. Child Neurol. Suppl. **109**, 8–14 (2007)
3. Wood, E., Rosenbaum, P.: The gross motor function classification system for cerebral palsy: a study of reliability and stability over time. Dev. Med. Child Neurol. **42**(5), 292–296 (2000)
4. Eliasson, A.C., et al.: The Manual Ability Classification System (MACS) for children with cerebral palsy: scale development and evidence of validity and reliability. Dev. Med. Child Neurol. **48**(7), 549–554 (2006)
5. Hidecker, M.J.C., et al.: Developing and validating the Communication Function Classification System for individuals with cerebral palsy. Dev. Med. Child Neurol. **53**(8), 704–710 (2011)
6. Rosenbaum, P., Gorter, J.W.: The 'F-words' in childhood disability: i swear this is how we should think! Child Care Health Dev. **38**(4), 457–463 (2012)
7. World Health Organization. Rehabilitation competency framework (2020). https://iris.who.int/bitstream/handle/10665/338782/9789240008281-eng.pdf?sequence=1. Accessed 27 Nov 2023
8. National Criteria for Referring People to Medical Rehabilitation 2022 Guide for Healthcare and Social Welfare Professionals and Those Working in Rehabilitation Services. Ministry of Social Affairs and Health. http://urn.fi/URN:ISBN:978-952-00-5423-6. Accessed 25 Nov 2023
9. Novak, I., et al.: State of the evidence traffic lights 2019: systematic review of interventions for preventing and treating children with cerebral palsy. Curr. Neurol. Neurosci. Rep. **20**, 1–21 (2020)
10. Miguel Cruz, A., Rios Rincon, A.M., Rodriguez Dueñas, W.R., Quiroga Torres, D.A., Bohórquez-Heredia, A.F.: What does the literature say about using robots on children with disabilities? Disabil. Rehabil. Assist. Technol. **12**(5), 429–440 (2017)
11. Rausch, A., Baur, H., Reicherzer, L., et al.: Physiotherapists' use and perceptions of digital remote physiotherapy during COVID-19 lockdown in Switzerland: an online cross-sectional survey. Arch. Physiotherapy **11**(1), 1–10 (2021)

12. Rintala, A., Päivärinne, V., Hakala, S., et al.: Effectiveness of technology-based distance physical rehabilitation interventions for improving physical functioning in stroke: a systematic review and meta-analysis of randomized controlled trials. Arch. Phys. Med. Rehabil. **100**(7), 1339–1358 (2019)
13. Cottrell, M.A., Galea, O.A., O'Leary, S.P., Hill, A.J., Russell, T.G.: Real-time telerehabilitation for the treatment of musculoskeletal conditions is effective and comparable to standard practice: a systematic review and meta-analysis. Clin. Rehabil. **31**(5), 625–638 (2017)
14. Hakala, S., Kivistö, H., Paajanen, T., et al.: Effectiveness of distance technology in promoting physical activity in cardiovascular disease rehabilitation: cluster randomized controlled trial, a pilot study. JMIR Rehabil. Assistive Technol. **8**(2), e20299 (2021)
15. Salminen, A., Hiekkala, S., Stenberg, J.: Etäkuntoutus. Kansaneläkelaitoksen julkaisuja. Juvenes Print, Tampere (2016)
16. Capecci, M., Ceravolo, M.G., Ferracuti, F., et al.: A Hidden Semi-Markov Model based approach for rehabilitation exercise assessment. J. Biomed. Inform. **78**, 1–11 (2018)
17. Scott Kruse, C., Karem, P., Shifflett, K., Vegi, L., Ravi, K., Brooks, M.: Evaluating barriers to adopting telemedicine worldwide: a systematic review. J. Telemed. Telecare **24**(1), 4–12 (2018)
18. Espinosa-Leal, L., Chapman, A., Westerlund, M.: Autonomous industrial management via reinforcement learning. J. Intell. Fuzzy Syst. **39**(6), 8427–8439 (2020)
19. Nazari, A., Höhn, S., Paikan, A., Ziafati, P.: Receptive language development diagnosis and tracking in conversational interactions with QTrobot for Autism. In: Proceedings of the First Workshop on Connecting Multiple Disciplines to AI Techniques in Interaction-Centric Autism Research and Diagnosis (ICARD 2023), pp. 12–16 (2023)
20. Peters, M.D., Godfrey, C., McInerney, P., Munn, Z., Tricco, A.C., Khalil, H.: Scoping reviews. In: JBI Manual for Evidence Synthesis, vol. 169, no. 7, pp. 467–473 (2020)
21. Arksey, H., O'Malley, L.: Scoping studies: towards a methodological framework. Int. J. Soc. Res. Methodol. **8**(1), 19–32 (2005)
22. Tricco, A.C., et al.: PRISMA extension for scoping reviews (PRISMA-ScR): checklist and explanation. Ann. Intern. Med. **169**(7), 467–473 (2018)
23. Munn, Z., et al.: What are scoping reviews? Providing a formal definition of scoping reviews as a type of evidence synthesis. JBI Evid. Synth. **20**(4), 950–952 (2022)
24. Martí Carrillo, F., et al.: Adapting a general-purpose social robot for paediatric rehabilitation through in situ design. ACM Trans. Hum.-Robot Interact. (THRI) **7**(1), 1–30 (2018)
25. Buitrago, J.A., Bolaños, A.M., Caicedo Bravo, E.: A motor learning therapeutic intervention for a child with cerebral palsy through a social assistive robot. Disabil. Rehabil. Assistive Technol. **15**(3), 357–362 (2020)
26. Martín, A., Pulido, J.C., González, J.C., García-Olaya, Á., Suárez, C.: A framework for user adaptation and profiling for social robotics in rehabilitation. Sensors **20**(17), 4792 (2020)
27. Gnjatović, M., et al.: Pilot corpus of child-robot interaction in therapeutic settings. In: 8th IEEE International Conference on Cognitive Infocommunications, pp. 253–258 (2017)
28. van den Heuvel, R.J., Lexis, M.A., de Witte, L.P.: Robot ZORA in rehabilitation and special education for children with severe physical disabilities: a pilot study. Int. J. Rehabil. Res. Internationale Zeitschrift fur Rehabilitationsforschung. Revue internationale de recherches de readaptation **40**(4), 353 (2017)
29. Suárez Mejías, C., et al.: Ursus: a robotic assistant for training of children with motor impairments. In: Pons, J., Torricelli, D., Pajaro, M. (eds.) Converging Clinical and Engineering Research on Neurorehabilitation. BIOSYSROB, vol. 1, pp. 249–253. Springer, Heidelberg (2013). https://doi.org/10.1007/978-3-642-34546-3_39
30. Borovac, B., Rakovic, M., Savic, S., Nikolic, M.: Design and control of humanoid robot MARKO: an assistant in therapy for children. In: Proceedings of the International Exploratory Workshop on New Trends in Medical and Service Robotics-MESROB (2014)

31. Chen, Y., Garcia-Vergara, S., Howard, A.M.: Effect of feedback from a socially interactive humanoid robot on reaching kinematics in children with and without cerebral palsy: a pilot study. Dev. Neurorehabil. **21**(8), 490–496 (2018)
32. Rahman, R.A.A., Hanapiah, F.A., Basri, H.H., Malik, N.A., Yussof, H.: Use of humanoid robot in children with cerebral palsy: the ups and downs in clinical experience. Procedia Comput. Sci. **76**, 394–399 (2015)
33. Kachmar, O., Kozyavkin, V., Ablikova, I.: Humanoid social robots in the rehabilitation of children with cerebral palsy. In: REHAB 2014 (2014)

Towards Seamless Communication for Sign Language Support: Architecture, Algorithms, and Optimization

Kei Yiang Lim[1], Ayan Priyadarshi[1], Nur Farah Nadiah[1], Jun Hao Jeff Lee[1], Jun Xiang Lau[1], Chyou Keat Lionel Chew[1], Peter ChunYu Yau[1], and Dennis Wong[2(✉)]

[1] School of Computing Science, University of Glasgow, Glasgow, UK
{2717838L,2717870P,2717943B,2717906L,2717836L,2717891C}@student.gla.ac.uk, PeterCY.Yau@glasgow.ac.uk
[2] Faculty of Applied Sciences, Macao Polytechnic University, Macao, China
cwong@mpu.edu.mo

Abstract. This study delves into the practical implementations of Computer Vision and Machine Learning within real-world contexts, specifically addressing the facilitation of communication between individuals with speech impairments and those without. The investigation focuses on deploying a learning model integrated with Computer Vision, designed to assimilate input data and generate user-friendly outputs. The refined model is subsequently adapted for seamless integration into over-the-counter transactions, streamlining consumer communication processes. A proposed solution, the Sign Assistance Ready App (SARA), is introduced in this report to address the identified communication gap. Throughout the ensuing sections, the application will be denoted as SARA for brevity and clarity.

Keywords: Computer Vision · Sign Language · Artificial intelligence · Machine Learning · Accessibility

1 Introduction

1.1 Background

Approximately 5% of the global population experiences hearing loss, classified as mild, moderate, severe or profound [1]. While those with mild to severe impairment can benefit from hearing aids, individuals classified as profoundly deaf find challenges in social scenarios [2] like airports where they have trouble communicating with ground staff or in the workplace where they encounter hurdles with potential employers who are sometimes unprepared to accommodate their needs and often overlooking their capabilities.

M. E. Auer et al. (Eds.): STE 2024, LNNS 1028, pp. 401–410, 2024.
https://doi.org/10.1007/978-3-031-61905-2_39

1.2 Objectives

Motivated by the aim of fostering inclusivity and providing opportunities for the deaf/mute community, this report seeks to identify methods for enhancing cohesiveness and collaboration between individuals with and without disabilities. The primary objective is aligned with the sustainable development goal of reducing inequalities. Through extensive research on existing software, artificial intelligence and machine learning models, the team recognizes the importance of interpersonal connections in shaping an individual's growth and aims to tackle this obstacle [3].

2 Literatures

2.1 Sign Language

Non-mute individuals rarely learn Sign Language (SL) [4], causing communication difficulties for those who are mute. This is because Sign Language is not widely taught in many education systems and most people rely solely on the help of translators. This solution is not always practical in most scenarios because of availability and is usually only used in business cases. This is where image recognition technology comes in, as it helps to automate the process of translating SL, eliminating the need for a human translator.

2.2 Computer Vision and Machine Learning for Gesture

Fig. 1. Google Translate recognising words and translating into English.

Computer Vision is the ability to allow computers to perceive the world, like how humans use their eyes [5]. As a powerful tool, computer vision is harnessed by many platforms for its applicability. One example of such application is Google Translate. In Fig. 1. it uses the mobile phone's camera to capture a photograph which will then utilise computer vision to recognise words and letters from the image [6]. Google Translates then processes the image, translates it into text of the same language, and finally translate it to another language of the user's choice. Computer vision uses neural network to recognise patterns for an object that the computer is looking at. However, it needs to be supplemented with training, which is to supply it with bulks of various images for the computer to find a distinctive pattern [7] (Fig. 2).

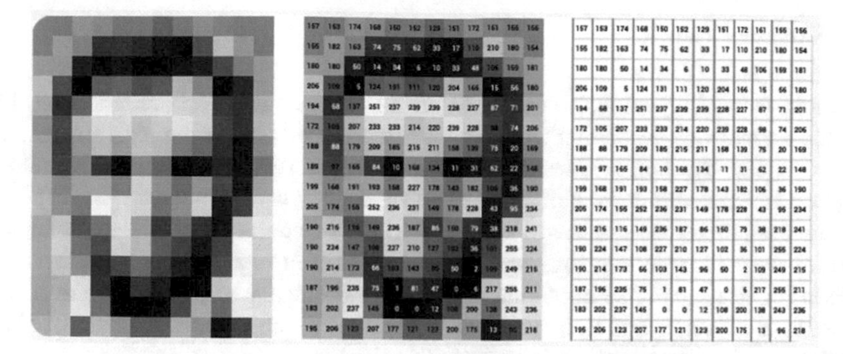

Fig. 2. Computer Vision labelling brightness levels of an image.

When an image is analysed, it is split into values of pixels based on brightness levels [7]. A big jump in values indicates a distinct feature in the image which helps the computer recognise different objects or features within the object. To identify these patterns in values, a neural network is typically required, such as Convolutional Neural Network [8].Once the model is sufficiently trained, the data generated would be of high accuracy for any images detected, producing a catered result to the end user.

2.3 Machine Learning for Gesture

Classifications: Machine learning has become a very prominent technology to include within various applications [9]. It is a subset of computer science that focuses on using models and algorithms to empower computers to make predictions. This can be done through training the machine learning model with past data to "teach" and reaffirm the model's learning about a subject [10].

CNN (Convolutional Neural Network): CNN adopts an algorithm which integrates different levels of classifier in a multi-layer stack. This algorithm's formulation is an adaptation from the feedforward neural network. Most of the filters that are used are 3 × 3 filters and follow similar rules such as having the same output size. It is a popular algorithm for Image classification and the use of CNN has produced a promising result in both ImageNet detection and localisation.

YOLOv3 Model (You Only Look Once v3): YOLOv3 is a deep learning model that is trained to identify objects in images and videos [11]. The YOLOv3 model is a single-shot detection (SSD) model, [12] it processes an entire image in one forward pass through the network. This makes it faster and more efficient compared to other object detection models. One of the key features of YOLOv3 is its ability to predict bounding boxes and class probabilities in a single pass, making it a popular choice for real-time object detection applications [13]. The model uses anchor boxes and anchor k-means clustering to predict multiple objects within an image, and it also uses a feature pyramid network to handle different object scales.

2.4 Existing Products

Wearable-Tech Gloves

Wearable technology gloves, created to aid in sign language communication, have become an innovative solution for instant interaction by converting sign gestures into speech or text through wireless connection to a smartphone providing a real-time communication solution for individuals who are deaf or hard of hearing [14, 15].

However, there are challenges such as possible inaccuracies caused by ambiguous hand movements in certain sign language letters as seen in Fig. 3 [16]. Moreover, the societal implications raise questions about how individuals with hearing impairments will need to adjust to this technology and the potential reinforcement of social biases, highlighting the necessity for education in this field.

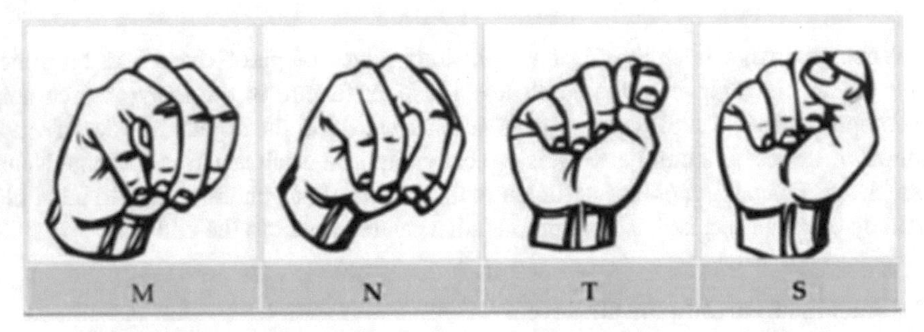

Fig. 3. Hand postures of letter M, N, T, S looks similar at first glance [16]

Sign Language Learning Apps

These applications make use of various features, such as instructional videos, illustrations, assessment, and engage users with games, to offer a comprehensive learning experience [17]. Some of the most well-known sign language learning apps include Duolingo, Cake, and LingoKids on ASL [18].

It is very convenient for users to learn different sign languages such as ASL and BSL at their own pace and on their own time [19]. In [17], the ability to offer tailored learning experiences based on an individual's abilities and progression enables a more productive and impactful method of education. Using engaging content such as media and games these apps make learning more engaging and enjoyable which could increase the effective learning experience [20]. However, these apps are still in their early stages and thus they lack the proper support, feedback [21], and engagement levels for it to be substantially effective which is its clear drawback.

3 Methodology

3.1 Cloud Architecture

In Fig. 4, it is a diagram depicting the cloud architecture we are proposing to use. It all will start from the Gesture Detector Device where it will be provided with an Application Programming Interface (API) for uploading and downloading.

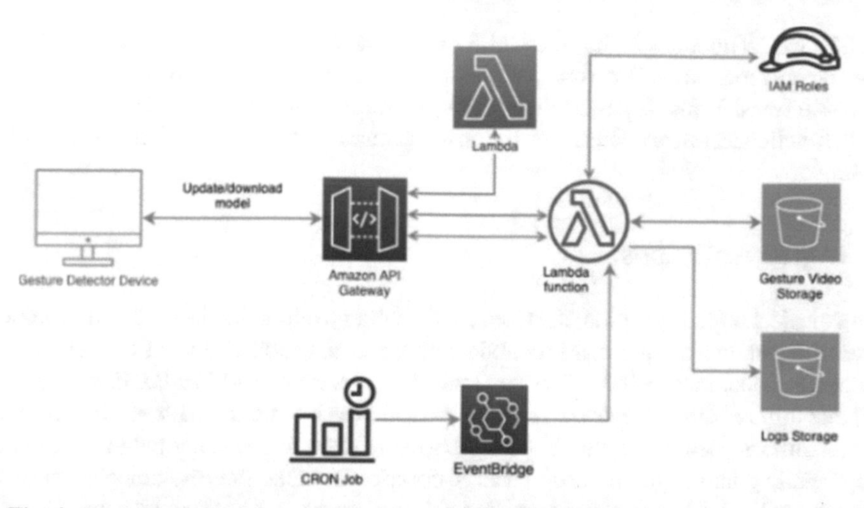

Fig. 4. Real-time handsign interpretation leveraging computational power on the Cloud.

The API will then be interfaced with Amazon API Gateway, to access the cloud server and this will also fetch new trained models daily. The API Gateway will then redirect to the Lambda functions where it will do various jobs. For example, one of the Lambda Function will be storing video data into an S3 Bucket and another Lambda Function will be storing all logs into another S3 Bucket. Cron Job will be the triggering point where it will start a Lambda Function to process or train the data. Lastly, IAM roles will be used to give authorization to the respective Lambda Functions that require it.

3.2 System Development

The team employs the OpenCV library to create a dataset for subsequent model training, comprising still images of the ASL alphabet and video files featuring commonly used ASL phrases. Before creation, a region of interest (ROI) within the camera view window is defined, bounded by x and y coordinates. OpenCV is utilized to map objects separated from the background as much as possible. After filtering, a contour filtering process is applied to extract the hands, eliminating unnecessary noise. The outcome is then saved to either test or train folders.

For the training/testing phase, an 80/20 split is implemented, and the chosen model is a Convolutional Neural Network. Parsing and training are executed using Keras and TensorFlow. Different parameter sets are tested through multiple iterations to determine the set to be further used in the development.

In terms of computer vision input, a webcam captures the vision input with the user positioned in front of the camera, ensuring that the upper body is within the frame. OpenCV captures, transfers, and parses on edge devices, followed by cleaning and processing the video before sending it to the image/video processing algorithm. This algorithm, using object detection, pose estimation, and deep learning, detects, tracks, and recognizes signs in real-time.

The model operates in the cloud, utilizing the server's resources to free edge devices from heavy processing. Once translation is complete, the program considers the top three probable translations. It passes them through an algorithm, powered by TensorFlow's NLP functionalities, to determine the most accurate translation when formed into a sentence.

4 Implementations

In the project implementation, the team used Docker to simulate the cloud environment. There are mainly 5 microservices container being used, which are the API Gateway, Data Upload Service, Fetch Model Service, Train Model Service and MinIO. Each service is callable through the API gateway, and they will also be interacting with MinIO for saving files. MinIO is used to simulate a local S3 bucket. The team have adopted to use python programming language and used various libraries to make the implementation of the project work. The libraries used are Flask, TensorFlow, MediaPipe, NumPy, OpenCV and Boto3.

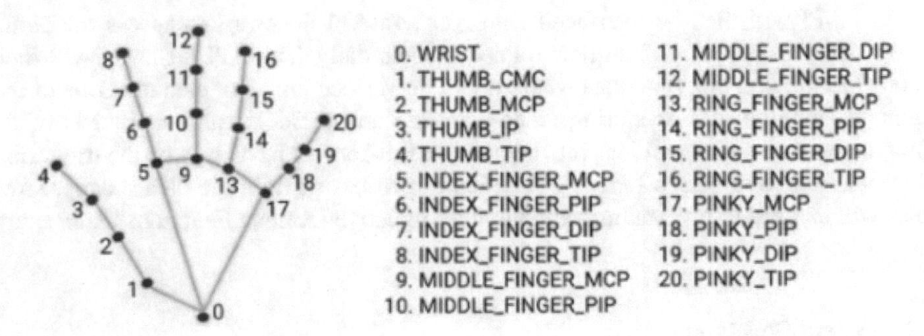

Fig. 5. Landmarks from MediaPipe

Flask is used as the main front-end client, along with OpenCV for communications with the webcam. MediaPipe is used to detect landmarks on the hands (Fig. 5). The team also used TensorFlow for prediction and supervised training of the model. For TensorFlow to work, data needs to be obtained, and the team used NumPy to gather data through manual recording. The data is then uploaded to MinIO. Each data processing is done through the NumPy, and the file is retrieved from MinIO. To access MinIO, Boto3 is required.

4.1 Models Training

The team comprising of five males and one female, aged 22 to 26, manually collected and trained local data on hand signs. Images of hand signs were captured at varying distances – approximately 50cm and 80cm from the camera and under well-lit conditions totalling 20 images for each sign covering both left and right hands. To eliminate noise affecting the capture of signs, the frame of the snapshot consisted only the member and a blank wall, enhancing the hand detection process. The above implementations are referenced from the "ActionDetectionforSignLanguage" GitHub library and YouTube videos by Nicholas Renotte or nicknochnack [22].

4.2 Architecture and Hyperparameters

The Convolutional Neural Networks used was initialised as a linear stack of layers using the *Sequential()* method made available by the *Keras* Deep Learning library. In total, there are nine layers within this model.

1D Convolutional Layers: The first convolutional layer accepts data input of twenty-one by three. It contains sixty-four filter. The second convolutional layer uses one-hundred and twenty-eight filters. Both convolutional layers utilise the *ReLU* activation function and uses has a kernel size of three. **Max Pooling Layer:** The max pooling layer the model uses to reduce spatial dimensions on input accepts a pool size of two.

Flattening Layer: A flattening layer was used to flatten data before directing data into the subsequent layers. **Dense Layers:** Two dense layers were used for this CNN model. The first dense layers utilise sixty-neurons while the second dense layer utilises thirty-two neurons. Both dense layers use the *ReLU* activation function. **Output Layer:** The output layer contains the same number of neurons as the number of pre-saved labels. A *softmax* function was also used to convert probabilities, ensuring that probability score sums to one.

5 User Study

5.1 Approach

The experimental design encompasses an examination of the non-functional requirements associated with the proposed solution. This entails a thorough assessment of accuracy, response time, and the complexity of the environment in which the solution operates. Specifics regarding the experiments will be expounded upon in the Research Settings and Participants section. Additionally, user studies constitute a pivotal evaluation method aimed at assessing the satisfaction levels of users with the solution. The particulars of this study will also be detailed in the Research Settings and Participants section.

5.2 Research Settings and Participants

The experimental design encompasses various aspects, including the measurement of accuracy, response time, and the level of complexity within the background environment. All experiments are conducted in a controlled setting with minimal interference.

For the accuracy experiment, a single researcher (User A) performs a series of sign languages as provided in a script, utilizing the proposed solution. Three sets of ten repetitions each are recorded for research purposes, documenting the occurrences of false positives, true positives, false negatives, and false positives.

Simultaneously, the environment complexity experiment is carried out in conjunction with accuracy testing. This involves conducting the experiment in three distinct background environments: a static white background, a background with three people walking past (representing a quiet environment), and a background with an uncontrolled number of people walking (representing a busy public environment). The accurate results are then analyzed in relation to the complexity of the background environment.

Additionally, the response time experiment is integrated into the environment complexity and accuracy testing. Using a stopwatch, the researcher measures the time between User A signing to the camera and the appearance of the predicted translation on the user interface. The response time is recorded for each iteration of the experiment.

5.3 Preparations

The preparations for the experiments include the procurement of test equipment, inviting participants and booking of test location. The environment used for the experiment will be a discussion room available in Singapore Institute of Technology, Nanyang Polytechnic Campus. Booking will be done through the school portal. Equipment used for this experiment includes a laptop with a webcam. This laptop will be used throughout the experiment and user study to prevent camera quality from skewing results. Participants for the user studies will include 10 random students from the Singapore Institute of Technology. Five males and five females' participants will be invited to ensure a gender-neutral experiment.

5.4 Setup

The experiment will be conducted in one session. This ensures that the researcher's performance will not vary due to confounding variables. The user study will also be conducted within one session. Each participant will be cordially invited into the discussion room one at a time. Participants will be requested for their consent. The user study will only proceed if participants agree to the information provided. Participants will only be allowed to perform the study once to prevent carryover and fatigue effects.

5.5 Data Collection and Analysis Procedures

In this experiment, data is collected to assess the accuracy of the sign language interpretation and the truthfulness of the predictions. The accuracy data is then organized into a matrix to determine the overall precision of the solution, providing a quantitative measure of its effectiveness. Similarly, for response time evaluation, the recorded data encompasses the time it takes for the translation to be displayed from the moment the user initiates signing. This measurement serves as an indicator of the solution's speed in providing timely translations. Considering the diverse and complex environments where the application is intended for use, we conduct experiments in various settings, including work environments and public spaces. This approach aims to test the adaptability and robustness of the application in real-world scenarios. To gauge user satisfaction, a survey is conducted, wherein participants are asked to rate their level of satisfaction with the procedures involved in using our application. The rating scale ranges from 1 to 5, with 1 representing 'Very Dissatisfied' and 5 indicating 'Very Satisfied'. This survey involves the same group of 10 student volunteers over a one-month period, coinciding with the implementation of our application.

6 Results

The results obtained from the experiments indicated an average response time of 37.4ms with the CNN implementation with 82.76% accuracy. In each set, the results of the response time for each increase as complexity increases. As for accuracy, the team used 40 epoch iterations and as seen in Fig. 6 CNN has a high accuracy to identify new patterns and that the model is not overfitting the training data.

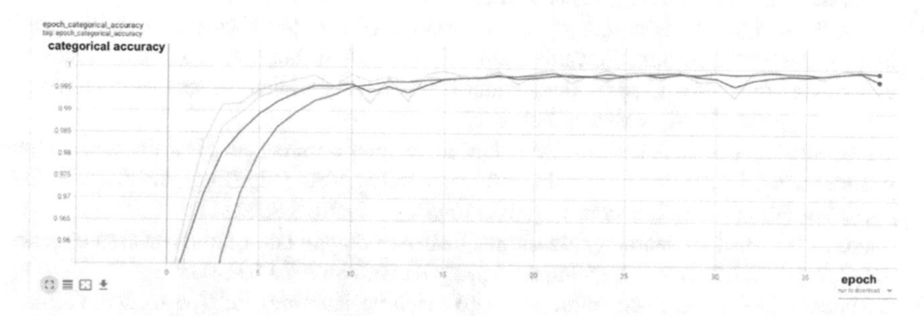

Fig. 6. CNN Epoch Categorical Accuracy Chart (Legend: Orange is training, Blue is testing)

7 Discussions and Conclusions

Based on the methodology aforementioned, the proposed solution, SARA, utilises machine learning and computer vision technologies to effectively translate hand sign languages to display the computed results. In this study, the team has explored the application of SARA and further tested its capabilities against our methodology. Nonetheless, there is scope for enhancing this solution. The team wishes to explore the possibility of utilising YOLOv3 model instead of CNN and test out the benchmarks against CNN to understand and experiment with its effectiveness and accuracy. With continued refinement, SARA has the potential to revolutionize and improve accessibility for the speech impaired in myriad environments.

Acknowledgements. The research is supported by the Macao Polytechnic University research grant (Project code: RP/FCA-05/2023).

References

1. "Deafness and hearing loss," Deafness and hearing loss (2021). https://www.who.int/newsroom/fact-sheets/detail/deafness-and-hearing-loss. Accessed 09 Feb 2023
2. "Difficulties the Deaf Face Every Day." Disability Experts of Florida. www.disabilityexpertsfl.com/blog/difficulties-the-deaf-face-every-day. Accessed 12 Feb 2023
3. "Deafness: The Invisible Disability." HandTalk. www.handtalk.me/en/blog/deafness-invisible-disability/. Accessed 12 Feb 2023

4. Nanda, C., Tuteja, T., Manimozhi, M.: Sign language recognition for deaf and mute people. Int J Pharm. Bio. Sci **2016**, 48–51 (2016)
5. Robin, J., Soni, M.R., Dubey, R.R., Datkhile, N.A., Kolap, J.: Computer vision for hand gestures. In: 2020 International Conference on Convergence to Digital World - Quo Vadis (ICCDW), Mumbai, India, pp. 1–4 (2020). https://doi.org/10.1109/ICCDW45521.2020.9318682
6. Schuster, M., Papineni, N.A.: How Google translate squeezes deep neural networks onto a phone. Google AI Blog (2015). https://ai.googleblog.com/2015/07/how-google-translate-squeezes-deep.html?m=1. Accessed 12 Feb 2023
7. Salek, S., Alharbi, A.: Everything you ever wanted to know about computer vision: here's a look why it's so awesome. Towards Data Science (2019). https://towardsdatascience.com/everything-you-ever-wanted-to-know-about-computer-vision-heres-a-look-why-it-s-so-awesome-e8a58dfb641e. Accessed 12 Feb 2023
8. Bantupalli, K., Xie, Y.: American sign language recognition using deep learning and computer vision. In: 2018 IEEE International Conference on Big Data (Big Data), Seattle, WA, USA, pp. 4896–4899 (2018). https://doi.org/10.1109/BigData.2018.8622141
9. Sarker, I.H.: Machine learning: algorithms, real-world applications and research directions. SN Comput. Sci. **2**, 160 (2021). https://doi.org/10.1007/s42979-021-00592-x
10. Baştanlar, Y., Özuysal, M.: Introduction to machine learning. In: Yousef, M., Allmer, J. (eds.) miRNomics: MicroRNA Biology and Computational Analysis. Methods in Molecular Biology, vol. 1107. Humana Press, Totowa (2014)
11. Mujahid, A., et al.: Real-time hand gesture recognition based on deep learning YOLOv3 model. Appl. Sci. **11**(9), 4164 (2021). https://doi.org/10.3390/app11094164
12. Redmon, J., Farhadi, A.: You only look once: unified, real-time object detection (2015). arXiv:1512.02325
13. Redmon, J., Farhadi, A.: YOLOv3: an incremental improvement (2018). arXiv:1804.02767
14. The Independent. Glove Translates American Sign Language into Speech (2020). [News article]. https://www.independent.co.uk/tech/glove-translates-american-sign-language-speech-a9598451.html
15. UCLA Newsroom. Glove Translates Sign Language to Speech (n.d.). [Press release]. https://newsroom.ucla.edu/releases/glove-translates-sign-language-to-speech
16. McGuire, R.M., Hernandez-Rebollar, J., Starner, T., Henderson, V., Brashear, H., Ross, D.S.: Towards a one-way American sign language translator. In: Sixth IEEE International Conference on Automatic Face and Gesture Recognition, Proceedings, Seoul, Korea (South), pp. 620–625 (2004). https://doi.org/10.1109/AFGR.2004.1301602
17. "Language Learning App Development" Octal Software. https://www.octalsoftware.com/blog/language-learning-app-development/. Accessed 11 Feb 2023
18. "Top Language Learning Apps Downloads" Statista. https://www.statista.com/statistics/1239522/top-language-learning-apps-downloads/. Accessed 11 Feb 2023
19. "Sign Language App" Healthline. https://www.healthline.com/health/sign-language-app. Accessed 11 Feb 2023
20. "Applying Mobile Technology to Enhance Language Learning" IEEE Xplore Digital Library. https://ieeexplore.ieee.org/document/5194206. Accessed 11 Feb 2023
21. Lee, P.L.: A review of sign language learning apps for the deaf community. Int. J. Inf. Educ. Technol. **7**(2), 127–132 (2017)
22. Solimine, N.J.: ActionDetectionforSignLanguage. GitHub. https://github.com/nicknochnack/ActionDetectionforSignLanguage. Accessed 22 Mar 2023

How Indirect and Direct Interaction Affect the Trustworthiness in Normal and Explainable Human-Robot Interaction

Truong An Pham and Leonardo Espinosa-Leal(✉)

Graduate School and Research, Arcada University of Applied Sciences,
Jan-Magnus Janssons plats 1, 00560 Helsinki, Finland
{truongan.pham,leonardo.espinosaleal}@arcada.fi

Abstract. Human-robot interaction (HRI) attracts significant attention from the public due to the ubiquity of robots in factories, restaurants, and even at home. However, the engagement of users in interacting with the robot is still a question mark due to the challenging trustworthiness. The trustworthiness becomes more complicated when discussing indirect interaction – humans observe the robot – and direct interaction – humans and robots may interact or not interact when being close to each other – in the robotic design. Several studies were conducted to analyze human trust in either indirect or direct aspects of robotic systems. The shortage of benchmarking indirect interaction and direct interaction initiates a significant gap in designing and developing a more subtle robotic system in complex scenarios that involve different stakeholders, such as users and observers, known as indirect users. In this study, we propose a novel guideline for evaluating such robotic systems in human-robot interaction. Particularly, we analyze differences between indirect and direct interaction about human trustworthiness in HRI. In addition, we also investigate the simulation methodology including virtual reality and video to evaluate a human-robot interaction scenario in both normal and explainable robotic systems by integrating a visual feedback module. By conducting quantitative and qualitative experiments, there is no significant difference between indirect and direct interaction in the trustworthiness of HRI. Instead, the explainable feature is recognized as the key factor in improving the trustworthiness of a robotic system.

Keywords: human-robot interaction · augmented reality · virtual reality

1 Introduction

Recently, we have witnessed the considerable emergence of several robotic applications. However, the ubiquity of the robots is extremely low, with limited applications such as vacuum robots. Trustworthiness and safety are two primary

M. E. Auer et al. (Eds.): STE 2024, LNNS 1028, pp. 411–422, 2024.
https://doi.org/10.1007/978-3-031-61905-2_40

reasons that prevent robots from completing other fine-grained tasks such as pick-and-place [13] and handover [24]. To the best of our knowledge, HRI can be categorized into direct interaction (first-person perspective) and indirect interaction (third-person perspective). In direct interaction scenarios [13,24], there are commonly some risks that robots may collide with the user. In contrast, with indirect interaction scenarios [27], there is only the risk that the robot may collide with objects or other people instead of the user. However, there is a limit to studying the difference between indirect and direct interaction with robots. Plus, explainable robotic systems with those two perspective types are still ambiguous.

To investigate the indirect and direct interaction in HRI, we design and implement an explainable robotic system that enhances human perception via spatial-augmented-reality feedback. We analyze trustworthiness through perceived safety, perceived risk, and understandability in first-person and third-person perspectives.

The rest of this paper is structured as follows. We introduce the background and the related work in Sect. 2. Section 3 presents the proposed robotic system and HRI scenarios. The evaluation protocol and result are described in detail in Sect. 4. In Sect. 5, we discuss the insights and limitations of the evaluation. Finally, we conclude the work in Sect. 6.

2 Background and Related Work

2.1 Trustworthiness in Human-Robot Interaction

Robotic trustworthiness is a characteristic or quality of a robotic system in human-robot interaction (HRI). So, trust is one of the important measures that can be used to evaluate an HRI system's trustworthiness. In [14], trust is defined as "the attitude that an agent will help achieve an individual's goals in a situation characterized by uncertainty and vulnerability". So, transparency or understandability of the robotic system plays a key role in shaping the way the user achieves the goal. Understandability aims to improve human trust as studied in [20]. In addition, the trust is built up on uncertainty and vulnerability situations, representing a connection with risks and safety. In [4,17], the authors define trust as the willingness to take risks under uncertain conditions. The trust [23] is strongly correlated with human risk perception. While risk represents the negative feeling of achieving the goal, perceived safety is also mentioned as one of the most essential elements to establish human trust in human-robot collaboration [6]. In summary, we aim to study the trustworthiness of HRI through four factors of trust, perceived risk, perceived safety, and understandability.

2.2 Explainable Human-Robot Interaction

As defined in [1], explainable artificial intelligence (XAI) refers to an artificial intelligence system that makes decisions and predictions more understandably and transparently to humans. As defined in [5], goal-driven XAI is a research domain aiming at building explainable agents and robots that are capable of explaining their behaviors. In HRI, there are several approaches to building an XAI system such as augmented reality (AR) to inform the robotic intention. In [25], through a teaching scenario, a human as a teacher can understand the reasons behind each step of a learner robot. With that augmented-reality-based XAI system, the trustworthiness of the robotic system is improved. However, the field of view of the AR headset, which is used for deploying reasoning feedback, is so narrow that it may not represent all information of the reasoning. So, in our study, spatial AR is used to break that limitation without requiring any AR headset. As a result, the user experience quality and performance of the user are improved.

2.3 First Person Perspective or Third Person Perspective in Human-Robot Interaction

First-person perspective interaction is an HRI case in which a person interacts directly with a robot. For instance, in handover tasks, a person tries to receive or give a certain object to the robotic arm as in [24]. Or in a pick-and-place scenario of industrial assembly setting [13], a robot collaborates with a person directly to complete the tasks. In contrast, third-person perspective interaction represents indirect HRI scenarios where humans observe the interaction. Due to the fuzzy benefit in HRI design, indirect interaction is under-investigated. In [27], the authors let participants interact with robots and record videos of those sessions. Later, those videos are shown to other participants. This is a standard approach for studying the trustworthiness difference between first-person perspective and third-person perspective in HRI. As studied in [15], trust calibration is essential to reduce trust evaluation ambiguity. However, a between-subject user study design lacks trust calibration because each participant experiences either a first-person or third-person perspective. Instead of designing a between-subject user study as [27], we implement a within-subject study, which allows participants to experience both conditions in a virtual-reality platform.

3 System Design and Scenario Implementation

3.1 System Design

In this study, indoor navigation is one of the most attractive challenges for robotic systems. To help the robotic system navigate successfully, human or obstacle avoidance is a must for a functional navigation robotic system. To implement that feature, the system is designed to process visual information that is one of the richest media in describing surrounding environments. Therefore, computer vision solutions are implemented for the robotic system in navigation scenarios to study the trustworthiness. However, computer vision processing tasks are too heavy to run on local mobile processors of robotic systems. Consequently, it causes a higher latency of the robotic feedback on the human-robot interaction (HRI). As also mentioned in [10,18], the latency issue is one of the challenges that affect the performance and experience when interacting with the robots. To overcome that challenge, we designed an edge-based robotic system in which computer vision tasks are offloaded to a nearby powerful server.

Edge-Offloading Robotic System. The proposed robotic system consists of two major modules, including a robotic module and a processing server module, as illustrated in Figure 1.

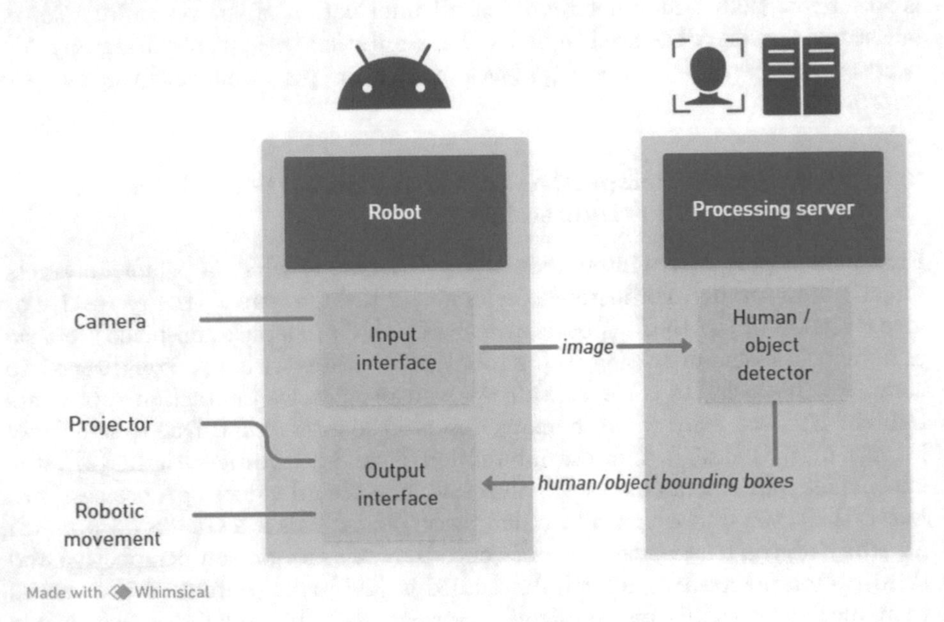

Fig. 1. System design

Robotic Module. This module is implemented with the robotic operating system (ROS). Plus, the module continuously captures the surrounding information through sensors such as cameras, haptic, audio, etc. In our study, we only focus on vision-based robotic systems with cameras. This module is used for capturing images through the eyes of the robot. The module controls robotic parts and a projector equipped with the robotic system. In this study, we integrate a virtual projector to visualize augmented-reality instructions to show the intent of the robot in movement. The arrow of robotic movement is projected on the floor as robotic movement intention as illustrated in Fig. 2(a).

Human and Object Detector Module. In HRI interface design, low latency and high accuracy are critical requirements of design guidelines for robotic systems. Therefore, YOLOv7 [26], which is a real-time and highly-accurate deep neural network (DNN) based detector, has been implemented in this module. To detect persons and objects, the YOLOv7 model is trained with MS COCO dataset [16].

Robot Controlling Protocol. gRPC protocol [7] is used for video streaming and feedback response. TCP-based protocol gRPC, which keeps the integrity of the image, is implemented to prevent computer vision algorithms from missing any part of surrounding information. As illustrated in Figs. 2(b) 2(c), the processing server detects persons and objects after receiving images from the robot. Then, it sends the command to control the projector and direction of movement based on the detected results.

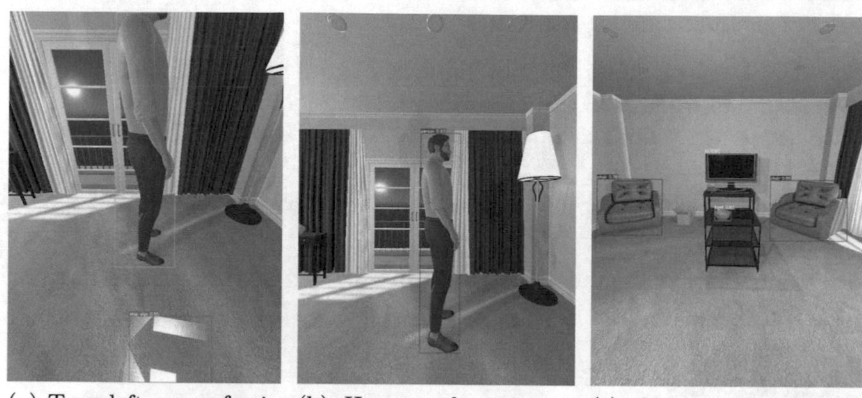

(a) Turn left arrow for informing the intention of robotic system (b) Human detection in processing server on edge (c) Object detection in processing server on edge

Fig. 2. Sample images that human-and-object detection in processing server on edge

3.2 Implementation

Based on the proposed design, we implement a robot model and simulation environment as follows.

Robot Model. There are several types of robots, such as humanoids, robotic arms, and navigation robots. Each type of robot is designed for a specific task. However, a humanoid robot is the most common robot used for studying trustworthiness due to its anthropology characteristics. In this study, to evaluate the trustworthiness, Reachy [2] is used for the implementation.

Simulation Environment. iGibson [22], Habitat [21], and AI2Thor [12] are state-of-the-art environments used for building a simulation environment as a testbed for studying the trustworthiness of the robotic system. Unity is a common platform for developing those virtual environments due to the cross-platform deployment, high fidelity, and physics engine features. So, in this study, robotic systems and scenarios are developed based on the Unity environment. Based on a room layout in AI2Thor, we designed and implemented multiple scenarios described in detail in Sect. 4. The robot is equipped with a camera to sense the environment and an equipped projector to project its movement intention on the floor. The projector is used as a visual feedback module.

4 Evaluation

To understand "How indirect and direct interaction affect the trustworthiness in normal and explainable human-robot interaction", we need to evaluate the validity of the following hypotheses.

- **H1.** The Trustworthiness level of indirect is significantly different from direct interaction in explainable human-robot interaction.
- **H2.** Explainable human-robot interaction significantly affects the sense of trustworthiness.
- **H3.** Trustworthiness awareness with the video approach is different from the virtual-reality simulation approach.

In this evaluation, we conducted the following user study protocol to clarify the hypotheses above.

4.1 Participants

We have recruited five persons to participate in the user study. They are invited without any specific criteria except that they need to be healthy enough to use virtual-reality glasses. All participants voluntarily took part in the user study. The following user study procedures are approved by our university via a research permit checking.

4.2 Scenario Design

In this study, we narrow down the scope of testing the trustworthiness of the robotic system into human and object avoidance for navigation tasks.

Robotic Movement Speed. In [9], the authors show that the aggressive driving style of an autonomous vehicle (AV) significantly affects the trust of pedestrians. That aggressive driving style consists of high speed, acceleration, and full stop. Therefore, in the testing scenarios, we only implement the high speed of movement and full stop for the robotic system to uncover the true feeling of trust from users.

Explainability: Different Explainable Level of Robot. Trustworthiness is also derived from the explainable robotic system. In this study, with the projector, spatial augmented reality as visual feedback is considered an explainable robotic approach. Therefore, we implement two conditions, including augmented reality (AR) and non-augmented reality, to check whether the visual feedback affects the trustworthiness of humans.

Perspective: Indirect vs. Direct Interaction. With direct interaction, the robotic system navigates in the direction of the participant with and without AR. In another way, the robotic system navigates in the direction of an avatar with and without AR with indirect interaction. Both scenario setups are shown in Figs. 3(a) 3(b).

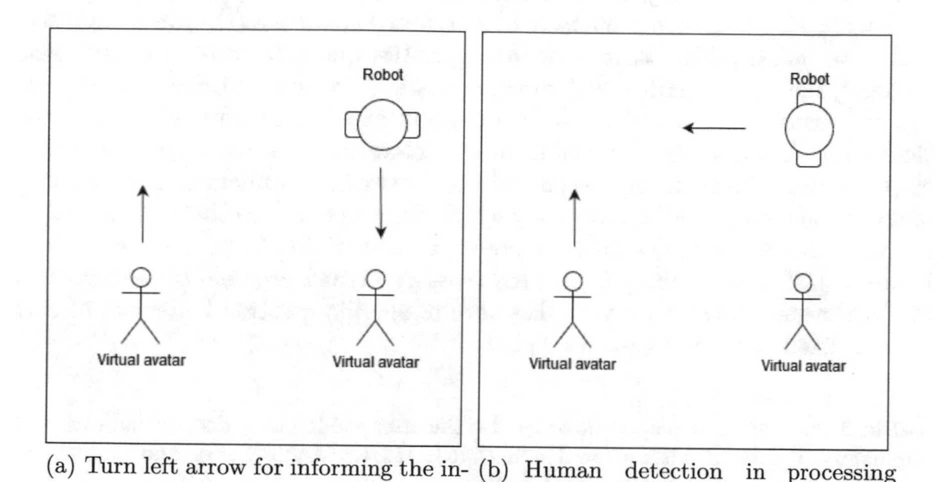

(a) Turn left arrow for informing the intention of robotic system

(b) Human detection in processing server on edge

Fig. 3. Direct vs. indirect interaction scenarios setup

Fidelity: Video. In the video, the robot navigates to confront the user with the first-person perspective. The view of the user is recorded in VR and shown on the TV for the participant. This scenario

4.3 Procedure

Participants attend the user study one-by-one. At first, information on the research was introduced to participants by a facilitator who is a co-author of this paper. Then, participants were given a consent form and signed them. Then, participants were informed of the experiment process, which includes three steps as follows. At step 1, participants' background information is collected via a pre-test questionnaire to understand their virtual reality and robotic knowledge and experience. At step 2, each participant experience 5 scenarios of the robotic system. The 5 scenarios was presented in Sect. 3. According to [8], "trust at the present moment is determined by trust at the previous moment" is one of the three properties to predict the trust in human-agent interaction. Therefore, the order of presenting the scenarios to participants may create a bias in the trustworthiness evaluation of scenarios. As a result, to avoid that order bias, we arranged the order of those scenarios differently for each participant. At first, the video scenario is placed at the beginning of scenarios sequence or at the end. We place an indirect scenario in one group and a direct scenario in another group to avoid the confusion of participants when they experience many scenarios with both conditions, which are augmented reality vs. non-augmented reality scenarios and indirect vs. direct scenarios. As a result, we have scenarios ordered as described in Table 1. Besides, in step 2, we also remind participants that this is a think-out-loud user study to let participants speak out their feelings if they want to. During step 2, to avoid memory loss on user experience (UX) in several scenarios, we asked participants to answer after-test questionnaires after the video scenario, indirect scenarios, and direct scenarios. So, the participants have two breaks during step 2 to reflect on their experience via answering after-test questionnaires. Additionally, the facilitator also plays the role of an observer in this step to explore more deeply insights of the survey questionnaires. This methodology is commonly used in user study which answers to subjective questionnaires are not reliable enough, such as remote user test in [11] or people with health issues in [19]. Plus, privacy is also the most concerned problem of participants. So, in the step 2, the observer takes note of specific reaction behaviors of user such as hand and body gestures, voice.

Table 1. Scenarios is placed in order for the user study. 1: Video; 2: indirect and augmented reality; 3: indirect and non augmented reality; 4: direct and augmented reality; 5: direct and non augmented reality

Participant index	Scenarios in order
P1	$1 \rightarrow 2 \rightarrow 3 \rightarrow 4 \rightarrow 5$
P2	$4 \rightarrow 5 \rightarrow 2 \rightarrow 3 \rightarrow 1$
P3	$1 \rightarrow 3 \rightarrow 2 \rightarrow 5 \rightarrow 4$
P4	$5 \rightarrow 4 \rightarrow 3 \rightarrow 2 \rightarrow 1$
P5	$5 \rightarrow 4 \rightarrow 2 \rightarrow 3 \rightarrow 1$

After the participant experiences all five scenarios, they attend a semi-structured interview to uncover more insights into the user experience on the robotic systems at step 3. This interview aims to obtain more explanation about the answers to the after-test questionnaire from step 2.

4.4 Apparatus

A processing server on edge is setup on a VR-ready PC with a powerful GPU (NVIDIA RTX 3080 with 12 GB VRAM). A living room with a 65-inch TV is arranged for video scenario setup. A VR valve index headset [3] is used for virtual reality (VR) setup.

4.5 Metrics

In Human Robot Interaction studies, metrics are categorized into two groups, including objective and subjective measures. With objective measures, they are commonly used for

Objective Measure Performance. Latency in video streaming and human/object detection is used to evaluate the performance of the robotic system.

Subjective Measure. We use post-test questionnaire to measure the trustworthiness through four ratings trust, safety, risk, and understandability. Plus, fidelity questions are also used to evaluate the effect of video and virtual reality on trustworthiness.

4.6 Results

Objective Measure: Performance. During the testing, the average video streaming latency is measured at 0.034 ms. Meanwhile, the processing server on edge can handle the human-and-object localization task with an average speed of 37.49 ms. In total, it takes approximately 37.52 ms for the robotic system to localize humans and obstacles and avoid the collision.

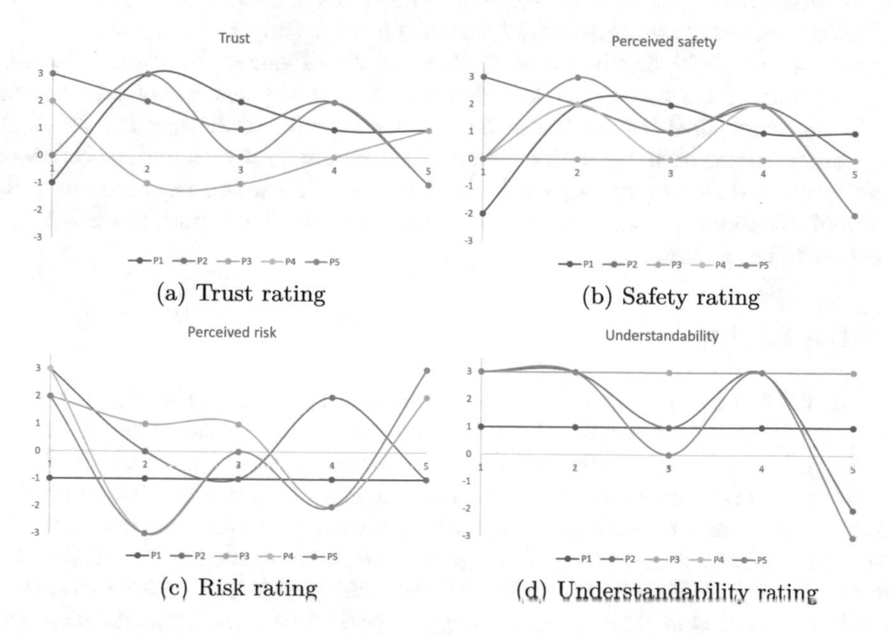

(a) Trust rating

(b) Safety rating

(c) Risk rating

(d) Understandability rating

Fig. 4. User experience rating via post-test questionnaires

Subjective Measures. Due to multi-condition experimental setup design, we use two-way ANOVA analysis to evaluate the effect of perspective condition (direct and indirect) and explainable robotic condition (with AR and non-AR).

Direct and Indirect Interaction. There is no value p in trust, safety, risk, and understandability rating is less than $\alpha = 0.05$ for the perspective condition and the interaction with the explainable robotic condition. Therefore, there is no clear effect of perspective conditions on human trustworthiness. Thus, with these experimental results, the hypothesis **H1** is not possible to be proven.

Explainable Robot via Augmented-Reality Feedback. All values $p = 0.01156$ in trust, $p = 0.00604$ in safety, $p = 0.0474$ in risk, and $p = 0.003$ understandability are less than $\alpha = 0.05$, so the null hypothesis is rejected. As a result, we can confirm that an explainable robot has a clear effect on trustworthiness. We can see the clear positive effect on the trust, safety, and understandability experience with two peak maxima in two scenarios 2 and 4 with augmented-reality feedback as shown in Fig. 4(a), 4(b), 4(d). Plus, participants P2, P4, and P5 usually emphasize the benefit of the augmented-reality feedback in showing the robotic intention in movement. Thus, hypothesis **H2** is true.

Fidelity in Video and Virtual Reality. Unlike perspective condition and explainable robotic condition, we verify the hypothesis **H3** by using one-way ANOVA analysis on scenario 1 and scenario 5 only. Only the understandability rating with $p = 0.02758$ gives a clear effect of the difference between video and virtual reality. That was also reflected easily through the answers of participants in the interview after testing the scenarios. For example, participants P3 and P5 only mention the high resolution as an advantage of VR over video. However, they mentioned that other ratings, including trust, safety, and risk, are not much different. Therefore, with these experimental results, the hypothesis **H3** is not possible to be proven.

5 Discussion

Firstly, the same room layout and same navigation path create memory bias for the participants. As a result, the feeling of safety and trustworthiness is high in the AR-or-non-AR condition and the indirect-or-direct condition. In future research, adding more room layouts and new navigation paths can solve the bias. Secondly, although 5 participants are an appropriate amount for the quantitative and qualitative user study, more participants are needed to strengthen the obtained insights. Finally, the designed scenarios are mainly passive situations in which participants barely do a thing. Therefore, the participants have free space and time to distinguish between reality and virtual worlds. So, it affects the trust and safety scale. In the next research, we aim to design some collaborative scenarios in which the robotic system will team up with the participants in completing a task such as dining table arrangement, furniture assembly, or lego playing to increase the immersion of the participant via occupying those free spaces and time.

6 Conclusion

In summary, in simulating the environments for human-robot interaction evaluation, three aspects, including vision perspective, explainable robot, and fidelity, are studied via a user study with 5 recruited participants. Firstly, an explainable robotic module has a positive effect on improving the trustworthiness of humans. Secondly, the effect of human perspective with first-person and third-person view is not clear enough. One of the reasons that has been discussed is the limitation on immersive scenarios and the small number of participants. Therefore, in the next research, we will focus on collaborative scenarios in which participants will feel more immersive when testing. Plus, more participants will be recruited to get more insights into the perspective effect on trustworthiness. Finally, with this next research, we can investigate more deeply the fidelity effect on trustworthiness also due to the higher immersive level for the participants.

References

1. Explainable AI. https://en.wikipedia.org/wiki/Explainable_artificial_intelligence. Accessed 29 Nov 2023
2. Reachy humanoid robot. https://www.pollen-robotics.com/. Accessed 11 Sept 2023
3. Valve index headset. https://store.steampowered.com/valveindex. Accessed 23 Nov 2023
4. Andras, P., et al.: Trusting intelligent machines: deepening trust within socio-technical systems. IEEE Technol. Soc. Mag. **37**(4), 76–83 (2018)
5. Anjomshoae, S., Najjar, A., Calvaresi, D., Främling, K.: Explainable agents and robots: results from a systematic literature review. In: 18th International Conference on Autonomous Agents and Multiagent Systems (AAMAS 2019), Montreal, Canada, 13–17 May 2019, pp. 1078–1088. International Foundation for Autonomous Agents and Multiagent Systems (2019)
6. Breton, L., Hughes, P., Barker, S., Pilling, M., Fuente, L., Crook, N.: The impact of leader-follower robot collaboration strategies on perceived safety and intelligence (2016)
7. gRPC: gRPC: a high performance open-source universal RPC framework (2020)
8. Guo, Y., Zhang, C., Yang, X.J.: Modeling trust dynamics in human-robot teaming: a Bayesian inference approach. In: Extended Abstracts of the 2020 CHI Conference on Human Factors in Computing Systems, pp. 1–7 (2020)
9. Jayaraman, S.K., et al.: Trust in AV: an uncertainty reduction model of AV-pedestrian interactions. In: Companion of the 2018 ACM/IEEE International Conference on Human-Robot Interaction, pp. 133–134 (2018)
10. Kaber, D.B., Riley, J.M., Zhou, R., Draper, J.: Effects of visual interface design, and control mode and latency on performance, telepresence and workload in a teleoperation task. In: Proceedings of the Human Factors and Ergonomics Society Annual Meeting, vol. 44, pp. 503–506. SAGE Publications, Los Angeles (2000)

11. Kaiser, J.N., Marianski, T., Muras, M., Chamunorwa, M.: Popup observation kit for remote usability testing. In: Proceedings of the 20th International Conference on Mobile and Ubiquitous Multimedia, pp. 233–235 (2021)
12. Kolve, E., et al.: AI2-THOR: an interactive 3D environment for visual AI. arXiv preprint arXiv:1712.05474 (2017)
13. Krenn, B., et al.: It's your turn!–A collaborative human-robot pick-and-place scenario in a virtual industrial setting. arXiv preprint arXiv:2105.13838 (2021)
14. Lee, J., Moray, N.: Trust, control strategies and allocation of function in human-machine systems. Ergonomics **35**(10), 1243–1270 (1992)
15. Lee, J.D., See, K.A.: Trust in automation: designing for appropriate reliance. Hum. Factors **46**(1), 50–80 (2004)
16. Lin, T.Y., et al.: Microsoft COCO: common objects in context. In: Fleet, D., Pajdla, T., Schiele, B., Tuytelaars, T. (eds.) ECCV 2014, Part V. LNCS, vol. 8693, pp. 740–755. Springer, Cham (2014). https://doi.org/10.1007/978-3-319-10602-1_48
17. Luhmann, N.: Trust and Power. Wiley, Hoboken (2018)
18. MacKenzie, I.S., Ware, C.: Lag as a determinant of human performance in interactive systems. In: Proceedings of the INTERACT 1993 and CHI 1993 Conference on Human Factors in Computing Systems, pp. 488–493 (1993)
19. Mäkelä, S., Bednarik, R., Tukiainen, M.: Evaluating user experience of autistic children through video observation. In: CHI 2013 Extended Abstracts on Human Factors in Computing Systems, pp. 463–468 (2013)
20. Muir, B.M.: Trust in automation: Part I. Theoretical issues in the study of trust and human intervention in automated systems. Ergonomics **37**(11), 1905–1922 (1994)
21. Savva, M., et al.: Habitat: a platform for embodied AI research. In: Proceedings of the IEEE/CVF International Conference on Computer Vision, pp. 9339–9347 (2019)
22. Shen, B., et al.: iGibson 1.0: a simulation environment for interactive tasks in large realistic scenes. In: 2021 IEEE/RSJ International Conference on Intelligent Robots and Systems (IROS), pp. 7520–7527. IEEE (2021)
23. Siegrist, M.: Trust and risk perception: a critical review of the literature. Risk Anal. **41**(3), 480–490 (2021)
24. Tian, L., He, K., Xu, S., Cosgun, A., Kulic, D.: Crafting with a robot assistant: use social cues to inform adaptive handovers in human-robot collaboration. In: Proceedings of the 2023 ACM/IEEE International Conference on Human-Robot Interaction, pp. 252–260 (2023)
25. Wang, C., Belardinelli, A.: Investigating explainable human-robot interaction with augmented reality. In: 5th International Workshop on Virtual, Augmented, and Mixed Reality for HRI (2022)
26. Wang, C.Y., Bochkovskiy, A., Liao, H.Y.M.: YOLOv7: trainable bag-of-freebies sets new state-of-the-art for real-time object detectors. In: Proceedings of the IEEE/CVF Conference on Computer Vision and Pattern Recognition, pp. 7464–7475 (2023)
27. Yanai, K., et al.: Evaluating human-robot interaction from inside and outside comparison between the first-person and the third-party perspectives (2018)

A Brain-Computer Interface for Controlling a Wheelchair Based Virtual Simulation Using the Unicorn EEG Headset and the P300 Speller Board

Oana Andreea Rusanu(✉)

Transilvania University of Brasov, Brasov, Romania
oana.rusanu@unitbv.ro

Abstract. The brain-computer interface (BCI) is a multidisciplinary research field at improving the biomedical engineering with innovator systems for helping people with neuromotor disabilities. This paper proposes a new solution to introduce the BCI as a portable, robust, and attractive technology right in the home environment of the users. The proposed application is related to using the mental commands for guiding a 3D model of a wheelchair by following different movement directions. The mental commands are triggered by the detection of the P300 evoked potentials from the Unicorn EEG headset provided by the GTEC Medical Engineering company. The wheelchair-based simulation is implemented in the LabVIEW programming environment by using the dedicated functions from the 3D Picture control. The control of the 3D wheelchair supposes to enable the corresponding operations of translating and rotating the elements from the 3D structure, as well as to accomplish its movement across different directions (forward, backward, turning left or right). The proposed software application involved the UDP based integration between LabVIEW and Unicorn P300 Speller. Moreover, the LabVIEW simulation provides multiple scenarios comprising different trajectories for guiding the wheelchair to reach the established destinations. The difficulty levels are increasing, and the users are required to achieve challenging tasks related to controlling the BCI by correctly transmitting the P300 based command. This paper presents three experimental sessions composed of three difficulty levels, such as: beginner (6 experiments with 2–3 commands each, intermediate (4 experiments with 7 commands each), and advanced (2 experiments with 13–14 commands each).

Keywords: Brain-Computer Interface · Unicorn P300 Speller · 3D Wheelchair

1 Introduction

The brain-computer interface (BCI) is a multidisciplinary research field aimed at improving the biomedical engineering with innovator systems for helping people with neuromotor disabilities. The BCI provides them the opportunity to replace natural pathways

M. E. Auer et al. (Eds.): STE 2024, LNNS 1028, pp. 423–435, 2024.
https://doi.org/10.1007/978-3-031-61905-2_41

(muscles and peripheric nerves) with artificial outputs for controlling both real and virtual devices only by thoughts. Therefore, the BCI technology includes various cognitive techniques for deciphering the mental intentions of the disabled users and one of the most frequently implemented modalities is based on detecting the P300 evoked biopotentials. Although the BCI systems that integrate de P300 Speller board are especially aimed at communication tasks, it is still possible to use the control signals determined by the P300 event related potentials for sending commands to enable the working of physical assistive devices or even to accomplish different virtual activities in a simulation-based environment.

This paper presents a software application for demonstrating, exploring, experimenting, providing a learning, and training tool for getting familiar with the brain-computer interface innovator technology that should be further used both by novice researchers in these field and disabled users. Considering the actual technical and challenges addressed by the brain-computer interface that is still encountering demanding intellectual (analysing and classifying the EEG signals) and financial resources, the pursue, attempts, and endeavours to improve the applications area or the experimental paradigms by involving new software or virtual ideas are welcome and may support both students and patients. Therefore, this paper proposes a new solution to introduce the brain-computer interface as a portable, robust, and attractive technology right in the home environment of the users. The proposed application is related to using the mental commands for guiding a 3D model of a wheelchair by following different movement directions depending on the implemented scenarios. The mental commands are triggered by the detection of the P300 evoked potentials from the Unicorn EEG headset provided by the GTEC Medical Engineering company.

The wheelchair-based simulation is implemented in the LabVIEW programming environment by using the dedicated functions from the 3D Picture control. The LabVIEW simulation provides multiple scenarios comprising different trajectories for guiding the wheelchair to reach established destinations by following different movement directions. The difficulty levels are increasing, and the users are required to achieve more challenging tasks related to controlling the brain-computer interface by correctly transmitting the P300 based command. The integration between the official Unicorn P300 Speller board software and the LabVIEW originally implemented simulation is enabled by the acquisition and processing of the UDP data packages embedding the P300 evoked response. The current research work involves the designing of a customized P300 Speller palette composed of both dark and flashing items that will be displayed as non-target and target symbols corresponding to BCI commands. The scope of the customized P300 Speller board is to provide a strong impact on the users' attention and eyesight to trigger an emphasized P300 evoked potential.

This paper presents three experimental sessions composed of three difficulty levels, depending on the number of required commands: 2–3 (beginner), 7 (intermediate), and 13–14 (advanced). There are also discussed the psychological and environmental implications on the performance, confidence, state of mind, and comfort of the users.

The structure of this scientific paper is based on the following sections: 1 - stating the Introduction, 2 - presenting the current status of the BCI research field involving the P300 evoked potentials for running 3D virtual simulations, 3 - describing the Hardware

System, 4 - revealing the Software system, 5 – discussing the experimental results, and 6 – concluding this article and envisioning the future research directions.

2 The Current Scientific Status of the P300 Based BCIs for Enabling Experimentation Involving 3D Virtual Simulations

Considering the futuristic input mental commands involved by the brain-computer interface, providing alternative solutions to the normal means of communication and interaction such as mouse, keyboard, joystick, or voice-controlled devices, it results in even more engaging and immersive experiences by integrating the detection of P300 event related EEG potentials with the virtual or augmented reality or the 3D simulations for developing and experimenting applications related to multiples areas: video games [1, 2], drone control [3], rehabilitation of neurological diseases [4], brain painting [5], mobile phones [6], and post-stroke recovery of motor functions [7, 8]. The P300 evoked potential is a positive deflection occurring during the EEG data acquisition with a delay of approximately 300 ms after a visually, auditory, or somatosensory stimulus was presented to the subject [10]. Being initially introduced by Farwell and Donchin in 1988 [9], the P300 based oddball paradigm showed different repeated stimuli included by two categories: target stimuli that were infrequently displayed and non-target stimuli that were frequently shown to the user. The P300 response is triggered by the infrequently flashing items. The P300 biopotentials are usually detected to the central and parietal cerebral regions by using scalp EEG sensors to the positions Fz, Cz, and Pz of The 10–20 International System [10].

3 The Hardware System – The GTEC Unicorn EEG Kit

The GTEC Unicorn EEG headset embedding 8 hybrid sensors was preferred for the current research work thanks to its high performance, convenience in terms of excellent signals quality, rapid setup and comfort, as well as usability considering the official P300 Speller software that provides the opportunity to detect the P300 evoked biopotentials and send the P300 response to external applications using UDP data communication protocol [11]. The Unicorn kit encourages creativeness by developing brain-computer interfaces using the already implemented machine learning algorithms and addressing to different categories of enthusiastic people (researchers, students, artists).

4 The Software System – LabVIEW Implementation of the Brain-Computer Interface Based on Unicorn P300 Speller

Figure 1 presents the graphical user interface of the LabVIEW software implementation aimed at the development of the P300 based brain-computer interface for running the virtual simulation consisting of controlling the 3D wheelchair through different movement directions (go forward/backward, turn left/right, rotate left/right) considering short and long distances. The primary phase is related to the implementation of the LabVIEW application by performing subsequent stages, such as: designing the 3D wheelchair by building the geometrical shapes components, controlling the 3D wheelchair by translating the internal structures both in a manual and automatic manner, programming a state-machine paradigm for switching between the movement commands triggered either by the buttons or UDP strings, acquiring and analyzing the received UDP data associated with the P300 response.

The secondary phase refers to the customization of a new palette with significant flashing and dark items using the official Unicorn P300 Speller provided by the GTEC Medical Engineering company.

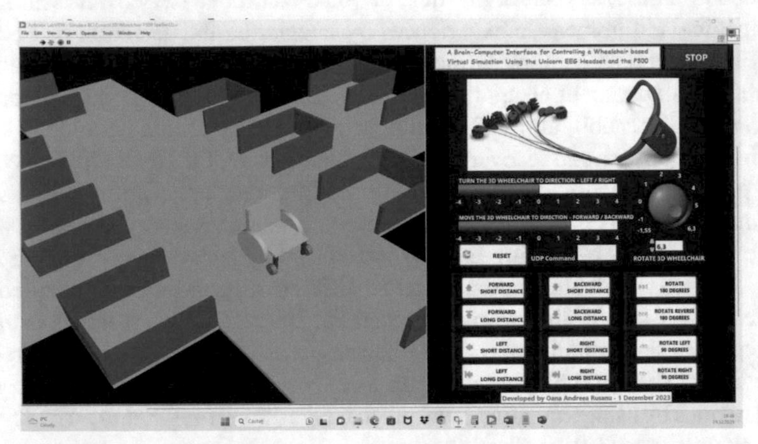

Fig. 1. The Graphical User Interface (Front Panel) of the proposed LabVIEW application for achieving the brain-computer interface based on using the P300 evoked EEG potentials for controlling the 3D wheelchair across different movement directions.

4.1 The LabVIEW Based Design of the 3D Virtual Wheelchair

According to Fig. 2, the LabVIEW based 3D design consisted in creating the structure of both the virtual wheelchair and the simulated environment providing the destinations and trajectories for different movement scenarios based on 12 experiments depending on the level (beginner, intermediate, or advanced) reached by the subject.

The wheelchair-based simulation is implemented in the LabVIEW graphical programming environment using the dedicated functions from the 3D Picture control palette. These functions are necessary for designing and animating the geometrical shapes composing the 3D model of the wheelchair. The design of the 3D wheelchair refers to setting the proper shapes, dimensions, and positions of each subsequent element. The control of the 3D wheelchair supposes to enable the corresponding actions of translating or rotating the elements from the structure of the wheelchair, as well as to accomplish its movement across different movement directions and distances.

The design of the simple 3D wheelchair supposed to create and link the following separate elements: the backrest, the seat, the left wheel, the right wheel, the left caster, the right caster, the left frame, and the right frame. The main LabVIEW function (virtual instrument) enabling the creation of a 3D element is called Create Object. The subsequent LabVIEW functions aimed at building the necessary 3D rectangular or circular geometrical shapes are known as Create Box and Create Cylinder. All the previously mentioned virtual instruments – Create Object, Create Box, Create Cylinder – are linked to the same Invoke Node assigned to the Drawable attribute. The input parameters of the Create Box function are the related length dimensions by the three axis (X, Y, Z) and the color that can be set using the Color Change function. The input parameters of the Create Cylinder virtual instrument are the dimensions of radius, height, detail as well as the preferred color.

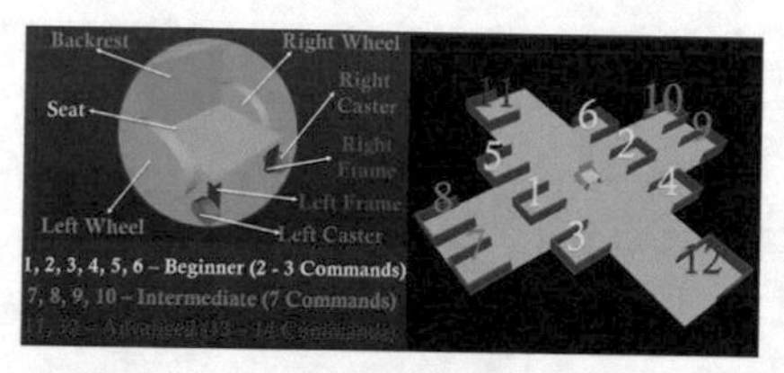

Fig. 2. The 3D structure of the virtual wheelchair and the environment simulated in LabVIEW.

Figure 3 presents the sequence of the Block Diagram consisting in using the functions from the 3D Picture Control Palette to create the elements composing the 3D wheelchair by setting the corresponding dimensions and geometrical shapes.

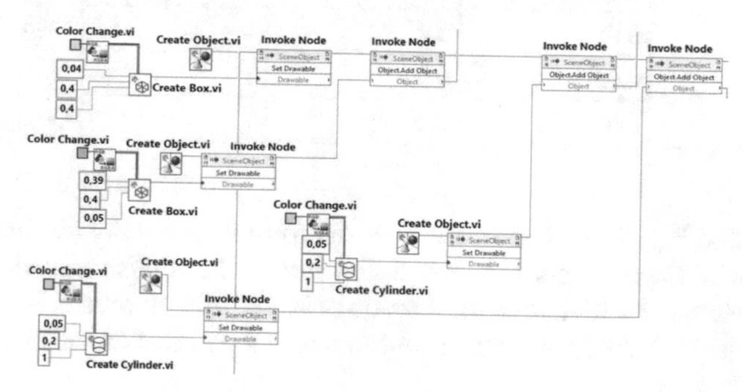

Fig. 3. The LabVIEW Block-Diagram comprising the functions provided by the 3D Picture Control palette used for designing and linking the backrest, the seat, the left and the right wheel from the structure of the virtual wheelchair.

Table 1 presents the input parameters of the Create Box LabVIEW function to design the backrest, the seat and the frames from the structure of the 3D wheelchair. Table 2 includes the input parameters of the Create Cylinder LabVIEW function used to design the wheels and the casters from the structure of the 3D wheelchair.

The proposed values from Table 1 and Table 2 were empirically estimated according to experimental results by visual checking the intermediate positions and manually adjusting the sizes of the related 3D elements. Considering the purpose of the researcher, the current LabVIEW virtual simulation supports further improvements, future updates, and additional geometrical elements to obtain more complex structure of the 3D wheelchair. Therefore, the specified values constitute a flexible approach.

Regarding the design of the LabVIEW based virtual environment aimed at stimulating the experimental scenarios based on running the movement commands, there were created the following 3D elements: the gray colored base floor platforms (one main plate of high size and four secondary plates of reduced dimensions), the green horizontal and vertical walls of different lengths.

Table 1. The Input Parameters of the Create Box function to design the 3D Wheelchair.

3D Element	Length X	Length Y	Length Z	Color
Backrest	0.04	0.4	0.4	Green
Seat	0.39	0.4	0.05	Yellow
Left Frame	0.2	0.04	0.03	Dark Blue
Right Frame	0.2	0.04	0.03	Dark Blue

Table 2. The Input Parameters of the Create Cylinder function to design the 3D Wheelchair.

3D Element	Height	Radius	Detail	Color
Left Wheel	0.05	0.2	1	Light Blue
Right Wheel	0.05	0.2	1	Light Blue
Left Caster	0.05	0.06	1	Orange
Right Caster	0.05	0.06	1	Orange

According to Fig. 2, the main base floor platform includes the destinations for the experiments with the numbers 1–2–3–4–5–6, whereas the similar secondary floor platforms comprise the locations for the experiments with the numbers 7–8–9–1–11–12. Table 3 presents the values set for the lengths on axis X–Y–Z of the main and secondary platforms.

Table 3. The Input Parameters of the Create Box function to design the 3D Base Platforms.

3D Element	Length X	Length Y	Length Z	Color
Main Base	0.05	4	4	Gray
Secondary Base	0.05	2	2.3	Gray

According to Fig. 2, the vertical and horizontal 3D walls are necessary to delimitate the destination places where the 3D wheelchair should be steered by the user during each of the 12 experimental scenarios depending on the difficulty level. The 3D walls are located to symmetrical positions. Likewise, the virtual walls are grouped together considering the similar sizes. Table 4 shows the lengths on axis X, Y, Z of the vertical 3D walls from the 12 experimental destinations. Table 5 includes the similar lengths on axis X, Y, Z of the horizontal 3D walls for all the 12 locations.

Table 4. The Input Parameters of the Create Box function to design the Vertical 3D Walls.

Experiments No.	Length X	Length Y	Length Z	Color
3, 4, 5, 6, 7, 9	0.4	0.04	0.8	Green
1, 2, 8, 10	0.4	0.04	0.75	Green
11, 12	0.4	0.4	1.2	Green

Table 5. The Input Parameters of the Create Box function to design the Horizontal Walls.

Experiments No.	Length X	Length Y	Length Z	Color
1, 2, 3, 4, 5, 6, 7, 8, 9, 10, 11, 12	0.4	1	0.05	Green

4.2 The LabVIEW Based Manual Control of the 3D Virtual Wheelchair

The control of the 3D wheelchair by different movement directions towards the virtual environment simulated in LabVIEW is enabled by a programmatic sequence (Fig. 4) of the following functions from the 3D Picture Control palette: Clear Transformation – Translate and/or Rotate Object – Scale Object.

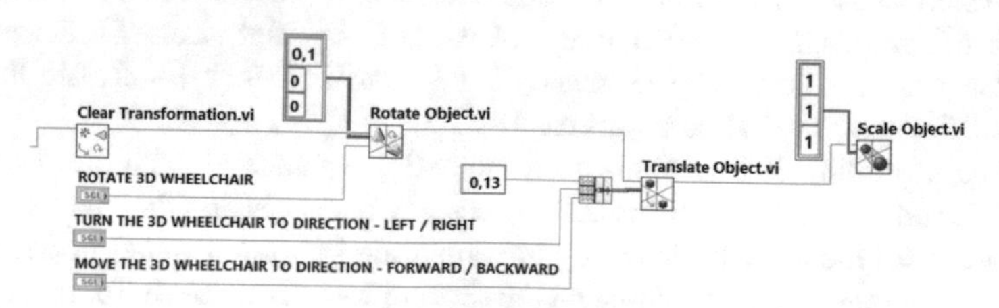

Fig. 4. The LabVIEW Block-Diagram comprising the functions provided by the 3D Picture Control palette used for controlling the 3D virtual wheelchair.

The Clear Transformation LabVIEW function accomplishes an initialization role and is necessary to modify the identity matrices characterizing the transformations (translations or rotations) applied to an object in the 3D scene. The Translate Object LabVIEW function achieves the movement of the 3D object to transversal directions of axis X or Y by applying a relative translation considering the current position of the 3D element.

The Rotate Object LabVIEW function performs the relative rotation of the 3D object depending on the set axis, the given angle and its current position. The minimum and maximum thresholds of both Translate and Rotate functions are empirically determined by the manual movement of the 3D wheelchair in the virtual environment depending on the entire dimensions of the main and secondary base floor platforms. The Scale Object function is necessary to display the individual elements in the 3D scene considering the factor set for the relative scaling from their current position. The input parameters of the Translate and Rotate Object functions constitute not only numeric constants to establish the permanent position of the elements in respect of the others or to keep the correct linking between 3D components, but also numeric controls (variable values) to enable the user to manually adjust the relative position of the 3D wheelchair to the movement directions: go forward, go backward, go to the left, go to the right, rotate to the left by 90°, rotate to the right by 90°, full rotate by 180°, and reverse rotate by 180°. In fact, the separate input parameters of the functions provided by the 3D Picture Control are either embedded by clusters of three numeric constants (single, 32-bit, real, 6-digit precision) or grouped using the Bundle function aimed at generating a cluster output. Otherwise, the input parameters of the Translate and Rotate Object are the local variables which allows the simultaneous access to the same LabVIEW data in multiple places from the Block-Diagram meaning that the user can both manually and automatically control the 3D wheelchair by moving it towards the virtual environment.

4.3 The LabVIEW State-Machine for Automatic Control of the Wheelchair

A state-machine programmatic paradigm was implemented in LabVIEW graphical environment to enable the control of the 3D wheelchair by different movement directions set either automatically by checking the UDP string associated with the P300 command or manually by selecting the virtual buttons or Boolean controls.

Figure 5 shows the Init state aimed at comparing the UDP command to the characters associated with the movement directions and running the next corresponding state: f – Forward Short Distance; F – Forward Long Distance; b – Backward Short Distance; B - Backward Long Distance; r – Right Short Distance; R – Right Long Distance; l – Left Short Distance; L – Left Long Distance; E – Rotate Left 90°; I – Rotate Full 180°; O – Rotate Full 180°; o – Rotate Reverse 180°.

Further, according to Fig. 6, the programmatic sequence of each previously mentioned state comprises the local variables necessary for accessing the input elements to manually control the 3D wheelchair by incrementing (adding a value) or decremented (subtracting a value) them depending on the short or long distance set for the movement command.

Figure 7 and Fig. 8 show the intermediate positions reached by the 3D wheelchair after sending the related command by running the state-machine paradigm for enabling its movement across the forward, backward, left, and right directions, both short and long distances, in the virtual environment.

To avoid the continuous or infinite movement of the 3D wheelchair, a Stop state is executed after the previously mentioned states. The Stop state includes the assignment of the string local variable of the UDP command to the empty string constant.

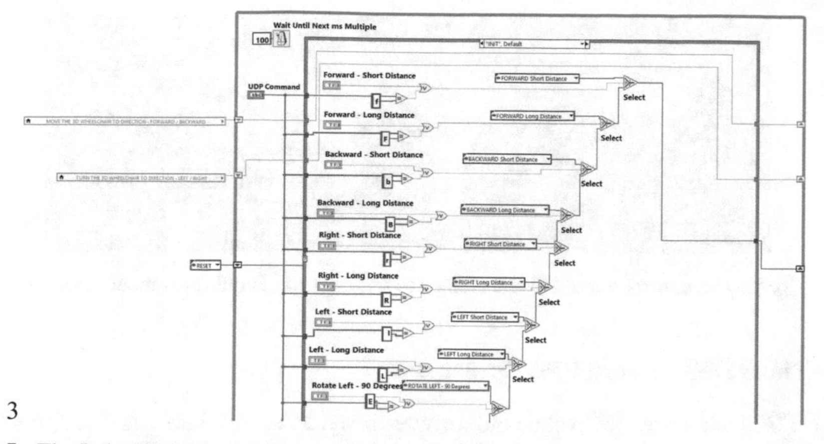

Fig. 5. The LabVIEW Block-Diagram with the *Initialization* Sequence of the State-Machine.

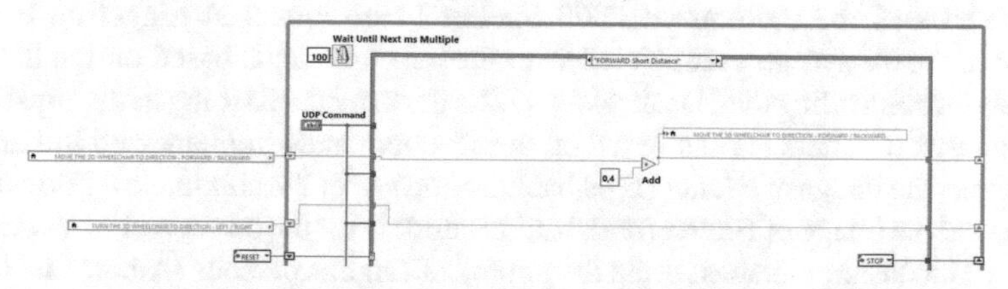

Fig. 6. The LabVIEW Block-Diagram with the *Forward Short Distance* State.

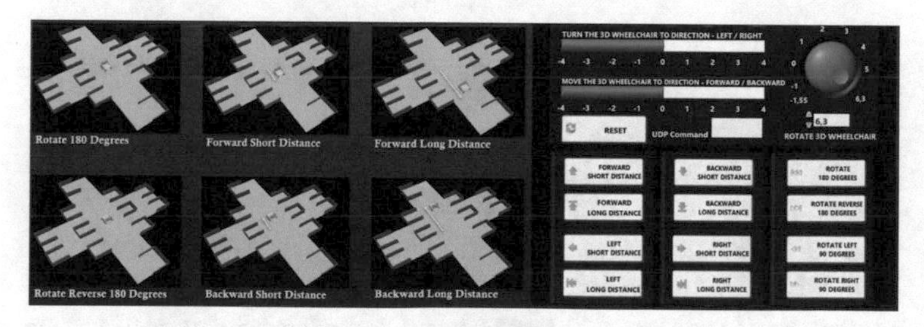

Fig. 7. The control of the 3D wheelchair to the Forward and Backward movement directions.

After Stop state is executed, the Init state is performed to check again the user selected command. Likewise, the state-machine includes the Reset state to reinitialize the position of the 3D wheelchair to the default values considering its placement towards axis X, axis Y, as well as a specific rotation axis and angle. The Reset state supposes the assignment of the numeric local variables for accessing the manual input elements to the default values (zero or maximum threshold) so that the 3D wheelchair is displayed in the center of the virtual experimental environment.

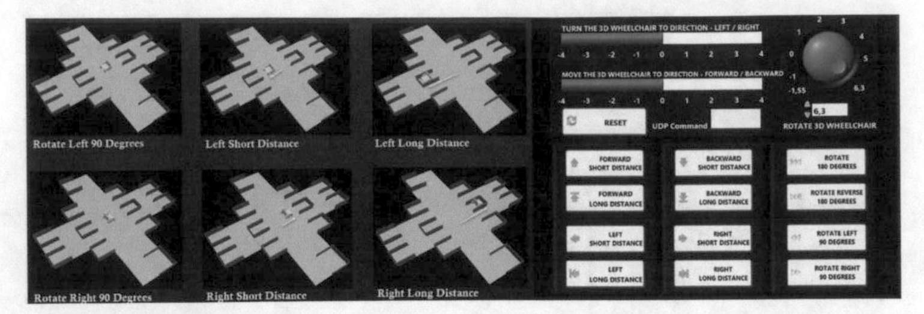

Fig. 8. The control of the 3D wheelchair to the Left and Right movement directions.

4.4 The GTEC Unicorn P300 Speller Board

The GTEC Unicorn P300 Speller paid software provides researchers with the opportunity to run both the official and customized design patterns comprising dark and flashing items for the experimentation of P300 controlled brain-computer interfaces. Figure 8 presents the two views of the customized P300 Speller board aimed at triggering the users' attention to the 12 symbols associated with the 12 commands based on the movement directions for controlling the 3D wheelchair. The dark or non-flashing items constitute the official images of letters chosen based on the 12 commands implemented in LabVIEW. For example, the image of F letter (light background) is for the command – Forward Short Distance and the image of F letter (dark background) is for the command – Forward Long Distance. The flashing images reveal the photos of famous persons (Albert Einstein, Mr. Been, Donald Trump) aimed at triggering a strong positive impact (curiosity, surprise, enthusiasm) on the users' emotional state (Fig. 9).

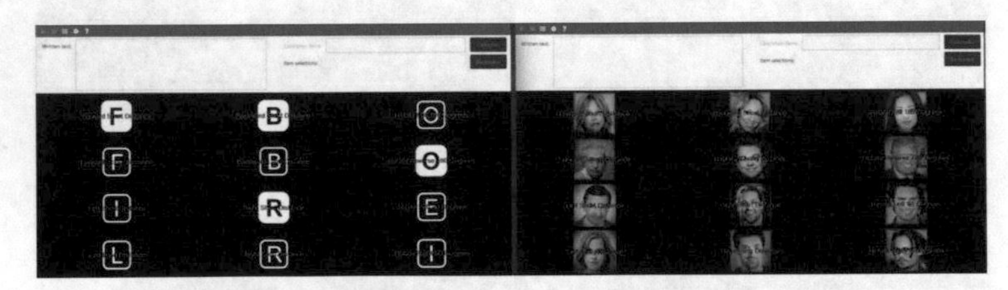

Fig. 9. The customized Unicorn P300 Speller board including the dark and flashing symbols associated with the commands based on movement directions for controlling the wheelchair.

5 Experimental Setup – Results and Discussions

The two YouTube links [12, 13] present the video live experimental session underlying the brain-computer interface application proposed by the current research paper. A single subject (girl, 31 years) was required to perform 12 experiments based on sending multiple P300 commands aimed at controlling the 3D wheelchair by different movement directions to reach the established positions from the virtual environment implemented in LabVIEW graphical programming environment. The 12 experiments were designed considering three levels of difficultness: beginner (experiments 1–2 based on executing two commands and experiments 3–4–5–6 based on performing three commands), intermediate (experiments 7–8–9–10 based on achieving 7 commands), and advanced (experiments 11–12 related to accomplishing 13–14 commands). The results obtained by the subject are described in the Table 6 (100% accuracy).

Table 6. Results obtained by performing 12 brain-computer interface experiments.

Experiment	Level	Commands	Results	Accuracy
1	Beginner	Rotate Left 90° Left Long Distance	Pass Pass	100%
2	Beginner	Rotate Right 90° Right Long Distance	Pass Pass	100%
3	Beginner	Forward Long Distance Rotate Left 90° Left Long Distance	Pass Pass Pass	100%
4	Beginner	Forward Long Distance Rotate Right 90° Right Long Distance	Pass Pass Pass	100%
5	Beginner	Backward Long Distance Rotate Left 90° Left Long Distance	Pass Pass Pass	100%
6	Beginner	Backward Long Distance Rotate Right 90° Right Long Distance	Pass Pass Pass	100%
7	Intermediate	Forward Short Distance Forward Short Distance Rotate Left 90° Left Long Distance Left Long Distance Left Short Distance Rotate Right 90°	Pass Pass Pass Pass Pass Pass Pass	100%
8	Intermediate	Backward Short Distance Backward Short Distance Rotate Left 90° Left Long Distance Lcft Long Distance Rotate Right 90°	Pass Pass Pass Pass Pass Pass	100%

(continued)

Table 6. (*continued*)

Experiment	Level	Commands	Results	Accuracy
9	Intermediate	Forward Short Distance Forward Short Distance Rotate Right 90° Right Long Distance Right Long Distance Right Short Distance Rotate Left 90°	Pass Pass Pass Pass Pass Pass Pass	100%
10	Intermediate	Backward Short Distance Backward Short Distance Rotate Right 90° Right Long Distance Right Long Distance Right Short Distance Rotate Left 90°	Pass Pass Pass Pass Pass Pass Pass	100%
11	Advanced	Rotate Reverse 180° Backward Long Distance Backward Short Distance Backward Short Distance Rotate Right 90° Right Short Distance Rotate Reverse 180° Backward Short Distance Backward Short Distance Backward Short Distance Rotate Left 90° Left Short Distance Left Short Distance Rotate Right 90°	Pass Pass Pass Pass Pass Pass Pass Pass Pass Pass Pass Pass Pass Pass	100%
12	Advanced	Forward Long Distance Forward Short Distance Forward Short Distance Rotate Left 90° Left Short Distance Rotate Full 180° Forward Short Distance Forward Short Distance Forward Short Distance Rotate Right 90° Right Short Distance Right Short Distance Rotate Left 90°	Pass Pass Pass Pass Pass Pass Pass Pass Pass Pass Pass Pass Pass	100%

6 Conclusions

This paper presented a simple prototype of a brain-computer interface for controlling a virtual simulation of a wheelchair using the GTEC Unicorn EEG headset and the P300 Speller board. The proposed software application involved the UDP based integration between the LabVIEW for developing a simulation of the 3D wheelchair and the Unicorn P300 Speller for providing the user interface to detect and analyze the EEG signals. The presented instrument enabled the experimentation of the BCI in the home environment showing that the portable EEG technology could be successfully leveraged to provide an attractive training, testing, and learning environment.

References

1. Cattan, G., Andreev, A., Visinoni, E.: Recommendations for integrating a P300-based brain-computer interface in virtual reality environments for gaming: an update. Computers **9**, 92 (2020). https://doi.org/10.3390/computers9040092
2. Hadjiaros, M., Neokleous, K., Shimi, A., Avraamides, M.N., Pattichis, C.S.: Virtual reality cognitive gaming based on brain computer interfacing: a narrative review. IEEE Access **11**, 18399–18416 (2023). https://doi.org/10.1109/ACCESS.2023.3247133
3. Kim, S., Lee, S., et al.: P300 brain-computer interface-based drone control in virtual and augmented reality. Sensors **21**(17), 5765 (2021). https://doi.org/10.3390/s21175765
4. Wen, D., Fan, Y., Hsu, S.-H., et al.: Combining brain–computer interface and virtual reality for rehabilitation in neurological diseases: a narrative review. Ann. Phys. Rehabil. Med. **64**(1), 101404 (2021). https://doi.org/10.1016/j.rehab.2020.03.015
5. McClinton, W., Laesker, D., Garcia, S., Caprio, D., Pinto, B., Andujar, M.: P300-based 3D brain painting in virtual reality. In: CHI Conference on Human Factors in Computing Systems (2019). https://doi.org/10.1145/3290607.3312968
6. Cattan, G.H., Andreev, A., et al.: A comparison of mobile VR display running on an ordinary smartphone with standard PC display for P300-BCI stimulus presentation. IEEE Trans. Games **13**, 68–77 (2021). https://doi.org/10.1109/TG.2019.2957963
7. Bulanov, V., Zakharov, A., Sergio, L., Lebedev, M.: Visuomotor transformation with a P300 brain-computer interface combined with robotics and virtual reality: a device for post-stroke rehabilitation (2021). https://doi.org/10.2139/ssrn.3811232
8. Baniqued, P.D.E., Stanyer, E.C., Awais, M., et al.: Brain–computer interface robotics for hand rehabilitation after stroke: a systematic review. J. NeuroEngineering Rehabil. **18**, 15 (2021). https://doi.org/10.1186/s12984-021-00820-8
9. Farwell, L.A., Donchin, E.: Talking off the top of your head: toward a mental prosthesis utilizing event-related brain potentials. Electroencephalogr. Clin. Neurophysiol. **70**, 510–523 (1988)
10. Philip, J.T., George, S.T.: Visual P300 mind-speller brain-computer interfaces: a walk through the recent developments with special focus on classification algorithms. Clin. EEG Neurosci. **51**(1), 19–33 (2020). https://doi.org/10.1177/1550059419842753
11. Rușanu, O.A.: A brain-computer interface-based simulation of vending machine by the integration between Gtec unicorn EEG headset and LabVIEW programming environment using P300 speller and UDP communication. In: Moldovan, L., Gligor, A. (eds.) Inter-Eng 2022. LNNS, vol. 605, pp. 836–849. Springer, Cham (2023). https://doi.org/10.1007/978-3-031-22375-4_68
12. YouTube Link. https://youtu.be/_KIFKc8NPp8. Accessed 5 Dec 2023
13. YouTube Link. https://youtu.be/I0Uyr3EHstg. Accessed 5 Dec 2023

Author Index

M. E. Auer et al. (Eds.): STE 2024, LNNS 1028, pp. 437–439, 2024.
https://doi.org/10.1007/978-3-031-61905-2

Printed in the United States
by Baker & Taylor Publisher Services